J. M. TEDDER
University of St. Andrews, Scotland

A. NECHVATAL
University of Dundee, Scotland

Basic Organic Chemistry

Part 2

1967
John Wiley and Sons
London New York Sydney Toronto

Reprinted 1972
Reprinted 1975

Library of Congress catalog card number: 67–17112
ISBN 0 471 85012 8

Printed in Great Britain
By Unwin Brothers Limited
The Gresham Press, Old Woking, Surrey, England
A member of the Staples Printing Group

Preface to Part 2

Like Part 1, this volume is derived from lecture notes prepared for the 2nd B.Sc. Class at the University of Dundee (formerly Queen's College of the University of St. Andrews), Scotland. As before, the lectures were dictated into a tape-recorder and the stencils prepared verbatim from the recording. The colloquial style of the stencilled sheets has been deliberately retained and, in the main, the book remains identical with the notes.

The lectures given to the second year class were intended to build on the previous year's work. The first four Chapters correspond to lectures given as "General Chemistry" and were intended to serve as an introduction to both organic and inorganic chemistry. We have retained this section as it forms a suitable foundation for the subsequent Chapters. Chapter 6 is intended to emphasize that thermodynamics is fundamental to all chemistry; it is in no sense intended to be an introduction to thermodynamics capable of standing by itself. The electronic theory of organic chemistry is discussed in Chapters 7 and 8 in the light of the earlier theoretical section. Chapters 9—22 develop the ideas of organic chemistry along the mechanistic lines of Part 1. Chapters 23 and 24 deal with stereochemistry in a fairly formal manner. Chapters 25 and 26 are concerned with structural determination, first by physical methods and then by chemical methods.

As in Part 1, we have attempted to develop the current ideas of organic chemistry and to show that the vast range of reactions undergone by the many different classes of compound can all be fitted in to the same general mechanistic picture. We should like, however, to add some words of caution. Some of the letters and comments we have had about Part 1 show that there are some teachers who wish to carry this approach far further than we think is justified at the present time. It is extremely important that the students should realize that mechanistic organic chemistry is nothing more than a rationalization of observed experimental facts. The student will quickly discover that, although mechanistic theory can suggest how two compounds may react together, it is very often unable to distinguish between two possible reaction paths. The days when the course of a new organic reaction can be predicted with certainty are still a long way off. We would also caution against the idea that if you understand mechanistic organic chemistry you do not need to remember facts. The whole point of the approach we have developed in these two books is to provide a scheme into which the facts can be fitted. Mechanistic organic chemistry without the experimental facts becomes pointless indeed.

Some teachers will be surprised that we delayed dealing with formal stereochemistry until Chapters 23 and 24 and we have had to fight very hard against the temptation to include little bits of formal stereochemistry as we went along. To do this would, we believe, have violated the most important of all teaching principles, namely, to deal with one concept at a time. We have found that by illustrating reaction mechanisms with diagrams and models representing the stereochemical situation correctly as we go along, the students welcome the material in Chapter 24 as pleasing confirmation of what they have already accepted. Thus, although formal stereochemistry is delayed until these two chapters, we have endeavoured throughout Parts 1 and 2 to emphasise the three-dimensional nature of chemistry as far as possible.

In a similar way some teachers may be surprised that we have delayed discussing physical methods until the penultimate chapter of the book. Again it was very tempting to introduce some aspects of spectroscopy earlier, but, as with formal stereochemistry, we believe such a course would have represented poor teaching. Only when students have enough knowledge of organic chemistry do they really appreciate the importance and value of spectroscopic methods. We have learned of two Universities where an over-early introduction of n.m.r. spectroscopy has had a disastrous effect on the students' interest and enthusiasm for organic chemistry. Conversely, we have found that when students have already acquired a considerable background knowledge of organic chemistry they find n.m.r. spectroscopy and other physical methods exciting and easy to assimilate. Above all, we believe that spectroscopic methods must be introduced into organic chemistry in conjunction with work in the laboratory.

We believe that organic chemistry has now reached the stage when mechanistic theory can replace compound classification as the main skeleton from which the various aspects of the subject can be developed. This is not the first book to attempt such an approach, but we feel that many previous authors have fallen between two stools by trying to make their books both textbooks for looking up facts and teaching books for developing ideas. We believe that this is no longer possible in one volume, and the present books are written on the assumption that students will either possess or have ready access to one of the many excellent conventional textbooks for seeking factual materials. Like Part 1, this volume is intended to be read.

Acknowledgements

Many of the ideas developed in the present volume came from our former teachers and from our colleagues over the years, and often what we believe to be original may, in fact, be derived from suggestions of others. We can only acknowledge our general debts to the wonderful colleagues with whom we have been fortunate to spend our working lives. Some help we can identify and we wish to thank: Professor J. N. Murrell and Dr. E. T. Stewart for reading and criticizing Chapters 1—4; Professor J. Leisten and Dr. P. A. H. Wyatt for initiating the ideas developed in Chapter 6, and Drs. C. R. Allen and P. G. Wright for criticising what we have written; Professor C. Rees for reading Chapter 22 and Dr. R. S. Cahn for considerable help with Chapters 23 and 24. An earlier version of the entire manuscript has been read by Professor T. S. Stevens, Dr. R. Brettle, and Dr. J. C. P. Schwartz, and their advice has proved invaluable. All through, our colleagues in Dundee have provided help and advice, and in thanking them all we would particularly like to mention Drs. J. Carnduff and W. M. Horspool. We must also acknowledge the tremendous help given to us at every stage by the London editorial staff of John Wiley & Sons. Our greatest debt is to the 2nd B.Sc. Class at Queen's College in the years 1965/66 and 1966/67. Their response to the original lectures provided the stimulus to produce the book.

April, 1967

J. M. T.
A. N.

[illegible]

[illegible]

[illegible]

[illegible]

[illegible]

Contents

CHAPTER 1

The Wave Equation

In Part 1 the chemistry of carbon compounds was treated in terms of the Lewis theory of valency. According to this theory a bond between two atoms is formed by the sharing of electrons to give electron pairs. Each atom tends to acquire an electron shell similar to that of the nearest noble gas. The shape of molecules is explained by allowing for the electrostatic repulsion between the electron-pair bonds. This simple picture adequately explained the valency of many elements and could readily be applied to describe the displacement reactions undergone by organic compounds. The theory was, however, inadequate to explain the rigidity of a carbon–carbon double bond and, moreover, failed to account for the properties of the acetate anion or the benzene nucleus. For the latter compounds the concept of resonance was introduced in Part 1, with a hint that it was derived from a more sophisticated theory. All modern theories of valency are based on quantum mechanics, but because full treatment of quantum mechanics involves fairly complex mathematics many students find themselves unable to comprehend the current valency theories. It is our present purpose to show that the basic ideas underlying the Schrödinger equation are quite simple and that a student without a strong mathematical background should be able to understand them.

We know from simple experiments involving interference and diffraction that visible light has wave characteristics. However, to understand the interaction of light and matter, chemists and physicists have come to the conclusion that light must also behave like a stream of particles. Visible light and indeed all electromagnetic radiation (e.g. wireless waves, X-rays, etc.) travels in a vacuum at a constant velocity c ($c = 3 \times 10^{10}$ cm sec^{-1}). The energy of the

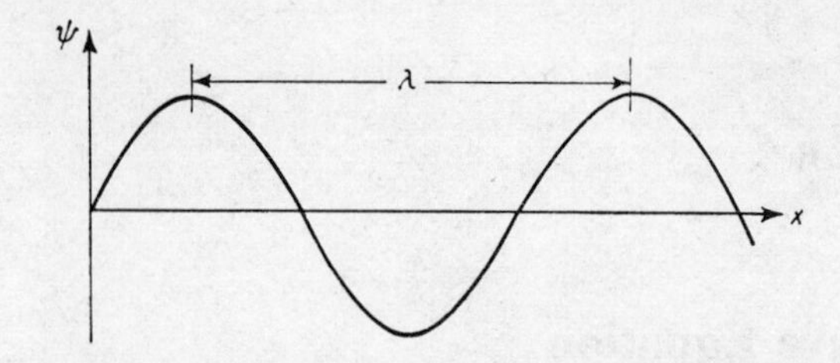

Figure 1.1. Frequency (ν) and wavelength (λ) are related to the velocity of light (c) by the simple relation $\lambda\nu = c$.

particles making up the light wave depends on the wavelength (or frequency) of the light (see Figure 1.1).

Experiment also shows that an atom cannot possess an arbitrary amount of energy but exists in discrete energy states, usually referred to as 'energy levels'. The change from the level E_1 to a higher energy level E_2 is accompanied by absorption of light, whereas the change from level E_2 to E_1 is associated with emission of light. Only light of a specific frequency is absorbed or emitted (see Figure 1.2) and this is expressed by the relation:

$$E_2 - E_1 = \boldsymbol{h}\nu$$

$\boldsymbol{h}$ is the universal constant known as Planck's constant. If the energy difference $E_2 - E_1$ is measured in ergs and the frequency ν is measured in $\sec^{-1}$, $\boldsymbol{h}$ must be expressed in erg sec ($\boldsymbol{h} \approx 6.63 \times 10^{-27}$ erg sec). The energy $\boldsymbol{h}\nu$ absorbed (or emitted) by an atom is called a quantum. The 'Quantum Theory' (Planck 1901) which recognized

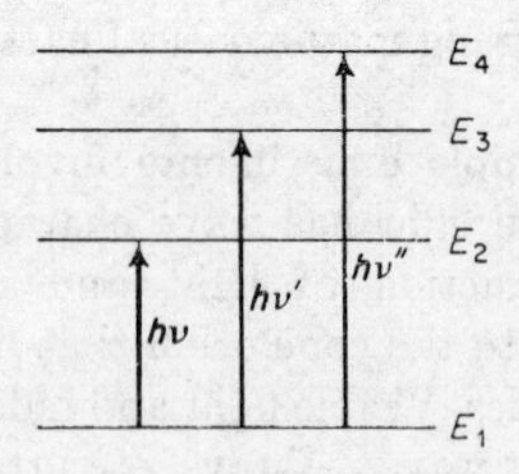

Figure 1.2. Relation of energy levels (E) to absorbed radiation.

that electromagnetic radiation exhibits both wave and particle characteristics and that only discrete energy levels are possible on the atomic scale marked the first great break from the concepts of classical physics.

In 1924 de Broglie suggested that electrons might also show dual character, exhibiting wave characteristics as well as particle characteristics, resembling the dual behaviour of electromagnetic radiation. Thus a particle with momentum p has an associated wavelength given by the expression

$$\lambda = \boldsymbol{h}/p$$

This concept was experimentally confirmed by the observation that a stream of electrons which passed through a metal foil gave a diffraction pattern analogous to that obtained with X-rays.

If an electron has wave characteristics it must be capable of description by a wave equation, in the same way as waves of light or sound or the vibration of a stretched string can be expressed in, or described by, a wave equation. A wave travelling in the direction of the x axis with wavelength λ and frequency ν has an amplitude of ψ given by the equation

$$\psi = A \sin 2\pi\left(\frac{x}{\lambda} - \nu t\right)$$

The equation for a standing wave, i.e. one in which the maxima and minima do not travel along, is given simply by

$$\psi = A \sin \frac{2\pi x}{\lambda}$$

Differentiating this equation twice gives the expression:

$$\frac{d^2\psi}{dx^2} = -\left(\frac{2\pi}{\lambda}\right)^2 A \sin \frac{2\pi x}{\lambda}$$

$$\therefore \frac{d^2\psi}{dx^2} + \frac{4\pi^2}{\lambda^2}\psi = 0$$

This is the 'wave equation' and is the basis of the whole theory. We introduce momentum from the de Broglie relation and the equation now becomes:

$$\frac{d^2\psi}{dx^2} + \frac{4\pi^2 p^2}{\boldsymbol{h}^2}\psi = 0$$

If we consider an electron moving freely in space but not subject to any external force, we can relate the energy of the electron solely to its kinetic energy $E = \frac{1}{2}mv^2$. Remembering that $p = mv$, we now have a wave equation in which the energy of the particle occurs, namely:

$$\frac{d^2\psi}{dx^2} + \frac{8\pi^2 mE}{h^2}\psi = 0$$

If the particle moves under the influence of an external force, we can separate the total energy into the kinetic energy T and the potential energy V, where

$$E = T + V$$

Our equation now becomes:

$$\frac{d^2\psi}{dx^2} + \frac{8\pi^2 m}{h^2}(E - V)\psi = 0$$

This is Schrödinger's equation for a particle in one dimension.

One of the most important features of the wave equation is that it predicts the existence of discrete energy levels. We can introduce this by considering the vibrations of stretched strings such as there are in a guitar or a piano. In these instruments the string is fixed at both ends and it can only vibrate giving the ground tone or a number of specific higher notes or overtones. It will not vibrate at other frequencies. This is depicted in Figure 1.3. This behaviour of a stretched string is described very simply by the equation given above for a standing wave:

$$\psi = A \sin \frac{2\pi x}{\lambda}$$

Now, because the string is fixed at both ends, we must apply to the equation restrictions called the 'boundary conditions'. These restrictions are that $\psi = 0$ when $x = 0$ and, $\psi = 0$ when $x = l$. These conditions are only met for certain values of λ, namely:

$$\frac{2\pi l}{\lambda} = n\pi$$

where n = any integer.

We therefore have a series of values of λ:

$$\lambda_1 = 2l, \quad \lambda_2 = \frac{2l}{2}, \quad \ldots, \quad \lambda_n = \frac{2l}{n}$$

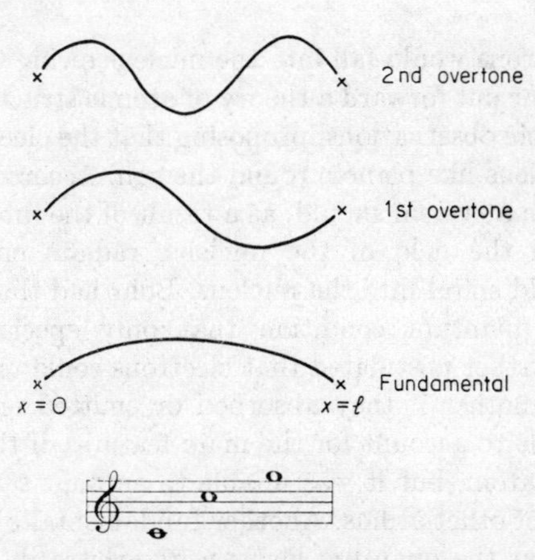

Figure 1.3. The fundamental and first and second overtones of, say, middle C.

Now let us return to consider a particle whose behaviour is described by the wave equation. If we replace λ by energy, using the de Broglie relation, we have:

$$E = \frac{1}{2}\,mv^2 = \frac{m}{2}\left(\frac{h}{m\lambda}\right)^2$$

and instead of wavelengths we have a series of energy levels:

$$E_1 = \frac{h^2}{8ml^2}, \quad \ldots, \quad E_n = \frac{h^2n^2}{8ml^2}$$

These are in fact, the energy levels for a particle restricted between 0 and l in a single dimension x. This is usually described as a particle in a 'one-dimensional box'. We thus see that as soon as we put boundary conditions on a particle described by the Schrödinger equation we restrict the energy of that particle to discrete levels.

We must next consider the structure of an atom. Rutherford's experimental observations led him to conclude that an atom consisted of a small, positively charged nucleus surrounded by electrons which sparsely occupy the rest of the atomic volume. Atoms are approximately 10^{-8} cm in diameter and the nucleus is of the order of 10^{-13} cm in diameter. According to classical mechanics the atom, as pictured by Rutherford, cannot be a static system because, if it were, the electrons would fall into the nucleus owing to electrostatic attraction. Bohr put forward a theory of atomic structure to account for spectroscopic observations, proposing that the electrons revolved round the nucleus like planets round the sun. According to classical mechanics such a system should, as a result of the interaction of the electrons with the field of the nucleus, radiate energy and the electrons should spiral into the nucleus. Bohr had thus to introduce an arbitrary quantum condition that only special orbits were allowed. He further postulated that electrons could only jump from one orbit to another if they absorbed or emitted energy $h\nu$. This theory was able to account for the main features of the spectrum of the hydrogen atom, but it was unable to account satisfactorily for the spectrum of other atoms. Another fundamental objection to the theory was that the quantum levels were arbitrarily introduced to fit experiment.

We have seen that discrete energy levels are predicted by the wave equation for a particle in a one-dimensional box. Let us apply the wave equation to the hydrogen atom. The hydrogen atom consists of an electron and a proton which attract; e is the electronic charge, their potential energy is $-e^2/r$ (provided the potential energy is taken to be zero when the proton and the electron are infinitely far apart); and the Schrödinger equation for the hydrogen atom is:

$$\frac{\partial^2\psi}{\partial x^2} + \frac{\partial^2\psi}{\partial y^2} + \frac{\partial^2\psi}{\partial z^2} + \frac{8\pi^2 m}{h^2}\left(E + \frac{e^2}{r}\right)\psi = 0$$

We need an equation in three dimensions, hence the variables x, y, and z. This equation is usually simplified as:

$$\nabla^2\psi + \frac{8\pi^2 m}{h^2}\left(E + \frac{e^2}{r}\right)\psi = 0$$

where

$$\nabla^2 = \frac{\partial^2}{\partial x^2} + \frac{\partial^2}{\partial y^2} + \frac{\partial^2}{\partial z^2}$$

We thus have an equation which accounts adequately for most of the experimental observations and can be solved to give the energy of an electron in an atom. Since the equation is a second-order differential equation (i.e. contains terms $\partial^2/\partial x^2$), any solution for ψ will contain two arbitrary constants for each dimension (i.e.

six in all). As with the stretched string, we need to know what boundary conditions to impose on ψ in order to eliminate this arbitrariness. We must, therefore, ask what is the significance of ψ. For the stretched string, ψ was the amplitude. We immediately had a clear physical picture of what it meant and could deduce that $\psi = 0$ at the ends of the string (i.e. the boundary conditions). However, the amplitude of an electron, which we are accustomed to think of as a discrete particle, is hardly a meaningful concept. In other forms of wave motion, such as light, the square of the amplitude is proportional to the intensity—the amount of energy per unit volume. For the electron we can interpret ψ^2 (or more correctly $\psi\psi^*$ if ψ is a complex function) as the 'amount of electron charge' per unit volume, interpreting the electron as a charge cloud distributed with density ψ^2 at any point. Since quantum-mechanical equations in general are based on considering electrons as point particles, a better interpretation (due to Born) is that ψ^2 is the probability of finding an electron in a unit volume. Thus we can determine the probability of finding an electron in a small volume about any point by substituting the coordinates of that point in the equation giving ψ^2 in terms of these coordinates.

We can now return to the hydrogen atom. The knowledge that ψ^2 is the probability distribution places certain restrictions on ψ. First, we note that the electron must be somewhere in space, hence

$$\int_{-\infty}^{\infty}\int_{-\infty}^{\infty}\int_{-\infty}^{\infty} \psi^2 \, dx \, dy \, dz = 1$$

which is usually written as:

$$\int \psi^2 \, dv = 1$$

where dv is the volume element $dx\, dy\, dz$, and the limits of integration are implicit. This equation is called the normalization requirement. The other conditions which we deduce from our interpretation of ψ^2 are: (i) ψ^2 and also ψ must be single valued since we cannot have two different probabilities of finding the electron at one place; (ii) since ψ^2 must obey the normalization condition, ψ must everywhere be finite; and (iii) the probability of finding an electron must vary from point to point in a continuous fashion. As in our example of the stretched string, applying these restrictions again gives us quantized energy levels. Moreover, the calculated values of the energy levels are in accord with those obtained for the hydrogen atom by spectroscopic experiments.

Experiment shows that both electromagnetic radiation and matter have dual wave–particle characteristics. Their behaviour can best be described by a wave equation in some experiments, and by an equation normally applying to discrete particles in other conditions. We do not, as yet, treat matter and radiation as different aspects of the same entity. An electron has charge and mass, whereas a photon has neither. The wave equation that we have found adequate for the electron is not the same wave equation that we use to describe light. Students should not let themselves be worried by this dual particle–wave relationship. We happily accept that the equation $d^2x/dt^2 = -gx/l$ describes the motion of a simple pendulum (see Figure 1.4). If we understand the meaning of the symbols we can predict the behaviour of the pendulum, but we can hardly say that the equation tells us what the pendulum looks like or would feel like. In the same way, the Schrödinger equation tells us how an electron behaves but gives us no help in trying to visualize what an electron is like. When we think of a discrete particle we are inclined to visualize a small billiard ball which we can see and feel. We can never hope to see or feel an electron, although we can perform experiments in which the effect of a single electron can be observed. Since it is impossible to see or feel an electron, we should not expect to be able to picture it correctly in ordinary concrete terms. Yet, because all our thought processes are derived from our everyday experience of concrete things it is often useful to picture the electron as a minute billiard

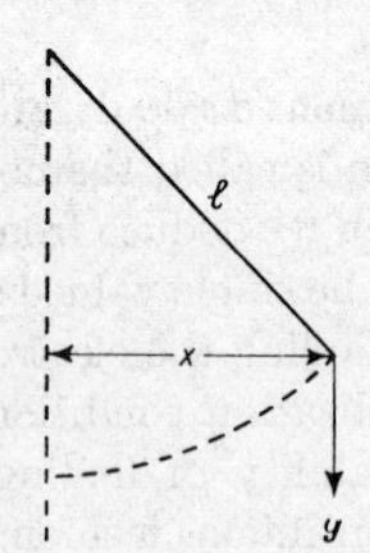

Figure 1.4. Motion of a simple pendulum.

ball. Pictures of this kind are an essential part of our chemical thoughts, but nevertheless we should remember that they are only pictures.

Problems

1. If an atom absorbs light whose wavelength is 5500 Å, what is the energy of the associated electronic transition?

2. What are the first three energy levels possible for a particle confined to one dimension x such that, when $x \leqslant 0$ and when $x \geqslant a$, there is an infinite potential (i.e. $V = \infty$), but between $x = 0$ and $x = a$, $V = 0$?

CHAPTER 2

Atomic Orbitals

Let us briefly reconsider the particle in a one-dimensional box. This phrase means that a particle of mass m is restricted to a line, its position on this line being determined by x. We introduce the boundary conditions that for x less than 0 or greater than l the potential energy is infinitely large, and between $x = 0$ and $x = 1$ the potential energy is zero; or, in mathematical terms, when $x < 0$ and $> l$, $V = \infty$; and when $0 < x < l$, $V = 0$.

The wave equation for a particle moving in one dimension, subject to no external force, is

$$\frac{\mathrm{d}^2\psi}{\mathrm{d}x^2} = -\frac{8\pi^2 mE}{\boldsymbol{h}^2}\psi$$

In Chapter 1 it was shown that applying the above boundary conditions then gives a series of solutions for the energy in the form:

$$E_n = \frac{\boldsymbol{h}^2 n^2}{8ml^2}$$

where n is any integer. We can call n a *quantum number* and it designates the particular energy level. In the case of the stretched string the wave function ψ giving the amplitude has direct physical meaning, but we have seen that for an electron it is ψ^2, the probability distribution, that has physical meaning. We represent this in Figure 2.1.

Notice that the greater the quantum number, the higher is the energy of the electron and the greater are the number of nodes (points or surfaces where the probability is zero). It would be possible to carry out a similar calculation for a particle in a two-dimensional box from which we should obtain two quantum numbers. If the box were a square box we should obtain pairs of solutions that had the same energies. Such non-identical solutions which have the same energy are said to be *degenerate*. Finally, we come to the case of a three-dimensional box and under these circumstances we should find three quantum numbers.

The equation for the hydrogen atom must, of course, be in three dimensions and, as already described, the potential energy is given by $-e^2/r$. If we attempt to solve the Schrödinger equation for the hydrogen atom using Cartesian coordinates the mathematics

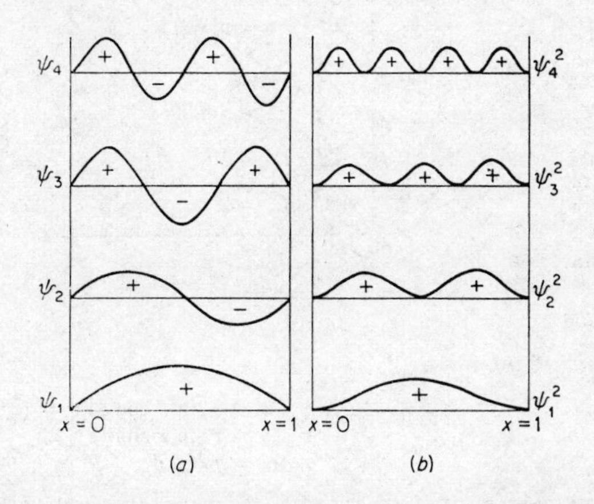

Figure 2.1. Representation of (*a*) ψ, the wave function and (*b*) ψ^2, the probability distribution, of an electron in a one-dimensional box.

becomes impossibly cumbersome. Instead, therefore, we use polar coordinates. The position of a point in three-dimensional space, instead of being defined by three distances, is defined by one distance and two angles (see Figure 2.2). The Schrödinger equation in polar coordinates looks very formidable:

$$\frac{\partial^2\psi}{\partial r^2} + \frac{2}{r}\cdot\frac{\partial\psi}{\partial r} + \frac{1}{r^2}\cdot\frac{\partial^2\psi}{\partial\theta^2} + \frac{\cos\theta}{r^2\sin\theta}\cdot\frac{\partial\psi}{\partial\theta} + \frac{1}{r^2\sin^2\theta}\cdot\frac{\partial^2\psi}{\partial\phi^2} + \frac{8\pi^2m}{h^2}\left(E + \frac{e^2}{r}\right)\psi = 0$$

We shall not go through the complete solution of this equation. It is important here that the wave function can be factorized into three functions, i.e. written as (function of r) $\times$ (function of θ) $\times$ (function of ϕ), the radial function involving r and two angular functions involving θ and ϕ. This deserves attention because in many books the 'shapes' of the wave functions are represented in terms of the angular functions only.

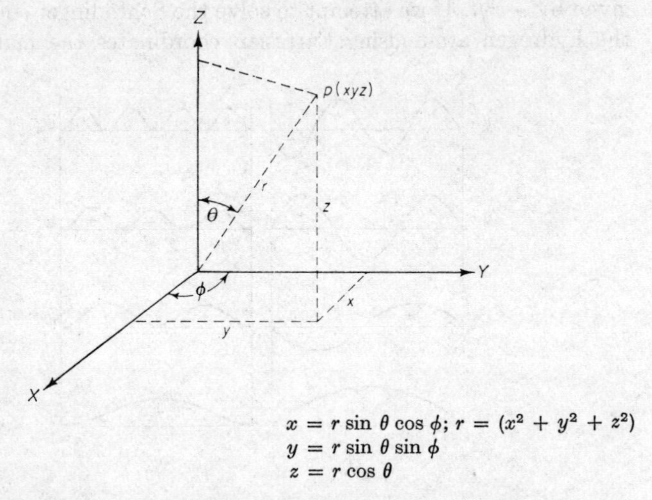

$$x = r \sin \theta \cos \phi ; \; r = (x^2 + y^2 + z^2)$$
$$y = r \sin \theta \sin \phi$$
$$z = r \cos \theta$$

Figure 2.2. Relation of Cartesian to polar coordinates.

We were able to represent solutions of the wave equation for a particle in a one-dimensional box in Figure 2.1. Clearly, when we come to do the same for solutions of the equation for a hydrogen atom we have the problem of representing the function in three dimensions. As discussed above, we have three quantum numbers which are represented by the letters n, l, and m (see Table 2.1). We need not trouble ourselves now with the names of the quantum

Table 2.1. Quantum numbers

Symbol	Name	Range
n	Principal quantum number	$1 \leqslant n$
l	Angular momentum quantum number	$0 \leqslant l \leqslant n - 1$
m	Magnetic quantum number	$-l \leqslant m \leqslant l$

numbers; all that concerns us is that assignment of appropriate values to the quantum numbers gives a specific solution to the wave equation. Roughly speaking, the principal quantum number n defines the 'size' of the electron's probability distribution, whereas quantum numbers l and m define the 'shape' and orientation. The solution with the lowest energy has quantum numbers $n = 1$, l and $m = 0$. It transpires that ψ is spherically symmetric for this solution and we can represent this in four ways. In the first of these (Figure 2.3*a*) we have drawn graphs of ψ and of ψ^2 against distance r defined by the polar coordinates in Figure 2.2. In Figure 2.3*b* we have drawn contours of ψ or ψ^2, the value in the centre being arbitrarily taken to be unity and the contours decreasing in value outwards from the centre. Figure 2.3*c* is a sketch representing the probability of finding the electron, the darker the 'cloud' the more likely we are to find the electron at that point. Figure 2.3*d* is a simplified form of Figure 2.3*b*, one contour being drawn to represent that most of the probability cloud of Figure 2.3*c* is inside this boundary surface.

According to Bohr's theory, the electron rotated in specific orbits around the nucleus and each orbit represented a different energy level. In quantum mechanics there is no orbit for the electron but only its probability distribution. This is called an *orbital*. The orbital of lowest energy ($n = 1$, $l = m = 0$) is called the 1*s* orbital. The next solutions of the Schrödinger equation for the hydrogen atom are four of equal energy, that is, four degenerate levels. The first of these ($n = 2$, $l = 0$, $m = 0$), like the 1*s* orbital, is spherically symmetric, but the remaining three ($n = 2$, $l = 1$, $m = 0$; $n = 2$, $l = 1$, $m = 1$; $n = 2$, $l = 1$, $m = -1$) have a node (i.e. ψ is zero) at the nucleus, and are not spherically symmetric but have different magnitudes in different directions. The orbitals with $l = 1$ are known as p orbitals; one way of indicating the shape of a p orbital is by means of a contour diagram (Figure 2.4). This orbital is axially symmetric round the z axis; ψ is zero on the nodal plane and positive on one side and negative on the other.

The directional variation of a p orbital may be illustrated by attempting to draw a single 'boundary' contour (Figure 2.5*a*). Often however, it is represented by a polar diagram. In a polar diagram we make use of the fact that the polar form of the Schrödinger equation can be factorized and we plot just the angular function. This is much easier to calculate and gives a useful approximate picture

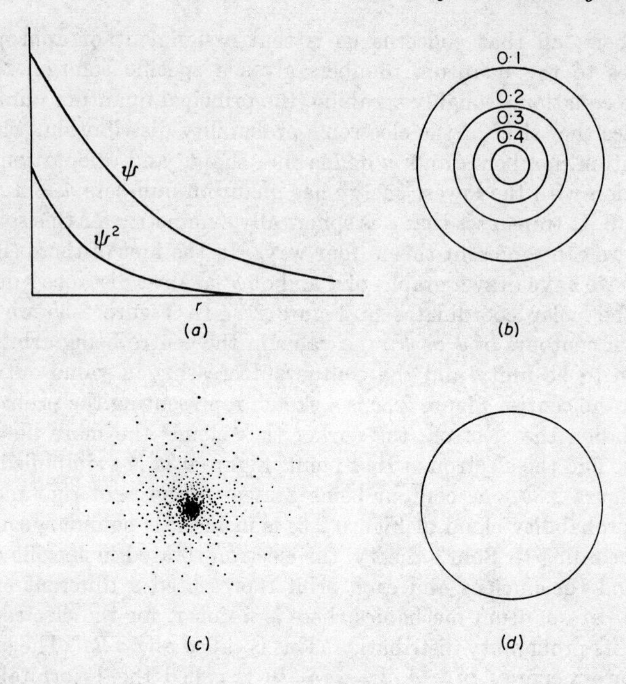

Figure 2.3. Four methods of representing a 1*s* atomic orbital.

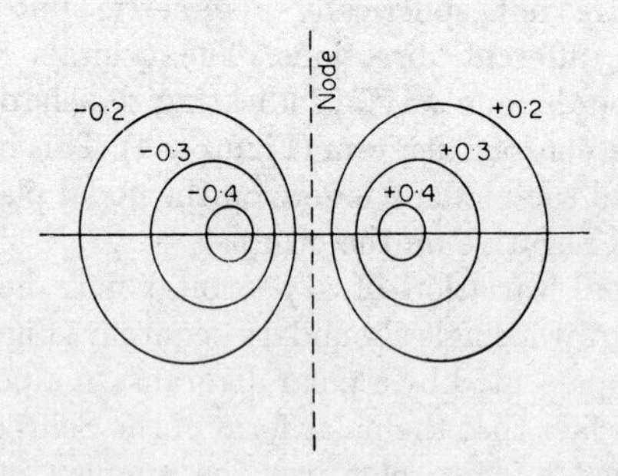

Figure 2.4. Contours for a 2*p* orbital of the hydrogen atom.

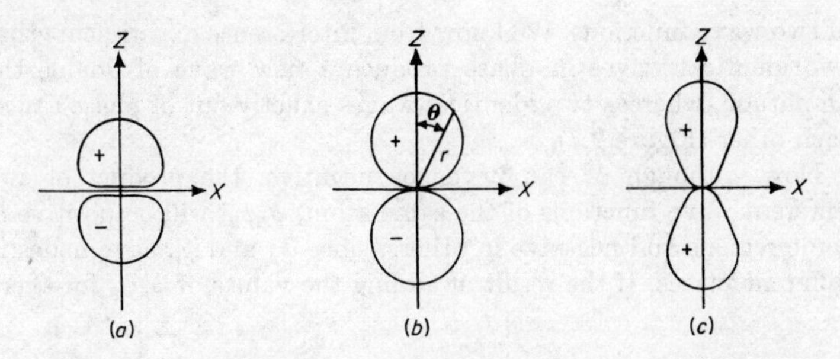

Figure 2.5. (*a*) Boundary surface for ψ, (*b*) polar diagram for ψ and (*c*) polar diagram for ψ^2.

(Figure 2.5*b*).* Another common representation is a figure of hour-glass shape (Figure 2.5*c*), which is a polar diagram for ψ^2; this seems inappropriate if some idea of 'shape' is to be conveyed. The + and − again represent the change of sign of ψ on opposite sides of the nodal surface. There are three degenerate p orbitals at right angles to each other. Figure 2.6 is a polar diagram representing their angular dependence.

The wave function ψ can be positive or negative, but the probability ψ^2 of locating the electron in a unit volume can never be negative. The sign of the wave function is only important for the interaction

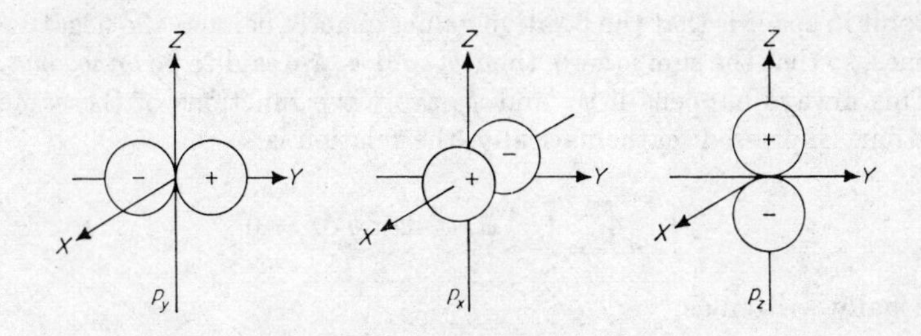

Figure 2.6. Polar diagrams representing the three $2p$ atomic orbitals.

* The polar diagram represents the path of a point along the radius vector r whose distance from the origin is equal to the angular function for θ defined by that vector.

of two wave functions. We know from interference experiments that two identical waves in phase produce a new wave of double the amplitude, whereas two identical waves exactly out of phase cancel each other (Figure 2.7).

Now, although ψ^2 can never be negative, the product of two different wave functions of the same atom, $\psi_A\psi_B$, will be positive in some regions and negative in others, since ψ_A and ψ_B have nodes in different places. If the result of adding the values of $\psi_A\psi_B$ for every

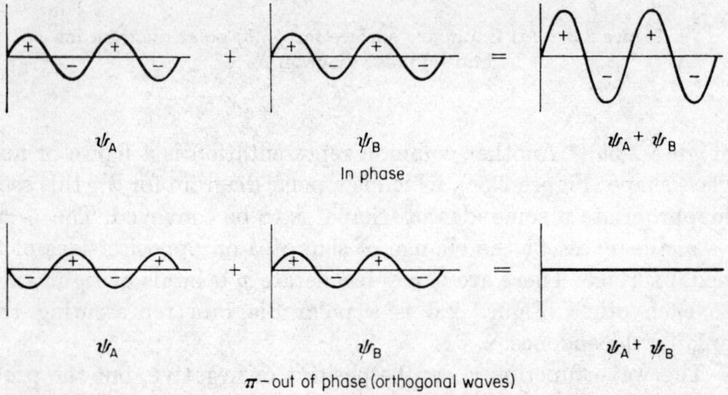

ψ_A + ψ_B = $\psi_A + \psi_B$

In phase

ψ_A + ψ_B = $\psi_A + \psi_B$

π - out of phase (orthogonal waves)

Figure 2.7. Cumulation of identical waves that are in phase (upper diagram) and π out of phase (lower diagram).

point in space is that the positive values exactly balance the negative ones, so that the sum is zero, then ψ_A and ψ_B are said to be *orthogonal*. This always happens if ψ_A and ψ_B are wave functions of the same atom. Expressed mathematically, the relation is:

$$\int_{-\infty}^{\infty}\int_{-\infty}^{\infty}\int_{-\infty}^{\infty} \psi_A\psi_B \, dx \, dy \, dz = 0$$

usually written as:

$$\int \psi_A\psi_B \, dv = 0,$$

where ψ_A and ψ_B are different wave functions.

We can solve (i.e. obtain solutions as simple expressions or *finite* series) the Schrödinger equation for the hydrogen atom, but with the equations for more complex atoms mathematical difficulties arise. In any system containing more than one electron interelectronic repulsion will appear in the potential-energy term and, since this involves the position of both electrons, it is impossible to deal with each separately. The complete equation, therefore, involves not only a complex potential-energy function but also six variables instead of three, that is, three coordinates for each electron. Direct integration of such an equation is not possible with existing mathematical techniques, and even for the helium atom the energy states are calculated by methods of successive approximation. The wave equation for a many-electron atom is normally handled by means of the *self-consistent field* (S.C.F.) *model*. The average distribution of one electron is so chosen that it represents the movement of that electron in the average field of all the others. We are not concerned here with the method, but only with the results. These results are a series of orbitals analogous to those of the hydrogen atom, except that the energy levels are further split. Whereas the single $2s$ and the three $2p$ orbitals in the hydrogen atom are degenerate, in atoms with more than one electron these separate into a $2s$ orbital which has a slightly lower energy than the three degenerate $2p$ orbitals. In the same way, the nine orbitals with the principal quantum number 3 are split into three levels. This is represented in Figure 2.8.

We can summarize the way in which the electronic structure of a many-electron atom is described as follows. Each electron is represented by a wave function called an atomic orbital and this orbital is found by a solution of the appropriate wave equation such that

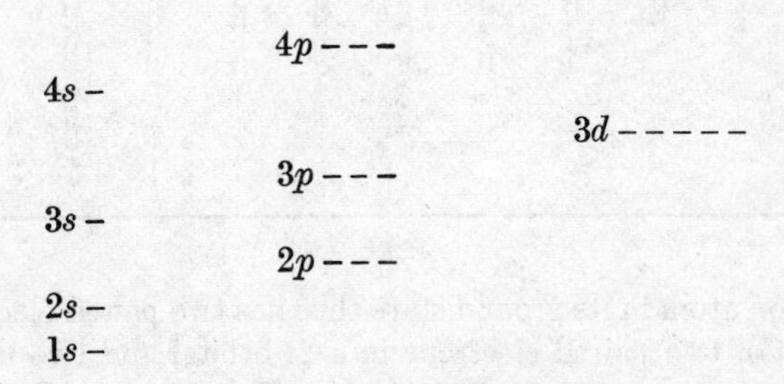

Figure 2.8. Energy levels (increasing upwards) of orbitals of a many-electron atom.

ψ^2 measures the probability density for this electron. Each atomic orbital is defined by a set of quantum numbers. These are analogous to the quantum numbers for the hydrogen atom: n the principal quantum number which defines the energy and size of the orbital, l and m which define its shape. We have to introduce a fourth quantum number which governs the spin of the electron and can have values of $+\frac{1}{2}$ or $-\frac{1}{2}$. Each atomic orbital has a specific energy level determined by the wave equation. An atom consists of a positively charged nucleus together with sufficient electrons exactly to balance the charge of the nucleus. We can thus build up the Periodic Table by filling up the orbitals, starting with those of the lowest energy and taking account of the Pauli exclusion principle which states that two electrons may never have exactly the same set of quantum numbers (this, in effect, means that only two electrons can fit into each orbital and then only if they have opposite spins). From hydrogen to boron we fill up each orbital in turn with a pair of electrons with opposing spins. When we reach carbon, however, we have the three degenerate $2p$ orbitals and in order to decide where the electrons go we need Hund's rules. These are that (1) electrons avoid the same orbital as far as possible and (2) two electrons occupying a pair of degenerate orbitals have their spins parallel in the lowest state of energy. We represent this in Table 2.2.

Table 2.2. Distribution of electrons in the first ten elements of the Periodic Table

	H	He	Li	Be	B	C	N	O	F	Ne
$1s$	↑	⇅	⇅	⇅	⇅	⇅	⇅	⇅	⇅	⇅
$2s$			↑	⇅	⇅	⇅	⇅	⇅	⇅	⇅
$2p_x$					↑	↑	↑	⇅	⇅	⇅
$2p_y$						↑	↑	↑	⇅	⇅
$2p_z$							↑	↑	↑	⇅

A carbon atom in its ground state thus has two paired electrons in a $1s$ orbital, two paired electrons in a $2s$ orbital and two unpaired electrons in two separate $2p$ orbitals. This is normally written $1s^2 2s^2 2p^2$.

Problems

1. What quantization exists for a particle in a constant potential space with no boundaries?

2. Plot $r = \cos\theta$ from $\theta = 0$ to $\theta = 2\pi$, (*a*) using Cartesian coordinates with r and θ as axes, and (*b*) using polar coordinates, with r as the radial vector.

CHAPTER 3

Molecular Orbitals and Chemical Bonds

Atoms combine to form molecules because their total energy is lowered in the process (cf. Part 1, Chapter 2). The dimensions of a molecule are fixed by that arrangement of the atoms which gives the lowest energy. The energy situation in a diatomic molecule can be represented by a diagram such as Figure 3.1.

The curve represents the sum of the electronic energy and the nuclear repulsion energy, plotted as a function of the internuclear distance r. When the atoms are far apart there is no force between them. As they come closer together there is a net attractive force (and consequent lowering of energy). The energy reaches a minimum when the internuclear distance is equal to r_0. This is the bond length of the molecule and represents the mean separation of the nuclei. If the atoms come closer together, a rapidly increasing repulsive force due to the internuclear repulsion becomes manifest. D is the electronic dissociation energy of the molecule.*

Chemical bonds are usually divided into two classes, ionic and covalent, an ionic bond being regarded as electrostatic in nature. In fact all bonds have some covalent character and there is no molecule for which the lowest-energy dissociation products are ions in the gas phase. The reason sodium chloride dissociates into ions in polar solvents is that the ions are bound much more tightly to the solvent (solvation) than is the undissociated molecule. There are currently two theories used to describe the covalent bond; the molecular-orbital theory and the valence-bond theory. Both theories explain why the Lewis electron-pairing scheme leads to stable bonds. Neither theory can really be understood without working through the mathe-

* The observed dissociation energy is slightly less than the electronic dissociation energy depicted in the diagram. This is due to the 'zero point' vibrational energy, which is small and need not concern us here.

matical arguments on which they are based, and at present we shall attempt no more than a very qualitative discussion.

The molecular-orbital theory is a logical extension of the treatment of electrons in atoms discussed above. If we can have atomic orbitals, then why not molecular orbitals? These molecular orbitals will extend over the whole molecule, and electrons can occupy these delocalized orbitals with the same restriction as for atoms, namely, that not more than two electrons can occupy any one orbital. For an atom there is a series of atomic orbitals, and electrons normally occupy those of the lowest energy. Similarly, there is a series of molecular orbitals and again the electrons will normally occupy

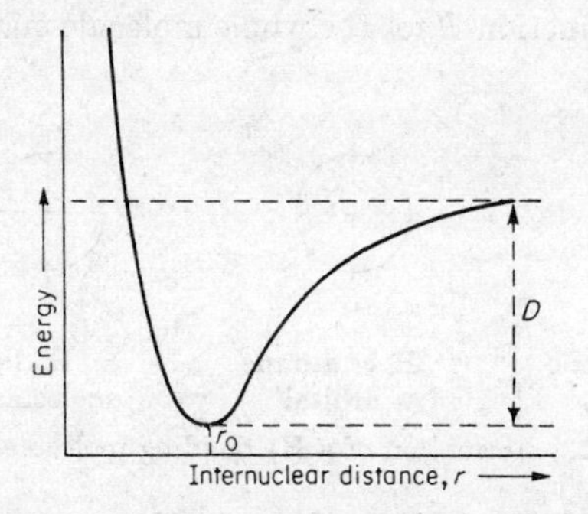

Figure 3.1. Energy curve for a diatomic molecule.

those of the lowest energy. The hydrogen molecule has a lower energy than two isolated hydrogen atoms because the lowest molecular orbital of the hydrogen molecule has a lower energy than the 1*s* orbital of the hydrogen atom.

The energy change associated with the formation of a bond between two hydrogen atoms is only 18% of their total energy. This suggests that the molecular orbitals of the hydrogen molecule are closely related to the atomic orbitals of the hydrogen atoms. Close to one nucleus an electron will be dominated by the potential field of that nucleus and a molecular orbital may resemble the atomic orbital. Thus as an approximation to the molecular orbital we may try some expression involving a sum of atomic orbitals. This is known as the 'linear combination of atomic orbitals' approximation to molecular orbitals (abbreviated to read L.C.A.O.).

Let us consider a diatomic molecule AB containing two electrons. When electron 1 is near nucleus A it will be described by an orbital similar to the atomic orbital of atom A which we shall call $\psi_A^{(1)}$. Similarly, when the same electron is near nucleus B it will be described by an orbital similar to an atom B's atomic orbital which we shall call $\psi_B^{(1)}$. The molecular orbital for this electron may then be expressed approximately as $(a\psi_A^{(1)} + b\psi_B^{(1)})$ where a and b are constants. We may construct an approximate molecular orbital for the second electron in a similar way, so that it would be written as $(a\psi_A^{(2)} + b\psi_B^{(2)})$. It can be shown that the complete wave function Ψ for any chemical species can be written reasonably accurately as a product of the individual orbitals (i.e. one-electron functions). Thus the total wave function Ψ for the whole molecule may be represented

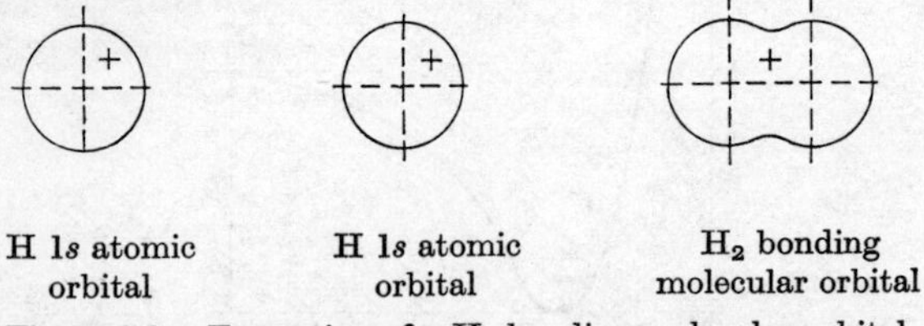

Figure 3.2. Formation of a H_2 bonding molecular orbital.

by the product of all the molecular orbitals:

$$\Psi = (a\psi_A^{(1)} + b\psi_B^{(1)})(a\psi_A^{(2)} + b\psi_B^{(2)})$$

$$\therefore\ \Psi = a^2\psi_A^{(1)}\psi_A^{(2)} + b^2\psi_B^{(1)}\psi_B^{(2)} + ab(\psi_A^{(1)}\psi_B^{(2)} + \psi_A^{(2)}\psi_B^{(1)})$$

The term $a^2\psi_A^{(1)}\psi_A^{(2)}$ represents both the electrons associated with nucleus A, i.e. A^-B^+. Similarly, the second term $b^2\psi_B^{(1)}\psi_B^{(2)}$ represents A^+B^-. For the hydrogen molecule, A is equal to B but, for any molecule where A is not equal to B, one or other ionic form will predominate and the molecule will have an uneven distribution of charge and will be polar. Pictorially we can represent the formation of the lowest molecular orbital of hydrogen from two hydrogen 1*s* orbitals as in Figure 3.2. The plus signs indicate the sign of the wave function. The linear combination of two 1*s* atomic orbitals leads to two molecular orbitals of the hydrogen molecule. Figure 3.2 shows the bonding orbital which is of lower energy than the 1*s* atomic orbital, and Figure 3.3 shows the antibonding orbital which

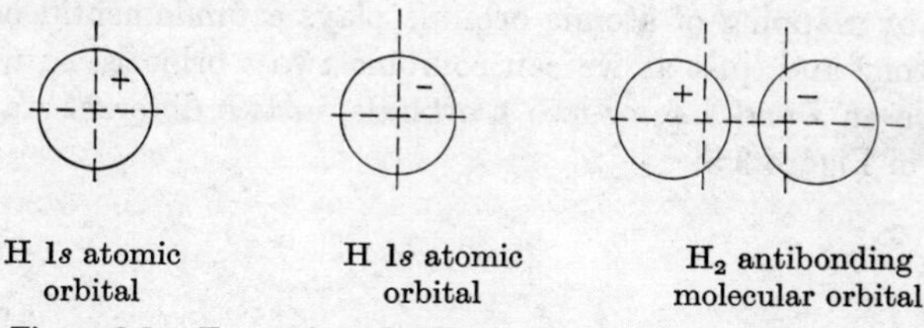

Figure 3.3. Formation of a H_2 antibonding molecular orbital.

is of higher energy than the 1*s* atomic orbitals and represents a repulsive state. The bonding orbital leads to a concentration of electrons between the nuclei. Pictorially, we can imagine the two positive nuclei as held together by the negative charge between them. On the other hand, in the antibonding orbital the internuclear space is relatively empty of electrons, so that the major effect is repulsion between the like charges of the two nuclei.

Bonds which are symmetric about the internuclear axis are called σ bonds. The relative energies of atomic orbitals are very conveniently represented on a correlation diagram (Figure 3.4). We distinguish the two molecular orbitals by denoting the bonding orbital as σ and the antibonding orbital as σ^*.

Bonds can be formed between orbitals of any two atoms provided that (i) all the atomic orbitals forming the bonds have components of the same symmetry with respect to the internuclear axis and (ii) the atomic orbitals do not differ too greatly in energy. The better the overlap of the atomic orbitals the stronger the bonds.

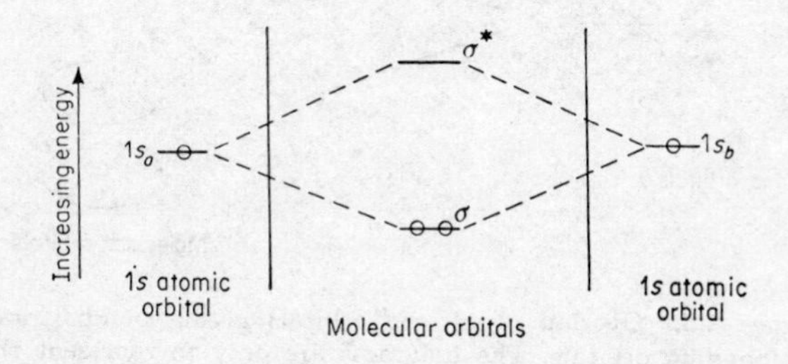

Figure 3.4. Correlation diagram for the hydrogen molecule.

The overlapping of atomic orbitals plays a fundamental part in the theory and, just as we can combine two *s* orbitals, so we can combine an *s* and a *p*, or two *p* orbitals, in two different ways, as shown in Figure 3.5.

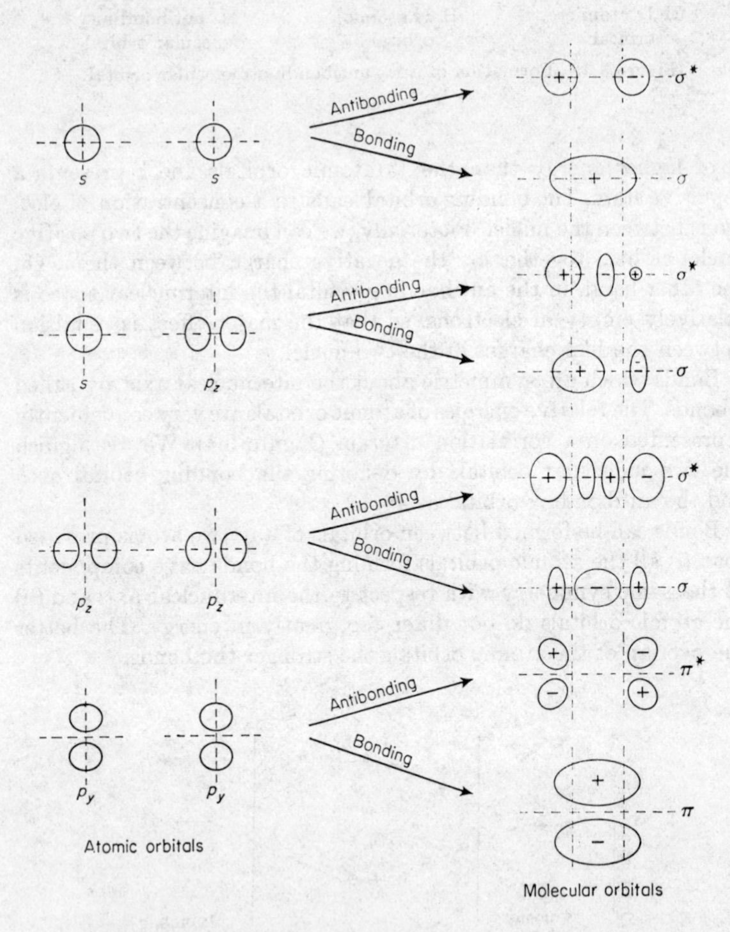

Figure 3.5. Overlap of atomic orbitals giving bonding and antibonding orbitals. The diagrams are only to represent the symmetry of the orbitals and are not intended to represent 'shape' accurately.

It will be noticed that two atomic orbitals always give rise to two molecular orbitals. The greater the lowering of energy in a bonding orbital, the greater the raising of the energy in the corresponding antibonding orbital. Molecular orbitals formed by combining two p_x (or two p_y) orbitals are antisymmetric to reflexion in a plane joining the two nuclei. These orbitals are called π orbitals. When there is negligible overlap, there is little interaction between the atomic orbitals. Such orbitals are referred to as non-bonding orbitals. For example, in the molecule Li_2, the 1*s* atomic orbitals act as non-bonding molecular orbitals, while the 2*s* atomic orbitals combine to give a bonding molecular orbital.

When we come to consider carbon, the situation is clearly much more complicated. In the ground state a carbon atom has filled 1*s* and 2*s* orbitals, and the two remaining electrons are unpaired in separate 2*p* orbitals. It is very difficult to see the connection between this ground-state electronic distribution and the observed fact that tetravalent carbon has four identical bonds distributed in a tetrahedral fashion. According to our molecular-orbital idea, however, there is no reason why our molecular orbitals should be directly related to the atomic orbitals of the individual atoms. What we need to do is to construct our molecular orbitals by taking a linear combination of the hydrogen 1*s* and the carbon 2*s* and 2*p* orbitals. The problem is, however, very difficult, so we usually approximate and borrow a technique which is really part of valence-bond theory which we shall be discussing later. If ψ_1 and ψ_2 are solutions of the wave equation for an isolated atom, corresponding to states of equal energy, then a linear combination of these solutions is also a solution of the wave equation. If we neglect the 1*s* atomic orbital as likely to be non-bonding in a carbon atom, we have to combine the 2*s* and three 2*p* orbitals to produce four identical 'hybrid' orbitals possessing tetrahedral symmetry, so that they have an arrangement consistent with empirical observation. Each 'hybrid' orbital must contain the same fraction of *s* and *p* character, i.e. each contains $\frac{1}{4}s$ character and $\frac{3}{4}p$ character. This means that the probability density ψ^2 for each orbital must contain $\frac{1}{4}s^2$ and a total of $\frac{3}{4}p^2$.

If we take the centre of the atom as the origin, then the corners of the tetrahedron can be labelled by their Cartesian coordinates as shown in the Figure 3.6. By symmetry the hybrid which points to the corner (1, 1, 1) must have equal weighting of p_x, p_y, and p_z.

This hybrid σ_1 must be

$$\sigma_1 = \tfrac{1}{2}(s + p_x + p_y + p_z)$$

Similarly we have

$$\text{for } (1, -1, -1), \quad \sigma_2 = \tfrac{1}{2}(s + p_x - p_y - p_z);$$
$$\text{for } (-1, 1, -1), \quad \sigma_3 = \tfrac{1}{2}(s - p_x + p_y - p_z);$$
$$\text{and,} \quad \text{for } (-1, -1, 1), \quad \sigma_4 = \tfrac{1}{2}(s - p_x - p_y + p_z)$$

Two atomic orbitals of the same atom are orthogonal and in the same way the four hybrid orbitals are orthogonal. These orbitals are

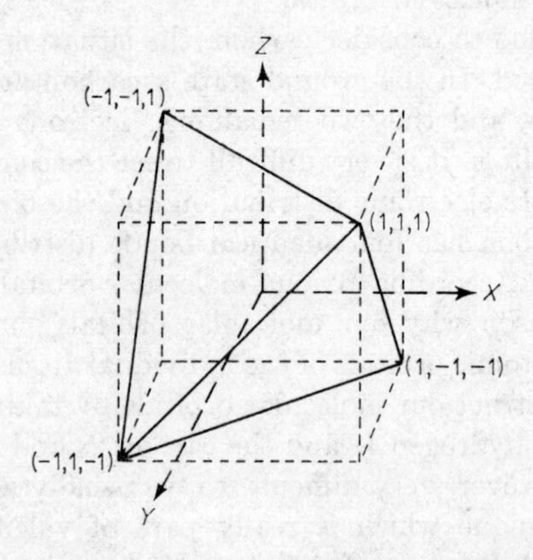

Figure 3.6. A tetrahedron set in a cube, such that its apices coincide with four of the corners of the cube. The centre of the tetrahedron is also the centre of the cube and the origin of the coordinates.

called sp^3 hybrid orbitals (denoting they are formed by the combination of an *s* orbital with three *p* orbitals). A hybrid orbital is of higher energy than the atomic orbitals from which it is derived, but the energy needed to promote an electron to a hybrid orbital is more than compensated by the energy gained in bonding. A contour diagram for an sp^3 hybrid is shown in Figure 3.7. We must always remember that hybridization is a mathematical device in our calculations. It is an essential first step in valence-bond calculations, but in molecular-orbital theory it is no more than a convenient step to simplify the calculation. Phrases such as 'the carbon atom then

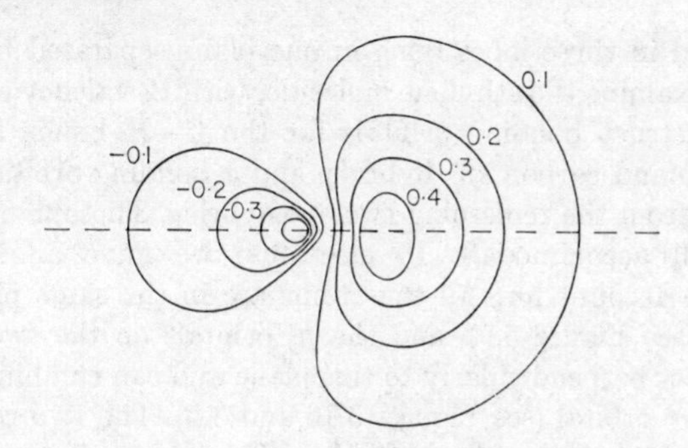

Figure 3.7. Contours for an sp^3 hybrid orbital.

rehybridizes . . .' occur frequently in the organic literature, suggesting that hybridization is a physical process, something like a car changing gear. We can only hope that our discussion has made it clear that this is not the case.

In methane we can consider the bond orbitals as formed by the overlap of the sp^3 hybrid orbitals of the carbon atom with the $1s$ orbitals of the hydrogen atoms (see Figure 3.8).

Combination of the $2s$, $2p_y$, and $2p_z$ orbitals gives us the sp^2 hybrid orbitals. These orbitals have their probability distribution

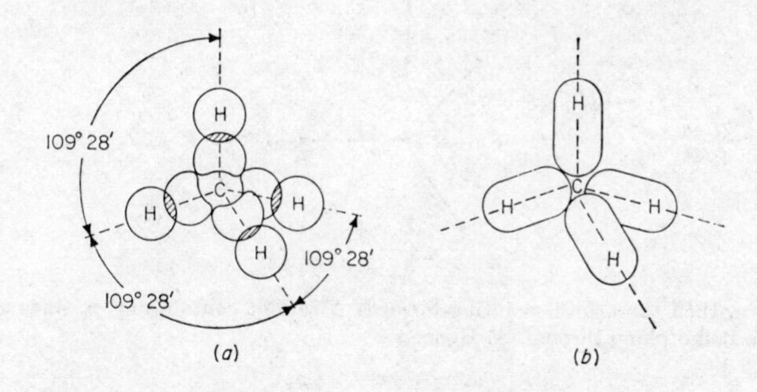

Figure 3.8. Diagrams very crudely representing (*a*) carbon sp^3 hybrid orbitals with overlapping $1s$ hydrogen orbitals and (*b*) the resulting bonding σ molecular orbitals.

concentrated in three lobes lying in one plane separated by 120°.*
If we now examine the ethylene molecule, with 12 valency electrons, we can construct bonding orbitals for the C—H bonds from the hydrogen 1s and carbon sp^2 hybrids, and a bonding orbital for the C—C bond from the remaining two sp^2 hybrids. These five bonding orbitals will accommodate 10 electrons; we have another two electrons to account for. All the atoms are in the same plane (the XZ plane; see Figure 3.9), and the p_y orbitals on the two carbon atoms project perpendicularly to this plane and can combine to give a π and a π^* orbital (see Figures 3.10 and 3.5). The two remaining electrons go into the bonding π orbital. This picture is entirely consistent with what we know about the chemistry of ethylene. The rigidity of ethylene leading to *cis*- and *trans*-isomers is readily accounted for by the angular dependence of the π bond. The chemical reactivity of ethylene differs from that of ethane because of the presence of π orbitals in the compound.

Finally, we have to consider what happens when we combine the 2s atomic orbital with just one 2p orbital. The resulting sp orbitals

* In olefins and aromatic hydrocarbons the carbon atom is joined to three other atoms which lie in a plane subtending angles of 120°. If these atoms lie on the XZ plane then the carbon p_y orbital cannot contribute to hybrid orbitals which point along the three bonds. If the hybrid orbitals are to be equivalent they must each contain $\frac{1}{3}s$ and $\frac{2}{3}p$ character. If we take axes as

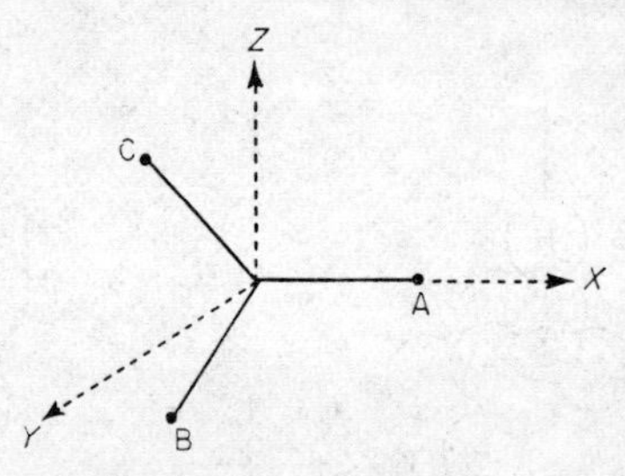

drawn, then the orbital pointing towards A cannot contain any p_z since this has a nodal plane through A. Hence,

$$\sigma_A = \sqrt{\tfrac{1}{3}}s + \sqrt{\tfrac{2}{3}}p_x$$

by symmetry arguments it is easy to show that the other two functions are:

$$\sigma_B = \sqrt{\tfrac{1}{3}}s - \sqrt{\tfrac{1}{6}}p_x + \sqrt{\tfrac{1}{2}}p_z$$
$$\sigma_C = \sqrt{\tfrac{1}{3}}s - \sqrt{\tfrac{1}{6}}p_x - \sqrt{\tfrac{1}{2}}p_z$$

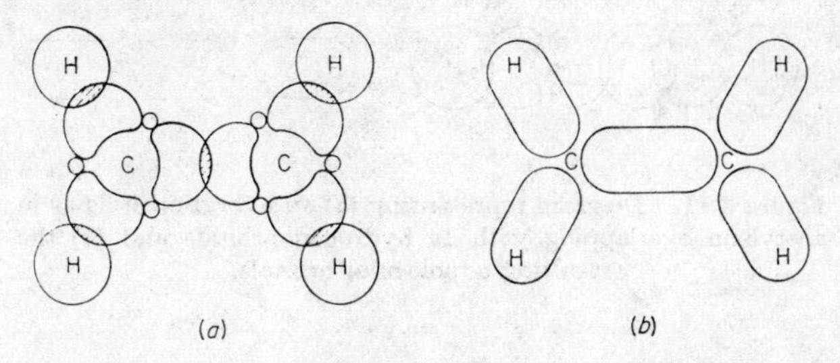

Figure 3.9. Diagram representing (*a*) carbon sp^2 hybrid orbitals in ethylene with overlapping 1*s* hydrogen orbitals and (*b*) the resulting bonding σ molecular orbitals.

project out from the carbon atom on opposite sides, perpendicular to the remaining two 2*p* orbitals.* We picture acetylene as being formed through the overlap of such *sp* orbitals (see Figure 3.11). Each carbon atom has two *p* orbitals with symmetry such that both can overlap those of the other carbon atom and form two π bonds. The net effect is that we have a π-electron density that is symmetrical around the carbon–carbon bond axis but is zero on the axis itself.

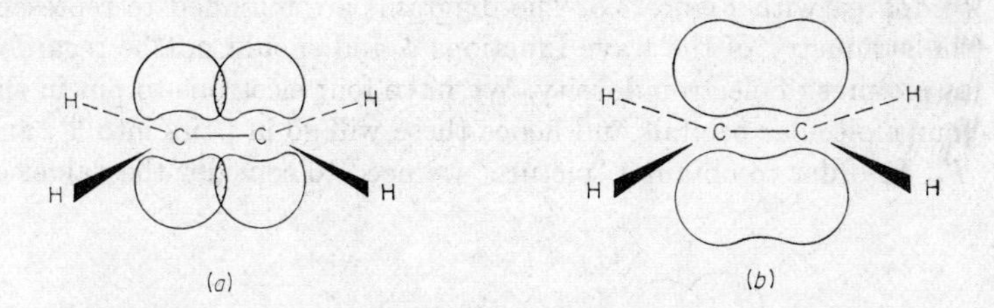

Figure 3.10. Diagram representing the $2p_y$ orbitals in ethylene, with the overlap (*a*) leading to π molecular orbitals [cf. Figure 3.5]. In the ground state the two electrons will go into the bonding π molecular orbital (*b*).

* In an *sp* orbital we require two equivalent orbitals projecting at opposite sides of the carbon atom. To be equivalent they must each contain $\frac{1}{2}s$ character and $\frac{1}{2}p$ character. Following the same procedure as for the sp^3 and sp^2 orbitals we can see that the two hybrid orbitals are defined by:

$$\sigma_1 = \sqrt{\tfrac{1}{2}}(s + p)$$
$$\sigma_2 = \sqrt{\tfrac{1}{2}}(s - p)$$

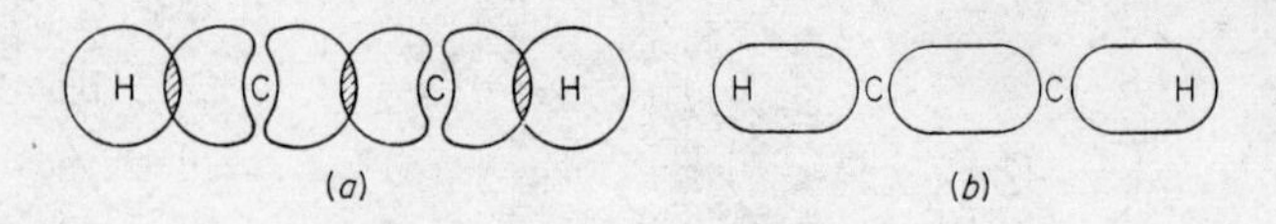

Figure 3.11. Diagram representing (*a*) *sp* hybrid orbitals as in acetylene overlapping with 1*s* hydrogen orbitals and (*b*) the resulting σ molecular orbitals.

A π bond is weaker than a σ bond because the overlap between atomic orbitals with π symmetry is less. π-Bonding is associated with a reduction of the carbon–carbon bond lengths from 1.54 Å in ethane to 1.35 Å in ethylene, and the presence of a second π bond in acetylene results in a further reduction in bond lengths to 1.21 Å.

In Part 1 (Chapter 13), when discussing the chemistry of butadiene, we came to the conclusion that a 1,3-diene behaved differently from molecules containing isolated ethylenic bonds. If we regard the carbon atoms in butadiene as sp^2 hybridized, then all the atoms lie in one plane and each carbon atom will have a $2p_y$ orbital perpendicular to this plane (see Figure 3.13). We can now combine these four $2p_y$ atomic orbitals and this will give us four π molecular orbitals. These are illustrated in Figure 3.14. Figure 3.14 should be compared with Figure 3.5. The diagrams are intended to represent the symmetry of the wave functions ψ and should not be regarded as pictures of electron density. We have four electrons to put in the four molecular orbitals and hence these will go in pairs into Ψ_1 and Ψ_2. In order to obtain a 'picture' we need to consider the values of

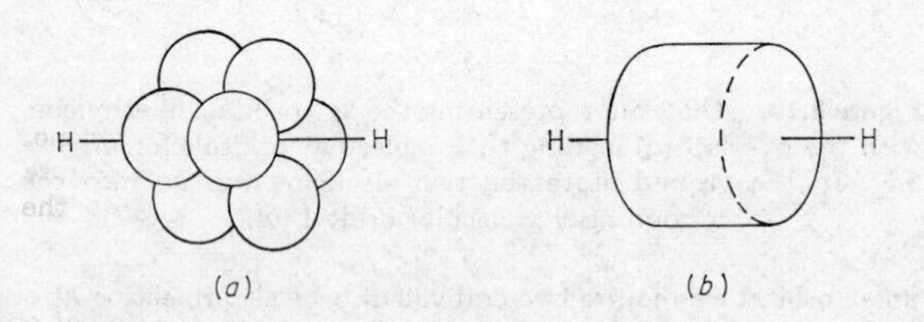

Figure 3.12. Diagram representing (*a*) the $2p_x$ and $2p_y$ atomic orbitals in acetylene with the overlap which leads to two π molecular orbitals (*b*). Notice that the π orbitals are cylindrically symmetric round the molecular axis.

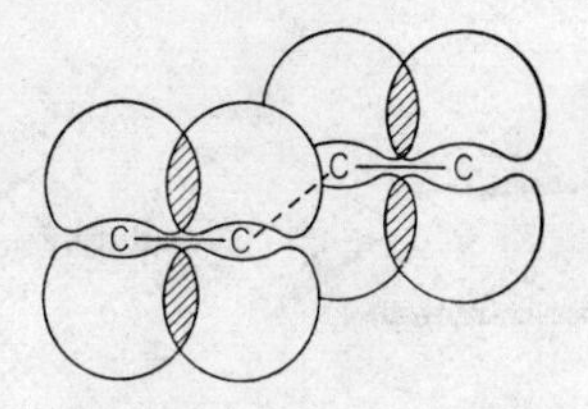

Figure 3.13. Diagram representing overlapping $2p_y$ atomic orbitals in butadiene.

Ψ^2 for the filled molecular orbitals, and Figure 3.15 is an attempt to represent the electron density clouds for ethylene and butadiene.

We see that the 1,2 and 3,4 bonds in butadiene 'look' very like the bond in ethylene and this suggests that our normal representation of butadiene with double bonds between the carbon atoms 1 and 2, and 3 and 4, and a single bond between carbon atoms 2 and 3 is a reasonable picture. However, if Figure 3.15 is correct, then there is some double-bond character associated with the bonds between carbon atoms 2 and 3; or, putting it another way, the π electrons are, to some extent, delocalized over the whole system even though they are concentrated in the 1,2 and the 3,4 bonds.

In Part 1 we went straight on from considering butadiene to consider the problem of cyclohexatriene and benzene. We found that there was a cyclic molecule C_6H_6, but that this molecule had none of

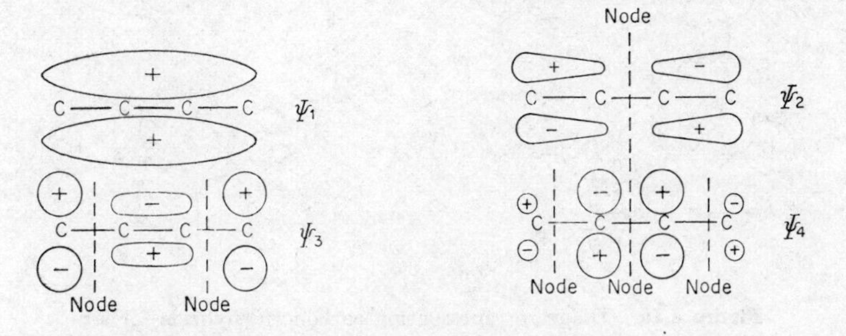

Figure 3.14. Four π molecular orbitals formed by combining the four $2p_y$ orbitals in butadiene. Energy increasing in the order $\Psi_1 < \Psi_2 < \Psi_3 < \Psi_4$.

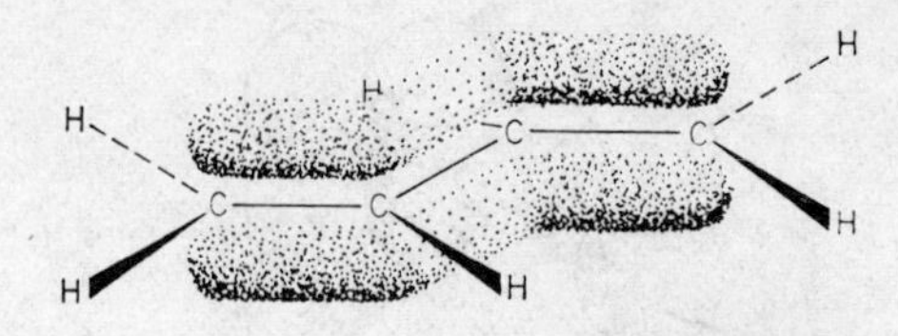

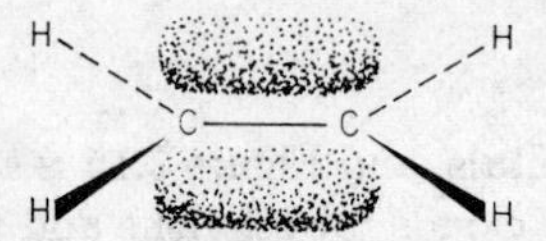

Figure 3.15. 'Electron density clouds' showing the electron distribution in the π orbitals for the ground state of ethylene and buta-1,3-diene perpendicular to the plane of the molecule.

the properties we should expect from cyclohexatriene. If we arrange six carbon atoms such that their sp^2 hybridized orbitals overlap to a maximum extent we find we must arrange these carbon atoms in a regular hexagon with an sp^2 hybrid pointing out at each apex, and these hybrids we can overlap each with a hydrogen 1*s* atomic orbital

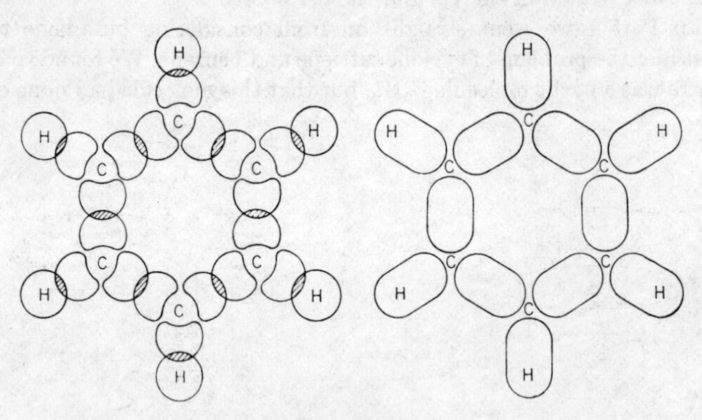

Figure 3.16. Diagram representing carbon sp^2 hybrids for six atoms arranged to provide maximum overlap between the carbon atoms. These orbitals, together with those shown overlapping with hydrogen 1*s* atomic orbitals form σ molecular orbitals.

Figure 3.17. Overlapping $2p_y$ atomic orbitals in benzene leading to six molecular π orbitals.

(see Figure 3.16). We now have the six carbon atoms and their six adjacent hydrogen atoms all in one plane. Perpendicular to this plane we have a $2p_y$ orbital associated with each carbon atom (see Figure 3.17). The six atomic orbitals will give us six molecular orbitals. Of the six molecular orbitals, four occur in two pairs of equal energy. We have depicted these orbitals in Figure 3.18.

Ψ_2 and Ψ_3 are of equal energy and are degenerate orbitals. The same is true of Ψ_4 and Ψ_5. We have six electrons to put into the six molecular orbitals, so that in the ground state two go into Ψ_1 and four into degenerate orbitals Ψ_2 and Ψ_3. If we want a 'picture' of the electron–density clouds we require to add together the values

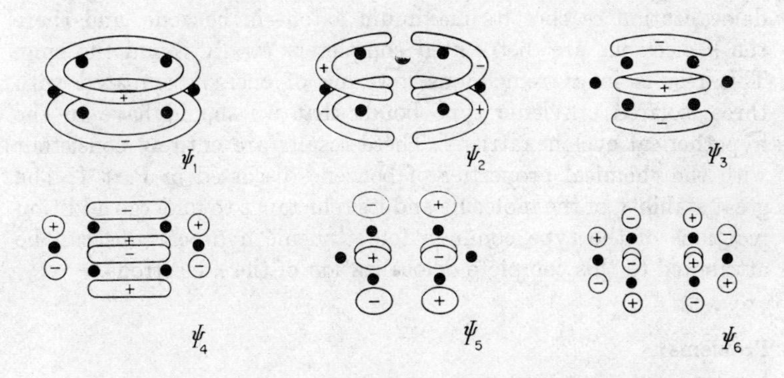

Figure 3.18. The six molecular π orbitals of benzene. Energy increasing $\Psi_1 < \Psi_2 = \Psi_3 < \Psi_4 = \Psi_5 < \Psi_6$.

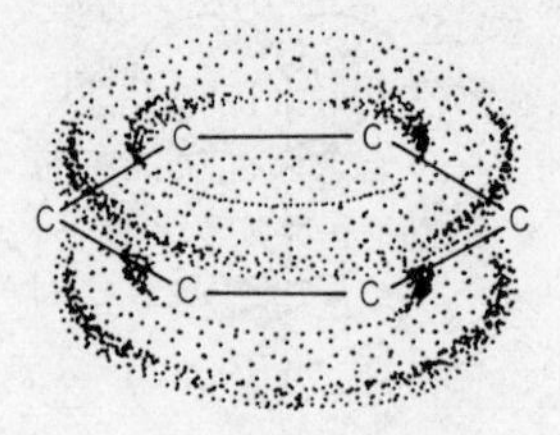

Figure 3.19. 'Electron density clouds' showing the electron distribution in the π orbitals of benzene in the ground state.

of $\Psi_1{}^2$ and $\Psi_2{}^2 + \Psi_3{}^2$. We have attempted to provide such a picture in Figure 3.19.

In Part 1 the reaction of hexa-1,3,5-triene with bromine was described and we reported that this yielded both 1,2-dibromohexa-3,5-diene and 1,6-dibromohexa-2,4-diene. Attention was drawn to the fact that no 3,4-dibromohexa-1,5-diene was formed; this suggested that a pair of conjugated double bonds is more stable than a pair of isolated double bonds. The quantum-mechanical calculations described qualitatively above give us energies of the orbitals as well as their shapes; and we find that four electrons in the first two molecular orbitals are delocalized to some extent and have a slightly lower energy than four electrons in two isolated ethylenic π orbitals. This is what we should expect if orbitals overlap to give a bonding interaction since this involves a lowering of energy. This delocalization reaches its maximum extent in benzene, and there the π electrons are distributed completely evenly round the ring. This results in a tremendous lowering of energy compared with three isolated ethylenic type bonds that we should have in the hypothetical cyclohexatriene. These results are entirely consistent with the chemical properties of benzene discussed in Part 1. The great stability of the molecule and its reluctance to undergo addition reactions of the type common for ethylenic hydrocarbons can be attributed to this complete delocalization of the π electrons.

Problems

1. Draw diagrams representing the possible interaction, bonding, non-bonding, or anti-bonding between orbitals of different atoms: (*a*) two 1*s* atomic orbitals; (*b*) a 2*s* atomic orbital with a 2*p* atomic orbital; (*c*) two 2*p* atomic orbitals. What kind of interaction is there between a 2*s* atomic orbital and a 2*p* atomic orbital of the *same* atom?

2. Consider possible electronic arrangements in the following compounds:

$$CH_2{=}CH{-}CH{=}CH_2$$
$$CH{\equiv}C{-}CH{=}CH_2$$
$$CH_3CH{=}C{=}CH_2$$

CHAPTER 4

The Valence-Bond Theory and Resonance Theory

We must next consider the second method by which we attempt to obtain suitable wave functions for molecules. The valence-bond theory is, in a sense, a direct translation of the Lewis theory into the language of quantum mechanics. The electrons in a molecule are still supposed to occupy atomic orbitals rather than molecular orbitals, but allowance is made for the fact that, if two atomic orbitals overlap one another, then we cannot be certain in which orbital an electron is to be found because electrons are indistinguishable from one another. Thus, if we consider a bond between two atoms A and B, that is formed by the pairing of two electrons with opposite spins—one from each atom—we can write a wave function for the system as follows:

$$\Psi = \psi_A^{(1)}\psi_B^{(2)}$$

where $\psi_A^{(1)}$ represents the electron 1 associated with atom A, and $\psi_B^{(2)}$ represents electron 2 associated with atom B. If the electrons changed places we would then have a new function:

$$\Psi' = \psi_A^{(2)}\psi_B^{(1)}$$

Because we cannot distinguish the electrons our true wave function must be the sum of these two wave functions with a numerical factor (c) allowing for the facts that we have only two electrons and that the total probability of finding any one electron somewhere in space is unity (i.e., the normalization condition; see Chapter 1). Our valence-bond wave function for the molecule AB is thus given by:

$$\Psi = c(\psi_A^{(1)}\psi_B^{(2)} + \psi_A^{(2)}\psi_B^{(1)})$$

This wave function gives us an electron distribution axially symmetrical about the AB bond, with the electrons concentrated between the two nuclei. This type of function gives a satisfactory result for a molecule such as hydrogen, but when dealing with a heteronuclear molecule such as hydrogen chloride we have to allow for the fact that the chlorine atom exerts a greater attraction for the electrons than a hydrogen atom does. Thus, if atom B in our original molecule attracts electrons more powerfully than A, we include a term allowing for a condition in which both electrons are on atom B, i.e. $\psi_B^{(2)}\psi_B^{(1)}$, which corresponds to the ionized molecule A^+B^-; and our wave function then becomes

$$\Psi = c'(\psi_A^{(1)}\psi_B^{(2)} + \psi_A^{(2)}\psi_B^{(1)}) + d\psi_B^{(2)}\psi_B^{(1)}$$

The relative magnitude of the coefficients c' and d will depend on the difference in the electron-attracting powers of the two nuclei. If our molecule AB is not very polar, the original symmetrical wave function will be fairly near to the truth and coefficient c' will be very much larger than coefficient d. On the other hand, in a highly polar molecule, coefficient d will become bigger.

Complete valence-bond calculations are usually very tedious, but we can see how this method led to the approximate resonance theory we described in Part 1. Let us first consider the hydrogen chloride molecule. A picture of the hydrogen chloride molecule in which there is no polarization is clearly unsatisfactory. Equally the hydrogen chloride molecule in the gas phase cannot be completely separated into a proton and a chloride anion. By resonance theory, therefore, we say that the molecule is a resonance hybrid of the two structures:

$$\text{H—Cl} \longleftrightarrow \text{H}^+ \quad :\text{Cl}^-$$

The double-headed arrow is to show that the ground state of the molecule is not correctly represented by either structure but is something intermediate between them. A mule is a hybrid—the result of crossing a horse with a donkey. It is not a horse one moment, and a donkey the next, but something quite different—a mule. Similarly, hydrogen chloride is a resonance hybrid of the two structures shown. We see that this argument is almost identical with that we used in constructing our valence-bond wave function.

We now extend this argument and in resonance theory regard a molecule as a hybrid of as many canonical forms as we can draw.

Let us take the 'classic' example of benzene. We can visualize the carbon atoms as being sp^2 hybridized, as we did in the molecular-orbital picture. We shall only consider the $2p_y$ atomic orbitals with π symmetry and consider the way in which we can pair these. We first obtain two regular 'cyclohexatriene' or Kekulé structures, by pairing orbitals on carbon atoms 1 and 2, 3 and 4, etc., or 6 and 1,

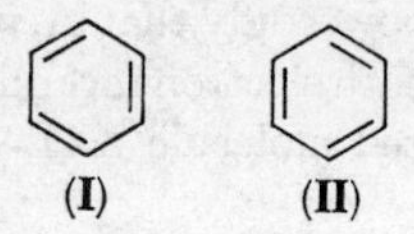

Kekulé structures for benzene.

2 and 3, etc. (see **I** and **II**). We could then pair orbitals on carbon atoms 1 and 4 and this would provide us with three additional structures (**III–V**), sometimes called 'Dewar' structures:

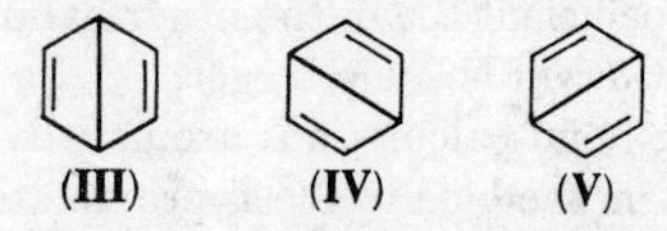

'Dewar' structures for benzene.

The complete wave function for HCl consisted of a linear combination of the wave functions for the covalent structures plus the wave function for the ionic structures. In a similar way, for benzene the wave function of the six π-electron system as the result of considering the two Kekulé structures and the three 'Dewar' structures is given by the equation:

$$\Psi = a\psi_{\mathrm{I}} + b\psi_{\mathrm{II}} + c\psi_{\mathrm{III}} + d\psi_{\mathrm{IV}} + e\psi_{\mathrm{V}}$$

Clearly, the coefficients a and b are equal on the one hand, and c, d, and e are equal on the other. If just a single Kekulé structure represented the true ground state of benzene its π-electron energy would be about the same as that of three ethylenes. We know that benzene is much more stable than this, and the wave function, which is a mixture of all the five structures, gives a lower energy than does a single structure. The 'Dewar' structures are clearly of higher energy than the Kekulé structures, since the overlap in the long bonds cannot be as good as the overlap in the bonds between adjacent carbon atoms. If we carry out the complete calculations we find the two Kekulé structures contribute about 85% of the delocalization energy, while a further 15% is contributed by the 'Dewar' structures. In order to obtain a complete valence-bond wave function for benzene we must, however, include ionic structures of the type **VI** and **VII**. These are high-energy structures and the contribution they make to the ground state is exceedingly small.

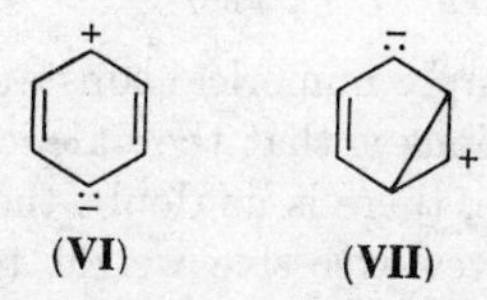

Complete calculations utilizing valence-bond theory are very tedious, but the qualitative ideas summed up in the theory of resonance that we outlined in Part 1 have proved extremely valuable to the organic chemist. We can define a set of rules for resonance to occur between two classical bond structures: (1) the nuclei must not be moved (except to adjust bond lengths), i.e. the structures differ only in electron distribution; (2) the energy of the two structures must be comparable (we have seen the ionic structures **VI** and **VII** contribute very little to the overall structure of benzene because they are of so much higher energy than the Kekulé structures); (3) the number of pairs of electrons with opposite spins must be the same in all the structures. Thus in hydrogen chloride we can have resonance between H:Cl with a covalent bond between the two atoms, and $H^{+}Cl{:}^{-}$ where the bonding electron pair is entirely on the chlorine atom; we cannot have resonance between these two structures and a structure H·Cl· in which there is an unpaired electron with parallel spins on each atom. This does not mean that we cannot have resonance between structures involving odd electrons; indeed, there certainly is resonance between two structures of the allyl radical, but notice the number of pairs of electrons with opposite spins is the same in both structures:

$$\dot{C}H_2—CH=CH_2 \longleftrightarrow CH_2=CH—\dot{C}H_2$$

Canonical structures representing 'resonance' in an allyl radical.

Valency Theories and the Organic Chemist

Valence-bond theory and molecular-orbital theory are nothing more than two methods for obtaining approximate solutions of wave equations for which it is impossible to obtain a complete solution. Let us consider our two expressions for the molecule AB:

$$\Psi_{\text{molecular orbital}} = a^2\psi_A^{(1)}\psi_A^{(2)} + b^2\psi_B^{(1)}\psi_B^{(2)} + ab(\psi_A^{(1)}\psi_B^{(2)} + \psi_A^{(2)}\psi_B^{(1)})$$

and

$$\Psi_{\text{valence bond}} = c(\psi_A^{(1)}\psi_B^{(2)} + \psi_A^{(2)}\psi_B^{(1)})$$

Since a, b, and c are simply numerical constants, the only difference between the two functions is that the valence-bond function omits ionic terms. In practice, there is no doubt that the L.C.A.O. molecular-orbital method gives excessive weight to ionic terms; and, to obtain better wave functions, molecular-orbital functions are modified in such a way as to reduce some of the emphasis of the ionic terms. Similarly, as we have seen, valence-bond wave functions for heteronuclear molecules can be improved by including terms to allow for the ionic contribution. Refining either molecular-orbital wave functions or valence-bond wave functions tends towards the same result. This is just as it should be; if these are both valid approximations, then their improvement should lead to the same end-product.

Quantitative calculation by the valence-bond theory is much more difficult than that by the molecular-orbital treatment. For this reason, far more work on complex organic molecules has been done by molecular-orbital theory than by valence-bond theory. A highly simplified form of molecular-orbital theory known as Hückel theory is so simple that a non-theoretically minded organic chemist can carry out the calculation for himself. Although valence-bond theory is far too difficult mathematically for the average organic chemist it does provide a theoretical basis for resonance theory. At the present time, organic chemists use qualitative pictures developed from both theories with gay abandon. There is no doubt that the use of resonance structures greatly assists the understanding of many organic reactions. On the other hand, a series of canonical forms describing a hybrid structure gives us no more of a ‘picture’ than ψ does of the electron. In order to provide this picture it is often helpful to draw ‘balloons’ and ‘sausages’ representing electron clouds or electron-cloud densities. We shall use both these qualitative pictures in subsequent chapters, so it is very important that the student should understand where these pictures come from. There is no doubt that a too superficial picture can lead to a great deal of misunderstanding, and many theoreticians may feel that this applies to the above discussion. The last chapters are an attempt to give as clear an insight into the basis of quantum mechanics applied to organic chemistry as is possible without recourse to mathematics. The reader will, it is hoped, be stimulated to read texts specially devoted to valence theories.*

Problem

Draw all the non-ionic canonical structures for anthracene, indicating which have the same energy.

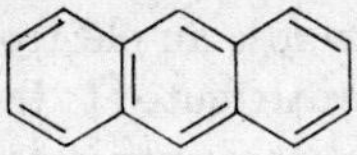

Anthracene (one Kekulé structure)

* These four chapters cover in more qualitative terms the same ground as the first five chapters in *Valence Theory* by J. N. Murrell, S. F. A. Kettle, and J. M. Tedder, John Wiley & Sons, Ltd., London, New York, and Sydney, 1965, to which the reader is directed for a more precise discussion.

CHAPTER 5

Electronegativity and Dipole Moments

In Chapter 1 of Part 1 the properties of lithium hydride, molecular hydrogen, and hydrogen fluoride were compared. Lithium hydride is salt-like and polarized in the form Li^+H^-; molecular hydrogen is non-polar and the electrons are shared equally between the two hydrogen atoms; and hydrogen fluoride is polarized in the form H^+F^-. The tendency of a lithium atom to lose an electron and the tendency of a fluorine atom to gain one was attributed to the stability of the inert-gas shell. We shall now look at these processes in somewhat more detail. The work required to remove an electron from a lithium atom, to yield a lithium cation and an electron, is known as the ionization potential and is denoted by I. Thus for any atom we have:

$$X \rightarrow X^+ + e^- \quad \Delta H = I_X$$

(I_X being the ionization potential, expressed in kcal atom^{-1} or in electron volts).

We should expect this process to be easiest for the alkali metals and (neglecting the noble gases) most difficult for the halogens. The experimentally determined values given in Table 5.1 show that this is so.

Table 5.1. Ionization potentials (ev) of the First Row of the Periodic Table

			H				He
			13.6				24.5
Li	Be	B	C	N	O	F	Ne
5.4	9.5	8.3	11.2	14.5	13.7	18.6	21.5

The energy released when a fluorine atom reacts with an electron to yield a fluoride anion is known as the electron affinity of fluorine and is denoted by A:

$$X + e^- \rightarrow X^- \quad -\Delta H = A_X$$

(A_X being the electron affinity, usually expressed in kcal atom^{-1}).

Experimental determination of electron affinities is very much more difficult than that of ionization potentials. The results are as we should expect, namely, that A_F for fluorine is very much larger than A_{Li} for lithium.

We have repeatedly assumed that, in a molecule such as hydrogen chloride HCl, electronic charge flows from the hydrogen atom to the chlorine atom. This is really saying that chlorine is more electron-attracting than hydrogen. The term used to describe the magnitude of this effect is *electronegativity*. Chlorine is said to be more electronegative than hydrogen. Clearly the flow of electronic charge from the hydrogen atom to the chlorine atom is related both to the ionization potentials I and the electron affinities A. R. C. Mulliken devised a scale of electronegativity by which the electronegativity χ of an atom is defined:

$$\chi_X^{(m)} = \tfrac{1}{2}(I_X + A_X)$$

The superscript (m) stands for Mulliken. The ionic character of a bond in X—Y is then determined by $|\chi_X - \chi_Y|$.

Because electron affinities are hard to measure, an approximate electronegativity scale has been constructed (in fact, the approximate scale predates the Mulliken scale). This is based on the concepts of valence-bond theory and employs a term Δ_{AB} which is the difference between the bond dissociation energy, D(A—B), and the expected covalent bond energy. This expected covalent bond energy is *assumed* to be the arithmetic mean of the bond dissociation energies, D(A—A), and D(B—B). The electronegativity difference between atoms A and B is then given by:

$$\sqrt{\Delta}_{AB} = |\chi_A^{(p)} - \chi_B^{(p)}|$$

The superscript (p) stands for Pauling, the originator of this scale. The values obtained by this method are related to the values obtained from the Mulliken scale by a numerical constant:

$$\chi_X^{(m)} = 2.78\chi_X^{(p)}$$

Table 5.2 lists Pauling electronegativities of the atoms of greatest interest to an organic chemist.

Table 5.2. Pauling's electronegativity scale $\chi^{(p)}$

			H 2.1			
Li 1.0	Be 1.5	B 2.0	C 2.5	N 3.0	O 3.5	F 4.0
Na 0.9	Mg 1.2	Al 1.5	Si 1.8	P 2.1	S 2.5	Cl 3.0
K 0.8					Se 2.4	Br 2.8

If, in a diatomic molecule such as hydrogen chloride, the chlorine atom takes a bigger share of the bonding electrons, it follows that the molecule must be polar, i.e. it must carry a negative charge at the chlorine end of the molecule and a corresponding positive charge at the hydrogen end of the molecule. If such a polar molecule is placed in an electric field we should expect it to become orientated in that field just as a bar magnet placed in a magnetic field becomes aligned with the magnetic field. The most common example of the latter phenomenon is a compass needle in the earth's magnetic field. The question now arises: can we actually observe this polar character of a molecule? The answer is, yes—by examining the dielectric properties of the substance. According to Coulomb's law, the force (f) between two charges q_1 and q_2 in a vacuum is given by:

$$f = q_1 q_2 / r^2$$

where r is the distance between the charges. The force between charges that are not in a vacuum is reduced and the equation has to be modified to:

$$f = q_1 q_2 / r^2 D$$

The constant D introduced in this equation is a property of the medium in which the charges are placed and is called the dielectric constant. The more polar the molecules of the medium, the larger the dielectric constant. Table 5.3 lists dielectric constants of some common liquids.

Table 5.3. Dielectric constant (liquids at 25°C)

CCl_4	1.5	$(CH_3)_2CO$	22
n-C_5H_{12}	1.8	CH_3OH	32
$(C_2H_5)_2O$	4.5	H_2O	79

The dielectric constant of a liquid is not solely dependent on the polar nature of the molecules, but it is noticeable in the above table that carbon tetrachloride and pentane (which according to our ideas should be completely non-polar molecules) have low dielectric constants whereas for example, acetone (which according to the ideas expressed in Part 1 should have a polar carbonyl bond) has a high dielectric constant.

Let us consider a substance in its vapour state placed in an electric field. If the molecules of that substance have polar bonds of the kind we have been discussing, they will become orientated in line with the applied field in so far as the thermal motion of the molecules makes this possible. In addition, each molecule will also have an induced polarization due to the electric field. Thus a non-polar molecule such as a hydrogen molecule will become polarized in the presence of an applied electrostatic field. The *dipole moment*, μ, of a molecule is defined as the permanent electric moment which exists in the absence of an applied field. This moment will be expressed in electrostatic units (e.s.u.) multiplied by the distance between the atoms in centimetres. The unit of electronic charge is 4.8×10^{-10} e.s.u. and the distance between atoms in a molecule is of the order of 10^{-8} cm. A molecular dipole is measured in Debye units, 1 Debye (D) $= 10^{-18}$ e.s.u. c.g.s. units.

The details of how dipole moments are measured need not concern us here, but we shall consider briefly the principles involved. If we keep the applied electrostatic field constant but vary the temperature, then the observed polarization, P, must be of the form

$$P = f(\alpha) + f'(\mu)$$

where $f(\alpha)$ is some function of the induced polarization and $f'(\mu)$ is some function of the permanent dipole. $f(\alpha)$ will be temperature-independent, but $f'(\mu)$ must vary inversely with the temperature since an increase in temperature will increase the molecular movement and reduce the extent to which the molecules become aligned in the applied field. Thus the dipole moment is obtained by measuring the total polarization (which is related to the dielectric constant) of a substance in the vapour phase over a range of temperatures and plotting the polarization against the inverse of the absolute temperature. The slope of the straight line so obtained is a numerical constant × a function of the dipole moment.

The relation between the measured dipole moment and electronegativity differences is shown for some simple diatomic molecules in Table 5.4.

Table 5.4. Electronegativity differences and dipole moments

	H—H	N—N	H—F	H—Cl	H—Br	H—I
$\chi_A^{(p)} - \chi_B^{(p)}$	0	0	1.9	0.9	0.8	0.4
μ (D)	0	0	1.9	1.0	0.8	0.4

In Part 1 considerable emphasis was laid on the increasing polar character of carbon–nitrogen, carbon–oxygen, and carbon–fluorine bonds. The dipole moments of some relevant compounds are listed below, though these are not completely valid comparisons since they are dipole moments for the whole molecule and involve bonds other than the C—X bond.

	CH_3—CH_3	CH_3—NH_2	CH_3—OH	CH_3—F
μ (D):	0	1.3	1.7	1.8

The dipole moment of a polyatomic molecule is made up of the vector sum of all the individual bond moments. Water, for example, has a dipole moment of 1.84 D. Note that if water were a linear

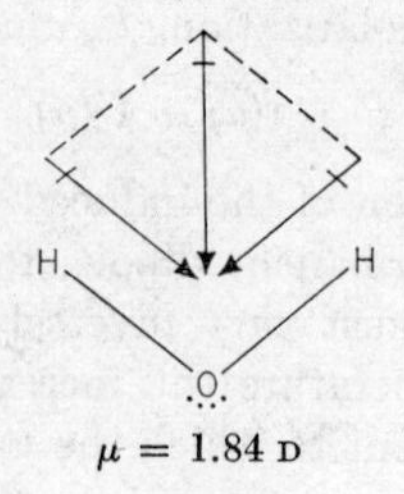

μ = 1.84 D

molecule the two hydrogen–oxygen bond moments would be in direct opposition to each other and the total dipole moment of water would be zero. The fact that water has a dipole moment is part of the evidence showing that the water molecule is bent. In the above diagram for water we have represented the dipole moment by an arrow with a cross on its tail. This representation is of great general utility and is really a shortened form of a positive charge and a negative charge with an arrow to show the movement of electrons.

$$\overset{+\ \longrightarrow\ -}{\text{H—Cl}} \quad \text{written as} \quad \overset{+\!\!\longrightarrow}{\text{H—Cl}}$$

Aromatic compounds are of particular interest because the direction of the bonds in the molecule are known and are fixed. Figure 5.1 shows the dipole moments of some monosubstituted benzenes. Notice that measurement of a dipole moment of a molecule gives no

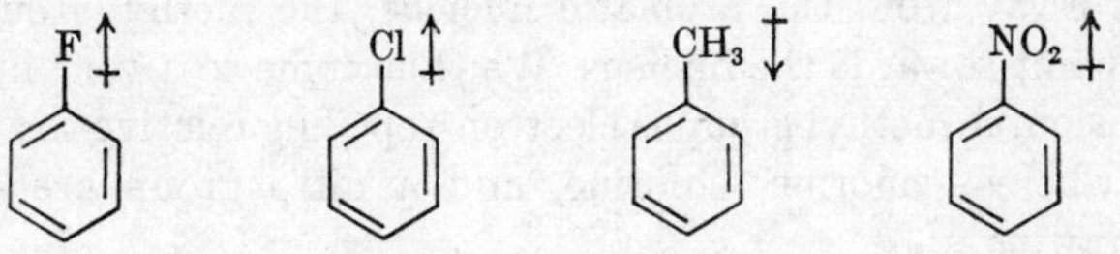

F Cl CH_3 NO_2

μ(D): 1.7 1.6 0.4 3.9

Figure 5.1. Some simple dipole moments.

indication of the direction of the dipole. In the compounds shown in Figure 5.1 we have diagrammatically represented the dipole pointing towards the fluorine atom in fluorobenzene, towards the chlorine atom in chlorobenzene, and towards the nitro group in nitrobenzene, but in toluene we have drawn the dipole pointing towards the benzene ring. We shall see the reason for doing this in just a moment. First let us look at the dipole moments of the three dinitrobenzenes (Figure 5.2) because they illustrate very clearly the way the total dipole moment of the molecule is made up of the vector sum of the individual bond moments. Now we can derive the direction of the molecular dipole in fluorobenzene and chlorobenzene from the previous discussion of the polarity of bonds. If we now look at Figure 5.3 we find that the moment of *p*-chloronitrobenzene is less than that of nitrobenzene (Figure 5.1). This must mean that the individual bond moments are in opposition, and as we have agreed

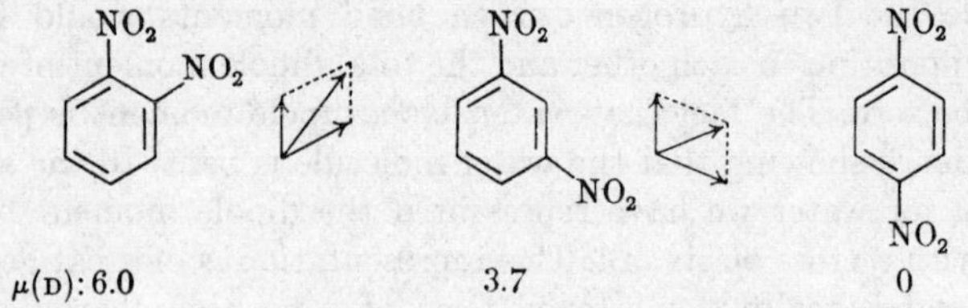

NO_2 NO_2 NO_2 NO_2 NO_2 NO_2

μ(D): 6.0 3.7 0

Figure 5.2. Vector sums of dipole moments.

that the carbon–chlorine bond moment is pointing towards the chlorine atom this tells us that the nitro group must be more electronegative than the carbon atoms in the benzene ring and that the dipole moment is again pointing away from the benzene ring. *p*-Nitrotoluene, on the other hand, has a larger dipole moment than nitrobenzene. This must mean that the bond moments are in the same direction. As we have just argued that the nitro-group dipole points away from the aromatic nucleus, the methyl-group dipole must point towards the nucleus. We thus come to a very important conclusion. A methyl group is electron-repelling relative to a benzene ring, whereas fluorine, chlorine, and a nitro group are electron-withdrawing.

The determination of dipole moments can also be of the greatest help in determining the structures of molecules, particularly of *cis*- and *trans*-isomers. For example, *trans*-dichloroethylene has no dipole moment, while *cis*-dichloroethylene has an appreciable one (see Figure 5.4).

To summarize this chapter: in Part 1 we suggested from our knowledge of inorganic chemistry that electrons in bonds between carbon atoms and other elements would not be equally shared. The

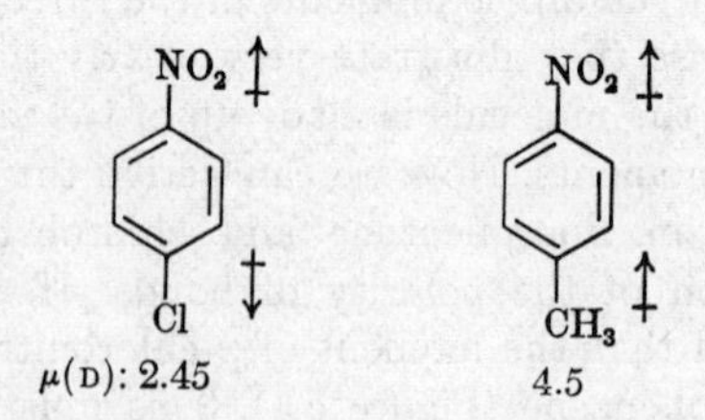

Figure 5.3. Opposing and reinforcing dipole moments.

present chapter has introduced the concept of electronegativity, such that the difference in electronegativity between two atoms gives a direct measure of the polarity of a bond formed between them. We have also found that this polarity of bonds is not a purely hypothetical concept but can actually be measured. In bonds between unlike atoms there is an electrostatic dipole moment, like the

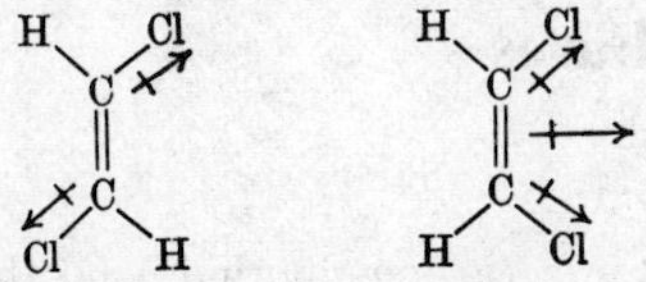

μ(D): 0 1.85

Figure 5.4. Dipole moments of *trans*- and *cis*-dichloroethylene.

magnetic moment of a bar magnet. In the next few Chapters we shall reconsider some of the reactions discussed in Part 1 in order to see how their course is affected by the presence of electronegative atoms or groups adjacent to a reaction site.

Problem

Predict the relative magnitude of the dipole moments of the following disubstituted benzenes:

1,2-dichlorobenzene; 1,3-dichlorobenzene; 1,4-dichlorobenzene; 2-chlorotoluene; 4-chlorotoluene; 1,3,5-trichlorobenzene [Chlorobenzene, μ = 1.6 D; toluene, μ = 0.4 D.]

CHAPTER 6

Electronegativity and the Dissociation Constants of Acids and Bases

We have seen that in a carbon–fluorine bond the fluorine tends to take a larger share of the electrons forming that bond. When we find experimentally that the dissociation constant of fluoroacetic acid is greater than that of acetic acid, it is not unreasonable to assume that the electron-withdrawing properties of the fluorine atom are in some way connected with the greater acidity of the fluoroacetic acid. It is tempting to suggest that, because the electrons in the fluorine–carbon bond are drawn towards the fluorine, the electrons in the carbon–carbon bond will, to a lesser extent, be drawn from the carbonyl-carbon to the carbon attached to the fluorine atom. If one continues this argument down the chain, from the carbon to the hydroxyl-oxygen, one finally comes to the oxygen–hydrogen bond. If the electron pair of this bond is drawn towards the oxygen, it will be

```
       H      O
       |     //
F<--+--C<--+--C
       |      \
       H       O<--+--H
```

easier for the hydrogen to separate as a proton. This would appear to provide an 'explanation' of the greater acidity of fluoroacetic than of acetic acid. However, we must proceed with caution. What we are apparently saying is that the electrostatic energy required to remove the (positive) proton away from the (negative) carboxylate anion is being reduced by the electronic pull of the fluorine atom. This in turn suggests that the enthalpy change associated with the ionization of fluoroacetic acid is more favourable than that in the dissociation of acetic acid. In fact, the situation is more complicated than this.

Let us go back to a much simpler system. Suppose we have two identical bulbs connected by a tap, and in one bulb we have the inert gas neon and in the other the inert gas helium, at equal low pressures. It is obvious what happens if we turn the tap: the two gases mix spontaneously and in a fairly short time there will be the same proportions of helium and neon molecules in both vessels. There is a spontaneous change in which no energy is absorbed or liberated and yet the change is not reversible—it is most improbable that the components would separate, regenerating the original situation. The mixed state has a greater probability of existence than the unmixed state: the driving force behind the mixing is the tendency for the system to attain the most probable, i.e. the most random state.

We can liken an exothermic process to a ball rolling down an inclined plane. However, we frequently meet endothermic processes. To take two simple examples: When we dissolve sodium nitrite in water the solution becomes cold; thus the sodium nitrite is dissolving even though heat has to be absorbed from the surroundings to complete the process. Similarly, ether, left on a watch-glass, will evaporate, but again heat will be absorbed from the surroundings and the watch-glass will become sufficiently cold for water to start to condense on its under surface. Both these processes involve an increase in randomness. Sodium nitrite changes from the ordered state in the crystal to a disordered state in solution. Similarly, the ether molecules in the liquid are in a more organized state than they are in the vapour. In both examples the driving force to obtain the most random state overcomes the enthalpy effect.

We are concerned with the ways in which energy is distributed in the system. A fixed amount of energy can be distributed among the molecules of, for example, a gas in a bulb, in W different ways. The more ways in which the energy is distributed (i.e. the larger W) the more random is the system. We can visualize two extreme distributions, one in which all the energy of the system is taken up by a single molecule, and another in which all molecules have identical energies. Both systems are highly ordered and both systems are extremely improbable. The most probable arrangement is the most random.

Thus there are two factors controlling a change which together we may call the 'net driving force': (1) the tendency to adopt the lowest possible energy and (2) the tendency to adopt the highest possible state of randomness. We can express this as an equation:

'Net driving force' = (The effect of the change in energy)
− (The effect of the change in randomness)

The 'net driving force' can be expressed in units of energy. If we are concerned with a chemical process at constant temperature and pressure, we call the 'net driving force' the change in *free energy* and denote it by the symbol ΔG, i.e.

'Net driving force' = The effect of the change in free energy (ΔG)

The free energy is a composite term involving energy (enthalpy) and some function of the randomness. Thus

$\Delta G = \Delta H - Xf$ (the randomness of the final state
− the randomness of the initial state)

Randomness we have associated with W, but W is only a number and ΔG is in units of energy, hence X must have units of energy. Statistical theory shows that $X = \boldsymbol{k}T$ ($\boldsymbol{k}$ = Boltzman's constant = $\boldsymbol{R}/\boldsymbol{N}$ where $\boldsymbol{R}$ is the gas constant and $\boldsymbol{N}$ is Avogadro's number; T = temperature in °K), and that f is a logarithmic function, i.e.

$$\Delta G = \Delta H - \boldsymbol{k}T\,\Delta(\ln W)$$

$\boldsymbol{k} \ln W$, a measure of the randomness of the system, is usually denoted by the symbol S and is given the name *entropy*. Hence

$$\underline{\Delta G = \Delta H - T\,\Delta S}$$

We must now return to fluoroacetic acid in water. ΔH and ΔS are quantities that we can determine by experiment. We find that ΔH is indeed more negative for fluoroacetic acid than it is for acetic acid; i.e. the ionization of the halogenoacetic acid is more exothermic than that of acetic acid, in line with the discussion at the beginning of this chapter. However, we also find in going from acetic to fluoroacetic acid there is a very significant change in ΔS and that the increased acidity of fluoroacetic acid at room temperature is to some extent due to a smaller entropy term. Some further examples are given in Table 6.1.

The dissociation constant of fluoroacetic acid is about 150 times that of acetic acid at room temperature. Let us consider a case

Table 6.1. Ionization data for halogenoacetic acids at 25°

	$10^5 K_a$	ΔG^0 (cal mole^{-1})	ΔH^0 (cal mole^{-1})	ΔS^0 (cal mole^{-1} degree^{-1})
CH_3COOH	1.754	6487	− 112	−22.1
CH_2FCO_2H	259.2	3527	−1390	−16.5
CH_2ClCO_2H	135.6	3911	−1123	−16.9
CH_2BrCO_2H	125.3	3958	−1239	−17.4
CH_2ICO_2H	66.80	4330	−1416	−19.3

where the dissociation constants are closer together. Chloroacetic acid ($10^5 K_a = 135.6$) is stronger than bromoacetic acid ($10^5 K_a = 125.3$) and we might, on the basis of the original argument, feel that this is what we should expect since chlorine is more electronegative than bromine. However, when we look at the values for ΔH and ΔS we find that ΔH is more negative for bromoacetic acid and that chloroacetic acid is only stronger at room temperature because it has a smaller entropy term. The simple argument has thus broken down and we might be tempted to argue that, although electronic effects can be used to explain large differences in dissociation constants, they should not be applied when the differences are small. There is, however, overwhelming experimental evidence that electronic effects can account for extremely small differences in dissociation constants, *so long as* the structures of the acids are similar, as in the case of chloroacetic and bromoacetic acid. Where then lies the fallacy? The fallacy lies right at the beginning where we assumed that the difference in the dissociation constants of acetic and fluoroacetic acid could be entirely attributed to the electrostatic energy necessary to remove a proton from the carboxylic acid and that this energy in turn was represented by the heat of ionization. The process is, in fact, extremely complicated. The undissociated acid is weakly bound to a number of solvent molecules, and in the same way the hydroxonium ion and the carboxylate anion have shells of solvent molecules weakly bound to them. The dissociation of acetic acid in water is thus not simply the transfer of a proton from acetic acid to a water molecule but a highly complex process involving the arrangement of a large number of molecules. Clearly, these changes involve changes in randomness, i.e. entropy, as much as changes in energy. Our false logic lay in trying to use a far too simple picture of the ionization process.

Instead of trying to use unjustifiably simple pictures of what are extremely complicated processes we must fall back on experiment. On the basis of a vast mass of accumulated experimental data we enunciate an '*empirical hypothesis*': '*A substituent in any charged species which tends to spread the charge in that species will lower its free energy.*' Thus, according to this hypothesis, fluoroacetic acid is a stronger acid than acetic acid because in the fluoroacetate anion the fluorine, by its electron attraction spreads the negative charge over more of the molecule. In the same way chloroacetic acid is stronger than bromoacetic acid because the more electronegative chlorine atom spreads the charge more effectively than does a bromine atom. We pay attention only to the free-energy changes and make no attempt to divide these into enthalpy and entropy changes. We shall find, as we come to consider not only equilibria but also reaction rates, that this hypothesis is one of very great generality. We must note, however, that the hypothesis as we have given it, is in terms of a substituent; we must, therefore, only apply the hypothesis inside a family of molecules with basically the same structure and differing only in the groups or atoms attached to this basic structure, as in the case of the four halogenoacetic acids. Provided we stick within these rules we can use the hypothesis meaningfully to interpret differences in dissociation constants of acids differing as little as from 2.8×10^{-5} to 3.1×10^{-5}.

Since we have found that we are concerned with free-energy changes it would be useful to have a direct relation between the free energy and the equilibrium constants that we determine. Such a relation is provided by the van't Hoff isotherm:

$$\Delta G^0 = -\boldsymbol{R}T \ln K$$

Thus an equilibrium constant is itself a measure of the free-energy change.

The concepts of free energy and entropy have been introduced in a qualitative fashion in this chapter in order that the reader should appreciate these ideas as fundamental to chemistry. The account here is quite inadequate for chemistry as a whole and is only intended to emphasize the significance of these quantities in organic chemistry. It is essential that students should realize the universality of thermodynamics and its importance in all branches of chemistry. *Proper understanding of the reactions of organic compounds is impossible without some knowledge of the basic principles of thermodynamics.*

Problem

'The replacement of the non-hydroxylic hydrogen atom of formic acid by an alkyl group would be expected to produce a weaker acid as the electron-donating inductive effect of the alkyl group will increase electron availability on the oxygen and so destabilize the anion.' Comment on the validity of this argument. Comment also on the use of the term 'electron-donating' with respect to an alkyl group.

CHAPTER 7

The Electronic Theory of Organic Chemistry

We shall now look at the dissociation constants of some acids and bases and, remembering the discussion of electronegativity (Chapter 5), see what interpretation we can put on the experimental results by means of the empirical hypothesis put forward in Chapter 6. Table 7.1 lists the dissociation constants of acetic acid, propionic acid, butyric acid, and pentanoic (n-valeric) acid together with their monochloro-derivatives, the chlorine atom always being at the end of the chain farthest from the carboxyl group.

Table 7.1. Dissociation constants ($10^5 K_a^{25}$) of some straight-chain acids and the corresponding ω-chloro-substituted acids

CH_3CO_2H	$CH_3CH_2CO_2H$	$CH_3[CH_2]_2CO_2H$	$CH_3[CH_2]_3CO_2H$
1.8	1.3	1.5	1.4
$ClCH_2CO_2H$	$Cl[CH_2]_2CO_2H$	$Cl[CH_2]_3CO_2H$	$Cl[CH_2]_4CO_2H$
160	8.2	3.0	1.9

Table 7.1 shows that the chloro-substituted acid is always stronger than the unsubstituted acid, although in the case of the pentanoic (valeric) acids the difference is insignificant. The ability of the chlorine atom to spread the negative charge decreases rapidly as the chain lengthens, i.e. the dipole of the carbon–chlorine bond exerts a very pronounced effect on the site adjacent to it, but its effect on sites more remote decreases very rapidly with increasing distance. This effect, which is basically electrostatic, is known as the *inductive effect* and signified as I_σ. The subscript σ is added because a dipole has a quite different effect on a molecule with π orbitals in the carbon chain. If the increase in acidity of chloroacetic acid can be attributed to spreading of the negative charge in the anion by the dipole of the

bond linking the electronegative chlorine atom, then any group linked by a polar bond will exert an inductive effect and affect the acidity of the substituted acetic acids. Table 7.2 lists the dissociation

Table 7.2. Dissociation constants ($10^5 K_a^{25}$) of some substituted acetic acids $X–CH_2CO_2H$

X = H	$CO_2C_2H_5$	OCH_3	CN	NO_2
1.8	22	34	300	4,800

constants of some such substituted acetic acids. All these groups contain atoms more electronegative than carbon (see Chapter 5), so that relatively they are all electron-attracting and all able to help to spread the negative charge in the acetate anion. Electron-attracting groups are denoted by $-I_\sigma$ (the minus sign indicating the withdrawal of electrons).*

In Chapter 5 it was shown that a methyl group is electron-repelling relative to the benzene ring. Many books say that acetic acid ($K_a^{25} = 1.8 \times 10^{-5}$) is a weaker acid than formic acid ($K_a^{25} = 1.7 \times 10^{-4}$) because of the electron-repelling properties of the methyl group. In terms of our hypothesis such a comparison is not justified. The introduction of a methyl group directly attached to the carboxyl group in place of a hydrogen atom is such a gross change in structure that the hypothesis cannot meaningfully be applied in this connexion, even though there is evidence from many other sources that a methyl group is in fact electron-repelling ($+I_\sigma$).

Let us now examine the effect of a substituent attached directly to an sp^2 carbon, i.e. to a carbon atom with π molecular orbitals emanating from it. Table 7.3 lists the dissociation constants of some *para*-substituted benzoic acids and anilines. In these examples, fluorine, which we know exerts an attracting effect on electrons

* There are two conventions used to indicate the direction of the inductive effect. The first, due to Robinson, takes note of the charge of an electron and denotes the attraction of electrons by $+I$. The second, due to Ingold, considers a group which attracts electrons as withdrawing them from the rest of the molecule and uses the reverse symbolism $-I$. We shall here follow the Ingold convention since it is used much more widely. However, as far as possible, we shall use the terms '*electron-attracting*' and '*electron-repelling*' and avoid symbols.

Table 7.3. Dissociation constants of some substituted benzene derivatives

$$X\text{-}C_6H_4\text{-}CO_2H + H_2O \rightleftharpoons X\text{-}C_6H_4\text{-}CO_2^- + H_3O^+$$

($10^5 K_a^{25}$ in H_2O)

X = H	F	OCH_3	NO_2	CH_3
6.3	7.3	3.4	37	4.2

$$X\text{-}C_6H_4\text{-}NH_2 + H^+ \rightleftharpoons X\text{-}C_6H_4\text{-}NH_3^+$$

$10^{10} K_b^{25}$ in 30% C_2H_5OH

X = H	F	OCH_3	NO_2
2.6	0.15	66	0.0001

relative to carbon, increases the acid strength of benzoic acid and decreases the basicity of aniline. The fluorine atom hinders the spread of charge in the anilinium cation, so raising its free energy and decreasing the ionization. The methyl group, which repels electrons, decreases the acidity of benzoic acid, as we should expect, but the effect of the methoxy group as a substituent is completely anomalous. Methoxybenzene, known as anisole, has an appreciable dipole moment and in view of the greater electronegativity of oxygen than of carbon this dipole must be pointing away from the benzene ring towards the oxygen atom. Thus a methoxy group would be expected to be electron-attracting and thereby to spread the charge in the anion of a substituted acid, so lowering its free energy. Table 7.2 shows that the methoxy group does markedly increase the dissociation constant of acetic acid. However, the *para*-methoxy group in benzoic acid lowers the dissociation constant and in the basic dissociation constants of aniline the methoxy group makes *p*-methoxyaniline (called *p*-anisidine) a much stronger base. We must conclude, therefore, that a picture of charge distribution depending entirely on the electronegativity of the atoms concerned breaks down when we come to consider conjugated molecules, i.e. molecules with π molecular orbitals.

We must now go back and consider the electron distribution in a

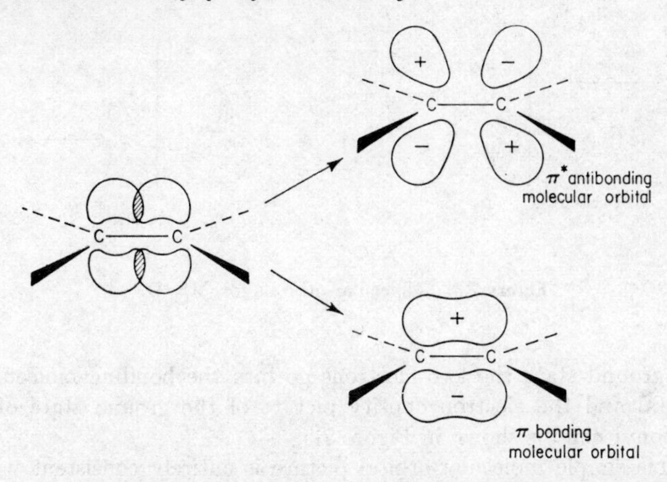

Figure 7.1. Molecular orbitals of >C=C<.

double bond in which one atom is more electronegative than the other. We have pictured a carbon–carbon double bond as being formed by the overlap of the two p orbitals on adjacent sp^2 hybridized carbon atoms (Figure 7.1). We can represent the relative energy of these orbitals in a correlation diagram (Figure 7.2). If one atom is more electronegative than the other, let us take for example the carbonyl bond, then, instead of a symmetrical arrangement of molecular orbitals, we obtain an unsymmetrical arrangement. The bonding orbital is closer in energy to the oxygen atomic orbital and the antibonding orbital is closer in energy to a carbon atomic orbital (Figure 7.3). This is diagrammatically represented in Figure 7.4. In

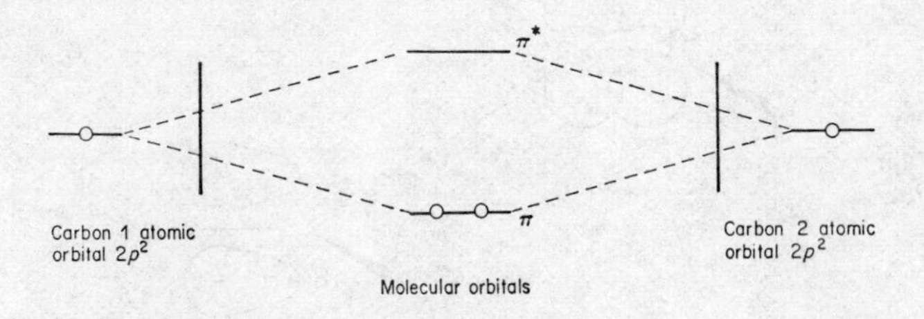

Figure 7.2. Correlation diagram for molecular orbitals in >C=C<.

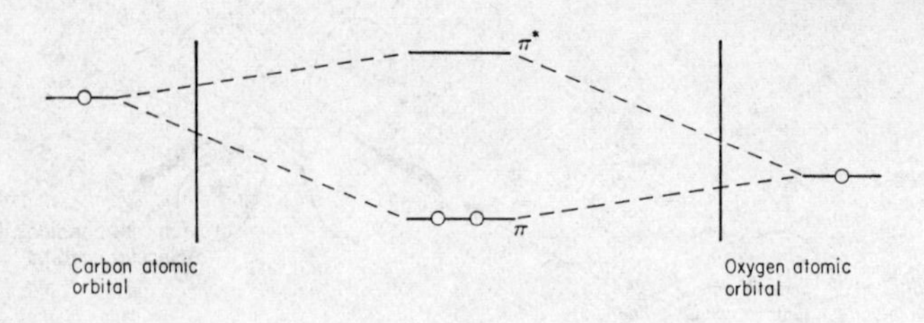

Figure 7.3. Molecular orbitals for >C=O.

the ground state the two electrons go into the bonding molecular orbital and the electron density picture of the ground state of a carbonyl bond is shown in Figure 7.5.

This simple molecular-orbital picture is entirely consistent with the ideas of the carbonyl bond developed in Part 1. The charge distribution in the bond is uneven and there is a definite concentration of negative charge on the oxygen atom. Now we must look at the effect of such an electronegative atom on the orbitals of a conjugated molecule, e.g. CH_2=CH—CH=CH—CH=O. For present purposes we are only concerned with the charge densities, and if we

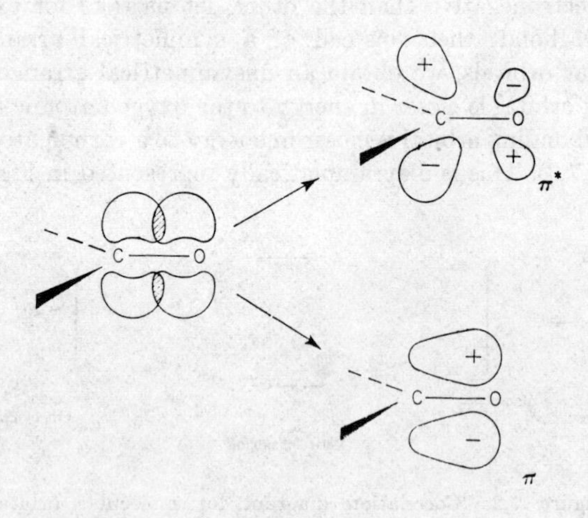

Figure 7.4. Electron distribution in >C=O.

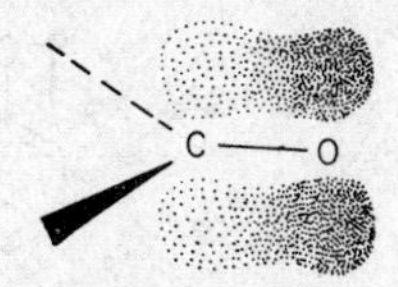

Figure 7.5. Electronic charge cloud for the ground state of a carbonyl bond.

carry out a simple molecular-orbital calculation on a molecule of this type we find that the electronegative oxygen atom has an effect on electron density in the π-orbitals all along the carbon chain. We find that there is a build-up of charge on one atom and a diminution of charge on the next one (see Figure 7.6). Notice that this effect is

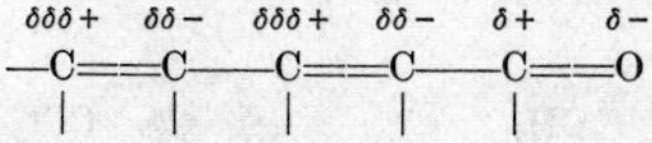

$$\overset{\delta\delta\delta+}{-\mathrm{C}}=\overset{\delta\delta-}{\mathrm{C}}-\overset{\delta\delta\delta+}{\mathrm{C}}=\overset{\delta\delta-}{\mathrm{C}}-\overset{\delta+}{\mathrm{C}}=\overset{\delta-}{\mathrm{O}}$$

Figure 7.6. The inductive effect of the electronegative oxygen atom on a conjugated chain.

due solely to the electronegativity of the oxygen atom and has nothing to do with the non-bonded electrons on that atom whose importance is discussed below. The same argument will apply to an aromatic ring. If we introduce an electronegative heteroatom (e.g. a nitrogen atom), the π-electron density will alternate around the ring.

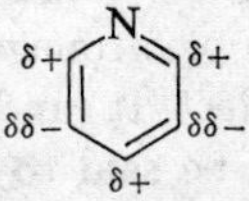

N
δ+ δ+
δδ− δδ−
δ+

π-Electron density in a heterocyclic molecule (pyridine)

π-Electron density in a heterocyclic molecule (pyridine).

When we come to consider a substitutent attached to carbon atom forming part of a conjugated system, we must investigate what effect the substituent will have on the relative electronegativity of the carbon atom to which it is attached. Let us examine two aromatic molecules, toluene and $\alpha\alpha\alpha$-trichlorotoluene (sometimes called benzotrichloride). We know from Chapter 6 that the methyl group is electron-repelling relative to an aromatic ring and we should expect

CH_3 (1) CCl_3 (1)

Toluene Benzotrichloride

the trichloromethyl group to be electron-attracting relative to the benzene ring. This, in effect, means that carbon atom 1 in toluene, which is having electrons pushed away from it, is more electropositive than a normal carbon atom, while carbon atom 1 in benzotrichloride, which is having electrons drawn towards it, will be more electronegative than a normal carbon atom. If we now consider the aromatic ring and regard carbon atom 1 as a heteroatom (i.e. different from the other atoms in the ring) then the charge distribution in the

CH_3: 1; 2 $\delta\delta-$; 3 $\delta\delta\delta+$; 4 $\delta\delta-$; 5 $\delta\delta\delta+$; 6 $\delta\delta-$

CCl_3: 1; 2 $\delta\delta+$; 3 $\delta\delta\delta-$; 4 $\delta\delta+$; 5 $\delta\delta\delta-$; 6 $\delta\delta+$

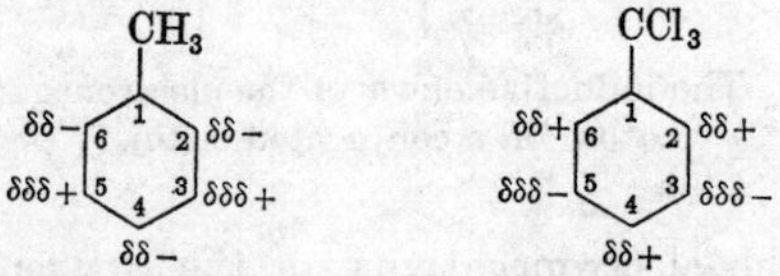

Figure 7.7. I_π effect: the charge distribution in the π orbitals of toluene and benzotrichloride due to the relative electronegativities of the CH_3 and CCl_3 groups.

π-orbitals of toluene and benzotrichloride will be as shown in Figure 7.7. An electronegative atom or group produces an alternating effect in the π-electron density along any conjugated chain to which it is attached. This effect is solely due to the electronegativity of the group concerned and is still an inductive effect. It can be denoted by I_π and the methyl group in toluene could be said to be exerting a $+I_\pi$ effect and the trichloromethyl group in benzotrichloride a $-I_\pi$ effect. (Unfortunately, few organic textbooks correctly describe the inductive effect in a conjugated molecule.)

A group will affect the electron distribution in a conjugated chain or a benzene ring purely through its relative electronegativity. It is now necessary to consider what additional effect may be associated with the orbitals of π symmetry in the substituent. In Part 1 we said that we could not distinguish between the two oxygen atoms in the acetate anion and the anion was described as a resonance

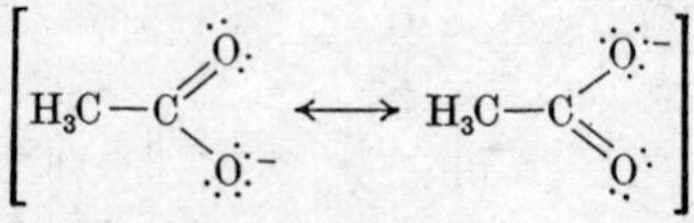

$$\left[H_3C-C\begin{matrix}\ddot{O}\!:\\ \ddot{O}\!:^-\end{matrix} \longleftrightarrow H_3C-C\begin{matrix}\ddot{O}\!:^-\\ \ddot{O}\!:\end{matrix} \right]$$

Figure 7.8. Resonance structures for the acetate anion.*

hybrid of two structures (see Figure 7.8). In terms of molecular-orbital theory we can picture this ion as having molecular orbitals with π symmetry spread over the two oxygen atoms and the carbonyl carbon atom, as depicted in Figure 7.9. The acidity of acetic acid is undoubtedly related to the fact that the charge is so effectively spread in the acetate anion. In Part 1 the acidity of phenol was attributed to the delocalization of the charge. According to our present hypothesis the benzene ring spreads the charge and so lowers the free energy of the phenoxide anion. In Part 1 this was depicted in terms of resonance theory (see Figure 7.10). Structures (*a*) and (*b*), with the negative charge on the oxygen atom, have the lowest energy. Here the negative charge is on the most electronegative atom and there is a stable benzene ring. However, in terms of valence-bond theory a more correct wave function would be a combination of the wave functions for all the possible structures, i.e. (*a*) and (*b*) together with (*c*), (*d*), and (*e*), and all these structures must contribute to the ground state.

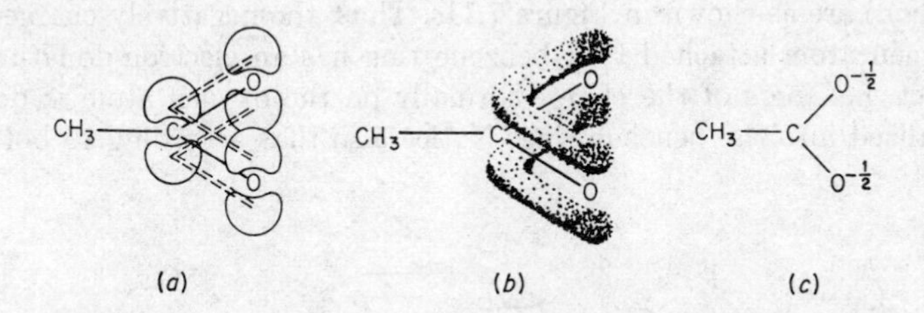

Figure 7.9. (*a*) Orbital interaction in the acetate anion; (*b*) π-electron 'clouds'; and (*c*) charge distribution as calculated by simple molecular-orbital theory.

* Dots representing non-bonding electron pairs are shown in formulae representing canonical forms enclosed in square brackets because these represent different pair-sharing schemes contributing to the resonance hybrid. Non-bonding electrons are not shown in other formulae, except in certain reaction sequences initiated by such an electron pair.

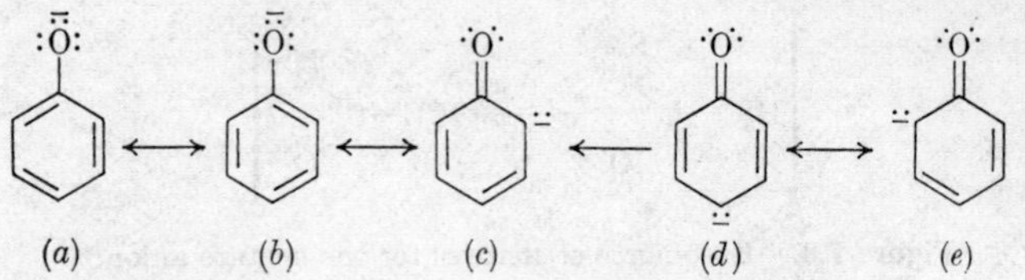

(*a*) (*b*) (*c*) (*d*) (*e*)

Figure 7.10. Resonance structures for the phenoxide anion.

In molecular-orbital theory the non-bonded electrons on the oxygen atom occupy 2*p* orbitals which have the same symmetry as the π orbitals of the ring, and the exocyclic oxygen atom must be included in the molecular orbitals for the whole molecule. The situation is very different from that of the carbonyl bond. In that case, we had a carbon atom with one electron in a *p* orbital and an oxygen atom with one electron in a *p* orbital. The two *p* atomic orbitals overlapped to give two π molecular orbitals, one bonding and one antibonding. The two electrons, one from the carbon and one from the oxygen, went into the bonding π molecular orbital. In the phenoxide anion there are seven *p* atomic orbitals which combine to give seven molecular orbitals. Three of these are bonding, three antibonding and one *non-bonding*. There are eight electrons in the anion; six of these go into the three bonding orbitals and the remaining two into the non-bonding orbital. The electron densities (if we neglect the greater electronegativity of oxygen compared with carbon) are as shown in Figure 7.11*c*. Thus, the negatively charged oxygen atom attached to a benzene ring has an electron-donating effect, i.e. some of the charge formally on the oxygen atom is delocalized into the benzene ring. Notice also that according to both

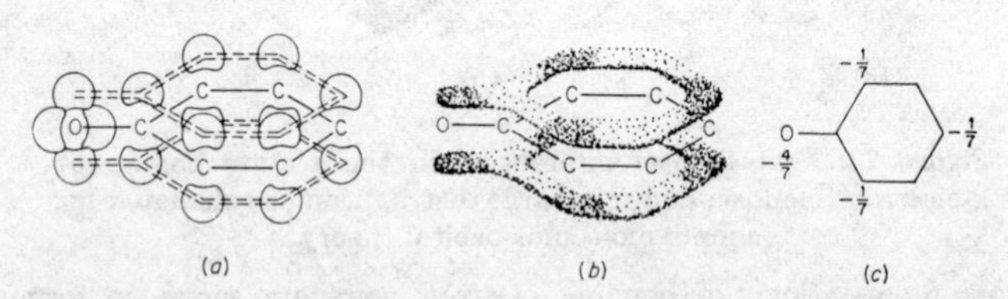

Figure 7.11. (*a*) Orbital interaction in the phenoxide anion; (*b*) π-electron-density clouds; and (*c*) charge distribution as calculated by simple molecular-orbital theory (neglecting the electronegativity of the oxygen atom).

theories this charge is distributed on the *ortho*- and *para*-positions of the benzene ring.

In Part 1 the reaction of primary amines with nitrous acid was discussed, and the formation of diazonium salts was described. With aliphatic compounds the diazonium salt is too unstable to exist and breaks down to yield the carbonium ion and nitrogen. But when the diazonium group is attached to a benzene ring, the diazonium salt is fairly stable. This was attributed to resonance stabilization. Structures (*a*) and (*b*) make the biggest contribution to the ground state of the diazonium cation, while (*c*), (*d*), and (*e*) make equal smaller contributions. According to the molecular-orbital picture we

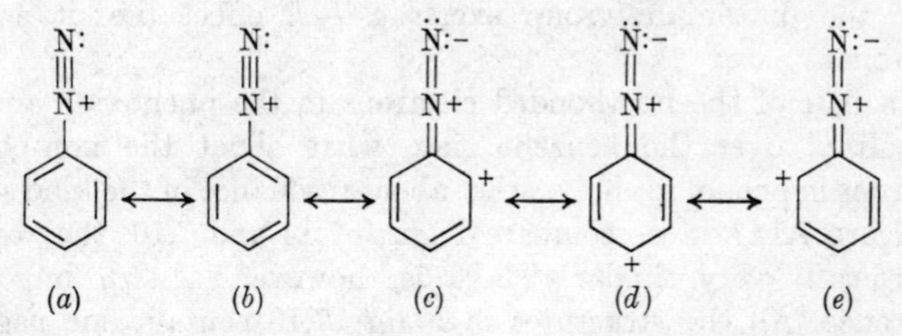

Figure 7.12. Resonance stabilization of the benzenediazonium ion.

now have an electron distribution which is the exact opposite of that in the phenoxide anion (effectively we have taken both electrons out of the non-bonding orbital and hence the charge distribution is the same but of the opposite sign) (see Figure 7.13). In the diazonium group there are low-lying antibonding orbitals of π symmetry which accept electrons from the ring. According to both resonance theory and simple molecular-orbital theory partial positive charges develop at the *ortho*- and *para*-positions. Since we can draw canonical forms illustrating this effect, we call it the *resonance* effect and denote it by the symbol R.* Thus, the oxygen atom with a negative charge, as

* The resonance effect has been called the electromeric effect; it has also been divided into an electromeric effect and a mesomeric effect, to distinguish between polarizability in an activated state and polarization in the ground state. It is now generally realized that this distinction is misleading rather than helpful, so we follow the current American practice and use the term resonance and the symbol R. Because of the confusion of signs used (i.e. $+R$ and $-R$), we shall so far as possible avoid the use of symbols and use the terms '*electron-donor*' and '*electron-acceptor*', thus distinguishing the resonance effect from the inductive effect which is due to electron-attractors and electron-repellers.

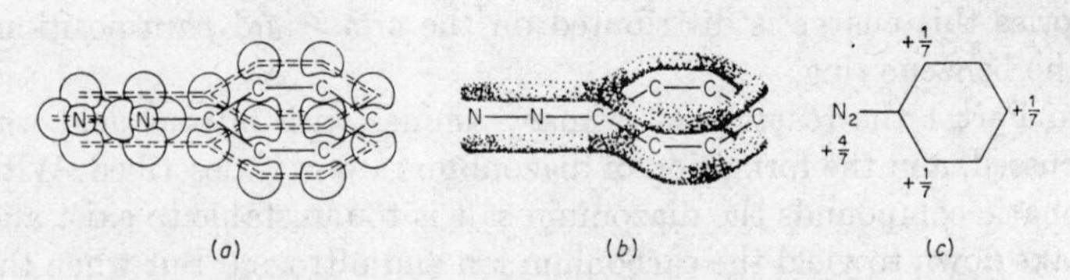

Figure 7.13. (*a*) Orbital interaction in the benzene diazonium cation; (*b*) π-electron-density clouds (there is additional π-electron density between the nitrogen atoms not shown); and (*c*) charge distribution.

in the phenoxide anion, exerts a $+R$ effect (i.e. it donates electrons), while the diazonium group exerts a $-R$ effect (i.e. it accepts electrons).

If a pair of the non-bonded electrons in the phenoxide anion is delocalized over the benzene ring, what about the non-bonded electrons in phenol itself, i.e. what about resonance of the kind shown in Figure 7.14? If we compare Figures 7.14 and 7.10, they appear superficially very similar. There is, however, a *very* important difference. All the structures in Figure 7.10 contain one negative charge. All that has happened in structures 7.10 (*c*), (*d*), and (*e*) is that the charge has been moved from the oxygen atom into the benzene ring. In Figure 7.14, however, we find that in (*c*), (*d*), and (*e*) we have created charge where there was none before; these dipolar structures are clearly of very high energy and will make relatively little contribution to the ground state.*

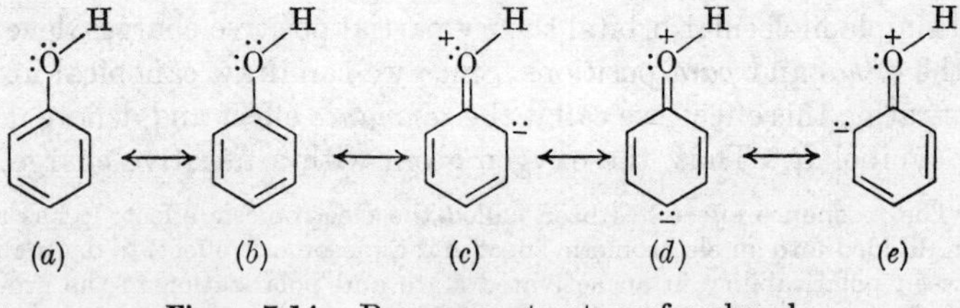

(*a*) (*b*) (*c*) (*d*) (*e*)

Figure 7.14. Resonance structures for phenol.

* A line joining two atoms represents two electrons *shared* between the atoms in question. In Figure 7.14 structures (*a*) and (*b*) the oxygen atom has eight electrons in its outer shell, but four of these are shared, two with the hydrogen atom and two with the carbon atom. The total negative charge associated with the oxygen atom remains the same as in an isolated neutral

Since nitrogen is less electronegative than oxygen one would expect structures analogous to those in Figure 7.14 (*c*), (*d*), and (*e*) to make a slightly bigger contribution to the ground state of the aniline molecule (see Figure 7.15). Aniline is a very much weaker base than methylamine and it is often suggested that this is because delocalization present in aniline is curtailed when a proton adds to the nitrogen atom to form the cation. Notice the important

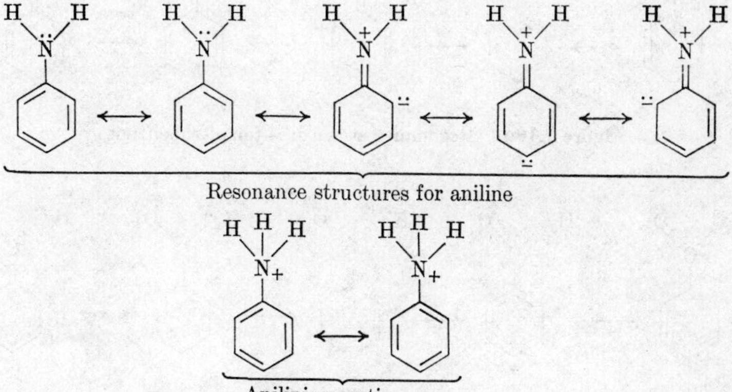

Resonance structures for aniline

Anilinium cation.
No additional resonance structures possible.

Figure 7.15.

difference between the diazonium cation and the anilinium cation: the former can accept electrons from the benzene ring, the latter cannot.

In discussing the diazonium cation we have used structures such as $[R{-}\overset{+}{N}{\equiv}N{:} \leftrightarrow \overset{+}{R}{=}\underset{\cdot\cdot}{\overset{+}{N}}{=}\overset{-}{N}{:}]$. This diagram really represents removing a pair of electrons from the molecular orbitals of the main part of the molecule and placing them into a low-lying antibonding

oxygen atom. In structures 7.14 (*c*), (*d*), and (*e*) there are still eight electrons in the outer shell of the oxygen atom but now six of these are shared, so that the total negative charge associated with the oxygen atom is one less than for an isolated oxygen atom. Thus the oxygen atoms in structures 7.14 (*c*), (*d*), and (*e*) carry a positive charge, and for a similar reason *ortho*- or *para*-carbon atoms carry a negative charge.

orbital of the diazonium group. We might thus expect any unsaturated group to be able to exert a similar effect, as shown in Figures 7.16*a*—*c*. In the canonical forms drawn for nitrobenzene there is a net transfer of electronic charge from the carbon atoms of the ring, to the oxygen atoms of the nitro group. Since oxygen

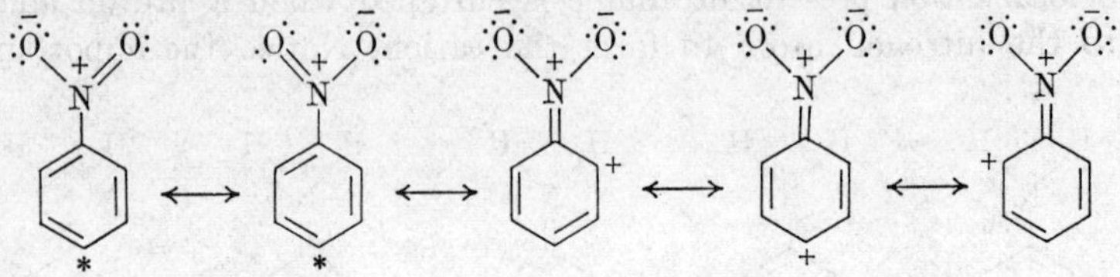

Figure 7.16*a*. Resonance structures for nitrobenzene

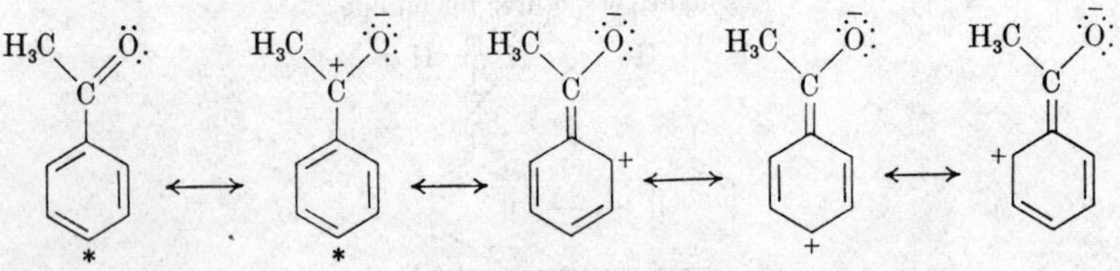

Figure 7.16*b*. Resonance structures for acetophenone.

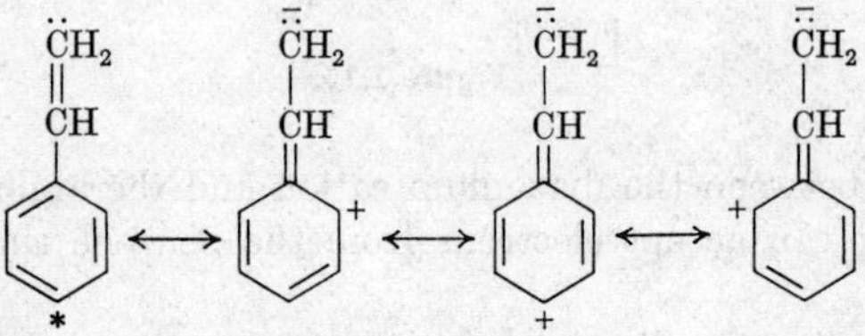

Figure 7.16*c*. *Possible?* Resonance structures for styrene ($CH{=}CH_2$ accepting electrons).

is more electronegative than carbon, this is a reasonable picture even if the more polar structures are of high energy. The same is true of the structures drawn for the carbonyl group. In styrene (phenylethylene) there is no reason why the electron should flow

* N.B. There is a second Kekulé structure corresponding to each of the formulae marked with an asterisk and these structures must be included in a complete resonance picture.

from the ring to the terminal carbon atom of the ethylenic group. We could equally well draw structures of the type shown in Figure 7.17. Thus the ethylenic group attached to a benzene ring neither accepts nor donates electrons in the ground state.

We began by considering the effect of the electronegativity of groups on the rest of the molecule. We then considered the effect that we should expect if the groups had orbitals of a symmetry such that they interact with π molecular orbitals of the rest of the molecule. We have called this kind of interaction a resonance effect. There are groups with filled non-bonding orbitals of π symmetry (—$\overline{\ddot{\underset{\cdot\cdot}{O}}}$:, —$\ddot{\underset{\cdot\cdot}{O}}H$, —$\ddot{N}H_2$, —$\ddot{\underset{\cdot\cdot}{Br}}$:, etc.) which donate electrons to the

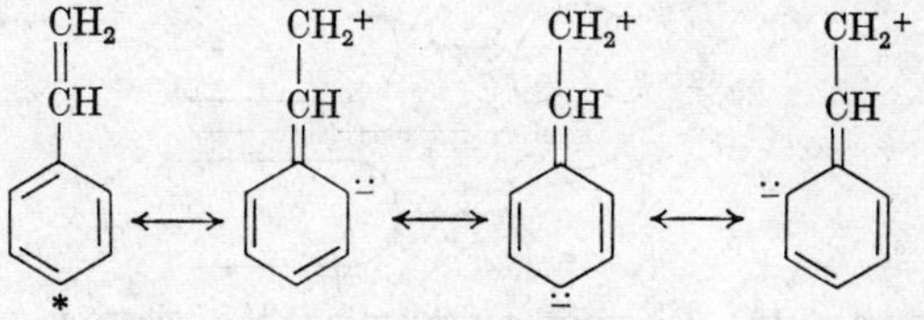

Figure 7.17. *Possible?* Resonance structures for styrene ($CH{=}CH_2$ donating electrons).

remainder of a conjugated molecule; and there are groups with vacant low-lying antibonding orbitals ($N{\equiv}\overset{+}{N}$—, O_2N—, RCO—, RO_2C—, etc.) which accept electrons from the remainder of a conjugated molecule. The extent of the donating effect ($+R$) will correlate with the ionization potential of the substituent, and the accepting effect ($-R$) will correlate with the electron affinities of the substituent. Table 7.4 shows the relative donating and accepting properties of some common groups.

Table 7.4. The resonance effects of groups

$+R$ groups (donors)	$^-O \ggg NH_2 > SH > OH > Cl$
$-R$ groups (acceptors)	$\overset{+}{N}{\equiv}N \ggg NO_2$; $SO_2R > RCO$; $CO_2R > CN$

Initially we considered the effect of the different electronegativities of various substituents and called this electronegativity effect an inductive effect. We then considered orbital interaction, which we called the resonance effect. It is possible to distinguish between resonance effects and inductive effects by the fact that canonical structures can be drawn to illustrate a resonance effect. This cannot be done for an inductive effect. A group exerting an inductive effect attracts or repels electrons, but a new electron-pairing scheme for the molecule cannot be drawn to depict this. A group exerting a resonance effect either accepts or donates an electron pair, and new electron-pairing schemes can be drawn to represent this.

There is, in addition, an I_π effect due to the non-bonded p electrons in a substituent such as the fluorine atom.* The fluorine atom will

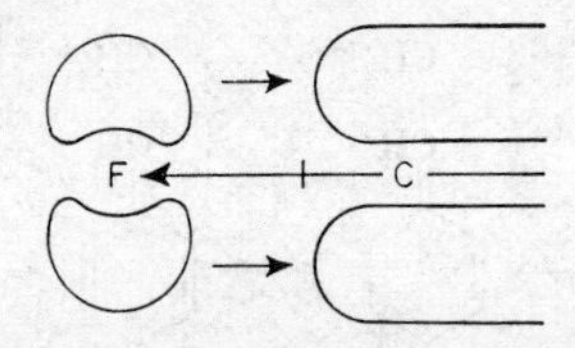

Figure 7.18. Diagram representing the electron repulsion between the p atomic orbitals of the fluorine atom and the π molecular orbitals of the remainder of the molecule.

not donate electrons by the $+R$ process. Fluorine is far too electronegative for this to be important in the ground state of an uncharged molecule. Thus there are the non-bonded p electrons of the fluorine atom adjacent to electrons in π orbitals in the remainder of the molecule (see Figure 7.18). There will be a repulsion between the two electron clouds (this will not be entirely electrostatic, but will be partly due to repulsion between electrons of the same spin), and we have already seen that a repulsion of π electrons results in an alternating polarization of charge all along the conjugated chain. This can be represented in Figure 7.19. Clearly oxygen atoms (as in phenol or anisole) and nitrogen atoms (as in aniline) have non-bonded p electrons which exert a similar $+I_\pi$ effect. In these molecules any $+I_\pi$ effect there may be is completely masked by the more powerful $+R$ effect discussed above. The repulsion between the non-bonded electrons of the substituent (e.g. fluorine) and the

* This paragraph deals with a more advanced concept and may be omitted at a first reading.

π electrons of the remainder of the molecule is only important in ground-state properties and for explaining fairly subtle differences in the electronic effects of the halogens, especially in the case of fluorine.

We began this discussion of the electronic effects of substituents in aromatic nuclei by considering the dissociation constants of some benzoic acids. Table 7.5 lists some more results. The first feature to

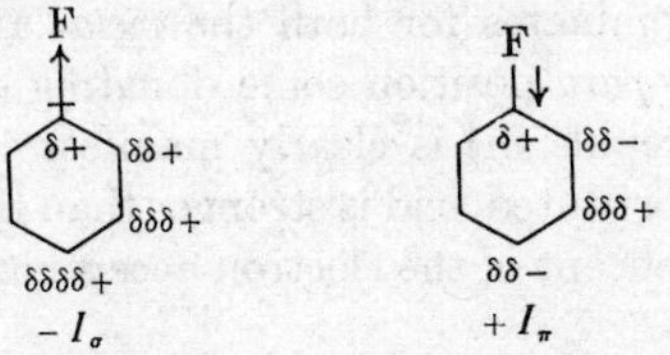

F δ+ δδ+ δδδ+ δδδδ+ $-I_\sigma$

F δ+ δδ− δδδ+ δδ− $+I_\pi$

Figure 7.19. The opposing $-I_\sigma$ and $+I_\pi$ effects of a fluorine atom. (The $-I_\sigma$ effect is transmitted through the σ orbitals and through space; see the beginning of this Chapter.)

Table 7.5. Dissociation constants ($10^5 K_a^{25}$) of some benzoic acids

Substituent	*ortho*	*meta*	*para*	Electronic effect
H	6.3	6.3	6.3	—
CH_3	12.4	5.4	4.2	$+I$
Cl	114	14.5	10.5	$-I$ and $+R$ (also $+I_\pi$)
NO_2	670	32.1	37.6	$-I$ and $-R$
CH_3O	8.1	8.2	3.4	$-I$ and $+R$

notice there is that any substituent in the *ortho*-position causes an appreciable increase in acid strength, even electron-repelling groups such as methyl and methoxyl groups which have a weakening effect in the *para*-position. A substituent in the *ortho*-position is adjacent to the carboxyl group: it will have a profound effect on the arrangement of the solvent molecules around the carboxyl group and the carboxylate anion. Such compounds, i.e. *ortho*-substituted benzoic acids, do not therefore come within the terms of our hypothesis and cannot meaningfully be discussed.

The next point to notice is that the methoxyl group has a weakening effect at the *para*-position and an acid-strengthening effect at the *meta*-position. This is in accord with the electronic theory as we have discussed it. The methoxyl group is a donor ($+R$), but owing to the electronegativity of the oxygen atom it is also an electron-attractor ($-I$). In the *para*-position the donating effect predominates and inhibits the spread of charge in the carboxylate anion. In the *meta*-position, the donating effect is unimportant and the electron-attracting effect predominates and spreads the charge in the carboxylate anion. In the chlorobenzoic acids the electron-attracting effect ($-I$) predominates for both the *meta*- and *para*-substituted acids, but at the *para*-position some donating effect ($+R$ or $+I_\pi$, due to electron repulsion) is clearly manifest. In the nitrobenzoic acids the *para*-substituted acid is stronger than the *meta*-substituted, in line with our concept of the electron-accepting properties ($-R$) of this substituent.

Problem

Predict the relative acid strengths of the following benzoic acids 3-nitro; 3-fluoro; 4-nitro; 3-ethoxy; 4-ethoxy; 4-fluoro; 3-methyl; 4-methyl.

CHAPTER 8

Directive Effects in Addition and Substitution Reactions

In Part 1, discussion of electrophilic addition to the ethylenic double bond, and of addition-with-elimination reactions of the benzene ring, was deliberately restricted to symmetrical ethylenes and unsubstituted benzene. Addition to an unsymmetrical double bond was considered for the case of the carbonyl group and here the polarization is so great that it was possible to say, on the basis of very simple ideas about the Periodic Table, that electrophiles would add to the oxygen and nucleophiles to the carbon atom. We must now consider the effect of substituents on addition to unsymmetrical ethylenes and the addition-with-elimination reactions of the substituted benzenes. But before we can do this we must briefly discuss the factors which control the rate of chemical reaction.

In the discussion of the reaction of the chlorine atoms with hydrocarbons in Chapter 2 of Part 1, diagrams were introduced in which energy was plotted against a coordinate representing the extent of reaction between two isolated reactant molecules, which was called the reaction coordinate. Figure 8.1 represents such a diagram for the reaction $X + YZ \rightarrow XY + Z$. As X approaches YZ there will be increasing repulsion between X and YZ. Eventually a stage is reached where the new bond between X and Y starts to be formed and the bond between Y and Z is being broken. If we consider the reverse reaction, starting with the products well separated, then as Z approaches YX there will be an increasing repulsion. Thus the activated complex occurs at a maximum on the energy/reaction coordinate curve.

The rate of the reaction is given by

$$\text{Rate} = k[X][YZ]$$

where k is the rate constant, and [X] and [YZ] represents the concentrations of the reacting species. If the temperature is raised the reaction is faster. Empirically, this increase in rate is found to obey the Arrhenius equation, where A and E are constants:

$$k = A\,e^{-E/RT}$$

$\boldsymbol{R}$ is the gas constant and T is the absolute temperature. The constant A is usually known as the 'A factor' and E as the 'activation energy'. The Arrhenius equation is an empirical equation and

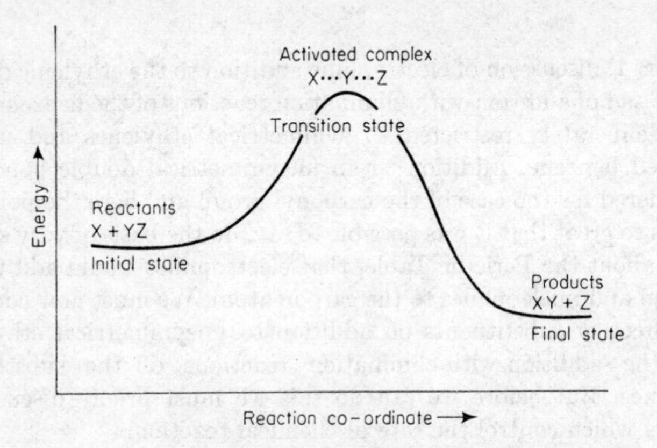

Figure 8.1. Energy/reaction coordinate diagram for a reaction X + YZ → XY + Z.

we shall now see how A and E can be interpreted in terms of known thermodynamic functions.

There is a theory called 'transition-state theory', according to which the activated complex is treated as though it were a molecule, normal in every respect except in regard to vibration along the reaction coordinate. X and YZ are regarded as being in equilibrium with the activated complex, which can break down to give the products or to regenerate X and YZ.

$$X + YZ \underset{k_{-1}}{\overset{k_1}{\rightleftharpoons}} \ddagger \xrightarrow{k_2} \text{Products}$$

The activated complex is usually denoted by the sign ‡. If it is assumed that there is a pre-equilibrium between X and YZ, then k_1

and k_{-1} must be very much greater than k_2. The overall rate of reaction can then be expressed by:

$$\text{Rate} = k_2 K^{\ddagger}[\text{X}][\text{YZ}] \quad K^{\ddagger} = k_1/k_{-1}$$

Statistical theory (the theory referred to in Chapter 6) shows that k_2 is a universal rate constant for all reactions:

$$k_2 = \boldsymbol{k}T/\boldsymbol{h} \quad \text{(the same for all reactions)}$$

where $\boldsymbol{k}$, as before (Chapter 6), is Boltzman's constant and is equal to $\boldsymbol{R}$, the gas constant, divided by N, Avogadro's number. T is the absolute temperature and $\boldsymbol{h}$ is Planck's constant.

Since $$\text{Rate} = k[\text{X}][\text{YZ}]$$

$$k = k_2 K^{\ddagger}$$

$$= \frac{\boldsymbol{k}T}{\boldsymbol{h}} K^{\ddagger}$$

Now $$\Delta G = -\boldsymbol{R}T \log K \quad \text{(see Chapter 6)}$$

Hence $$k = \frac{\boldsymbol{k}T}{\boldsymbol{h}} \mathrm{e}^{-\Delta G\ddagger/RT}$$

or, since $$\Delta G = \Delta H - T\Delta S \quad \text{(see Chapter 6)},$$

$$k = \frac{\boldsymbol{k}T}{\boldsymbol{h}} \mathrm{e}^{\Delta S\ddagger/R}\, \mathrm{e}^{-\Delta H\ddagger/RT}$$

Thus, according to transition-state theory, it is the *free-energy* difference which matters and which determines the rate of a reaction. We also see that the Arrhenius activation energy is related to the enthalpy change from reactants to activated complex and that the Arrhenius A factor is not entirely temperature-independent and contains an entropy term. These results are consistent with intuitive expectation. In Chapter 2 of Part 1, fluorination, chlorination, and bromination were compared and it was shown that the differences in rates could be explained in terms of differences of activation energy. It was emphasized very strongly that this argument could only be applied to similar reactions, and we now see that this argument will only apply if $\Delta S^{\ddagger}$ is constant for the reaction series under consideration.

Having discussed the factors which control reaction rates in general we can now turn to the influence of substituents in addition reactions. Let us start by considering the electrophilic addition of a compound such as hydrogen chloride to an unsymmetrical olefin:

$$RCH{=}CH_2 + H{-}X \longrightarrow R\overset{+}{C}H{-}CH_3 + X^- \longrightarrow RCHXCH_3$$

We can depict this process in an energy/reaction coordinate diagram (Figure 8.2). Notice that this is a two-stage reaction and the overall

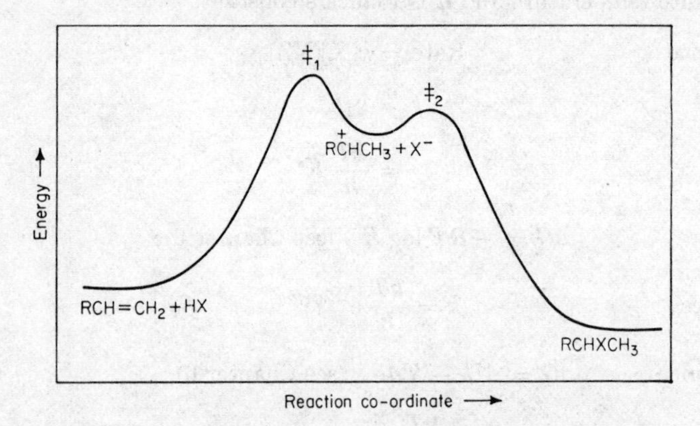

Figure 8.2. Energy/reaction coordinate diagram for a reaction $RCH{=}CH_2 + HX \rightarrow RHXCH_3$.

rate depends on the nature of the first activated complex ($\ddagger_1$). Unfortunately, we know very little about the nature of this activated molecule. We can, therefore, either talk about the properties of the reactants, particularly $RCH{=}CH_2$, or about the products of the initial reaction which in this case is the initial addend, i.e. $R\overset{+}{C}HCH_3$. In the first approach, which we might call 'the isolated molecule approach', we have to consider what effect the substituent R has on the electron distribution in the ground state of the initial molecule and we then expect that H^+ would attack the site of the highest electron density. There are two difficulties with this approach. The first is that it assumes that the electronic effects of R in the ground state are the same as the electronic effects when H^+ approaches the molecule. Though this may sometimes be so, there is no reason to suppose it will always be true. The second objection to the isolated-molecule approach is that it assumes the rate of reaction to depend only on the electronic distribution, whereas we have seen that it really depends on the free-energy change. If one is comparing a series of similar reactions occurring by identical mechanisms the isolated-molecule approach may not be too unsatisfactory, but clearly we ought to avoid it if we can.

The alternative approach is to look at the nature of the products, in this case the initial carbonium ion $R\overset{+}{C}HCH_3$. As this is clearly a high-energy species it is not unreasonable to assume that it is nearer to the transition state in electronic energy and possibly also nearer in free energy. When we come to consider the nature of this addend an important feature will be the extent to which the electrons can be delocalized. We have seen the tremendous lowering of energy associated with electron delocalization in benzene compared with three separate ethylenic bonds. We have also seen the effect of electron delocalization in stabilizing the acetate anion compared with the ethanolate anion. We next have to consider what the electronic effects of the substituent will be. In order to do this we invoke a similar hypothesis to that put forward when considering the dissociation constants of halogenoacetic acids. '*A substituent in any charged species which tends to spread the charge in that species will lower its free energy.*' We shall now apply these arguments to a few specific examples.

Directive Effects in Electrophilic Addition to Olefinic Double Bonds

We shall take the addition of hydrogen bromide to styrene as our first example:

$$C_6H_5{-}CH{=}CH_2 + HBr \longrightarrow \textit{either}\ C_6H_5{-}CH_2\overset{+}{C}H_2\ (a)$$

$$\textit{or}\ C_6H_5{-}\overset{+}{C}HCH_3\ (b)$$

It can be seen that addition of a proton to the ethylenic carbon atom nearest the benzene ring gives species (*a*) in which the positive charge is situated on the terminal carbon atom. The original delocalization between the ethylenic double bond and the benzene ring has been lost and the positive charge is completely localized on the terminal carbon atom. In the alternative intermediate (*b*), in which the proton adds to the terminal carbon atom of the ethylenic double bond, it is true that the delocalization between the ethylenic double bond and the benzene ring has been lost but we have a new form of delocalization depicted in (i)–(v). Although structures (iii), (iv), and

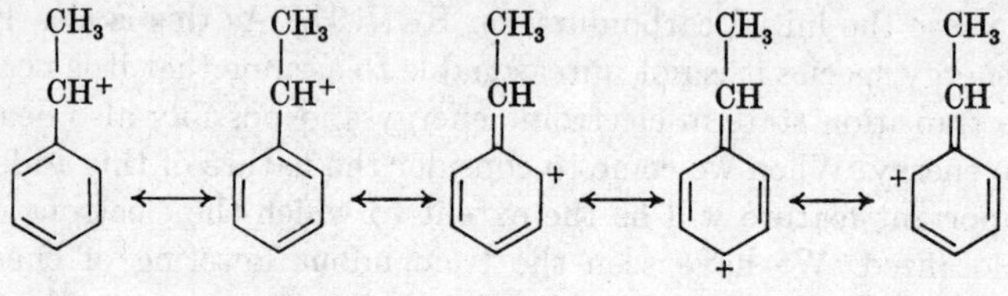

CH_3 CH_3 CH_3 CH_3 CH_3

CH^+ CH^+ CH CH CH

(i) $\longleftrightarrow$ (ii) $\longleftrightarrow$ (iii) $\longleftrightarrow$ (iv) $\longleftrightarrow$ (v)

(v) are of higher energy than structures (i) and (ii), they nevertheless must contribute to the overall electron distribution. This resonance results in considerable spreading of the charge as well as electron delocalization, and thus addition of the proton to the terminal carbon atom should be greatly favoured. This is entirely in accord with experiment.

As a second example we shall take the addition of hydrogen fluoride to methyl vinyl ether:

$$CH_3OCH{=}CH_2 + H{-}F \longrightarrow \begin{cases} \textit{either } CH_3OCH_2CH_2^+ & (a) \\ \textit{or } CH_3O\overset{+}{C}HCH_3 & (b) \end{cases}$$

In the intermediate (*a*), where the proton has been added to the substituted ethylenic carbon atom, any delocalization there may have been in the initial molecule between the $2p$ atomic orbital of the oxygen atom and the π molecular orbital of the ethylenic double bond has been removed. The positive charge is sited exclusively on the terminal carbon atom. In the alternative intermediate (*b*), where the proton has been added to the terminal carbon atom of the ethylenic double bond, it is again true that any delocalization in the initial molecule has been destroyed, but delocalization is possible between the carbon atom carrying the positive charge and the oxygen atom:

$$\left[CH_3-\ddot{\underset{..}{O}}-\overset{+}{C}H-CH_3 \longleftrightarrow CH_3-\overset{+}{\underset{..}{O}}=CH-CH_3\right]$$

So here some electron delocalization remains and the positive charge is spread. Thus, we should predict that this is substantially the more favourable product and that addition to a vinyl ether will give exclusively the α-halogeno-ether, e.g. $CH_3OCHFCH_3$.

In both examples so far, limited electron delocalization is possible in the initial addition product and we should, therefore, predict that both these reactions proceed more rapidly than the similar addition to ethylene. This is in accord with experimental observation.

We will now consider the addition of hydrogen chloride to propene.

$$CH_3CH{=}CH_2 + H{-}Cl \longrightarrow \begin{cases} \textit{either } CH_3CH_2CH_2^+ \ (a) \\ \textit{or } CH_3\overset{+}{C}HCH_3 \ (b) \end{cases}$$

In neither of these addition products is there any orbital of π symmetry and there is no possibility of electron delocalization. Decision as to their relative stabilities must, therefore, be made entirely on the basis of the charge distribution. Now a methyl group repels electrons (we can assume that an ethyl group will also repel electrons in a similar manner). When the positive charge is on the central carbon atom, as in (*b*), the two flanking methyl groups with their electron-repelling effect, tend somewhat to neutralize this positive charge, so that the overall effect is to spread the positive charge. In the other case (*a*), where the positive charge is located on a terminal carbon atom, it is true that the ethyl group may be as electron-repelling as the methyl group but there is only one electron-repelling group attached to C^+ compared with two in case (*b*). We therefore predict that the proton will add preferentially to the terminal carbon atom, so that the chief product will be $CH_3CHClCH_3$. We also predict, on the basis of the same argument, that this reaction would proceed slightly more rapidly than the similar addition to ethylene itself. Both these predictions are in accord with experiment.

For our next example we shall consider the hydration of 1,1,1-trifluoropropene by sulphuric acid:

$$CF_3CH{=}CH_2 + H{-}OSO_3H \longrightarrow \begin{array}{l} \textit{either } CF_3CH_2CH_2^+ \ (a) \\ \textit{or } CF_3\overset{+}{C}HCH_3 \ (b) \end{array}$$

The trifluoromethyl group has a powerful electron-attracting effect, so that in the ion (*b*), where the positive charge is adjacent to the trifluoromethyl group, the electronic effect of the CF_3 group would intensify this positive charge (i.e. according to our hypothesis this would increase the free energy of the ion). In the alternative ion (*a*), the CF_3 group is further removed from positive charge and, although it will still exert an electron-withdrawing effect, its influence will be much less. Thus we should expect trifluoropropene to react very much less readily than ethylene and to yield, after hydrolysis, $CF_3CH_2CH_2OH$ (any student who was unfortunate enough to have heard of Markovnikov's rule, please forget it!).

Let us now consider the addition of a hydrogen halide to ethyl acrylate:

$$C_2H_5O_2CCH{=}CH_2 + H{-}Cl \longrightarrow \begin{array}{l} \textit{either } C_2H_5O_2CCH_2\overset{+}{C}H_2 \ (a) \\ \textit{or } C_2H_5O_2C\overset{+}{C}HCH_3 \ (b) \end{array}$$

Because of the electronegativity of oxygen a carboxyl group is strongly polarized:

$$\overset{\delta+}{C}{=}\overset{\delta-}{O} \quad \left(\text{i.e. } C{=}\ddot{O}\!: \longleftrightarrow \overset{+}{C}{-}\ddot{\underset{..}{O}}{}^{-}\right)$$

In ethyl acrylate the carbonyl bond has orbitals of π symmetry, but we cannot use these to delocalize the positive charge (try drawing the appropriate canonical forms). In structure (*b*) we have a carbonium ion at position 2, next to the carbonyl-carbon atom which already carries an incipient positive charge. Thus, in forming the ion (*b*) we concentrate the positive charge instead of spreading it. In structure (*a*) there is no such concentration and we should therefore predict that the product of such a reaction would be the β-chloro-ester, i.e. $ClCH_2CH_2CO_2C_2H_5$. There is some delocalization possible in ethyl acrylate between the ethylenic double bond and the carbonyl double bond, but no delocalization at all is possible in the initial addition product. We therefore expect this reaction to occur much less readily than with ethylene. All our expectations are met in experiment.

As a final example, we shall consider the addition of hydrogen chloride to vinyl chloride:

$$Cl{-}CH{=}CH_2 + H{-}Cl \longrightarrow \begin{array}{l} \textit{either } ClCH_2CH_2^+ \ (a) \\ \textit{or } Cl\overset{+}{C}HCH_3 \ (b) \end{array}$$

The predominant effect of the substituent chlorine atom in the ground state of the ethylene molecule will be one of electron-withdrawal, owing to the electronegativity of the chlorine atom. However, a chlorine atom has filled atomic orbitals ($2p_z$ and $2p_y$). These orbitals will interact only very slightly in the ground state with the π molecular orbitals of the ethylenic double bond, i.e. according to resonance theory, a structure $^+\ddot{C}l{:}{=}CH{-}\ddot{C}H_2^-$ makes only a small contribution to the ground state of vinyl chloride. In the addition product, however, the molecule carries a positive charge, and the electronegativity of the chlorine which prevents it from acting as a donor in the ground state is not sufficient to prevent it from acting as a donor when adjacent to a carbonium ion [as in (*b*)], i.e.:

$$[{:}\ddot{C}l{-}\overset{+}{C}HCH_3 \longleftrightarrow {:}\overset{+}{C}l{=}CHCH_3]$$

No such delocalization is possible with the other intermediate (*a*), so we should predict that addition of hydrogen chloride to vinyl chloride will produce 1,1-dichloroethane, Cl_2CHCH_3, but that, because of the electronegativity of the chlorine atom, this reaction should occur less rapidly than with ethylene. Notice again how chlorine, which exerts a negligible $+R$ effect in the ground state of a non-polar molecule, will donate electrons when attached to a molecule with a positive charge. The same argument applies with greater force to fluorine.

Directive Effects in Aromatic Addition-with-Elimination Reactions, so-called "Electrophilic Aromatic Substitution"

Chapter 14 of Part 1 describes how electrophiles add to the benzene nucleus and how, instead of adding an anion, the resulting carbonium ion ejects a proton so that the product is a substituted benzene. Care was taken not to consider the reaction of any substituted benzene, but we can see from the above discussion that electrophilic addition-with-elimination reactions will involve a situation analogous to electrophilic addition to olefins. The substitution process can be depicted as follows:

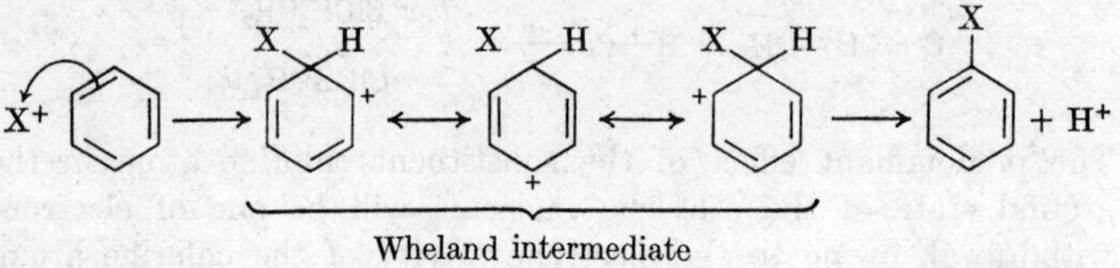

Wheland intermediate

where $X^+ = NO_2^+$, etc.

We shall first consider the effect of donors, that is, groups with filled orbitals of π symmetry which can interact with the π orbitals of the benzene ring (i.e. $+R$ groups).

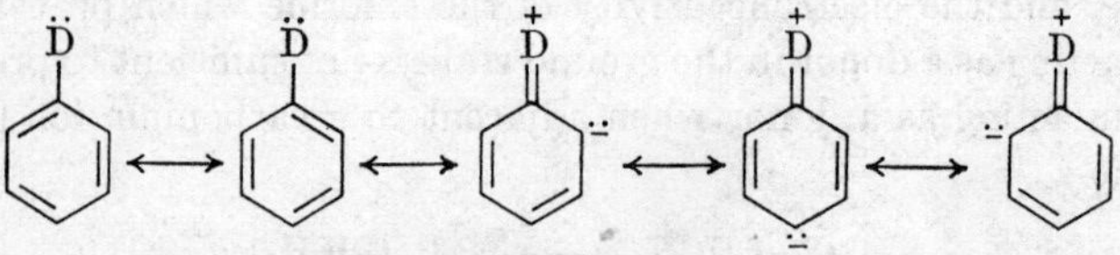

Canonical forms for benzene containing a donor $(+R)$ substituent in the ground state.

If we write out the canonical forms for the Wheland intermediates for benzene nuclei substituted by a donor group D, we find that when the electrophile is attached to a position *ortho* or *para* to the donor group we can extend the conjugation. Thus a donor group stabilizes the Wheland intermediate for attack at the *ortho*- and *para*-positions and, in addition, it spreads the positive charge, so that on both counts we should expect *ortho*- and *para*-substitution to be favoured in a benzene nucleus substituted by a donor group. We can represent the relative donating properties of some common donor groups as follows:

$$^-O \ggg NH_2;\ NR_2 > OH > OR > CH{=}CH_2 > Cl;\ F > CH{=}CHNO_2$$

Notice that the vinyl group, which, in Chapter 7, we said neither donated nor accepted electrons in the ground state of the styrene molecule, acts as a donor in the Wheland intermediate. Similarly, the chlorine atom and the fluorine atom, which have negligible donor effects in the ground state of chlorobenzene and fluorobenzene, act as donors in the positively charged Wheland intermediates; however, their electronegativity is such that, although *ortho/para*-substitution is observed, the reaction is much slower than with benzene itself. The final group in the above list, the nitrovinyl group,

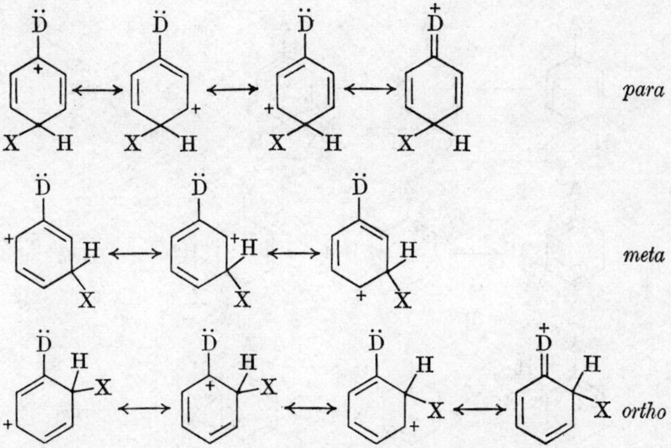

Canonical forms for the three possible Wheland intermediates when a benzene nucleus substituted by a donor is attacked by an electrophile.

we might at first sight expect to behave as an acceptor. However, the nitro group is a less deep sink of electrons than the positive charge encountered in the Wheland intermediate, so that, like the halogens, a nitrovinyl group directs attack into the *ortho*- and *para*-positions although it has a retarding effect on nitration. All the other donors not only direct the entering group to *ortho/para*-positions but also accelerate the nitration.

We must now turn to the effect of an acceptor group, A, as a substituent in a benzene ring:

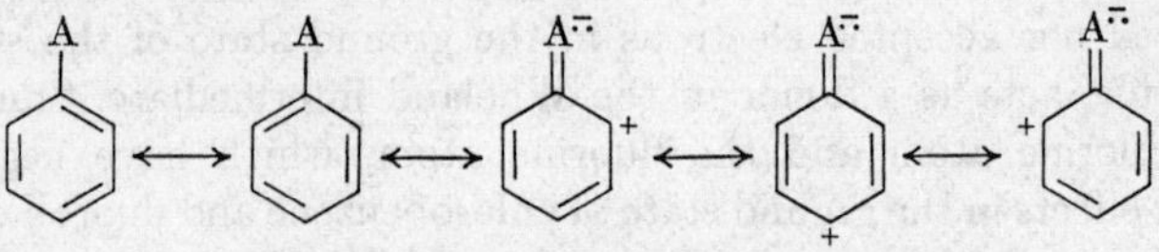

Canonical forms for benzene containing an acceptor ($-R$) substituent in the ground state.

Acceptors are unsaturated groups with low-lying antibonding orbitals. We can depict the canonical forms for the three possible Wheland intermediates as shown.

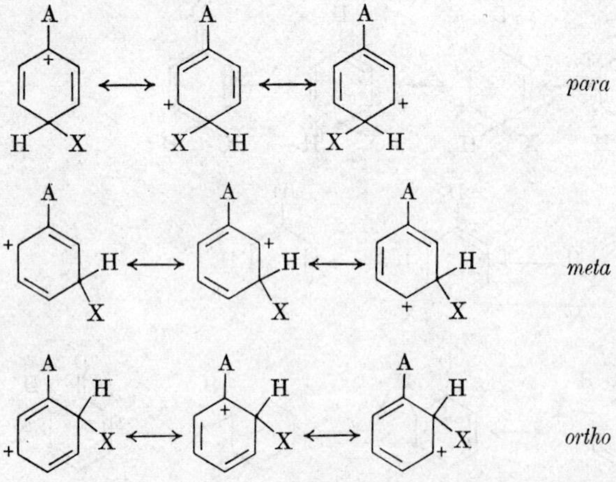

para

meta

ortho

Canonical forms for the three possible Wheland intermediates when a benzene nucleus substituted by an acceptor is attacked by an electrophile.

Notice that in the Wheland intermediate the delocalization of electrons between the acceptors and a benzene ring is destroyed and that, no matter whether the electrophile attacks at the *ortho*-, *para*-, or *meta*-position, no delocalization involving the acceptor is possible in the Wheland intermediate. We predict at once, therefore, that all acceptors have a retarding influence on addition-with-elimination reactions. We then have to consider what additional electronic effect they may have. Let us look at the electronic properties of the common acceptor groups by drawing their possible canonical forms.

$$\left[-\overset{+}{N}(=O)-O^{-} \leftrightarrow -\overset{+}{N}(-O^{-})=O\right];$$

$$\left[-S(=O)_2-OR \leftrightarrow -\overset{+}{S}(=O)(-O^{-})-OR \leftrightarrow -\overset{+}{S}(=O)(-O^{-})-OR\right] > \left[-C(R)=O \leftrightarrow -\overset{+}{C}(R)-O^{-}\right];$$

$$\left[-C(OC_2H_5)=O \leftrightarrow -\overset{+}{C}(OC_2H_5)-O^{-}\right] > [-C\equiv N \leftrightarrow -\overset{+}{C}=\overset{-}{N}:]$$

Common acceptor substituents ($-R$).

The formulae show that in all the common acceptors the atom attached to the aromatic nucleus carries an incipient positive charge. We thus conclude that all acceptors will greatly deactivate the aromatic nucleus to electrophilic addition-with-elimination reactions and will have the same directing influence as a non-accepting electronegative group, i.e. a normal electron-attractor or $-I$ group. Before considering acceptor substituents further, let us therefore consider groups whose effect depends solely on their relative electronegativity.

We can represent the charge distribution in the Wheland intermediates as follows:

S, $+\frac{1}{3}$, $+\frac{1}{3}$, $+\frac{1}{3}$, H, X	S, $+\frac{1}{3}$, $+\frac{1}{3}$, H, X, $+\frac{1}{3}$	S, H, $+\frac{1}{3}$, X, $+\frac{1}{3}$, $+\frac{1}{3}$
para-Substitution (I)	*meta*-Substitution (II)	*ortho*-Substitution (III)

Charge distribution in Wheland intermediates in which the substituent S cannot conjugate with the rest of the molecule.

Let us consider toluene where the substituent is the methyl group. The methyl group repels electrons and thus partially neutralizes the positive charge which develops at the 1-position of the Wheland intermediate when attack is at the 4(*para*)-position (I) or the 2(*ortho*)-position (III). This is, of course, exactly analogous to the argument we used for the addition of hydrogen chloride to propene. On the other hand, electron-attracting groups such as CCl_3 or $^{+}N(CH_3)_3$ pull electrons away from the position to which they are attached and therefore intensify the positive charge in the 1-position of the Wheland intermediates involved in attack at the *ortho*- or *para*-position (I or III).

On the basis of our hypothesis, therefore, an electron-repelling group such as methyl will, by spreading the charge, lower the free energy of the Wheland intermediates involved in *ortho*- and *para*-attack and will direct substitution into the *ortho*- and *para*-positions at a rate slightly greater than that of unsubstituted benzene. Electron-attracting groups tend to concentrate the charge at the 1-position in the Wheland intermediates involved in *ortho*- or *para*-attack, so that in these molecules *meta*-substitution will be preferred.

We can now return to consider the common accepting groups. The incipient positive charge on the atom adjacent to the benzene

Table 8.1. Orientating effects of substituents in the benzene ring

Type of group	Examples	Effect
Donors $(+R)$	^{-}O, NH_2, NR_2, OH, OR, SH, $CH{=}CH_2$	*ortho*/*para*-Directing, strongly activating
Electron-repelling $(+I)$	CH_3, $(CH_3)_2CH$	*ortho*/*para*-Directing, weakly activating
Electron-attracting donors $(+R, -I)$	Cl, F, $CH{=}CHNO_2$	*ortho*/*para*-Directing, weakly deactivating
Electron-attracting $(-I)$	CCl_3, CF_3, $(CH_3)_3N^{+}$,[a]	*meta*-Directing, deactivating
Acceptors $(-R)$	NO_2, SO_2OR, SO_2R, CO_2R, COR, CN	*meta*-Directing, strongly deactivating

[a] Groups carrying a positive charge will be particularly deactivating to electrophilic attack by positively charged species such as the nitronium ion. This purely electrostatic effect is felt more strongly at the *meta*, than at the *para*-position. The overall effects are that the molecule is very unreactive to attack by positive ions and that the *meta*- and *para*-positions are attacked about equally fast.

ring means that all these groups behave like the electron-attracting groups, e.g. like the trichloromethyl group, in their directing properties. The important difference is that an acceptor group will be more deactivating than simple electron-attracting groups because of the loss of delocalization in the formation of the Wheland intermediate. We can summarize all our conclusions as in Table 8.1.

Table 8.2 gives some idea of the magnitude of these effects for nitration. Selectivity varies substantially for different addition-with-elimination reactions. Sulphonation has a selectivity similar to that

Table 8.2. Orientation and approximate relative rates of nitration of substituted benzenes $X{-}C_6H_5$

	X	Orientation (%)			Relative rate[a]
		o	*m*	*p*	($C_6H_6 = 1$)
$+R$	CH_3O	45 (72)	—	55 (28)	10^3
$+I$	CH_3	58	5	37	10
$-I, +R$	Cl	30	1	69	10^{-2}
$-I$	CCl_3	7	65	28	10^{-3}
$-R$	NO_2	6	93	1	10^{-4}

[a] The relative-rate data are no more than intelligent guesses. There are accurate values for similar compounds, e.g. the four halogenobenzenes, but nitrobenzene is nitrated in a concentrated sulphuric acid medium and anisole in a solution of acetic acid, so that direct comparison is impossible. The medium and conditions can also have a big effect on orientation; the figures for anisole are for reaction in acetic acid at 65° and those in parenthesis are for reaction in acetic anhydride at 0°.

of nitration, but halogenation is much more selective. Selectivity depends, in fact, a great deal on the conditions and on the details of the mechanism. The generalizations in Table 8.1 remain true for all electrophilic addition-with-elimination reactions though the magnitude of the directing effects varies.

Directive Effects in Nucleophilic Aliphatic Substitution

Nucleophilic aliphatic substitution is initiated by attack of an electron-donating group, which usually carries a negative charge (Nu^-). We might, at first sight, predict that an electron-attracting group attached to the carbon atom undergoing substitution would enhance the rate of substitution. However, the reaction is completed by the ejection of an anion, which would be facilitated by an electron-repelling group:

$$\bar{Nu}: + \underset{H\ \ H\ \ X}{\overset{R}{C}} \longrightarrow \overset{\delta-}{Nu}\cdots\underset{H\ \ H}{\overset{R}{C}}\cdots\overset{\delta-}{X} \longrightarrow \underset{Nu\ \ H\ \ H}{\overset{R}{C}} + :X^-$$

Overall, therefore, aliphatic nucleophilic substitution is not particularly sensitive to the presence of electron-repelling or electron-attracting groups adjacent to the substituted carbon atoms. Substituents do have a tremendous effect in changing the mechanism of the reaction, but for the moment we are considering, solely, those reactions which occur by the bimolecular (S_N2) mechanism.

Not all nucleophilic substitutions take place without the formation of charge; in particular, the formation of a quaternary ammonium ion by nucleophilic substitution results in the formation of charged species, and in this reaction we should expect polar substituents to have a very big effect:

$$R_3N: + \underset{H\ \ H\ \ I}{\overset{R'}{C}} \longrightarrow R_3\overset{+}{N}CH_2R' + I^-$$

The rates of quaternization of some *para*-substituted *N*,*N*-dimethylanilines with methyl iodide have been studied in nitrobenzene solution; the results are reproduced in Table 8.3, which shows

Table 8.3. Rates of quaternization of *para*-substituted *N*,*N*-dimethylanilines in nitrobenzene solution

$$p\text{-X–}C_6H_4\text{–}N(CH_3)_2 + CH_3I \rightarrow p\text{-X–}C_6H_4\text{–}\overset{+}{N}(CH_3)_3I^-$$

X	k (10^3 l.mole^{-1} sec^{-1})	E (kcal mole^{-1})
CH_3O	5.6	11.7
CH_3	2.5	12.3
H	0.8	12.8
Br	0.26	13.7
Cl	0.21	13.9

excellent accord with the ideas we have developed. However, the substituents are in the nucleophile, and results relating to the effect of substituents adjacent to the substituted carbon atom are limited and hard to interpret; this is largely because the substituents likely to have a big effect on the rate cause, instead, a change in mechanism. The effect of substituents on nucleophilic substitution occurring by ionization or by the unimolecular mechanism (S_N1) will be discussed in Part 3.

Problems

1. At which position (or positions) would you expect benzenediazonium chloride to couple with phenol in alkaline solution and with *N,N*-dimethylaniline.

2. Predict the orientation of mononitration of the following:

$C_6H_5NO_2$; $C_6H_5COCH_3$; C_6H_5Cl; $C_6H_5OCH_3$; *p*-nitrotoluene; resorcinol dimethyl ether [1,3-$(CH_3O)_2C_6H_4$].

3. Predict the products from the addition of hydrogen fluoride to:

$C_6H_5CH{=}CH_2$; $CH_3CH{=}CH_2$; $CF_3CH{=}CH_2$; $CH_3COCH{=}CH_2$; $CH_3C{\equiv}CCH_3$

CHAPTER 9

Carbanions

The carbon–hydrogen bonds in organic molecules are in general non-polar, and carbon does not readily form compounds in which a carbon atom carries a positive or a negative charge. We have seen that delocalization of electrons greatly increases the stability of a molecule. It would, therefore, be reasonable to ask whether it is possible to have a carbon anion (called a *carbanion*) in a molecule in which the electrons were extensively delocalized. This delocalization would result in spreading of the negative charge which, according to our hypothesis, would lower the free energy of the species and hence increase its stability in solution. The triphenylmethide ion, formed by the reaction of triphenylmethane with sodamide (or by the reaction of sodium with triphenylmethyl chloride) is an example of just such a stabilized anion:

$$(C_6H_5)_3CH + Na^+NH_2^- \longrightarrow \underset{\text{Sodium triphenylmethide}}{(C_6H_5)_3C^-Na^+} + NH_3 \qquad ...(1)$$

There are no less than forty-four canonical forms which can be drawn for the triphenylmethide anion: three are as depicted here; there are

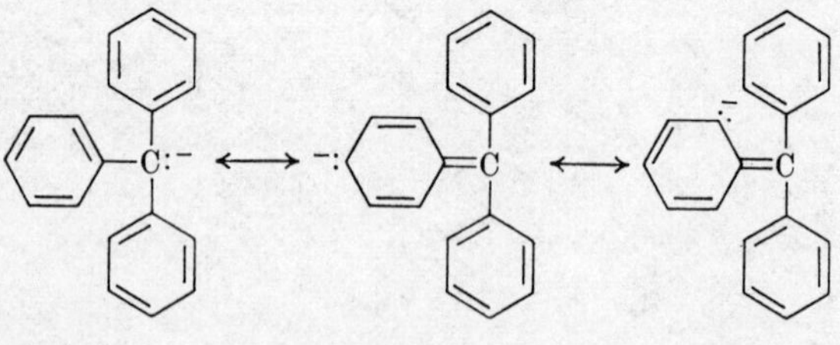

seven additional others in which the negative charge is centred on other carbon atoms, and thirty-four more corresponding to alternative arrangements of bonds in the Kekulé benzene rings.

Triphenylmethane is an exceedingly weak acid and, although the triphenylmethide ion is stable, sodium triphenylmethide is hydrolysed instantly by water and even by such a weak acid as ethanol. If in the canonical forms of a carbanion the negative charge is distributed over more electronegative atoms such as oxygen, this should enhance its stability compared with that of an ion in which the negative charge is distributed only over carbon atoms. For example, β-diketones behave as very weak acids because, in the anions, the negative charge is distributed over five atoms including two oxygen atoms:

$$H_3C{-}C({=}O){-}CH_2{-}C({=}O){-}CH_3 \xrightarrow{NaOC_2H_5}$$

$$\left[H_3C{-}C(-\ddot{O}:^-){=}CH{-}C({=}O){-}CH_3 \longleftrightarrow H_3C{-}C({=}O){-}\bar{C}H{-}C({=}O){-}CH_3 \longleftrightarrow H_3C{-}C({=}O){-}CH{=}C(-\ddot{O}:^-){-}CH_3 \right]$$

$$+ Na^+ + C_2H_5OH \qquad \text{...(2)}$$

As the formulae indicate, treating the diketone (acetylacetone; pentane-2,4-dione) with a solution of sodium ethoxide in alcohol results in the formation of the sodium salt of the diketone. In other words, acetylacetone is more acidic than ethanol. The anion is drawn as a resonance hybrid of three possible canonical forms, two of which have the negative charge centred on an oxygen atom. According to molecular-orbital theory there are molecular orbitals of π symmetry associated with all five atoms involved in the conjugation (see Figure 9.1).

If a β-diketone is more acidic than ethanol we may reasonably ask: how acidic are simple carbonyl compounds such as acetone or ethyl acetate? Clearly, they are far less acidic than the β-dicarbonyl compound. Nonetheless, it is reasonable to suppose that they will ionize to a very limited extent in the presence of a strong base. In Part 1, discussion of the reactions of aldehydes with alkali was restricted to aldehydes with no hydrogen atom attached to the α-carbon atom. Such aldehydes underwent the Cannizzaro reaction [e.g. (3) and (4)]. This reaction involves a hydride shift. But carbonyl

$$2\,HCHO \xrightarrow{NaOH} CH_3OH + HCO_2H \qquad ...(3)$$

$$2\,C_6H_5CHO \xrightarrow{NaOH} C_6H_5CH_2OH + C_6H_5CO_2H \qquad ...(4)$$

compounds with hydrogen on the α-carbon atom, such as acetaldehyde, will ionize on treatment with a strong base, forming an anion:

$$CH_3C(H){=}O \underset{}{\overset{Base}{\rightleftharpoons}} \left[H_2C{=}C(H){-}\ddot{O}{:}^{-} \longleftrightarrow H_2\bar{C}{-}C(H){=}O \right] + BH^{+} \qquad ...(5)$$

Notice that acetaldehyde will be only very weakly acidic, and the concentration of the anion will be very small. We know that nucleophiles add to the carbon atom of the carbonyl group, and clearly the anion of acetaldehyde is a strong nucleophile. The anion adds to un-ionized acetaldehyde as fast as it is formed [reaction (6)]:

$$H_3C(H)C{=}O + {}^{-}CH_2{-}C(H){=}O \rightleftharpoons H_3C{-}CH(O^{-}){-}CH_2{-}C(H){=}O \overset{BH^{+}}{\rightleftharpoons} \underset{\text{Aldol}}{CH_3CH(OH)CH_2CHO} \qquad ...(6)$$

Although the equilibrium between the anion and the free aldehyde lies very much on the side of the aldehyde, this does not prevent the overall reaction going to completion. The product (aldol) is still an aldehyde and the reaction can go further, so that polymerization

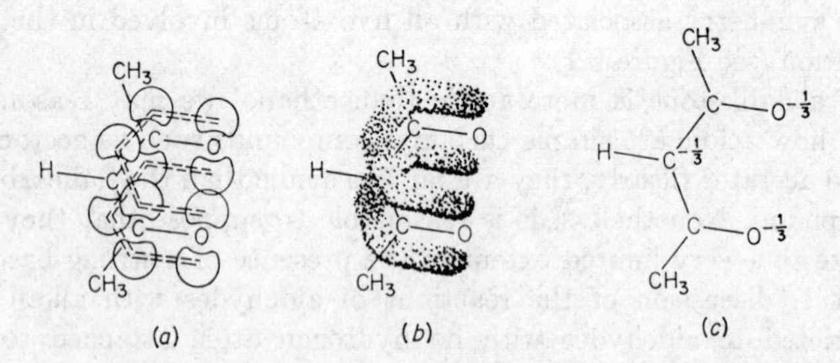

Figure 9.1. The pentane-2,4-dione anion: (*a*) orbital interaction; (*b*) 'π-electron clouds'; and (*c*) charge distribution.

is possible. Equally, the reaction is reversible and aldol can be depolymerized. In practice, the preparation of aldol by treatment of acetaldehyde with aqueous alkali is extremely difficult, not only because further polymerization occurs, but also because aldol itself loses water very readily [reaction (7)].

$$H_3C{-}CH(OH){-}CH_2{-}CHO \xrightarrow{\ddot{X}^-} \underset{\text{Crotonaldehyde}}{H_3C{-}CH{=}CH{-}CHO} + HX + OH^- \qquad ...(7)$$

The reaction is suitably called the *aldol condensation*, and this name is used in a general sense for condensation reactions of this type. Acetone, in the presence of a weak base, undergoes a similar reaction (8).

$$H_3C{-}CO{-}CH_3 \xrightarrow{Ba(OH)_2} \left[H_2\ddot{C}^-{-}CO{-}CH_3 \longleftrightarrow H_2C{=}C(CH_3){-}\ddot{O}{:}^- \right]$$

$$(H_3C)_2C{=}O + {}^-H_2C{-}CO{-}CH_3 \longrightarrow (H_3C)_2C(O^-){-}CH_2{-}CO{-}CH_3 \xrightarrow{H^+}$$

$$(CH_3)_2C(OH){-}CH_2{-}CO{-}CH_3 \qquad ...(8)$$

The product of this condensation, sometimes called diacetone alcohol, although more stable than aldol, readily loses water to yield 4-methylpent-3-en-2-one (mesityl oxide):

$$(CH_3)_2C(OH)\text{—}CH_2\text{—}COCH_3 \xrightarrow{:X^-} (CH_3)_2C{=}CH\text{—}COCH_3 + HX + OH^- \quad ...(9)$$

Mesityl oxide

Very often it is impossible to isolate the β-hydroxy carbonyl compound because loss of water occurs too readily and the unsaturated carbonyl compound is all that can be isolated after such a reaction.

By using a carbonyl compound with no α-hydrogen atom it is possible to obtain condensation products from two different compounds. For example, treating a solution of benzaldehyde and acetone with alkali results in a high yield of the unsaturated ketone, benzylideneacetone (4-phenylbutenone):

$$C_6H_5CHO + CH_3COCH_3 \xrightarrow{NaOH} C_6H_5CH{=}CHCOCH_3\ (78\%) \quad ...(10)$$

Benzylideneacetone

In the presence of an excess of benzaldehyde the reaction proceeds further, to yield dibenzylideneacetone (1,5-diphenylbutadien-3-one):

$$C_6H_5CHO + C_6H_5CH{=}CHCOCH_3 \xrightarrow{NaOH} C_6H_5CH{=}CHCOCH{=}CHC_6H_5\ (94\%) \quad ...(11)$$

Dibenzylideneacetone

In the stabilized anions that we have discussed so far, the carbonyl group behaves as an acceptor. It would be reasonable to ask whether compounds in which other powerful acceptor groups are attached to methylene groups form anions in a similar fashion. The answer is that they do, and even a comparatively weak acceptor such as a cyanide group is able to stabilize a carbanion sufficiently to permit it to undergo the type of reaction just described; e.g. reactions (12) and (13):

$$C_6H_5CH_2CN + NaOC_2H_5 \longrightarrow$$
$$[C_6H_5\ddot{C}H^-\text{—}C{\equiv}\ddot{N} \longleftrightarrow C_6H_5CH{=}C{=}\ddot{N}:^-] + Na^+ + C_2H_5OH \quad ...(12)$$

$$C_6H_5CHO + H\bar{C}(C_6H_5)CN \longrightarrow C_6H_5CH(O^-)-CH(C_6H_5)CN \xrightarrow{HX}$$

$$C_6H_5CH(OH)-CH(C_6H_5)CN \longrightarrow C_6H_5CH{=}C(C_6H_5)CN + HX + OH^-$$

...(13)

The nitro group is a more powerful acceptor than the cyanide group and it is more effective in stabilizing carbanions:

$$\left[RCH_2-\overset{+}{N}(=O)-O^- \longleftrightarrow RCH_2-\overset{+}{N}(-O^-)=O\right] \xrightarrow{NaOC_2H_5}$$

$$\left[R\bar{C}H-\overset{+}{N}(=O)-O^- \longleftrightarrow R\bar{C}H-\overset{+}{N}(-O^-)=O \longleftrightarrow RCH{=}\overset{+}{N}(-O^-)_2\right] \quad ...(14)$$

Condensation reactions of the anions of aliphatic nitro compounds are very common, and we illustrate just two examples (15) and (16) [here and henceforth C_2H_5 is abbreviated to Et].

$$C_6H_5CHO + CH_3NO_2 \xrightarrow{NaOEt} C_6H_5CH{=}CH_2NO_2 \quad ...(15)$$

1-Nitro-2-phenylethylene
(β-Nitrostyrene)

$$3\,HCHO + CH_3NO_2 \xrightarrow{NaOEt} (HOCH_2)_3CNO_2 \quad ...(16)$$

Carboxylic acids and their derivatives contain carbonyl groups and, although carboxylic acids themselves react with bases to form merely the carboxylate anion, their esters form unstable carbanions [reaction (17)] analogous to the anions from the aldehydes and ketones:

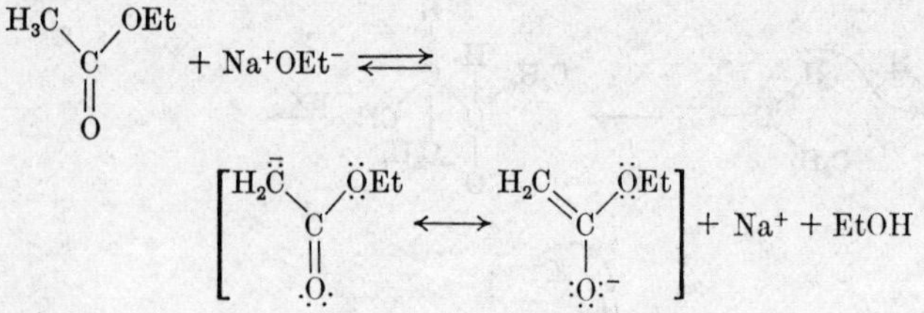

$$H_3C{-}C({=}O){-}OEt + Na^+OEt^- \rightleftharpoons \left[H_2\bar{C}{-}C({=}O){-}OEt \longleftrightarrow H_2C{=}C(O^-){-}OEt \right] + Na^+ + EtOH \qquad ...(17)$$

Just as the anion derived from acetaldehyde reacts with unchanged acetaldehyde, so the anion derived from ethyl acetate will react with unchanged ethyl acetate [cf. (18)].

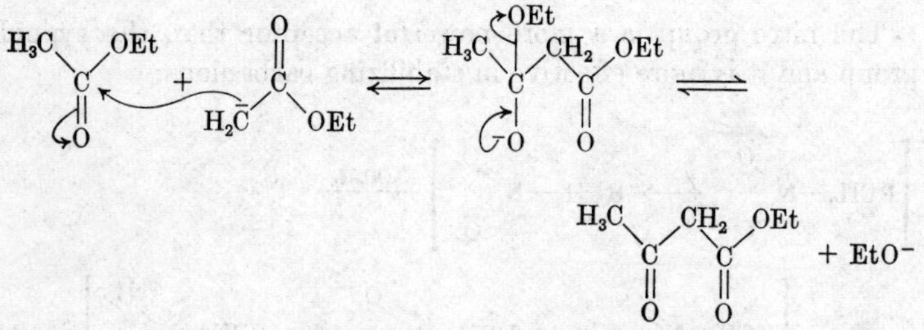

$$H_3C{-}C({=}O){-}OEt + H_2\bar{C}{-}C({=}O){-}OEt \rightleftharpoons H_3C{-}C(O^-)(OEt){-}CH_2{-}C({=}O){-}OEt \rightleftharpoons H_3C{-}C({=}O){-}CH_2{-}C({=}O){-}OEt + EtO^-$$

Ethyl acetoacetate

...(18)

At the beginning of this Chapter we considered the anion derived from a dicarbonyl compound and gave acetylacetone as an example. Clearly, ethyl acetoacetate is also a β-dicarbonyl compound and in the presence of sodium ethoxide it will be largely converted into its stable anion [cf. (19)]. Provided, therefore, that we have one mole

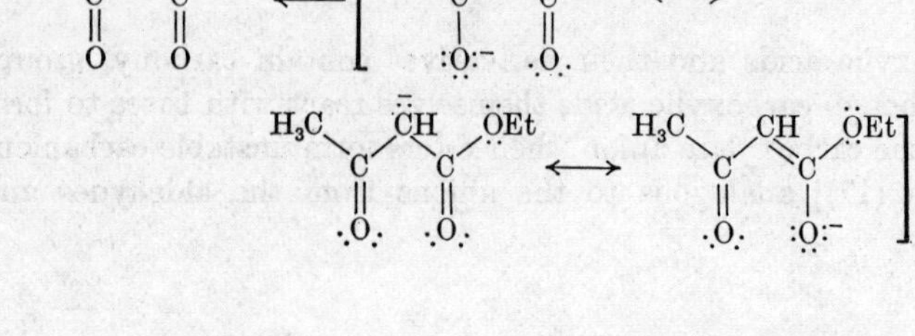

$$H_3C{-}C({=}O){-}CH_2{-}C({=}O){-}OEt \overset{^-OEt}{\rightleftharpoons} \left[H_3C{-}C(O^-){=}CH{-}C({=}O){-}OEt \longleftrightarrow H_3C{-}C({=}O){-}\bar{C}H{-}C({=}O){-}OEt \longleftrightarrow H_3C{-}C({=}O){-}CH{=}C(O^-){-}OEt \right] \qquad ...(19)$$

of sodium ethoxide, the reaction, reversible at every stage, will yield the sodium salt of ethyl acetoacetate.

This self-condensation of esters is known as the *Claisen ester condensation*. It is a general reaction of considerable practical importance. Esters with two α-hydrogen atoms can be condensed with sodium ethoxide in the same way as ethyl acetate itself:

$$RCH_2CO_2Et \xrightarrow{NaOEt} RCH_2CO\overset{\displaystyle R}{\overset{|}{C}}HCO_2Et \qquad ...(20)$$

Esters with only one α-hydrogen atom condense very reluctantly with sodium ethoxide and it is better to use a much stronger base; a convenient strong base is sodium triphenylmethide which we discussed at the beginning of this Chapter:

$$R_2CHCO_2Et \xrightarrow{(C_6H_5)_3C^-Na^+} R_2CHCO\underset{\displaystyle R}{\overset{\displaystyle R}{C}}CO_2Et \qquad ...(21)$$

Esters with no α-hydrogen atom cannot, of course, form an anion and do not undergo such self-condensation. They can however, be used in cross-condensations:

R_3CCO_2Et ↛ No self condensation—but used in cross reactions

By a cross-reaction we mean the reaction between two different esters. For example, if we take one mole of an ester with no α-hydrogen atom, e.g. ethyl trifluoroacetate, with one mole each of ethyl acetate and sodium ethoxide, the corresponding keto ester is obtained in very high yield [reaction (22)].

$$CF_3C(=O)OEt + H_2\bar{C}C(=O)OEt \rightleftharpoons CF_3C(OEt)(O^-)CH_2C(=O)OEt \longrightarrow CF_3C(=O)CH_2C(=O)OEt + OEt^- \qquad ...(22)$$

Ethyl trifluoroacetate

Ethyl benzoate will react in a similar way [cf. (23)].

$$C_6H_5C(=O)OEt + {}^-C(CH_3)_2CO_2Et \xrightleftharpoons{(C_6H_5)_3C^-Na^+} C_6H_5C(O^-)(OEt)C(CH_3)_2CO_2Et \longrightarrow C_6H_5COC(CH_3)_2CO_2Et + OEt^- \quad ...(23)$$

Other examples include ethyl formate, which will also acylate other esters [cf. (24)].

$$HC(=O)OEt\ \text{(Ethyl formate)} + {}^-CHRCO_2Et \rightleftharpoons HC(O^-)(OEt)CHRCO_2Et \longrightarrow HCOCHRCO_2Et + {}^-OEt \quad ...(24)$$

Ethyl oxalate, the diester of the simplest dibasic acid, oxalic acid, $HO_2C—CO_2H$, behaves similarly [cf. (25)]. The reactivity of ethyl

$$EtO_2C—C(=O)OEt\ \text{(Diethyl oxalate)} + {}^-CHRCO_2Et \rightleftharpoons EtO_2C—C(O^-)(OEt)—CHR—CO_2Et \longrightarrow EtO_2C—CO—CHR—CO_2Et \quad ...(25)$$

oxalate becomes particularly interesting when it is allowed to condense with the diester of another dibasic acid, as shown in (26).

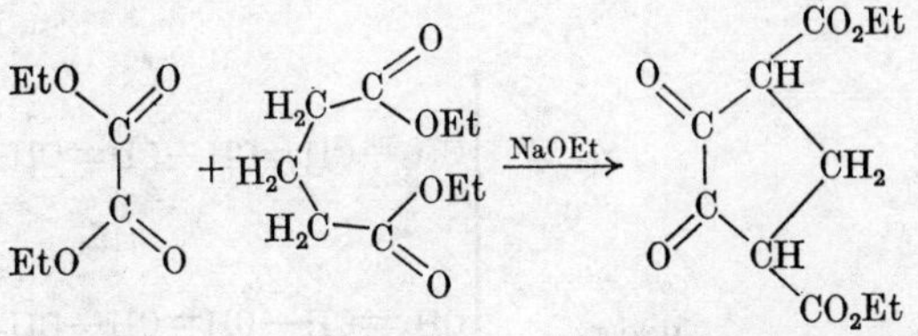

$$(EtO)OC{-}CO(OEt) + H_2C(CH_2CO_2Et)_2 \xrightarrow{NaOEt} \text{cyclo}[O{=}C{-}C({=}O){-}CH(CO_2Et){-}CH_2{-}CH(CO_2Et)] \quad \text{...(26)}$$

Diethyl oxalate — Diethyl glutarate — Diethyl 1,2-dioxo-cyclopentane-3,5-dicarboxylate

Looking at the formula of an ester of a dibasic acid raises the question: will such esters 'bite their own tails'? Such cyclizations can be brought about particularly readily with the esters of adipic and pimelic acid which yield derivatives of cyclopentanone and cyclohexanone [reaction (27)]. Three- and four-membered rings *cannot* be prepared in this manner. This cyclic condensation of the esters of dibasic acids is known as the *Dieckmann condensation*.

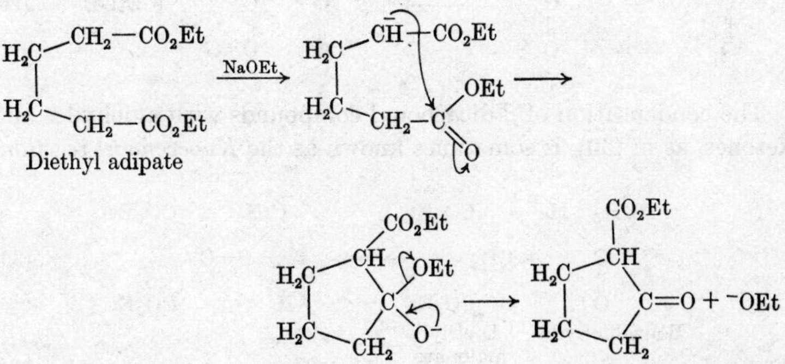

$$H_2C(CH_2{-}CO_2Et){-}H_2C(CH_2{-}CO_2Et) \xrightarrow{NaOEt} H_2C(\bar{C}H{-}CO_2Et){-}H_2C(CH_2{-}C({=}O)OEt) \longrightarrow$$

Diethyl adipate

$$\text{cyclo}[H_2C{-}CH(CO_2Et){-}C(OEt)(O^-){-}CH_2{-}H_2C] \longrightarrow \text{cyclo}[H_2C{-}CH(CO_2Et){-}C{=}O{-}CH_2{-}H_2C] + {}^-OEt$$

Ethyl cyclopentanone 2-carboxylate

...(27)

Unsaturated esters will also form carbanions, as in reaction (28).

$$\mathrm{CH_3CH{=}CH{-}CH{=}CH{-}C({=}O)OEt} \xrightarrow{\mathrm{NaOEt}}$$

$$\left[\mathrm{\ddot{C}H_2{-}CH{=}CH{-}CH{=}CH{-}C(\ddot{O}Et){=}\ddot{O}{:}} \longleftrightarrow \mathrm{CH_2{=}CH{-}CH{=}CH{-}CH{=}C(\ddot{O}Et){-}\ddot{O}{:}^{-}}\right]$$

$$\xrightarrow{\mathrm{HCO_2Et}} \mathrm{H(O{=})C{-}CH_2{-}CH{=}CH{-}CH{=}CH{-}C(OEt){=}O} \qquad \text{...(28)}$$

If esters can condense with themselves, it is clear they should also condense with other carbonyl compounds, and, similarly, other carbonyl compounds should react with the anions of esters. For example, acetylacetone (see p. 91) can be prepared by the condensation of acetone with ethyl acetate in the presence of sodium ethoxide [cf. (29)].

$$\mathrm{H_3C{-}C({=}O){-}OEt} + \mathrm{H_3C{-}C({=}O){-}CH_3} \xrightarrow{\mathrm{NaOEt}} \mathrm{H_3C{-}C({=}O){-}CH_2{-}C({=}O){-}CH_3} + \mathrm{EtOH} \qquad \text{...(29)}$$

The condensation of β-dicarbonyl compounds with aldehydes and ketones, as in (30), is sometimes known as the *Knoevenagel reaction*.

$$\underset{\text{Benzaldehyde}}{\mathrm{C_6H_5{-}C({=}O){-}H}} + \underset{\text{Diethyl malonate}}{\mathrm{CH_2(CO_2Et)_2}} \longrightarrow \mathrm{(C_6H_5)(H)C{=}C(CO_2Et)_2} \qquad \text{...(30)}$$

We have not detailed the electron shifts involved in this reaction as they are identical with those involved in the self-condensation of acetaldehyde or acetone. All the reactions discussed in this Chapter so far involve the addition of a nucleophile to the carbonyl double bond [see (31)].

$$\ddot{\text{Nu}}^- + R(R')C{=}O \rightleftharpoons \text{Nu}{-}CR(R'){-}O^- \quad \ldots(31)$$

What about the other common reaction of a nucleophile, namely, its displacement reaction (32) with an alkyl halide? If we treated an

$$\ddot{\text{Nu}}^- + RCH_2X \longrightarrow \text{Nu}CH_2R + X^- \quad \ldots(32)$$

alcoholic solution of acetone or ethyl acetate with sodium ethoxide and ethyl bromide, the principal product of the reaction would be diethyl ether, as in (33). This is because acetone and ethyl acetate are much weaker acids than ethanol and the concentration of the carbanions derived from them is very small.

$$\text{EtO}^- + CH_3CH_2Br \longrightarrow \text{EtO}CH_2CH_3 + Br^- \quad \ldots(33)$$

Ethoxide ions, will of course, also add to the carbonyl bonds, when an ester condensation or an aldol-type condensation is taking place. Unlike the nucleophilic displacement this kind of reaction (34) is reversible and does not affect the overall reaction sequence. The

$$\text{EtO}^- + R_2C{=}O \rightleftharpoons \text{EtO}{-}CR_2{-}O^- \quad \ldots(34)$$

anions derived from β-dicarbonyl compounds are much more stable than the ethoxide ion; alternatively one could say that acetylacetone, ethyl acetoacetate, and diethyl malonate are more acidic than ethanol (cf. Table 9-1). All the β-dicarbonyl compounds that are more

Table 9.1. Relative acidities of some very weak acids

Acid		K_a
$(C_6H_5)_3CH$		$\sim 10^{-31}$
$(CH_3)_2CO$	Increasing acid strength ↓	$\sim 10^{-20}$
C_2H_5OH		$\sim 10^{-18}$
$CH_2(CO_2Et)_2$		5×10^{-14}
$CH_3COCH_2CO_2Et$		7×10^{-11}
$CH_3COCH_2COCH_3$		1.5×10^{-9}

acidic than ethanol react with sodium ethoxide, yielding the carbanion which acts like other nucleophiles and displaces halogen from an alkyl halide, as in the sequence (35). The butylmalonic ester

$$EtO{-}C({=}O){-}CH_2{-}C({=}O){-}OEt \xrightarrow{NaOEt} \left[EtO{-}C(O^-){=}CH{-}C(=O){-}OEt \longleftrightarrow EtO{-}C(=O){-}\overline{C}H{-}C(=O){-}OEt \longleftrightarrow EtO{-}C(=O){-}CH{=}C(O^-){-}OEt \right]$$

$$(EtO{-}C(=O))_2CH^- + H_2C(C_3H_7){-}Br \longrightarrow (EtO{-}C(=O))_2CH{-}C_4H_9 + Br^- \qquad ...(35)$$

thus produced still has a hydrogen atom α to two carbonyl groups and, if treated with further sodium ethoxide and a different alkyl halide, again undergoes a displacement reaction. We can summarize these consecutive steps as in (36).

$$CH_2(CO_2Et)_2 + C_4H_9Br \xrightarrow{Na^+\ ^-OEt} C_4H_9CH(CO_2Et)_2 \xrightarrow[Na^+\ ^-OEt]{C_2H_5I} (C_4H_9)(C_2H_5)C(CO_2Et)_2 \quad ...(36)$$

Besides acetylacetone, other common and important β-dicarbonyl compounds which can be alkylated in this fashion are ethyl acetoacetate [see (37)] and ethyl cyanoacetate [see (38)]. Mechanistically

$$\underset{\text{Ethyl acetoacetate}}{H_3C{-}C(=O){-}CH_2{-}C(=O){-}OEt} \xrightarrow{\text{Base}} \left[H_3C{-}C(-\ddot{O}:^-){=}CH{-}C(=\ddot{O}){-}\ddot{O}Et \longleftrightarrow H_3C{-}C(=\ddot{O}){-}\ddot{C}H^-{-}C(=\ddot{O}){-}\ddot{O}Et \longleftrightarrow H_3C{-}C(=\ddot{O}){-}CH{=}C(-\ddot{O}:^-){-}\ddot{O}Et \right] \quad ...(37)$$

$$\underset{\text{Ethyl cyanoacetate}}{N{\equiv}C{-}H_2C{-}C(=O){-}OEt} \xrightarrow{\text{Base}} \left[:\ddot{N}^-{=}C{=}HC{-}C(=\ddot{O}){-}\ddot{O}Et \longleftrightarrow :N{\equiv}C{-}H\ddot{C}^-{-}C(=\ddot{O}){-}\ddot{O}Et \longleftrightarrow :N{\equiv}C{-}HC{=}C(-\ddot{O}:^-){-}\ddot{O}Et \right] \quad ...(38)$$

these compounds behave in the same way as diethyl malonate [see (39)].

$$\underset{\text{Anion of ethyl acetoacetate}}{(EtO{-}C{=}O)(CH_3{-}C{=}O)CH^-} + CH_3I \longrightarrow (EtO{-}C{=}O)(CH_3{-}C{=}O)CH{-}CH_3 + I^- \quad ...(39)$$

These alkylations are of great synthetic value, but full discussion of the use of, particularly, diethyl malonate and ethyl acetoacetate must await discussion of syntheses in general (Chapter 15). We shall just note here two reactions (40) and (41) of diethyl malonate with

$$\begin{matrix} CH_2Br \\ | \\ CH_2Br \end{matrix} + 2CH_2(CO_2Et)_2 + 2NaOEt \longrightarrow \begin{matrix} CH_2CH(CO_2Et)_2 \\ | \\ CH_2CH(CO_2Et)_2 \end{matrix} \qquad \ldots(40)$$

$$\begin{matrix} CH_2Br \\ | \\ CH_2Br \end{matrix} + CH_2(CO_2Et)_2 + 2NaOEt \longrightarrow \begin{matrix} CH_2 \\ | \\ CH_2 \end{matrix}\!\!>\!C(CO_2Et)_2 \qquad \ldots(41)$$

ethylene dibromide and see how by altering the concentrations it is possible to change the products. Clearly, such reactions cannot yield exclusively one product, as indicated by the equations. Nonetheless, using ethylene dibromide, diethyl malonate, and sodium ethoxide in the proportions shown, W. H. Perkin, Junior, was able to isolate from the two separate reactions reasonable yields of the two products shown.

The anions from β-dicarbonyl compounds can be readily oxidized, the result being a coupling reaction (42). The usual oxidizing agent is molecular iodine.

$$2\ \begin{matrix} O{=}C(CH_3) \\ \quad \bar{C}H \\ O{=}C(OEt) \end{matrix} \xrightarrow{-2e} \begin{matrix} O{=}C(CH_3) & & (H_3C)C{=}O \\ \quad CH & — & CH \quad \\ O{=}C(OEt) & & (EtO)C{=}O \end{matrix} \qquad \ldots(42)$$

At the beginning of our discussion of the reactions of carbanions as nucleophiles in displacement reactions, we ruled out the carbanions derived from monocarbonyl compounds because they were less acidic than ethanol, so that the alkoxide ion was always a more reactive nucleophile than the carbanion. If, however, we can prepare these carbanions without the presence of another anion, they will undergo displacement reactions. One way of doing this is to treat the carbonyl compounds with sodamide, as shown in the sequence (43). This reaction has frequently been used to alkylate cyclohexanone [see (44)].

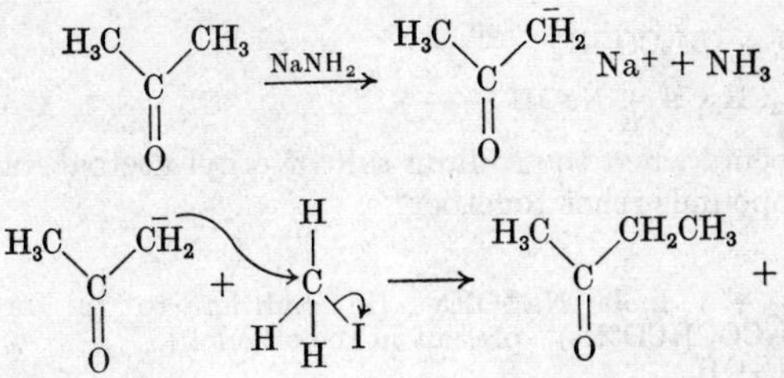

$$H_3C{-}CO{-}CH_3 \xrightarrow{NaNH_2} H_3C{-}CO{-}\bar{C}H_2\ Na^+ + NH_3$$

$$H_3C{-}CO{-}\bar{C}H_2 + H_3C{-}I \longrightarrow H_3C{-}CO{-}CH_2CH_3 + I^- \qquad ...(43)$$

$$\text{Cyclohexanone} \xrightarrow{NaNH_2} \text{(2-carbanion of cyclohexanone)}\ Na^+ \xrightarrow{CH_2{=}CHCH_2Br} \text{2-(}CH_2CH{=}CH_2\text{)cyclohexanone} \qquad ...(44)$$

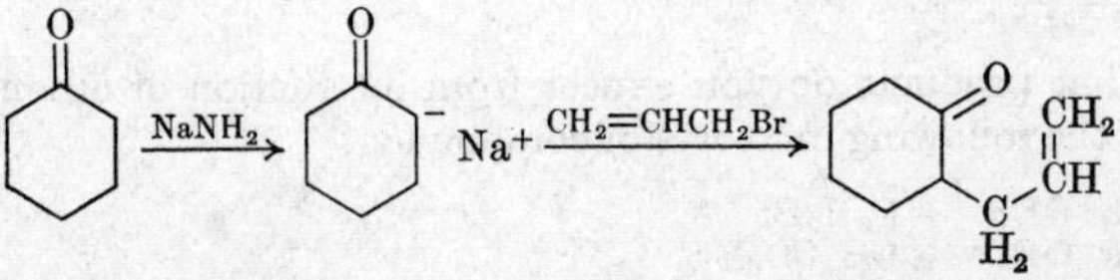

The use of sodamide is not restricted to carbonyl compounds; suitable nitriles will behave in the same fashion, as indicated in the sequence (45).

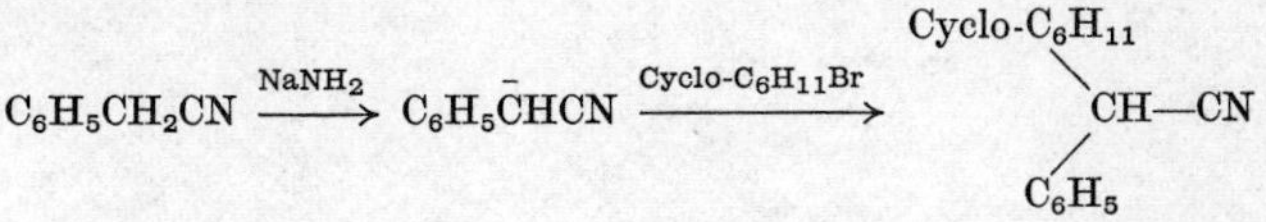

$$C_6H_5CH_2CN \xrightarrow{NaNH_2} C_6H_5\bar{C}HCN \xrightarrow{\text{Cyclo-}C_6H_{11}Br} (\text{Cyclo-}C_6H_{11})(C_6H_5)CH{-}CN \qquad ...(45)$$

Carbanions are highly reactive species completely different from the familiar stable anions of inorganic chemistry. We have seen, however, that they are important intermediates in a wide variety of reactions. We shall, later, when discussing synthetic sequences, return to many of the reactions described in the present Chapter.

Problems

1. Which of the following compounds form anions when treated with sodium triphenylmethide [$(C_6H_5)_3C^-Na^+$]?

CH_3COCH_3; $CH_3CH_2COCH_2CN$; $C_6H_5CH_2CN$; $C_6H_5CH_2CH_3$; $CH_2(CO_2CH_3)_2$; $CH_3COCH_2CO_2Et$; $(CH_3)_3CH$; $CH_3CH_2NO_2$.

Put the anions in an approximate order of stability, explaining your order of precedence.

2. Predict the outcome of the following reactions:

$CH_3CH_2CH_2CO_2Et + Na^+OEt^- \xrightarrow[\text{reflux}]{C_2H_5OH}$

$CH_3CO_2Et + CF_3CO_2Et + Na^+OEt^- \xrightarrow[\text{reflux}]{C_2H_5OH}$

$CH_3CHO + \text{aqueous } NaOH \longrightarrow$

$C_6H_5CHO + CH_3COCH_3 \xrightarrow{NaOH}$

$ClCH_2CH_2CH_2CN + NaOH \longrightarrow$

3. What occurs when the sodium salt of ethyl acetoacetate and the following compounds react together?

(*a*) C_2H_5I
(*b*) $(CH_2Br)_2$ + 1 mole Na^+OEt^- [in addition to the original $Na^+(CH_3COCHCO_2Et)^-$ present in the ethanol.]
(*c*) $I_2 + C_2H_5OH$
(*d*) C_6H_5Br

4. What products do you expect from interaction of ethanolic solutions of the following mixtures of compounds?

(*a*) $Na^+OEt^- + C_3H_7Br$
(*b*) $Na^+OEt^- + CH_3CO_2Et$
(*c*) $Na^+OEt^- + C_3H_7Br + CH_3CO_2Et$
(*d*) $Na^+OEt^- + C_3H_7Br + CH_3COCH_2CO_2Et$

CHAPTER 10

Carbonium Ions

Formation

In the last Chapter we showed that carbanions are very unstable when the negative charge is isolated on a single carbon atom but that moderately stable carbanions can be formed when the charge is spread over several atoms. For example, although the methyl carbanion is extremely reactive and unstable, the triphenylmethyl carbanion can be prepared by the reaction of sodamide with triphenylmethane and the resulting sodium triphenylmethide is quite a stable substance. We attributed this increased stability of the triphenylmethyl anion to the large number of possible canonical forms which spread the negative charge over the whole molecule. Clearly, we can have an identical number of canonical forms [see (1)] for the triphenylcarbonium ion, Ph_3C^+.*

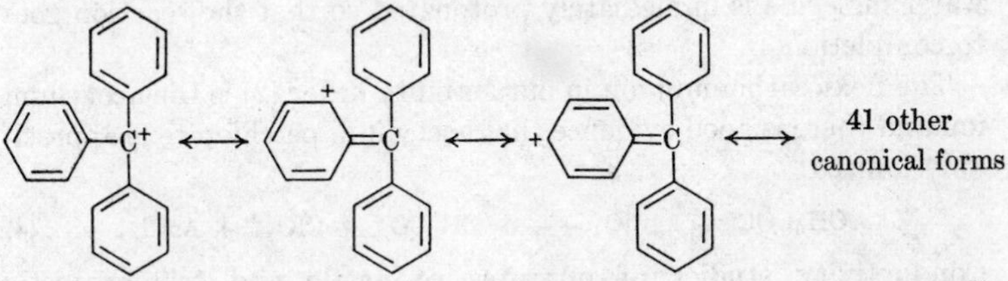

C^+ ⟷ C ⟷ C ⟷ 41 other canonical forms ...(1)

Triphenylcarbonium salts can be prepared and are quite stable. Just as carbanions can be stabilized by conjugation with electron-accepting groups and by electron-attractors, so carbonium ions can

* Ions containing positively charged carbon are called 'carbonium ions' generically, i.e. as a class. In addition, 'carbonium' is used as the name for the methyl cation, H_3C^+, and as a basis for naming its substitution products.

be stabilized by electron-donors and electron-repellers. Some common carbonium ions are shown in decreasing order of stability (2).

$$(C_6H_5)_3C^+,\quad (CH_3\overset{+}{C}{=}\ddot{O}{:} \longleftrightarrow CH_3C{\equiv}\overset{+}{O}{:}),\quad CH_3{-}\underset{CH_3}{\overset{CH_3}{\overset{|}{\underset{|}{C^+}}}},$$

$$H{-}\underset{CH_3}{\overset{CH_3}{\overset{|}{\underset{|}{C^+}}}},\quad CH_3CH_2^+,\quad CH_3^+ \qquad \ldots(2)$$

Carbanions are formed by treating an organic compound with a powerful base and exist in basic media. Carbonium ions are formed by treating organic molecules with strong acids (especially Lewis acids) and exist in acidic media. The triphenylcarbonium ion is formed by dissolving triphenylmethanol in sulphuric acid [reaction (3)]. In this reaction the oxonium ion formed by the protonation of

$$(C_6H_5)_3C{-}OH + H_2SO_4 \rightleftharpoons (C_6H_5)_3C{-}\overset{+}{O}\begin{smallmatrix}\diagup H\\ \diagdown H\end{smallmatrix}\ HSO_4^- \rightleftharpoons$$

$$(C_6H_5)_3C^+ + H_2O + HSO_4^- \qquad \ldots(3)$$

the oxygen atom in triphenylmethanol breaks down to yield a triphenylcarbonium ion and a water molecule. In sulphuric acid the water molecule is immediately protonated so that the reaction goes to completion.

The next carbonium ion in our stability order (2) is the acetylium ion and there is good evidence that acetylium perchlorate is appreciably ionized:

$$CH_3COCl + AgClO_4 \longrightarrow CH_3CO^+ + ClO_4^- + AgCl \qquad \ldots(4)$$

Conductivity studies of mixtures of acetic and trifluoroacetic anhydrides show that the unsymmetrical anhydride which is formed is ionized to a very small extent into acetylium and trifluoroacetate ions, as in (5).

$$CH_3CO{\cdot}O{\cdot}COCF_3 \rightleftharpoons CH_3CO^+ + {}^-OCOCF_3 \qquad \ldots(5)$$

Treatment of an alkyl halide with aluminium chloride, although not producing free ions, produces a complex with a very polar bond [see reaction(6)].

$$CH_3CH_2Cl + AlCl_3 \longrightarrow CH_3CH_2^+ \cdots ClAlCl_3^- \qquad ...(6)$$

Free carbonium ions can only exist in highly acidic media, but there is very good evidence that they occur as transient intermediates in many reactions. For example, when discussing the addition of the hydrogen halides to olefins in Part 1 (Chapter 6) and again in Chapter 8 (this Part), we assumed that the first step was the addition of a proton, yielding a carbonium ion, as in (7). Another reaction in which

$$(H_3C)_2C{=}CH_2 \xrightarrow{H-X} (H_3C)_2\overset{+}{C}-CH_3 + X^- \qquad ...(7)$$

transient carbonium ions may occur is that of nitrous acid with a primary amine in acidic media. Initially a diazonium salt is formed. Aromatic diazonium salts are moderately stable compounds, but most aliphatic diazonium salts break down immediately to yield a carbonium ion and nitrogen [reaction (8)].

$$C_2H_5{-}NH_2 + HNO_2 \xrightarrow{HCl} C_2H_5{-}\overset{+}{N}{\equiv}N\ X^- \longrightarrow C_2H_5^+ + N_2 + X^- \qquad ...(8)$$

When discussing nucleophilic substitution reactions of the alkyl halides in Part 1 we described how the reaction is usually a bimolecular process (9) involving attack by a nucleophile on the carbon atom carrying the halogen atom. However, we also described how, with a

$$\bar{N}u + {-}\overset{|}{\underset{|}{C}}{-}X \longrightarrow Nu{-}\overset{|}{\underset{|}{C}}{-} + \bar{X}: \qquad ...(9)$$

few compounds, particularly tertiary alkyl halides, the reaction was independent of the concentration of the nucleophile and must therefore involve the initial ionization of the alkyl halide to yield a carbonium ion transiently [see (10)].

$$\overset{|}{\underset{|}{C}}{-}X \xrightarrow{slow} C^+ + X^- \xrightarrow{fast} \text{Products} \qquad ...(10)$$

Reactions

The first reaction of carbonium ions we must consider is their reaction with nucleophiles. This will be very rapid and result in the formation of a new bond between the positively charged carbon atom and the electron pair of the nucleophile. For example, the triphenylcarbonium ion reacts instantly with water to produce triphenylmethanol, and with methanol to produce methyl triphenylmethyl ether [reactions (11)].

$$(C_6H_5)_3C^+ \quad HSO_4^- \begin{cases} \xrightarrow{H_2O} (C_6H_5)_3COH + H_2SO_4 \\ \xrightarrow{CH_3OH} (C_6H_5)_3COCH_3 + H_2SO_4 \end{cases} \quad \text{...(11)}$$

Similarly, the ethylium ion produced by the reaction of aqueous nitrous acid with ethylamine reacts with the solvent water to produce ethanol, as in (12). As we saw in Part 1 (Chapter 5), *tert*-

$$C_2H_5{-}NH_2 + HNO_2 \xrightarrow{HX} [C_2H_5{-}\overset{+}{N}{\equiv}N] \longrightarrow [C_2H_5^+] \xrightarrow{OH_2} C_2H_5OH + H^+ \quad \text{...(12)}$$

butyl chloride (2-chloro-2-methylpropane) reacts with aqueous alcohol by a unimolecular process to give both *tert*-butyl alcohol (2-methylpropan-2-ol) and isobutene (2-methylpropene) (see annexed scheme). In the first reaction the carbonium ion reacts with water molecules to produce the alcohol (an S_N1 reaction). In the other reaction, however, a proton is eliminated (an *E*1 reaction). This is another way in which a carbonium ion can stabilize itself, namely, by the elimination of a proton.

$$(CH_3)_3C{-}Cl \xrightarrow[\text{C}_2\text{H}_5\text{OH 4 parts (slow)}]{\text{H}_2\text{O 1 part}} (CH_3)_3C^+ + Cl^-$$

$$(CH_3)_3C^+ \xrightarrow{\text{(Fast) } E1} CH_2{=}C(CH_3)_2$$

$$(CH_3)_3C^+ \xrightarrow{S_N1 \text{ (Fast)}} CH_3{-}C(CH_3)_2{-}OH$$

Carbonium ions are electrophiles, and cationic polymerization [cf. (13)] was described in Part 1 (Chapter 6). Now, although carbonium

$$\text{R—CH=CH}_2 + \text{X}^+ \longrightarrow \overset{+}{\text{R}}\text{CHCH}_2\text{X} \xrightarrow{\text{RCH=CH}_2} \text{XCH}_2\underset{}{\overset{\text{R}}{\text{C}}}\text{HCH}_2\overset{\text{R}}{\text{C}}\text{H}^+ \xrightarrow{\text{RCH=CH}_2}$$

$$\text{X—CH}_2\overset{\text{R}}{\text{C}}\text{HCH}_2\overset{\text{R}}{\text{C}}\text{HCH}_2\overset{\text{R}}{\text{C}}\text{H}^+ \longrightarrow \text{etc.} \quad \text{...(13)}$$

ions often cause polymerization of olefins, addition without polymerization can occur, as illustrated by reaction (14).

$$\text{RCH=CH}_2 + (\text{CH}_3\text{CO})(\text{CF}_3\text{CO})\text{O} \longrightarrow \text{RCH(OCOCF}_3\text{)—CH}_2\text{COCH}_3 \quad \text{...(14)}$$

Carbonium ions add also to the aromatic nucleus, by the usual addition-with-elimination reactions (electrophilic aromatic substitution). The reactivity of the various carbonium ions and their relative stability go hand in hand. Thus, the most stable carbonium ion in our list, the triphenylcarbonium ion, attacks only very reactive aromatic compounds such as phenols [see (15)]. The next most

$$(\text{C}_6\text{H}_5)_3\text{C}^+ + \text{C}_6\text{H}_5\text{—OH} \longrightarrow (\text{C}_6\text{H}_5)_3\text{C—C}_6\text{H}_4\text{—OH} + \text{H}^+ \quad \text{...(15)}$$

stable, the acetylium ion, is more reactive: it reacts with activated aromatic nuclei, for example, anisole, to yield 4-methoxyacetophenone [see (16)]. Contrary to what is said in many textbooks the

$$\underset{\text{Acetylium perchlorate}}{\text{CH}_3\text{CO}^+\text{ClO}_4^-} + \underset{\text{Anisole}}{\text{C}_6\text{H}_5\text{—OCH}_3} \longrightarrow \underset{\text{4-Methoxyacetophenone}}{\text{CH}_3\text{CO—C}_6\text{H}_4\text{—OCH}_3} + \text{HClO}_4 \quad \text{...(16)}$$

acetylium ion does not react directly with benzene. The complex, formed between acetyl chloride and aluminium chloride, is much more electrophilic and, in effect, the benzene nucleus behaves as a nucleophile, displacing the aluminium tetrachloride anion from the complex. A method of formulating this is shown in (17), where the dotted line shows the range of delocalization of electrons.

$$C_6H_6 + CH_3\overset{\cdots}{C}(=O)\cdots AlCl_4 \longrightarrow [C_6H_6(H)(COCH_3)]^+ \; AlCl_4^- \xrightarrow{H_2O} C_6H_5COCH_3 \quad \text{...(17)}$$

Direct alkylation of benzene by olefinic hydrocarbons in the presence of a strong acid is a very important reaction. If isobutene is bubbled through a mixture of sulphuric acid and benzene, for example, *tert*-butylbenzene is formed (18). Such alkylations are often

$$(H_3C)_2C{=}CH_2 + C_6H_6 \xrightarrow{H_2SO_4} H_3C{-}C(CH_3)_2{-}C_6H_5 \quad \text{...(18)}$$

done by using hydrogen fluoride as the acid catalyst; see, for example, reaction (19).

$$C_6H_6 + CH_2{=}CH{-}CH{=}CH_2 \xrightarrow{HF} C_6H_5CH_2CH{=}CHCH_3 \quad \text{...(19)}$$

In all the reactions involving carbon chains that we have discussed so far, the chain can be lengthened (for example, by nucleophilic substitution involving the cyanide anion, carbanions, or Grignard reagents), or it can be broken (for example, by oxidation of olefinic double bonds, as in the Barbier–Wieland step-by-step degradation of a carbon chain). Unless such lengthening or shortening occurs, the carbon chain has remained unchanged in the reaction; however, in the chemistry of certain carbonium ions, the carbon chain can rearrange during a reaction. We shall begin by considering a fairly simple example. When the glycol called pinacol (2,3-dimethylbutane-2,3-diol) is treated with strong acid such as sulphuric acid, it loses a molecule of water to yield a carbonium ion. In scheme (20) we have drawn a curved arrow from the carbon–oxygen bond to the protonated oxygen atom; this indicates that a molecule of water is going

$$(CH_3)_2C(OH)\text{—}C(OH)(CH_3)_2 \xrightarrow{H^+} (CH_3)_2C(\ddot{O}H)\text{—}C(CH_3)_2(OH_2^+) \longrightarrow \text{H}_3\text{C}(\text{HO})\overset{+}{C}\cdots CH_3 \cdots C(CH_3)_2\cdots OH_2 \qquad \ldots(20)$$

Activated complex

to separate. From the non-protonated oxygen atom we have drawn a curved arrow from the lone pair on the oxygen atom into the carbon–oxygen bond. Previously we used curved arrows in this way, but notice also that we have drawn a curved arrow from the bond between one of the methyl groups and carbon atom 2 to carbon atom 3, from which the water molecule is being eliminated. This arrow still represents the transfer of a pair of electrons, so that the methyl group is migrating and these electrons form the new carbon–carbon bond. We have also attempted to represent the activated complex—the dotted lines represent the bonds being formed and broken. When the water molecule finally departs, we are left with a stabilized oxonium ion which can lose a proton to yield the ketone called pinacalone (3,3-dimethylbutanone) [see (21)].

$$\left[H_3C\text{—}C(=\overset{+}{\ddot{O}}\text{—}H)\text{—}C(CH_3)_3 \longleftrightarrow H_3C\text{—}\overset{+}{C}(\text{—}\ddot{O}{:}\text{—}H)\text{—}C(CH_3)_3 \right] \xrightarrow{-H^+} H_3C\text{—}C(=O)\text{—}C(CH_3)_3 \qquad \ldots(21)$$

Pinacolone

This kind of reaction is known as a molecular rearrangement; the rearrangement of glycols by acids is a general reaction.

$$\underset{(a)}{(C_6H_5)_2C(OH)\text{—}CH_2OH} \xrightarrow{H^+} (C_6H_5)_2CHCHO \xleftarrow{H^+} \underset{(b)}{C_6H_5CH(OH)\text{—}CH(OH)C_6H_5} \qquad \ldots(22)$$

H atom migrates — Phenyl group migrates

If the groups R in a glycol $HOCR^1R^2\text{—}CR^3R^4OH$ are not all identical, the question arises which of them will migrate. The answer lies, as usual, in the need to spread the positive charge in the first step of the reaction: the proton adds to the oxygen atom best able to spread the charge. As a result, the two glycols (22*a*) and (22*b*) give the same product. Similarly, in reaction (23) the group which

$$(p\text{-CH}_3\text{OC}_6\text{H}_4)(\text{C}_6\text{H}_5)\underset{\text{HO}}{\text{C}}-\underset{\text{OH}}{\text{C}}(\text{C}_6\text{H}_4\text{OCH}_3\text{-}p)(\text{C}_6\text{H}_5) \xrightarrow[\text{group migrates}]{p\text{-Methoxyphenyl}} (\text{CH}_3\text{O}-\text{C}_6\text{H}_4)_2(\text{C}_6\text{H}_5)\text{C}-\underset{\text{O}}{\underset{\|}{\text{C}}}-\text{C}_6\text{H}_5 \qquad \text{...(23)}$$

migrates is that which can most readily donate electrons. We are not concerned at present about the details of the pinacol rearrangement: what is important is that carbonium ions undergo this kind of reaction.

We have discussed how the treatment of a primary amine yields a diazonium salt and that aliphatic diazonium salts are unstable and lose nitrogen to yield a carbonium ion. As a result we may expect

$$\text{H}_3\text{C}-\underset{\text{OH}}{\overset{\text{CH}_3}{\text{C}}}-\underset{\text{NH}_2}{\text{C}}(\text{CH}_3)_2 \xrightarrow{\text{HNO}_2} \text{H}_3\text{C}-\underset{\text{OH}}{\overset{\text{CH}_3}{\text{C}}}-\underset{\text{N}^+\equiv\text{N}}{\text{C}}(\text{CH}_3)_2 \longrightarrow$$

$$\text{H}_3\text{C}-\underset{:\ddot{\text{O}}\text{H}}{\overset{\text{CH}_3}{\text{C}}}-\overset{+}{\text{C}}(\text{CH}_3)_2 + \text{N}\equiv\text{N} \longrightarrow \text{CH}_3\text{COC}(\text{CH}_3)_3 + \text{H}^+ \qquad \text{...(24)}$$

amine alcohols to undergo a rearrangement (24) analogous to that of pinacol. This reaction has been carried out with an optically

$$C_6H_5-C(C_6H_5)(OH)-{}^*CH(CH_3)(NH_2) \xrightarrow{HNO_2} C_6H_5CO\overset{*}{C}H(CH_3)(C_6H_5) \quad ...(25)$$

Still optically active

active amino alcohol. The fact that the resultant ketone is optically active shows that the reaction (25) is a concerted process.

We considered above how *tert*-butyl chloride ionized in aqueous alcohol, yielding a transient carbonium ion which either reacted with a nucleophile (in the case of water to yield *tert*-butyl alcohol) or lost a proton (to yield isobutene). Neopentyl halides are very unreactive in nucleophilic substitution reactions, and when the reaction does occur it is by a unimolecular process; presumably, a bimolecular process is retarded because the bulky *tert*-butyl group hinders the approach of the attacking nucleophile towards the halogen-substituted carbon atom; the important feature of this reaction (26), however, is that only rearranged products are obtained. Compare this sequence very carefully with the sequence shown earlier for the reaction of *tert*-butyl chloride in aqueous ethanol.

$$(H_3C)_3C-CH_2-Br \xrightarrow[\substack{C_2H_5OH \\ (slow)}]{Na^{+}\,{}^{-}OC_2H_5} (H_3C)_2\overset{+}{C}\cdots CH_3\cdots CH_2 \longrightarrow (H_3C)_2\overset{+}{C}-CH_2CH_3$$

Neopentyl bromide (1-bromo-2,2-dimethylpropane)

$$(H_3C)_2\overset{+}{C}-CH_2CH_3 \xrightarrow{S_N1} C_2H_5O-C(CH_3)_2-CH_2CH_3$$

$$(H_3C)_2\overset{+}{C}-CH_2CH_3 \xrightarrow{E_1} (H_3C)_2C=CHCH_3$$

...(26)

We must next consider why these rearrangements occur. Notice that if, in the last case, no rearrangement had occurred we should have obtained the very unstable primary carbonium ion: rearrangement resulted in the formation of the much more stable tertiary carbonium ion. This tendency to form the most stable carbonium ion is quite general; thus, dehydration of a branched-chain alcohol under acidic conditions also results in rearrangement. This is illustrated in scheme (27). The starting alcohol in this reaction can be

$$(CH_3)_2C(CH_3)\text{-}CH(CH_3)OH \xrightarrow[\text{Heat}]{(CO_2H)_2} (CH_3)_2C(CH_3)\text{-}CH(CH_3)OH_2^+ \longrightarrow (CH_3)_2\overset{+}{C}\text{-}C(CH_3)_2H$$

$$\longrightarrow \underset{(a)}{(CH_3)_2C{=}C(CH_3)_2}\ \text{(5 parts)} + \underset{(b)}{(CH_3)_2CH\text{-}C(CH_3){=}CH_2}\ \text{(2 parts)} \quad ...(27)$$

N.B. No $(CH_3)_3CCH{=}CH_2$ is formed

prepared by the reduction of pinacalone (see p. 115), and the formation of symmetrical tetramethylethylene (27*a*) amounts to reversal of the pinacol rearrangement and so is called a 'retro-pinacol rearrangement'. The driving force of the retro-pinacol rearrangement is the formation of a tertiary rather than a secondary carbonium ion; in the pinacol rearrangement the ion that would be formed if the water molecule were eliminated without rearrangement would already be a tertiary carbonium ion, but a tertiary carbonium ion is less stable than the oxonium ion formed as a result of rearrangement.

The examples of molecular rearrangement that we have represented so far may seem remote from the reactions of ordinary molecules, but this is not so. Any reaction involving a carbonium ion intermediate is liable to result in rearrangement. A simple example is the treatment of primary aliphatic amines with nitrous acid. Unfortunately, very many elementary books say that this reaction yields the corresponding alcohol. In the case of ethanol, where no rearrangement is possible, this is true; but for the next member of the series, namely propylamine the reaction yields propene and a mixture of normal and isopropyl alcohol in which the latter predominates [see (28)].

$$CH_3CH_2CH_2NH_2 \xrightarrow{HNO_2} \underbrace{CH_3CH{=}CH_2;\quad (H_3C)_2CHOH;}_{\text{Major products}} \text{ and some } CH_3CH_2CH_2OH \quad (28)$$

The isomerization of hydrocarbons to yield more branched products was discussed in Part 1 (Chapter 17) in relation to the petroleum industry. For example, when normal butane is treated with aluminium chloride at room temperature an equilibrium mixture (29) is produced containing 80% of the branched isomer.

$$CH_3CH_2CH_2CH_3 \underset{}{\overset{AlCl_3}{\rightleftharpoons}} (H_3C)_2C(H)CH_3 \quad (80\% \text{ at } 25°) \qquad ...(29)$$

Actually, if the butane is completely pure, it does not react at all with aluminium chloride. Traces of olefin or alcohol must be added to form the first carbonium ion. The reaction then proceeds by a chain process (30).

$$\underset{\text{Olefin or alcohol}}{RH} \xrightarrow{AlCl_3} \underset{\text{Carbonium ion}}{R^+}$$

$$CH_3CH_2CH_2CH_3 + R^+ \rightleftharpoons CH_3\overset{+}{C}HCH_2CH_3 + RH$$

$$(H_3C)(H)\overset{+}{C}-CH_3CH_2 \rightleftharpoons H_3C-C(CH_3)(H)-\overset{+}{C}H_2 \rightleftharpoons H_3C-\underset{+}{C}(CH_3)-CH_3$$

$$(CH_3)_3C^+ + CH_3CH_2CH_2CH_3 \rightleftharpoons (CH_3)_3CH + CH_3\overset{+}{C}HCH_2CH_3 \qquad (30)$$

We may mention one more, very practical example. The reaction of alkyl halides with aluminium chloride to yield ionic compounds was described at the beginning of this Chapter, and the use of these ionic compounds in the alkylation of the benzene nucleus (the Friedel–Crafts reaction) was reported in Chapter 14 of Part 1. However, it will be clear from our latest discussion that, if we used a primary alkyl halide in the Friedel–Crafts reaction, some rearrangement would almost certainly occur; and this is verified in practice. For example, some normal propylbenzene is formed from normal propyl chloride, benzene, and aluminium chloride, but the predominant product is isopropylbenzene [reaction (31)].

$$CH_3CH_2CH_2Cl + C_6H_6 \xrightarrow{AlCl_3} C_6H_5CH(CH_3)_2 \text{ (Main product)} + C_6H_5CH_2CH_2CH_3 \text{ (Minor product)} \quad ...(31)$$

Summary

Carbonium ions are in general, unstable species but they occur as intermediates in a wide variety of reactions. Carbonium ions are powerful electrophiles and will add to olefinic double bonds and to the aromatic nucleus. They react, of course, extremely rapidly with any nucleophile. If a carbonium ion cannot react instantly with a nucleophile, it may stabilize itself by elimination of a proton or by rearrangement of the carbon skeleton in such a way to produce a more stable carbonium ion. If we compare this Chapter with the Chapter on carbanions we notice that we have not discussed the carbonium ions formed by adding a proton to a carbonyl compound, as in (32), though some reactions of this type were discussed in

$$R_2C{=}O + H^+ \rightleftharpoons \left[R_2C{=}\overset{+}{O}{-}H \longleftrightarrow R_2\overset{+}{C}{-}\ddot{O}{-}H \right] \quad ...(32)$$

Chapter 7 of Part 1. Effectively the proton accelerates nucleophilic addition to the carbonyl double bond and, in particular, we discussed the importance of such intermediates in the formation of acetals from ketones and in the acid-catalysed hydrolysis of esters. Many of the condensations described in the last Chapter that involved carbanions formed by treating carbonyl compounds with bases can also be completed in acidic media. For example, acetone condenses with itself to yield diacetone alcohol and eventually mesityl oxide in the presence of a mineral acid [sequence (33)]; this involves the enol form of acetone which is further discussed in Chapter 12. Although many of the condensations involving carbanions with basic catalysts will also proceed *via* carbonium ions with acidic

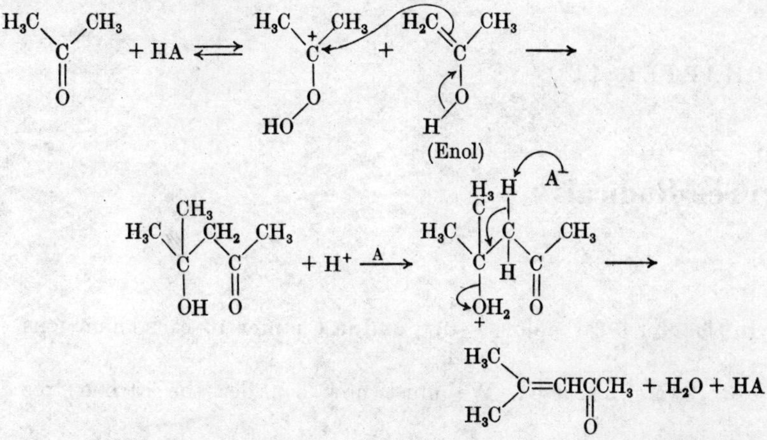

$$\underset{\text{}}{(H_3C)_2C{=}O} + HA \rightleftharpoons (H_3C)_2\overset{+}{C}{-}O{-}H + \underset{\text{(Enol)}}{H_2C{=}C(CH_3){-}O{-}H} \longrightarrow$$

$$(H_3C)_2C(OH){-}CH_2{-}C(=O){-}CH_3 + H^+ \xrightarrow{A} (H_3C)_2C(\overset{+}{O}H_2){-}CH_2{-}C(=O){-}CH_3 \longrightarrow$$

$$(H_3C)_2C{=}CHCCH_3(=O) + H_2O + HA \qquad \text{...(33)}$$

catalysts, the carbanion route is usually better from a practical point of view and involves fewer side reactions.

Carbonium ions and carbanions are, in general, extremely reactive species; their importance lies in their occurrence as intermediates in reaction processes. Carbonium ions are likely in highly acidic media and in the presence of acids. Carbanions, on the other hand, are likely in basic media and in the presence of strong bases.

Problems

1. Predict the outcome of the following reactions:

(*a*) $(CH_3)_2C{=}CH_2 + H_2SO_4 \longrightarrow$
(*b*) $(CH_3)_2C{=}CH_2 + H_2SO_4 + C_6H_6 \longrightarrow$
(*c*) $(CH_3)_2C{=}CH_2 + CH_3CO_2H + (CF_3CO)_2O \longrightarrow$

2. Complete the following reaction sequences, all of which involve carbonium ions:

(*a*) $(C_6H_5)(p\text{-}NO_2C_6H_4)C(OH){-}C(OH)(C_6H_5)(CH_3) + H_2SO_4 \longrightarrow$
(*b*) $(CH_3)_3CCH_2Br + AgNO_3 + H_2O \longrightarrow$
(*c*) $CH_3CH_2CH_2NH_2 + HNO_2 \longrightarrow$
(*d*) $CH_3CH_2CH_2Cl + CH_3C_6H_5 \xrightarrow{AlCl_3}$

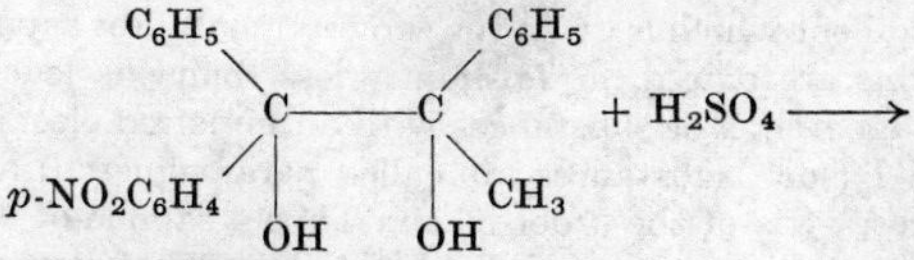

CHAPTER 11

Free Radicals

In Chapter 9 carbanions $-\overset{|}{\underset{|}{\text{C}}}{:}^{-}$, and in Chapter 10 carbonium ions $-\overset{|}{\underset{|}{\text{C}}}{}^{+}$, were discussed. We must now consider the carbon free radical $-\overset{|}{\underset{|}{\text{C}}}\cdot$. A free radical has no charge but has an unpaired electron. The presence of an unpaired electron has an important effect on the magnetic properties of a molecule.* There is a very big difference in the chemical behaviour of anions and cations on the one hand, and of radicals on the other. In a solution of sodium chloride in water, there is a very high concentration of sodium cations (Na^+) and chloride anions (Cl^-). The positively and negatively charged ions do not come together and combine because they are 'solvated', that is, bound to solvent molecules. In the same way, we can have high concentrations of carbanions, e.g. triphenylmethide anions in liquid ammonia. In fact, it is possible to prepare

* If two magnetic poles p_1 and p_2 are separated by a distance r in a medium X, the force between the poles is given by

$$F = \frac{1}{\mu_X} \frac{p_1 p_2}{r^2}$$

The quantity μ_X is the magnetic permeability of substance X (not to be confused with the dipole moment which is given the same symbol). For substances in which all the electrons are paired, μ_X is slightly less than one (such substances are called diamagnetic). For substances with an unpaired electron, μ_X is slightly greater than 1 (such substances are called paramagnetic); and for ferromagnetic substances, μ_X is of the order of 10^3. Thus a sample of a paramagnetic substance is drawn into a magnetic field and a sample of diamagnetic substance tends to be pushed out. This property is sometimes used to demonstrate the presence of stable free radicals.

solid sodium triphenylmethide, in which there is considerable ionic character in the crystal:

$$(C_6H_5)_3CH + NaNH_2 \longrightarrow (C_6H_5)_3C^- + Na^+ + NH_3$$

Similarly, we can have very high concentrations of the triphenylmethyl cation in concentrated sulphuric acid:

$$(C_6H_5)_3COH + 2H_2SO_4 \longrightarrow (C_6H_5)_3C^+ + HSO_4^- + (H_3O^+ + HSO_4^-)$$

and we can have a largely ionic, crystalline triphenylmethyl perchlorate. We have attributed the stability of the triphenylmethide anion and the triphenylmethyl cation to the fact that the charge is spread over the whole molecule and there are large numbers of canonical forms. Clearly there is an exactly equal number of canonical forms for the triphenylmethyl radical.

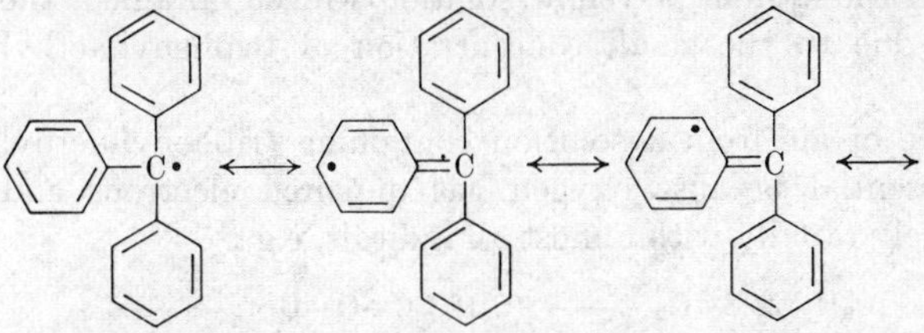

C• ⟷ =C ⟷ =C ⟷ and 41 other canonical forms.

Let us not worry, just for the moment, how the triphenylmethyl radical could be prepared but consider what would happen if we had it in solution. A normal carbon–carbon single bond is formed as a result of the sharing of two electrons between the two carbon atoms, so that in solution we should expect triphenylmethyl radicals to combine to form hexaphenylethane. Herein lies the great difference between ions and radicals. Triphenylmethide anions are stable in liquid ammonia solution; they are solvated and can be present in very high concentrations. Triphenylmethyl cations are stable in concentrated sulphuric acid solution in which they are solvated, and we can have very high concentrations of them. But triphenylmethyl radicals, even though stabilized, can never exist in high concentration because they are not solvated; hence they combine. Recent work shows that the central carbon atom of one radical couples with the para-position of one of the benzine rings of the other radical. This dimer is only dissociated into triphenylmethyl radicals to a very limited extent at room temperature when dissolved in organic solvents:

$$\text{Dimer} \rightleftarrows (2C_6H_5)_3C\cdot$$

$$K_c^{25} = 2 \times 10^{-4}$$

Ethane, on the other hand, shows no tendency to split into two methyl radicals at normal temperatures, and temperatures must be well in excess of 500° before any such reaction takes place. The dimer dissociates mainly because the triphenylmethyl radical is so highly stabilized. In addition, the bulky phenyl groups make the carbon–carbon single bond of the molecule longer than normal and hence weaker. The dimer can be prepared in solution by treating triphenylmethyl chloride (sometimes called trityl chloride) with metallic silver or zinc dust. Provided air is excluded from the system, a yellow solution will be obtained, the colour being due to the small concentration of triphenylmethyl radicals.

The absence of air from a solution containing triphenylmethyl radicals is essential because oxygen has unpaired electrons and reacts extremely rapidly with almost all radicals, e.g.:

$$(C_6H_5)_3C\cdot + O_2 \longrightarrow (C_6H_5)_3C{-}O{-}O\cdot$$

$$(C_6H_5)_3C{-}O{-}O\cdot + (C_6H_5)_3C\cdot \longrightarrow (C_6H_5)_3CO{-}OC(C_6H_5)_3$$

The reason why oxygen behaves in this fashion can readily be seen by drawing a correlation diagram of the type used in Chapters 3 and 7.

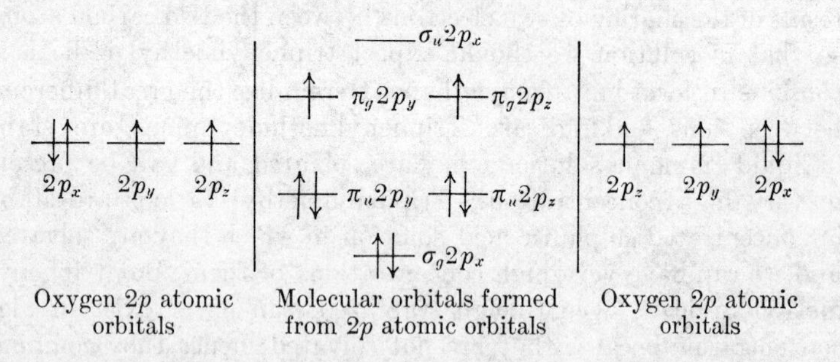

The diagram shows that in molecular oxygen two electrons have to go into a degenerate pair of antibonding orbitals of π-symmetry. According to Hund's rules one electron will go into each of the degenerate orbitals with parallel spins. This is called a triplet state. Molecular oxygen has unpaired electrons, just as a carbon radical has, and readily forms a bond with the carbon atom carrying the odd electron. Nitric oxide has one electron less than molecular oxygen and of necessity one electron must be unpaired. Nitric oxide likewise reacts extremely rapidly with radicals to form the corresponding nitroso compound:

$$(C_6H_5)_3C\cdot + NO \longrightarrow (C_6H_5)_3C{-}NO$$

Nitroso compound

Triphenylmethyl radicals are exceptional examples of carbon free radicals because of their very high stability. Nonetheless, interesting information can be obtained from them. For instance, the effect of substituents on the stability of radicals can to some extent be assessed by preparing derivatives of hexaphenylethane with substituents in the benzene ring, and then measuring the extent of dissociation, e.g.:

$$2(p\text{-}R\text{–}C_6H_4)_3C\cdot \rightleftarrows \text{“dimer”}.$$

It is found that both donor and acceptor groups stabilize the radical; presumably through canonical forms of the type shown.

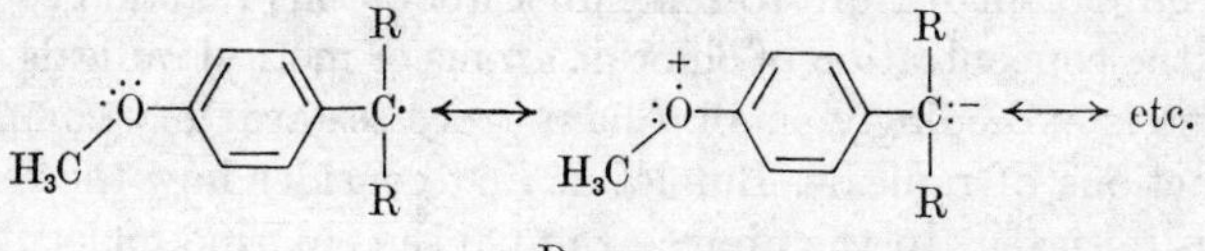

Donor

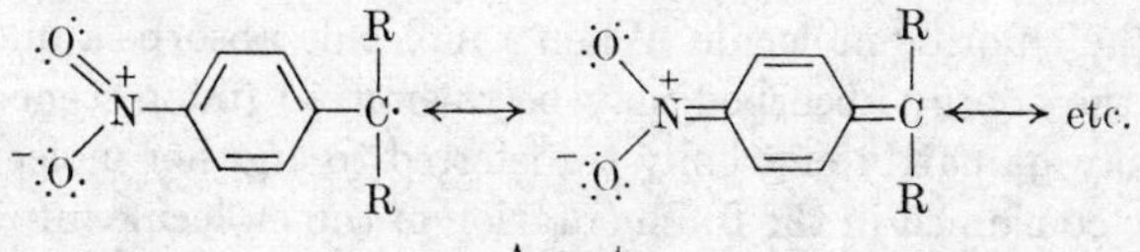

Acceptor

It is important to realize that the stable radicals are not the most common kind of radical. As we shall see below, most organic radicals exist only as transient intermediates in reaction processes. There is, however, a very limited number of radicals so stable that they do not dimerize at all and can even be prepared in crystalline form. One example of such a species is the diphenylpicrylhydrazyl radical:

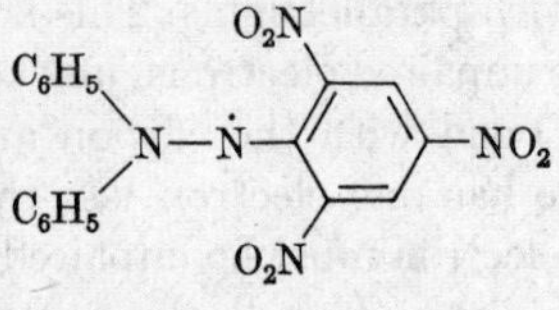

$$(C_6H_5)_2N{-}\dot{N}{-}C_6H_2(NO_2)_3$$

Diphenylpicrylhydrazyl

Although ethane does not dissociate into two methyl radicals at any normal temperatures, methyl radicals are formed at high temperatures, i.e. 600° and above, but various other complex reactions also occur. In Chapter 2 of Part 1 we discussed the chlorination and bromination of alkanes, including methane, and we described how methyl radicals were formed by abstraction of a hydrogen atom from methane by a chlorine atom, in a chain reaction.

$$Cl_2 \xrightarrow[\text{or } \Delta H]{h\nu} 2\,Cl\cdot \qquad \text{Initiation}$$

$$\left.\begin{array}{l} Cl\cdot + CH_4 \longrightarrow CH_3\cdot + HCl \\ CH_3\cdot + Cl_2 \longrightarrow CH_3Cl + Cl\cdot \end{array}\right\} \text{Chain-propagating steps}$$

$$\left.\begin{array}{l} Cl\cdot + Cl\cdot + M \longrightarrow Cl_2 + M \\ CH_3\cdot + CH_3\cdot \longrightarrow C_2H_6 \\ CH_3\cdot + Cl\cdot \longrightarrow CH_3Cl \end{array}\right\} \text{Chain-terminating steps}$$

The important feature to remember about this reaction sequence is that the concentration of chlorine atoms or methyl radicals at any one time is exceedingly small. Chain processes are very common in the reactions of radicals. But let us first consider how the reaction chain is initiated. In the above example the chlorine molecule may be dissociated into chlorine atoms either by the action of light or by heating the chlorine molecule. When a molecule absorbs a quantum of light, the energy absorbed may be re-emitted (fluorescence); the energy may be partly or wholly transferred to another molecule; or it may be consumed in the fragmentation of the molecule into atoms or radicals. The two most common ways of preparing radicals are either this photochemical process or heating a molecule which has a weak bond. In the case of chlorination, either process may be used.

Radicals may also be prepared by one-electron transfer from carbanions or carbonium ions. Thus the triphenylmethyl cation may be reduced to the triphenylmethyl radical by vanadous chloride:

$$(C_6H_5)_3C^+ + V^{2+} \longrightarrow (C_6H_5)_3C\cdot + V^{3+}$$

In a similar way the carbanion may be oxidized. Oxidation and reduction reactions may also produce radicals from neutral molecules. Cumyl hydroperoxide ($\alpha\alpha$-dimethylbenzyl hydroperoxide) is decomposed by ferrous salts into alkoxyl radicals and hydroxyl anions:

$$C_6H_5C(CH_3)_2{-}O{-}OH + Fe^{2+} \longrightarrow C_6H_5C(CH_3)_2{-}O\cdot + Fe^{3+} + OH^-$$

A similar reaction occurs with hydrogen peroxide itself, to yield hydroxyl radicals (so-called Fenton's reagent). Reactions involving one-electron transfers will be discussed again in Chapter 19.

The simplest organic radical, the methyl radical, can be prepared from azomethane which can be decomposed into two methyl radicals and nitrogen by either heat or light:

$$CH_3{-}N{=}N{-}CH_3 \xrightarrow[\text{or } h\nu]{\Delta H} 2\,CH_3\cdot + N_2$$

Acetone, photolysed by a medium-pressure mercury arc, also yields a methyl radical and an acetyl radical. At temperatures above 80°, the acetyl radical breaks down to give a methyl radical and carbon monoxide:

$$(CH_3)_2C{=}O \xrightarrow{h\nu} CH_3\cdot + CH_3CO\cdot$$

$$CH_3CO\cdot \xrightarrow{>80^\circ} CH_3\cdot + CO$$

Peroxides, where the oxygen–oxygen bond is particularly weak, are very frequently used as sources of radicals, and *tert*-butyl peroxide breaks down when heated, to give two alkoxyl radicals which decompose to acetone and a methyl radical:

$$(CH_3)_3C{-}O{-}O{-}C(CH_3)_3 \xrightarrow{\Delta H} 2\,CH_3{-}C(CH_3)_2{-}O\cdot \longrightarrow (CH_3)_2C{=}O + CH_3\cdot$$

Next, we must consider what happens to a radical when it has been formed. If, for example, we photolyse azomethane in the gas phase, an important reaction will be radical combination:

$$CH_3\cdot + CH_3\cdot \longrightarrow C_2H_6$$

We emphasized that the radical concentration will always be very low indeed, and so the reactive methyl radical will take part in many other reactions. Just as a chlorine atom abstracts a hydrogen atom from methane, so a methyl radical may abstract the hydrogen atoms from other organic molecules and, in this case, the only other organic molecule present is azomethane.

$$CH_3\cdot + CH_3N{=}NCH_3 \longrightarrow CH_4 + \cdot CH_2N{=}NCH_3$$

The new radical derived from the azomethane then breaks down to give a series of complex products. Methyl radicals may also add to the nitrogen–nitrogen double bond:

$$CH_3\cdot + CH_3N{=}NCH_3 \longrightarrow (CH_3)_2N{-}\dot{N}CH_3 \xrightarrow{CH_3\cdot} (CH_3)_2N{-}N(CH_3)_2$$

Extensive studies have been made of the reactions of methyl radicals with other organic molecules by decomposing either acetone or azomethane in the presence of these molecules, particularly in the gas phase. In the first step, the photolysis of acetone is studied and the ratio of methane to ethane formed is measured. The reaction is then repeated in the presence of another organic molecule, say, butane. When no butane is present, all the methane formed is the result of methyl radicals abstracting hydrogen atoms from acetone; but when butane is present, more methane will be formed because methyl radicals abstract hydrogen atoms from butane. By measuring the new methane: ethane ratio it is possible to determine the rate of attack of methyl radicals on the butane molecule. Very extensive studies of this type have been carried out.

We have now described four reaction that radicals undergo. First, we have radical combination, including recombination of triphenylmethyl radicals to yield hexaphenylethane and of methyl radicals to yield ethane, and of chlorine atoms to yield molecular chlorine. We then have the decomposition of radicals to yield a new radical and a stable molecule, including decomposition of the acetyl radical to a methyl radical and carbon monoxide and of the *tert*-butoxyl radical to acetone and a methyl radical. Next we have hydrogen-abstraction reactions, called radical-transfer reactions because a new radical is formed; reactions of this type that we have discussed include the reaction of chlorine atoms with methane to yield methyl radicals, and of methyl radicals with butane to yield butyl radicals. Finally, we have the addition reaction, of which we have described only one so far, namely, addition of a methyl radical to azomethane.

$$X\cdot + Y\cdot \longrightarrow X\text{—}Y \qquad \text{Radical combination}$$
$$(\text{e.g. } CH_3\cdot + CH_3\cdot \longrightarrow C_2H_6)$$
$$YZ\cdot \longrightarrow Y\cdot + Z \qquad \text{Radical fragmentation}$$
$$(\text{e.g. } CH_3CO\cdot \longrightarrow CH_3\cdot + CO)$$
$$X\cdot + Y\text{—}Z \longrightarrow X\text{—}Y + Z\cdot \qquad \text{Radical transfer}$$
$$(\text{e.g. } CH_3\cdot + C_4H_{10} \longrightarrow CH_4 + C_4H_9\cdot)$$
$$X\cdot + Y\text{=}Z \longrightarrow XY\text{—}\dot{Z} \qquad \text{Radical addition}$$
$$(\text{e.g. } CH_3\cdot + CH_3N\text{=}NCH_3 \longrightarrow (CH_3)_2N\text{—}\dot{N}CH_3)$$
$$X\cdot + Y\text{—}Z\text{—}\dot{W} \longrightarrow XY + Z\text{=}W \qquad \text{Radical disproportionation}$$
$$(\text{e.g. } CH_3\cdot + CH_3CH_2\cdot \longrightarrow CH_4 + CH_2\text{=}CH_2)$$

We have not so far described radical disproportionation, in which two radicals interact to yield two stable molecules (one of which must be unsaturated). For example, two ethyl radicals may combine to form butane or disproportionate to yield ethane and ethylene. Experiment shows combination to be favoured over disproportionation by a factor of over 7:1 in this case.

Notice that only in radical–radical reactions do we obtain non-radical products. Radical–radical reactions are usually extremely fast, but they depend on the square of the radical concentration. In most radical reactions the radical concentration is extremely low, so that instead of radical-combination predominating, a chain process involving transfer and/or addition is observed.

Transfer reactions were discussed briefly in Chapter 2 of Part 1. The reactivity of an atom or radical is largely determined by the thermochemistry involved. Thus the weakest bond is usually broken first, so that from an alkane a tertiary hydrogen atom is abstracted more readily than a secondary hydrogen atom, which in turn is abstracted more readily than a primary hydrogen atom. A fluorine atom reacts very rapidly and unselectively with an alkane, and the reaction is extremely exothermic. Methyl radicals are reactive but are much more selective, the reaction in this case being only slightly exothermic. Bromine atoms, on the other hand, react slowly and are very selective; the reaction in this case is endothermic (Table 11.1). Directive effects in radical-transfer reactions are discussed later in the present Chapter.

Turning now to consider radical-addition reactions in more detail, we recall some of these reactions discussed in Chapter 6 of Part 1.

Table 11.1. Relative rates of hydrogen-atom abstraction from alkanes by F·, CH_3·, and Br· at 25°

	CH_3—	CH_2<	CH<
F·	1	1.2	1.4
CH_3·	1	35	800
Br·	1	100	1,600

For example, we discussed the radical addition of a normally ionic molecule such as hydrogen bromide to ethylene:

$$\left.\begin{array}{l} \text{R—O—O—R} \xrightarrow{\text{heat}} 2\,\text{RO}\cdot \\ \text{A peroxide} \\ \text{RO}\cdot + \text{HBr} \longrightarrow \text{ROH} + \text{Br}\cdot \end{array}\right\}\text{Initiation}$$

$$\left.\begin{array}{l} \text{Br}\cdot + CH_2{=}CH_2 \longrightarrow BrCH_2CH_2\cdot \\ BrCH_2CH_2\cdot + \text{HBr} \longrightarrow BrCH_2CH_3 + \text{Br}\cdot \end{array}\right\}\text{Chain propagation}$$

A reaction that is at first sight unlikely is the addition of the very inert carbon tetrachloride to ethylene, but this can also be initiated by peroxide. The reaction is carried out in an autoclave, which is partly filled with carbon tetrachloride and a very small amount of benzoyl peroxide is added. The autoclave is then charged with ethylene at quite a high pressure and is heated to 100° at which temperature the benzoyl peroxide breaks down to initiate the reaction.

$$\left.\begin{array}{l} (C_6H_5CO_2)_2 \xrightarrow{\text{heat}} C_6H_5CO_2\cdot \longrightarrow C_6H_5\cdot + CO_2 \\ C_6H_5\cdot + CCl_4 \longrightarrow CCl_3\cdot + C_6H_5Cl \end{array}\right\}\text{Initiation}$$

$$\left.\begin{array}{l} CCl_3\cdot + CH_2{=}CH_2 \longrightarrow CCl_3CH_2CH_2\cdot \\ CCl_3CH_2CH_2\cdot + CCl_4 \longrightarrow CCl_3CH_2CH_2Cl + CCl_3\cdot \end{array}\right\}\text{Propagation}$$

Additions of this kind can also be carried out by photochemical initiation; for example, bromotrichloromethane adds to ethylene in a similar way. This reaction can be initiated with peroxides, as for carbon tetrachloride, but more readily by light in either the gas or the liquid phase. (Why, in this reaction sequence, does the 1,1,1-trichloropropyl radical abstract the bromine atom from the bromotrichloromethane and not a chlorine atom?)

$$\mathrm{CCl_3Br \xrightarrow{h\nu} CCl_3\cdot + Br\cdot} \qquad \text{Initiation}$$

$$\left.\begin{array}{l}\mathrm{CCl_3\cdot + CH_2{=}CH_2 \longrightarrow CCl_3CH_2CH_2\cdot} \\ \mathrm{CCl_3CH_2CH_2\cdot + CCl_3Br \longrightarrow CCl_3CH_2CH_2Br + CCl_3\cdot}\end{array}\right\} \text{Chain propagation}$$

Chlorination of 1-chlorobutane yields a mixture of all the possible dichlorobutanes (i.e. 1,1-, 1,2-, 1,3-, and 1,4-dichlorobutane). When 1-bromobutane is chlorinated, a rather interesting phenomenon is observed: four products are obtained; namely, 1-bromo-4-chloro-, 1-bromo-3-chloro-, and 1-bromo-1-chlorobutane together with 1,2-dichlorobutane; but no 1-bromo-2-chlorobutane is isolated.

$$\mathrm{BrCH_2CH_2CH_2CH_3 \xrightarrow[h\nu]{Cl_2} BrCH_2CH_2CH_2CH_2Cl,\ BrCH_2CH_2CHClCH_3,}$$
$$\mathrm{BrClCHCH_2CH_2CH_3,\ and\ ClCH_2ClCHCH_2CH_3}$$

The reason for this is that the $\mathrm{BrCH_2\dot{C}HCH_2CH_3}$ radical breaks down to yield but-1-ene and a bromine atom. This reaction is, of course, reversible because a bromine atom will add to the carbon–carbon double bond:

$$\mathrm{BrCH_2\dot{C}HCH_2CH_3 \rightleftharpoons Br\cdot + CH_2{=}CHCH_2CH_3}$$

However, under the conditions of the chlorination, it is far more likely that a chlorine atom will add to the double bond than that the minute amount of the bromine atoms present will be effective.

In a somewhat similar way, chlorination of an ether in the gas phase yields none of the α-chloro-product because the radical breaks down:

$$\mathrm{RCH_2OCH_3 + Cl\cdot \longrightarrow R\dot{C}HOCH_3 + HCl}$$
$$\mathrm{R\dot{C}H{-}O{-}CH_3 \longrightarrow RCHO + CH_3\cdot}$$

This instability of radicals can be used to prepare a specific product. For example, if we chlorinate or brominate propene, we should expect to obtain the results of both addition and substitution (see Part 1, Chapter 6). Now from what is said above, the addition of a chlorine atom to the double bond of propene will be reversible (i). On the other hand, the abstraction of a hydrogen atom from the methyl group of propene will not be reversible (iii). The chloropropyl radical reacts with molecular chlorine to yield the dichloropropane (ii), and the allyl radical reacts with molecular chlorine to yield allyl chloride (iv), both these reactions being chain-propagating steps.

$$\left.\begin{array}{ll} Cl\cdot + CH_3CH{=}CH_2 \rightleftharpoons CH_3\dot{C}HCH_2Cl & \text{(i)} \\ CH_3\dot{C}HCH_2Cl + Cl_2 \longrightarrow CH_3CHClCH_2Cl + Cl\cdot & \text{(ii)} \end{array}\right\} \text{Addition}$$

$$\left.\begin{array}{ll} Cl\cdot + CH_3CH{=}CH_2 \longrightarrow HCl + \underset{\text{Allyl radical}}{\overbrace{CH_2CHCH_2}^{\cdot}} & \text{(iii)} \\ \overbrace{CH_2CHCH_2}^{\cdot} + Cl_2 \longrightarrow \underset{\text{Allyl chloride}}{CH_2ClCH{=}CH_2} + Cl\cdot & \text{(iv)} \end{array}\right\} \text{Allylic substitution}$$

By carrying the reaction out at high temperatures, where the chloropropyl radical breaks down rapidly, it is possible to steer the reaction in such a way that allyl chloride is almost the sole product. This reaction is important because allyl chloride is used in the industrial manufacture of glycerol. A similar use is made of this type of reaction involving a compound called *N*-bromosuccinimide, which is a very useful brominating agent. *N*-Bromosuccinimide is usually slightly impure, containing traces of molecular bromine or of substances which will react with the imide to yield molecular bromine. If a compound such as ethyl crotonate, *trans*-$CH_3CH{=}CHCO_2Et$, is refluxed in carbon tetrachloride with *N*-bromosuccinimide and the reaction mixture is irradiated with ultraviolet light, the traces of molecular bromine dissociate into bromine atoms and these react with the ethyl crotonate in the same way as chlorine atoms react with propene; i.e. they will add and will also abstract a hydrogen atom:

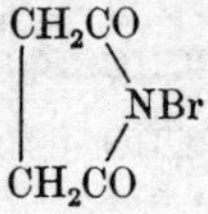
CH2CO
NBr
CH2CO
N-Bromosuccinimide

$$Br\cdot + CH_3CH{=}CHCO_2Et \rightleftharpoons CH_3CHBr\dot{C}HCO_2Et$$

$$Br\cdot + CH_3CH{=}CHCO_2Et \longrightarrow \overbrace{CH_2CHCH}^{\cdot}CO_2Et + HBr$$

Because the concentration of molecular bromine is extremely low, there is very little probability that the brominated radical will react with molecular bromine; instead, it dissociates to yield again ethyl crotonate and a bromine atom. However, once a hydrogen atom has been abstracted from ethyl crotonate, the resulting allylic radical is relatively stable and is likely to remain present until it has the opportunity of reacting with a bromine molecule; and the hydrogen bromide formed in this step reacts instantly with *N*-bromosuccinimide to yield molecular bromine and succinimide. Thus the function

$$\begin{array}{l} CH_2CO \\ | \quad \backslash \\ | \quad\; NBr \\ | \quad / \\ CH_2CO \end{array} + HBr \longrightarrow \begin{array}{l} CH_2CO \\ | \quad \backslash \\ | \quad\; NH \\ | \quad / \\ CH_2CO \end{array} + Br_2$$

$$\overbrace{CH_2CH\dot{C}HCO_2Et} + Br_2 \longrightarrow BrCH_2CH{=}CHCO_2Et + Br\cdot$$

of *N*-bromosuccinimide in these reactions is to provide a constant, very low concentration of molecular bromine. *N*-Bromosuccinimide is used extensively in synthetic organic chemistry for preparing compounds with a bromine atom in the allylic position to a carbon–carbon double bond.

We have not yet discussed the reactions of free radicals with aromatic compounds. These are complicated and we shall treat them only very briefly. One of the most convenient sources of phenyl radicals is benzoyl peroxide, which can be decomposed by heat to yield initially benzoylperoxy radicals, which break down to a phenyl radical and carbon dioxide:

$$C_6H_5COO{-}OOCC_6H_5 \longrightarrow 2\,C_6H_5CO_2\cdot \longrightarrow C_6H_5\cdot + CO_2$$

If this reaction is carried out in benzene, the phenyl radical adds, as we might expect, to the benzene nucleus. This produces a new radical which may be dehydrogenated by the reaction of a further phenyl radical or may dimerize to yield a tetraphenyl derivative which is readily oxidized in the air. All substitutents in the benzene ring are *ortho-para*-directing for radical attack, and this is what we should expect if we remember that both acceptors and donors stabilize the triphenylmethyl radical.

$$C_6H_5\cdot + C_6H_6 \longrightarrow \text{(H)(}C_6H_5\text{)}C_6H_5\cdot \xrightarrow{\text{Combination}} \text{(H)(}C_6H_5\text{)}C_6H_5{-}C_6H_5\text{(H)(}C_6H_5\text{)}$$

$$\xrightarrow{C_6H_5\cdot\ \text{Disproportionation}} C_6H_5{-}C_6H_5 + C_6H_6$$

$$\xrightarrow{\text{Disproportionation}} C_6H_5{-}C_6H_5 + \text{(H)(}C_6H_5\text{)}C_6H_4\text{(H)(}C_6H_5\text{)}$$

Another source of the phenyl radical is *N*-nitrosoacetanilide:

$$C_6H_5NH_2 \longrightarrow C_6H_5NHCOCH_3 \longrightarrow C_6H_5N(N{=}O)C({=}O){-}CH_3 \longrightarrow$$

$$C_6H_5N{=}N{-}O{-}COCH_3 \longrightarrow C_6H_5\cdot + N_2 + \cdot OCOCH_3$$

Radicals occur as intermediates in many important chemical processes, from combustion to photosynthesis. Because they exist only transiently they have in the past received less study than ionic intermediates.

Directive Effects in Free-radical Reactions

Since radicals are, in general, uncharged species we might at first sight expect that radical reactions would be unaffected by polar substituents. In practice this is not the case. Free-radical substitution of an aliphatic compound initially involves hydrogen abstraction, in an overall chain-reaction process:

$$X{-}Y \xrightarrow{1} X\cdot + Y\cdot \quad \text{Initiation}$$

$$\left.\begin{aligned} R{-}H + X\cdot &\xrightarrow{2} R\cdot + HX \\ R\cdot + YX &\xrightarrow{3} RY + X\cdot \end{aligned}\right\} \text{Chain propagation}$$

The reaction which determines the site of substitution is reaction 2 in the above sequence, the hydrogen-abstraction step. Thus, for chlorination of propane ($X = Y = Cl$), it is the relative ease of abstraction of hydrogen from the central and the terminal atoms that determines the proportions of propyl and isopropyl chloride in the product.

Let us next examine the energy/reaction co-ordinate diagram, using the method discussed already in Chapter 2 of Part 1. We noted in the paragraphs above, that radical-transfer reactions can be considered in terms of thermochemistry, i.e. in terms of the strengths of the bonds broken and made. The stronger the R—H bond the lower the reactants are on the energy/reaction coordinate diagram, so that strengthening the R—H bond increases the activation energy E. On the other hand, increasing the H—X bond energy corresponds to lowering the position of the products on the energy/reaction coordinate diagram (Figure 11.1), so that an increase in the H—X bond strength decreases the activation energy (E). We then have two additional forces to consider, namely, the repulsion between the starting molecule RH and the initial

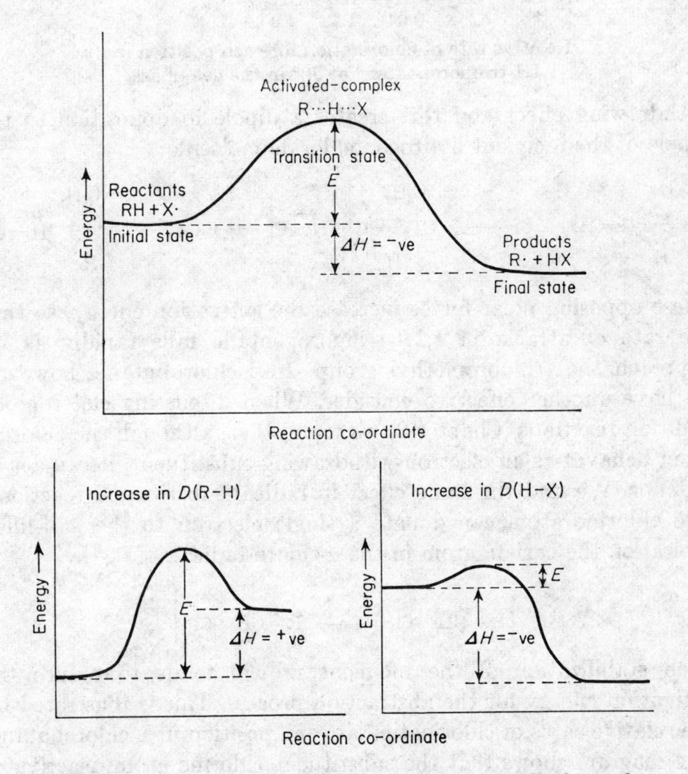

Figure 11.1. Reaction coordinate diagrams for radical attack.

radical X· and the repulsion between the new radical R· and the new molecule HX. If, in the first instance, we consider chlorination, then since atoms are non-polar, the presence of polar groups in R can neither increase nor decrease the repulsion between the chlorine atom and the molecule RH as the atom approaches. However, the repulsions between the products of the reaction, that is between the radical R· and the molecule HY (i.e. hydrogen chloride), can be greatly affected by polar substituents in R. The relative rates of chlorination of the four possible positions in 1,1,1-trifluoropentane are illustrated. The trifluoromethyl group has a powerful electron-

$$\underset{}{CF_3}—\underset{0.04}{CH_2}—\underset{1.2}{CH_2}—\underset{4.0}{CH_2}—\underset{1}{CH_3}$$

Relative rate of chlorination for each position in 1,1,1-trifluoropentar at 70° in the gas phase.

withdrawing effect and this creates a dipole in opposition to the dipole of the incipient hydrogen chloride molecule:

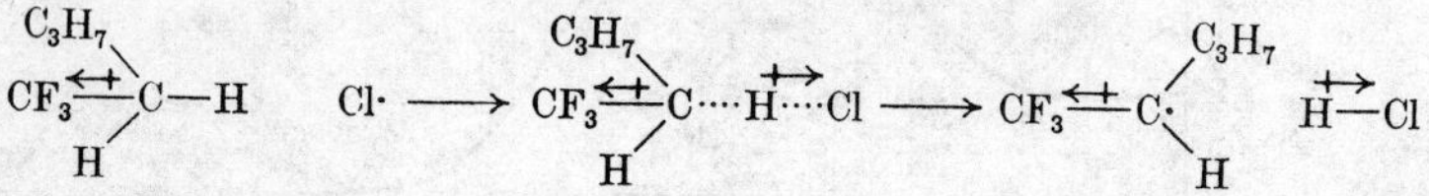

$$CF_3 \overset{\leftarrow\!+}{—} C(C_3H_7)(H)—H \quad Cl\cdot \longrightarrow CF_3 \overset{\leftarrow\!+}{—} C(C_3H_7)(H)\cdots \overset{+\!\rightarrow}{H \cdots Cl} \longrightarrow CF_3 \overset{\leftarrow\!+}{—} \dot{C}(C_3H_7)(H) \quad \overset{+\!\rightarrow}{H—Cl}$$

These opposing polar forces increase the activation energy, so that the rate of attack on 1,1,1-trifluoropentane falls rapidly as we approach the trifluoromethyl group. In 1-chlorobutane, however, we have another effect to consider. When discussing electrophilic addition reactions (Chapter 8) we found that, although the chlorine atom behaves as an electron-withdrawing substituent, it can act as a donor. We find the same effect in radical substitution reactions. The chlorine atom can donate a single electron to the half-filled orbital on the carbon atom in the α-chloro-radical:

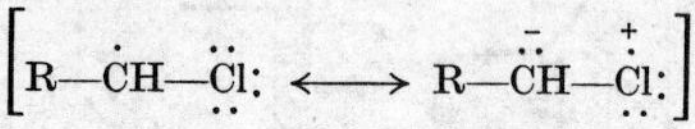

$$\left[R—\dot{C}H—\ddot{\underset{..}{Cl}}: \longleftrightarrow R—\overset{-}{C}H—\overset{+}{\dot{\underset{..}{Cl}}}:\right]$$

Such stabilization of the incipient radical tends to reduce the activation energy for the abstraction process. This is illustrated by the relative rates of chlorination at each position in 1-chlorobutane. The diagram shows that the substituent chlorine atom deactivates

$$Cl—\underset{0.8}{\overset{\alpha}{C}H_2}—\underset{2}{\overset{\beta}{C}H_2}—\underset{4}{\overset{\gamma}{C}H_2}—\underset{1}{\overset{\delta}{C}H_3}$$

Relative rates of chlorination at various positions of 1-chlorobutane in the gas phase at 70°.

the β-position by a factor of two, but compared with the other terminal position (the δ-position) has very little effect on the rate of attack at the α-position. Thus chlorination is retarded by the polar effect at the β-position, but, although the polar effect is stronger at the α-position, this is almost completely off-set by the stabilization of the incipient radical.

We can now summarize the directive effects in chlorination: Electron-attracting groups exert a retarding influence on the α-carbon atom to which they are attached, and a weakly retarding influence on the β-carbon atom. Substituents with orbitals of π-symmetry which can donate electrons to the incipient radical have an accelerating influence at the α-position.

When we consider bromination, we must remember that the hydrogen abstraction step is endothermic, so that factors affecting bond strength (or the stability of the incipient radical) will be more important than in chlorination. On the other hand, the dipole moment of hydrogen bromide is smaller than that of hydrogen chloride, so that the polar effects will be less pronounced. In fluorination, the reaction is very exothermic and very unselective. But the dipole moment of hydrogen fluoride is large, so that polar forces should have some effect, even in spite of the general unselectivity of the reaction. These opposing effects are well illustrated by comparing the relative rates of chlorination, bromination, and fluorination of 1-fluorobutane, shown in Table 11.2.

Table 11.2. The relative rates of halogenation at each site of 1-fluorobutane

	F—$\overset{\alpha}{CH_2}$—	$\overset{\beta}{CH_2}$—	$\overset{\gamma}{CH_2}$—	$\overset{\delta}{CH_3}$
Fluorination at 20°	0.3	0.8	1	1
Chlorination at 78°	0.9	1.7	4	1
Bromination at 150°	10	9	90	1

Notice that the rate of fluorination at the α-position is less than at the δ-position. The rates of chlorination at the α-position is almost the same as at the δ-position. The rate of bromination at the α-position is appreciably greater than that at the δ-position. Whereas in electrophilic substitution of the benzene nucleus, the orientating effect of a substituent remains the same regardless of the nature of the electrophile, in radical substitution of aliphatic compounds the position most reactive to attack by one radical may be the least reactive to attack by another.

We have discussed the radical reactions directly in terms of activation energy and have not invoked our hypothesis concerning free energy. This is because these reactions can be studied in the gas phase in the absence of a solvent and all the relative rates that we have discussed are known to be principally due to the changes in the activation energy and not to the entropy term. An important point of this discussion of radical substitution in aliphatic compounds is that the concepts developed in Chapter 8 for ionic reactions are still applicable.

In aromatic radical substitution, the mechanism is, as we have indicated, very complex, and it is hard to be certain which step is rate-determining. We have seen that both donor and acceptor groups stabilize the triphenylmethyl radical (see page 123). It is not surprising, then, that free-radical phenylation of both anisole ($CH_3OC_6H_5$) and nitrobenzene ($O_2NC_6H_5$) occurs preferentially at the *para*-position.

The orientation of free-radical addition to an olefin is determined by the strength of the bond formed. Thus a radical will add preferentially to, say, the CH_2 end of ethyl acrylate because the bond so formed is stronger than the bond obtained by addition to the $CHCO_2Et$ end of the molecule:

$$X\cdot + CH_2{=}CHCO_2Et \longrightarrow XCH_2\dot{C}HCO_2Et \quad \text{mainly}$$
$$+\ \dot{C}H_2CHXCO_2Et \quad \text{trace}$$

We can distinguish between the strengths of the two bonds in a qualitative fashion by considering what happens when the bond is broken, thus:

$$X{-}CH_2{-} \longrightarrow X\cdot + \dot{C}H_2{-} \qquad \text{(A)}$$

$$X{-}\overset{|}{C}H{-}\underset{\underset{O}{\|}}{C}OEt \longrightarrow X\cdot + \left[{-}\dot{C}H{-}\underset{\underset{\cdot\ddot{O}\cdot}{\|}}{C}{-}OEt \longleftrightarrow {-}CH{=}\underset{\underset{:\ddot{O}\cdot}{|}}{C}{-}OEt \right] \qquad \text{(B)}$$

In the fission (B) the carbon free radical is 'resonance-stabilized', whereas in the former case it is not. Therefore, the former process would be expected to be less favoured, i.e. the $X{-}CH_2$ bond is stronger than the X—$CHCO_2Et$ bond. It has often been suggested that it is 'resonance stabilization' of the initial addend radical which is important: recent work has shown that this is not the case.

Notice that many polar molecules add to olefins by a radical process in exactly the opposite way to which they add heterolytically.

$$RO\cdot + HBr \longrightarrow ROH + Br\cdot$$

$$CH_3CH{=}CH_2 + Br\cdot \longrightarrow CH_3\dot{C}HCH_2Br$$

$$CH_3\dot{C}HCH_2Br + HBr \longrightarrow CH_3CH_2CH_2Br + Br\cdot \quad \text{Chain propagation}$$

$$\text{i.e.}\quad \mathbf{CH_3CH{=}CH_2 + HBr \xrightarrow[\text{or peroxide}]{\text{Light}} CH_3CH_2CH_2Br} \quad \textbf{Free-radical addition}$$

$$(H_3C)(H)C{=}CH_2 + H{-}Br \longrightarrow (H_3C)(H)C^{+}{-}CH_3 + \bar{B}r \longrightarrow (CH_3)(H_3C)(H)C{-}Br$$

$$\text{i.e.}\quad \mathbf{CH_3CH{=}CH_2 + HBr \xrightarrow{\text{Dark}} CH_3CHBrCH_3} \quad \textbf{Ionic addition}$$

Free-radical reactions are particularly important in the study of 'directive effects'. The reactions can be carried out over a temperature range in the gas phase without the complications of a solvent. Thus, measured activation energies can be directly related to bond strengths and polar forces, without invoking any *ad hoc* hypothesis.

Problems

1. What products do you expect from the reaction of 1-phenylprop-1-ene ($C_6H_5CH{=}CHCH_3$) and bromine under the following conditions?

(*a*) In CCl_4 solution in the dark (1 mole of Br_2)
(*b*) In the presence of $FeBr_3$ (2 moles of Br_2)
(*c*) In boiling CCl_4 solution in the presence of ultravoilet light (excess of bromine present at the start)
(*d*) In boiling CCl_4 solution, in the presence of ultraviolet light to which 1 mole of bromine is added very slowly so that the concentration of molecular bromine is always very small.

2. Predict the products from a reaction in which CF_2ClBr and ethylene are heated together in the presence of a small amount of benzoyl peroxide. Write out a full reaction scheme, distinguishing between initiation, propagation, and termination steps.

CHAPTER 12

Tautomerism

(a) Prototropy

In Chapter 9 the stability of anions derived from β-dicarbonyl compounds was attributed to the spreading of the negative charge. For example, the anion of ethyl acetoacetate was depicted as follows:

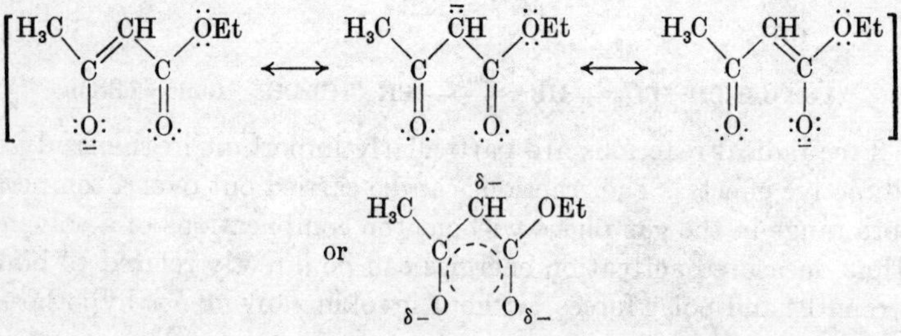

```
[ H3C    CH    ÖEt          H3C    C̈H    ÖEt          H3C    CH    ÖEt  ]
    \   //  \  /               \   /  \   /               \   /  \\  /
     C       C       <--->      C       C       <--->      C       C
     |       ||                 ||      ||                 ||      |
    :Ö:     .O.                .O.     .O.                .O.     :Ö:
[                                                                       ]

                              δ−
                    H3C       CH      OEt
                       \     /  \    /
              or        C ˙ ˙ ˙ ˙ C
                        |         |
                     δ− O         O δ−
```

If this anion is treated with a proton acid there appear to be at least three sites to which the proton might become attached. Two of these sites are more probable than the third.

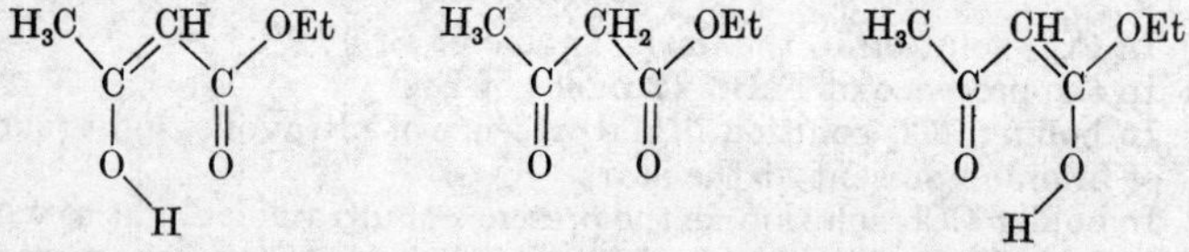

```
 H3C    CH    OEt          H3C    CH2    OEt          H3C    CH    OEt
    \   //  \  /              \   /   \   /              \   /  \\  /
     C       C                 C       C                  C       C
     |       ||                ||      ||                 ||      |
     O       O                 O       O                  O       O
      \                                                           /
       H                                                         H
    (a)                           (b)                     (c) least likely
```

Since oxygen is more electronegative than carbon, we might expect that the proton would preferentially become attached to the oxygen atom. On the other hand, carbon–hydrogen bonds do not readily become ionic and are presumably thermodynamically more stable than oxygen–hydrogen bonds. Clearly we cannot make predictions with certainty and we must proceed to experiment. If the sodium salt of ethyl acetoacetate is treated with anhydrous hydrogen chloride in light petroleum solution at −78°, the product has structure (*a*):

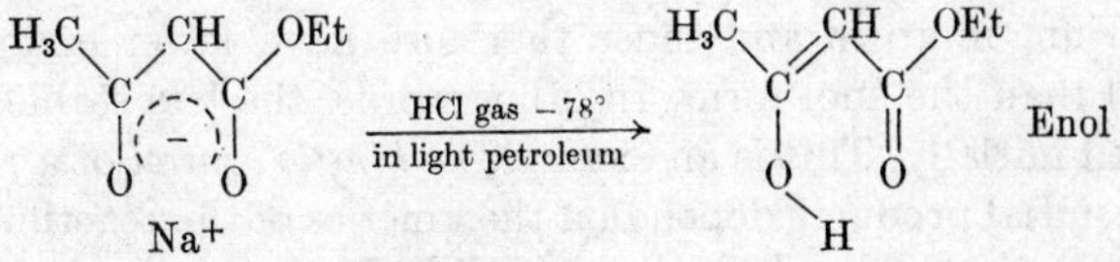

H_3C CH OEt, C C, O O, Na^+ —(HCl gas −78°, in light petroleum)→ H_3C CH OEt, C C, O O, H Enol

If, on the other hand, a solution of ethyl acetoacetate in light petroleum is cooled to −78°, a crystalline substance is obtained which differs from that obtained by treating the sodium salt with hydrogen chloride, and this other substance can be shown to have structure (*b*):

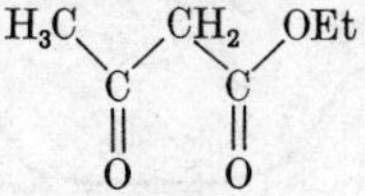

H_3C CH_2 OEt, C C, O O

Keto-form
(crystallized from
petroleum at −78°)

Thus there are two different compounds, which can be converted into each other by the transfer of a proton. The transformation can be catalysed by bases through the anion or with acids via the cation: we shall discuss its mechanism in detail shortly. In the presence of catalytic amounts of acid or base there is an equilibrium between the two forms and in the pure liquid the equilibrium proportions are keto: enol = ca. 12:1. If we try to separate the two forms by dis-

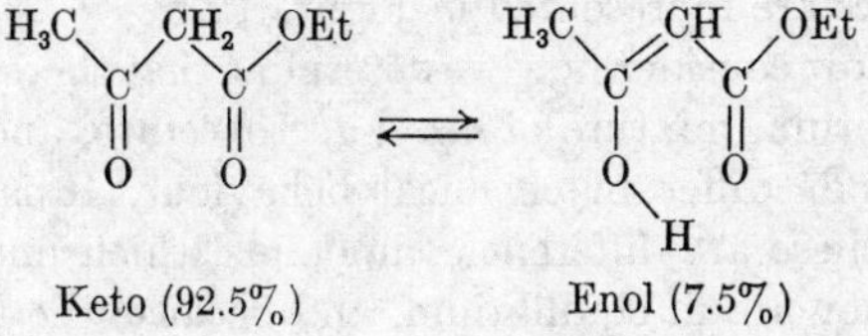

H_3C CH_2 OEt, C C, O O ⇌ H_3C CH OEt, C C, O O, H

Keto (92.5%) Enol (7.5%)

tillation, ordinary glass is sufficiently basic to catalyse the change. However, if a quartz vessel is used, separation becomes possible, the enol form being the more volatile, for in the absence of a catalyst the change is very slow indeed.

In the above discussion we have introduced an extremely important idea. We have said that treatment of the anion with acid leads to the enol form, but that the keto-form predominates in the equilibrium mixture and must therefore have lower energy (free energy) than the enol form. In other words, the less stable form is obtained initially. This is an example of *kinetic control* of a reaction, i.e. the initial products depend on the kinetics of the reaction. When equilibrium is allowed to be established, the most stable form predominates, and this is called *thermodynamic control.* Quite often the product of a chemical reaction is not thermodynamically

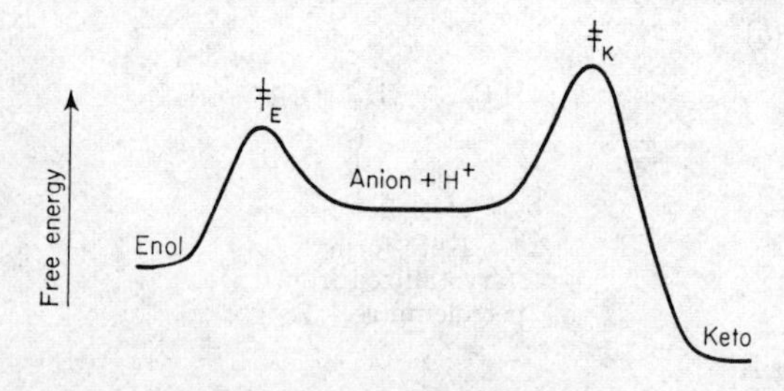

Figure 12.1. Diagram representing the equilibrium situation in ethyl acetoacetate. The symbol ‡ represents the activated complex.

the most stable compound but that which is formed most rapidly, which emphasizes the danger of trying to predict the course of a reaction entirely on the basis of relative stability of the products. The energetics are represented in Figure 12.1.

We must now consider how we determine that there is 7.5% of enol in the equilibrium mixture of ethyl acetoacetate, and also how the two compounds differ in chemical behaviour. It is important to realize that these are different compounds which undergo different reactions. They are in equilibrium, and because the transformation from one form to the other is catalysed by both acids and bases, it is difficult to distinguish the reactions of one form from those of the other. A particularly obvious feature of the enol form is that it has a carbon–carbon double bond; so, if bromine is added to a cold solution of the ester, in slight excess, the enol form reacts very rapidly with the bromine, whereas the keto-form is unaffected:

$$H_3C(OH)C{=}CH{-}C(=O)OEt + Br_2 \longrightarrow H_3C{-}C(=O){-}CHBr{-}C(=O)OEt + HBr$$

If some other compound that reacts extremely rapidly with bromine is now added to take up the excess of bromine before the keto form has an opportunity to change into the enol form, it is possible to estimate the concentration of enol present. The compound normally used to extract the excess of bromine is 2-naphthol:

$$Br_2 + \text{2-Naphthol} \xrightarrow{\text{Fast}} \text{1-Bromo-2-naphthol} + HBr$$

The amount of bromine that has added to the ester is estimated by treating the solution with sodium iodide and acid. The hydrogen iodide formed reduces the ethyl bromoacetoacetate to the starting ester, hydrogen bromide, and molecular iodine:

$$CH_3COCHBrCO_2Et + 2\,HI \longrightarrow CH_3COCH_2CO_2Et + HBr + I_2$$

and the iodine is then estimated with thiosulphate solution in the ordinary way. In addition to this chemical method for determining the concentration of enol in equilibrium mixtures of carbonyl compounds, there are various spectroscopic methods. Table 12.1 lists the enol content of a variety of mono- and di-carbonyl compounds estimated by the chemical method described.

Notice the very much higher enol content of the β-dicarbonyl compounds, especially acetylacetone. This is probably due, in part, to hydrogen bonding in the enol form. These β-dicarbonyl compounds

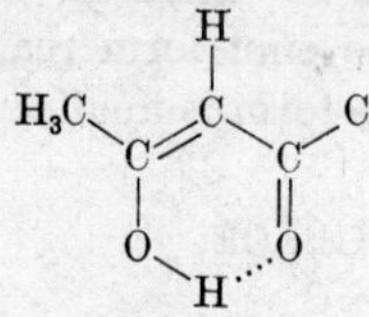

Hydrogen bonding in acetylacetone

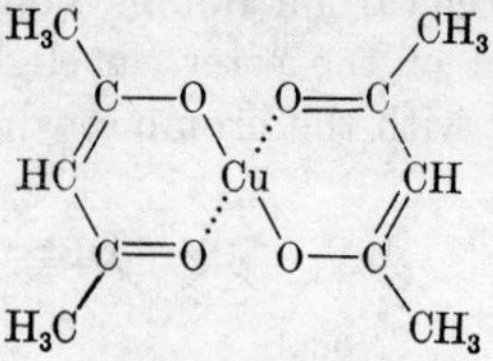

Copper(II) chelate with ethyl acetoacetate

form very stable chelates with transition-metal ions. Since the interconversion of enol and keto-compounds is catalysed both by acids and bases, a carbonyl compound usually shows the reactions of both the enol and the keto-form. Even with compounds such as acetone, where the extent of enolization is very small, the enol may react so much more rapidly than the keto-form that the only reaction we may observe is that of the enol compound. A particularly important illustration of this phenomenon is provided by the halogenation of ketones. Lapworth (1904) found that, in the presence of acids, the

Table 12.1. The enol content of some carbonyl compounds (pure liquids)

	Carbonyl compound	Enol (%)
Acetone	$CH_3COCH_3 \rightleftharpoons CH_2{=}C(OH){-}CH_3$	0.00025
Biacetyl	$CH_3COCOCH_3 \rightleftharpoons CH_2{=}C(OH){-}COCH_3$	0.0056
Cyclohexanone	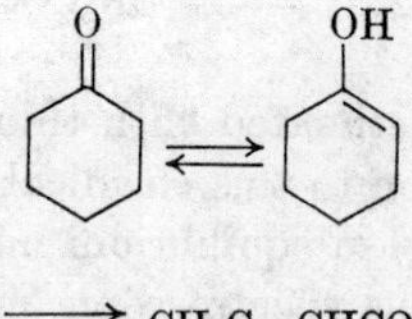 cyclohexanone $\rightleftharpoons$ cyclohex-1-en-1-ol (O / OH)	0.020
Ethyl acetoacetate	$CH_3COCH_2C_2H_5 \rightleftharpoons CH_3C(OH){=}CHCO_2C_2H_5$	7.5
Acetylacetone	$CH_3COCH_2COCH_3 \rightleftharpoons CH_3C(OH){=}CHCOCH_3$	80

rate of bromination of acetone is independent of the concentration of bromine. This requires that there be some slow rate-determining step before the bromination. It is, of course, the acid-catalysed enolization:

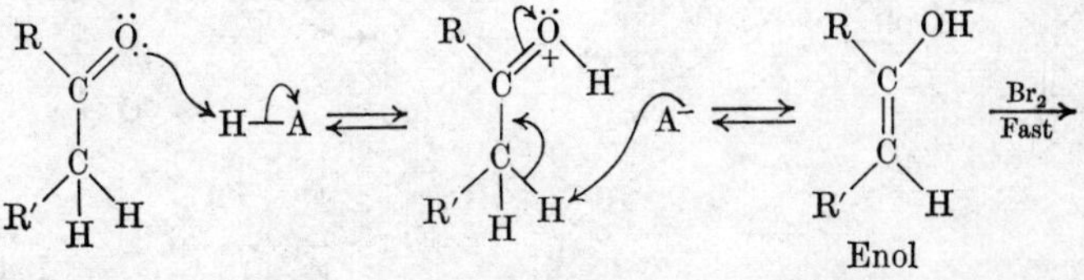

Since the interconversion of keto and enol forms can be base-catalysed, we should expect halogenation, e.g. iodination or bromination, to be catalysed by bases as well as by acids, and indeed this is so. In base-catalysed halogenation it is the enolate anion rather than the free enol which reacts:

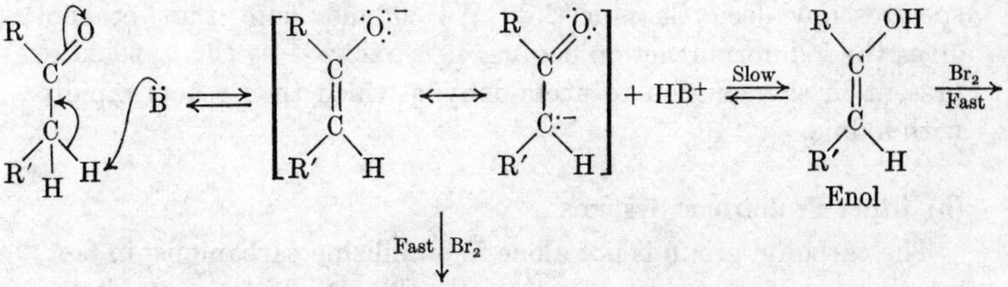

Notice, that if R′ is an electron-attracting substituent, it facilitates the base-catalysed enolization by making it easier for the proton on the same carbon atom to separate; on the other hand, it hinders acid-catalysed enolization by reducing the basicity of the carbonyl group. This can have a considerable effect on the course of the reaction. For example, in the base-catalysed bromination of acetone the first isolable product is 1,1,1-tribromoacetone. In other words, once one hydrogen atom has been replaced by a bromine atom the reaction becomes faster. On the other hand, in acid-catalysed bromination of acetone, the monobromo-compound can be isolated. The base-catalysed iodination of a methyl ketone is known as the iodoform reaction because the triiodomethyl ketone decomposes to give iodoform, a yellow crystalline substance.

$$\mathrm{RCOCH_3} \xrightarrow[\mathrm{I_2}]{\mathrm{OH^-}} \mathrm{R(^-O)C{=}CH_2} + \mathrm{I{-}I} \longrightarrow \mathrm{RCOCH_2I} \xrightarrow[\mathrm{I_2}]{\mathrm{OH^-}} \mathrm{R(^-O)C{=}CHI} + \mathrm{I{-}I} \longrightarrow$$

$$\mathrm{RCOCHI_2} \xrightarrow[\mathrm{I_2}]{\mathrm{OH^-}} \mathrm{R(^-O)C{=}CI_2} + \mathrm{I{-}I} \longrightarrow \mathrm{RCOCI_3} \xrightarrow{\mathrm{OH^-}} \mathrm{R(CI_3)(OH)C{-}O^-} \longrightarrow$$

$$\mathrm{RCOOH} + \mathrm{CI_3^-} \longrightarrow \mathrm{RCO_2^-} + \underset{\text{Iodoform}}{\mathrm{CHI_3}}$$

The formation of iodoform when a compound is treated with an alkaline solution of iodine was formerly used as a diagnostic test for the presence of a CH_3CO group. In most circumstances the presence of such a grouping can nowadays be detected with greater certainty by a combination of infrared and nuclear magnetic resonance spectroscopy (see Chapter 25). We should note that ethanol gives the iodoform reaction because it is oxidized by the hypoiodite present in the solution to acetaldehyde which then reacts rapidly with iodine.

(b) Other Prototropic Systems

The carbonyl group is not alone in stabilizing carbanions; in fact, most acceptor groups are effective (see Chapter 9), for example the nitro-group.

$$\left[\mathrm{R{-}\overset{-}{C}H{-}\overset{+}{N}(=O)(O^-)} \longleftrightarrow \mathrm{R{-}\overset{-}{C}H{-}\overset{+}{N}(O^-)(=O)} \longleftrightarrow \mathrm{R{-}CH{=}\overset{+}{N}(O^-)(O^-)}\right]$$

$$\text{or}\quad \mathrm{R{-}\overset{\delta-}{C}H{\cdots}\overset{+}{N}(O^{\delta-})(O_{\delta-})}$$

Phenylnitromethane is a very pale yellow liquid which dissolves only slowly in alkali. If the alkaline solution is treated with acetic acid a colourless solid is precipitated. This is the *aci*-form which corresponds to the enol form of a ketone.

$$C_6H_5{-}CH_2{-}\overset{+}{N}(=O){-}O^- \rightleftharpoons C_6H_5{-}CH{=}\overset{+}{N}(OH){-}O^-$$

Nitro-form (pale yellow liquid) — *aci*-Form (colourless solid)

The *aci*-form is strongly acidic and releases carbon dioxide from aqueous sodium hydrogen carbonate. Phenylnitromethane is a liquid, and the concentration of the *aci*-form in the pure liquid at room temperature is negligible. The concentration of the *aci*-form of nitromethane at room temperature is 0.00001%.

With aliphatic nitroso-compounds the reverse is true and the form corresponding to the enol is the only form present under normal conditions. If, for example, we treat ethyl methyl ketone with nitrous acid, nitrosation of the reactive methylene group takes place but the nitroso-compound rearranges to the corresponding monoxime of biacetyl (butane-2,3-dione):

$$CH_3COCH_2CH_3 \xrightarrow{HNO_2} CH_3COCH(NO)CH_3 \longrightarrow CH_3COC(=N{-}OH)CH_3$$

(Notice that this product has the same structure as the oximes prepared by treating a carbonyl group with hydroxylamine.) True aliphatic nitroso-compounds are known, but these have no hydrogen atoms on the carbon atom to which the nitroso-group is attached.

Weak acceptors such as the cyanide group do not exist in tautomeric forms ($RC{\equiv}C{-}NH_2$ does not exist, but RCH_2CN does).

(c) Enol Acetates and Enol Ethers

Enol esters and ethers can be made by additions to acetylene. In Part 1, when describing the hydration of acetylene to form acetaldehyde, we rather glossed over the fact that the initial product of this reaction is the enol of acetaldehyde, namely, vinyl alcohol:

$$HC{\equiv}CH \xrightarrow[H_2SO_4/H_2O]{Hg^{2+}} (H_2C{=}CHOH) \longrightarrow CH_3CHO$$

Vinyl alcohol

If, instead of being carried out in aqueous sulphuric acid, the reaction is carried out in the presence of sulphuric acid and acetic acid, then vinyl acetate can be prepared:

$$HC{\equiv}CH + CH_3CO_2H \xrightarrow[H_2SO_4,\ CH_3CO_2H]{Hg^{2+}} \underset{\text{Vinyl acetate}}{CH_2{=}CHOCOCH_3}$$

Hydrolysis of vinyl acetate yields acetaldehyde and acetic acid. (Vinyl acetate is also prepared directly from ethylene by a mechanistically complicated industrial process and is used as a monomer to be polymerized with other substituted ethylenes in certain plastics.)

In our discussion of carbonium ions we described how acetic acid dissolved in trifluoroacetic anhydride to yield an unsymmetrical anhydride which partly ionized into acetylium cations and trifluoroacetate anions:

$$CH_3COOH + (CF_3CO)_2O \longrightarrow CH_3CO{-}O{-}COCF_3 + CF_3CO_2H$$

$$CH_3CO{-}O{-}COCF_3 \rightleftharpoons CH_3CO^+ \quad ^-OCOCF_3$$

Although the concentration of ions is very low indeed, the acetylium ion is extremely reactive and will add to an acetylenic triple bond. The product of such a reaction is the enol trifluoroacetate of a β-diketone. Trifluoroacetate esters are easily hydrolysed and treatment with an excess of methanol yields methyl trifluoroacetate and the enol which, of course, is in equilibrium with the β-diketone.

$$C_4H_9C{\equiv}CH + CH_3COOH + (CF_3CO)_2O \longrightarrow C_4H_9\underset{\displaystyle OCOCF_3}{\underset{|}{C}}{=}CHCOCH_3$$

$$\xrightarrow[\text{excess}]{CH_3OH} C_4H_9COCH_2COCH_3 + CF_3COOCH_3$$

Vinyl ethers can also be prepared directly from acetylene and alcohols in the presence of alkali at high temperatures and pressures:

$$HC{\equiv}CH + ROH \xrightarrow[150\text{-}180°]{KOH} H_2C{=}CHOR$$

They break down readily in acid to give the corresponding alcohol and acetaldehyde.

We began this Chapter by considering stabilized carbanions such as that from ethyl acetoacetate. We saw that they reacted with a proton to yield the enol. On the other hand, we know that these carbanions will react with alkyl halides by displacement, which results in *C*-alkylation. This reaction is initiated by the carbanion and hence the most nucleophilic centre in the carbanion will be the effective part of the molecule in attacking the relatively inert alkyl halide. However, we have seen that a proton becomes attached to

$$H_3C{-}C(=O){-}\overset{-}{C}H{-}C(=O){-}OEt \xrightarrow{RCH_2Br} H_3C{-}C(=O){-}CH(CH_2R){-}C(=O){-}OEt + Br^- \quad C\text{-Alkylation}$$

$$H_3C{-}C(=O){-}\overset{-}{C}H{-}C(=O){-}OEt \xrightarrow[\text{(pyridine solution)}]{CH_3COCl} H_3C{-}C(OOCCH_3){=}CH{-}C(=O){-}OEt + Cl^- \quad O\text{-Acylation}$$

the place of the highest electron density, namely, the oxygen atom. Thus, if we have a powerful electrophile, we should expect it likewise to attack the site of highest electron density. It is possible, in accord with this idea, to make enol esters by treating the carbanion with very reactive reagents. For example, treating ethyl acetoacetate with acetyl chloride in pyridine yields the enol acetate by so-called *O*-acylation.

It is possible to make enol esters even of such unstable enols as that derived from acetone, by using exceedingly reactive reagents. Ketene ($CH_2{=}C{=}O$) (see Chapter 21), which reacts very rapidly with hydroxy-compounds to produce the corresponding acetate esters, reacts with acetone in the presence of acids to yield the corresponding enol acetate:

$$H_3C{-}C(=O){-}CH_3 \underset{}{\overset{H^+}{\rightleftharpoons}} H_2C{=}C(OH){-}CH_3 \xrightarrow{CH^2{=}C{=}O} H_2C{=}C(OCOCH_3){-}CH_3$$

We have just described how all these enol esters and ethers can be hydrolysed to give a ketone (or aldehyde).

(d) Enamines

α,β-Unsaturated primary and secondary amines do not exist, i.e. the corresponding imines have no tendency to tautomerize:

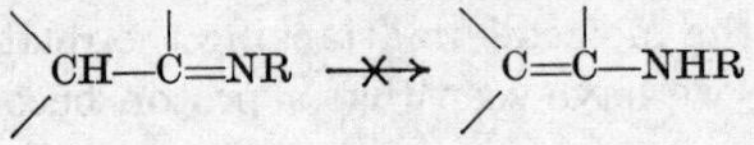

$$\mathrm{>CH{-}\overset{|}{C}{=}NR} \not\rightarrow \mathrm{>C{=}\overset{|}{C}{-}NHR}$$

but vinylamines (enamines) without hydrogen on the nitrogen atom can be prepared. For example, the heterocyclic secondary amine pyrrolidine condenses with cyclic ketones, e.g. cyclohexanone, to yield such a compound:

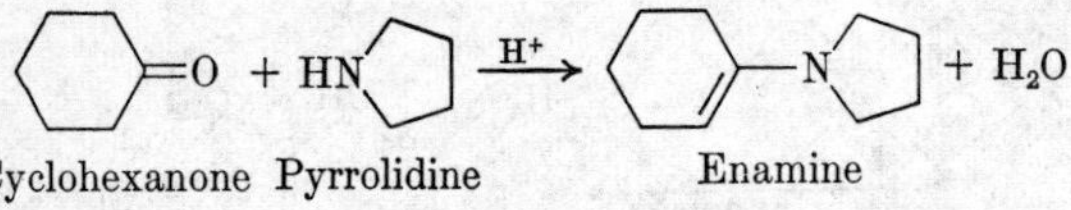

Cyclohexanone + Pyrrolidine $\xrightarrow{H^+}$ Enamine + H_2O

Enamines are interesting because they undergo *C*-alkylation under very mild conditions. Remember how ammonia or an amine reacts with an alkyl halide by donating its lone pair of electrons. In much the same way, an enamine may displace a halide ion from an alkyl halide although this displacement (which we might call a conjugate displacement) actually involves the β-carbon atom:

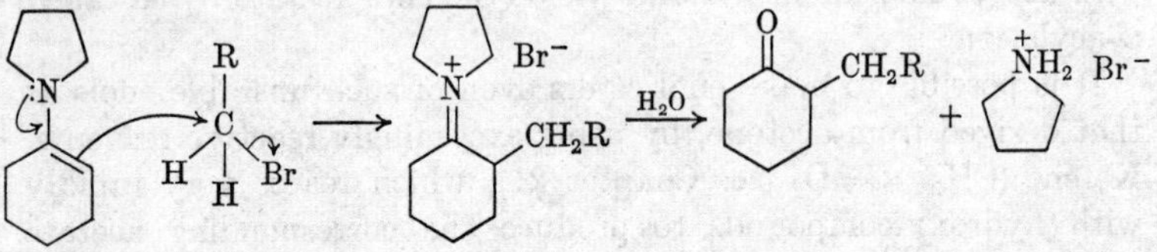

Enamine + RCH_2Br $\longrightarrow$ iminium salt (Br^-, CH_2R) $\xrightarrow{H_2O}$ 2-(CH_2R)-cyclohexanone + pyrrolidinium ($\overset{+}{N}H_2$ Br^-)

To appreciate this reaction it is very important to compare it with the reaction of a saturated amine with an alkyl bromide. The enamine reaction, besides being interesting, is also useful because the cyclohexanone molecule has been alkylated under very mild conditions. When discussing the reactions of carbanions we described the alkylation of cyclohexanone by use of sodamide as the base to provide the carbanion. Clearly, many molecules would decompose if treated with such a powerful reagent, and the use of enamines provides an alternative method.

(e) Anionotropy

We have seen that a proton or an electrophile may attack the resonance-stabilized carbanion in more than one place, i.e. may lead to *C*-alkylation or *O*-acylation, depending on the conditions. In a somewhat similar way carbonium ions and oxonium ions may be attacked in more than one place by an anion. The product of such a reaction may be such that it appears that the anion has migrated from one end of the molecule to the other. For example, when but-2-en-1-ol is treated with a catalytic amount of acid in aqueous media, it is converted into an equilibrium mixture in which but-3-en-2-ol predominates (70%):

$$\underset{(70\%)}{CH_3{-}\underset{\underset{\ddot{O}H}{|}}{CH}{-}CH{=}CH_2} \xrightarrow{H^+} CH_3{-}\underset{\underset{\overset{+}{O}H_2}{|}}{CH}{-}CH{=}CH_2 \;\; \bar{O}H \rightleftharpoons \underset{(30\%)}{CH_3{-}CH{=}CH{-}CH_2OH} + H_2O$$

If the double bond can conjugate with an aromatic ring, such anionotropy, as it is called, can result in almost complete conversion:

$$C_6H_5\underset{\underset{OH}{|}}{CH}{-}CH{=}CH{-}CH_3 \underset{}{\overset{H^+}{\rightleftharpoons}} C_6H_5CH{=}CH{-}\underset{\underset{OH}{|}}{CH}{-}CH_3$$

In a conjugated carbon chain, the anion may attack at any point along the chain, but is most likely to do so at an end, normally the least substituted end. Thus, it may appear that the anion has moved as many as six carbon atoms down a chain:

$$R{-}\underset{\underset{HO}{|}}{\overset{\overset{CH_3}{|}}{C}}{-}CH{=}CH{-}\overset{\overset{CH_3}{|}}{C}H{-}CH{=}CH{-}CH{=}CH_2 \xrightarrow{H^+} R{-}\underset{\underset{H_2\overset{+}{O}}{|}}{\overset{\overset{CH_3}{|}}{C}}{-}\ldots{-}CH{=}CH_2 \xrightarrow{OH^-} R{-}\overset{\overset{CH_3}{|}}{C}{=}CH{-}CH{=}CH{-}\overset{\overset{CH_3}{|}}{C}{=}CH{-}CH_2OH$$

Anionotropy is less common than prototropy, but it will be clear from our discussion that the two phenomena are closely related.

Problems

1. 1,3-Diphenylpropane-1,3-dione (dibenzoylmethane) was treated with sodium in ether and then with benzoyl chloride. The ether solution was washed with water and evaporated to yield a crystalline compound *A*, $C_{22}H_{16}O_3$, which melted at 240° in a soft-glass tube but at 155° in a quartz tube. Acidification of the aqueous washings gave an isomeric compound *B* which melted at 240–245° in either soft-glass or quartz tubes. *A* dissolved readily in a small quantity of alcohol; on standing this solution precipitated crystals of the less soluble compound *B*. Suggest structures for *A* and *B* and account for the observations described.

2. Bromination of acetone in aqueous solution is accelerated by both acids and bases but is independent of the concentration of bromine. Explain these observations.

CHAPTER 13

Nucleophilic Addition

Nucleophiles do not add to hydrocarbon olefinic double bonds (see Chapter 6 of Part 1). On the other hand, nucleophiles add readily to the carbonyl double bond and we attributed this to the electron-attracting properties of the oxygen atom (Chapter 7 of Part 1). When discussing, in Chapter 8 of the present Part, the directive effects in addition and substitution we saw that electron-attracting groups retard the addition of electrophiles to olefinic double bonds. The question we now must consider is whether nucleophilic addition will occur at a double bond surrounded by electron-attracting groups.

The most obvious electron-attracting groups with which to surround a carbon–carbon double bond are fluorine atoms. Alkoxide anions are typical nucleophiles, and experimentally we find that indeed they add quite readily to fully fluorinated olefins:

$$CH_3O^- + CF_3CF{=}CF_2 \longrightarrow CF_3\overset{-}{C}F{-}CF_2OCH_3 \xrightarrow{H{-}OCH_3} CF_3CHF{-}CF_2OCH_3 + CH_3O^-$$

Chlorofluoro-olefins react similarly:

$$ClFC{=}CF_2 + EtOH + Na^+\ {}^-OEt \xrightarrow[20^\circ]{2\frac{1}{2}\ hr.} F{-}CClH{-}CF_2OEt$$

$$Cl_2C{=}CF_2 + C_6H_5O^-K^+ \longrightarrow CCl_2HCF_2OC_6H_5$$

Another typical nucleophilic anion is the cyanide anion* and it reacts similarly:

$$^{-}CN + CF_2{=}CFCl \longrightarrow \begin{matrix} CF_2CN \\ | \\ CFCl^- \end{matrix} \xrightarrow{H-CN} \begin{matrix} CF_2CN \\ | \\ CHFCl \end{matrix} \xrightarrow{H_2SO_4} CHFClCF_2CO_2H$$

Although ammonia carries no negative charge we saw in Part 1 (Chapter 3) how it behaves as a strong nucleophile, taking part in displacement reactions with alkyl halides and adding to the carbonyl double bond. When ammonia adds to a fluoro-olefin the initial product undergoes further transformations such as those shown:

$$H_3\ddot{N} + C_2F_5CF{=}CF_2 \longrightarrow C_2F_5\bar{C}F{-}CF_2\overset{+}{N}H_3 \longrightarrow C_2F_5CHF{-}CF_2\ddot{N}H_2 \longrightarrow C_2F_5CHF{-}CF{=}\overset{+}{N}H_2\ F^- \xrightarrow{-HF}$$

$$C_2F_5CHF{-}CF{=}\ddot{N}H \longrightarrow C_2F_5CHF{-}C{\equiv}\overset{+}{N}{-}H\ F^- \xrightarrow{-HF} CF_3CF_2CHFCN\ (20\%)$$

These examples show that a carbon–carbon double bond flanked by sufficient electron-attracting groups undergoes addition reactions with nucleophiles. On the other hand, of course, the powerful electron-attracting groups inhibit electrophilic addition to the double bond.

Since electron-attractors activate a carbon–carbon double bond to attack by a nucleophile, we should expect that an electron-acceptor would do so more effectively. Experimentally we find alkoxides add readily to unsaturated ketones and esters, as illustrated here for a vinyl ketone and for ethyl acrylate.

* In all these reactions the addition occurs at the CF_2 group. This is because the negative charge is spread more effectively (there is considerable I_π repulsion between the filled non-bonding p orbital of the negatively charged carbon atom and the filled non-bonding p orbitals of an adjacent fluorine atom).

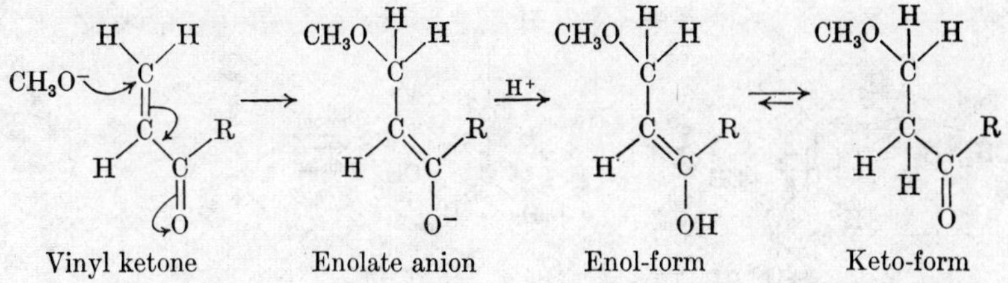

CH_3O^- + Vinyl ketone $\longrightarrow$ Enolate anion $\xrightarrow{H^+}$ Enol-form $\rightleftharpoons$ Keto-form

Vinyl ketone Enolate anion Enol-form Keto-form

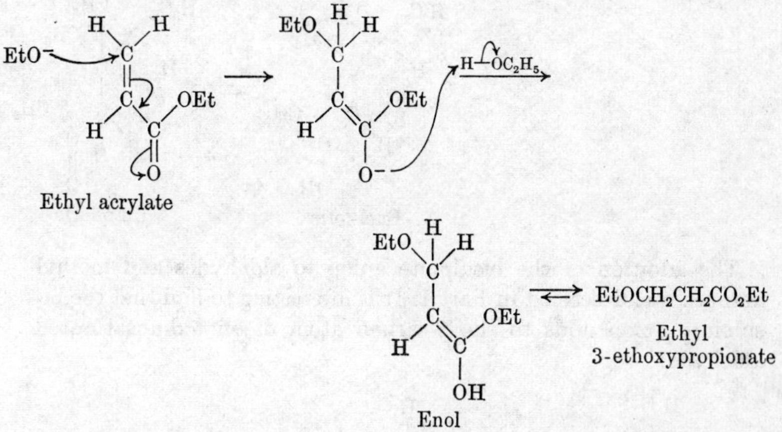

EtO^- + Ethyl acrylate $\longrightarrow$ $\xrightarrow{H-OC_2H_5}$ Enol $\rightleftharpoons$ $EtOCH_2CH_2CO_2Et$

Ethyl acrylate

Enol

Ethyl
3-ethoxypropionate

The cyanide ion, ammonia, and amines will all add similarly (see below).

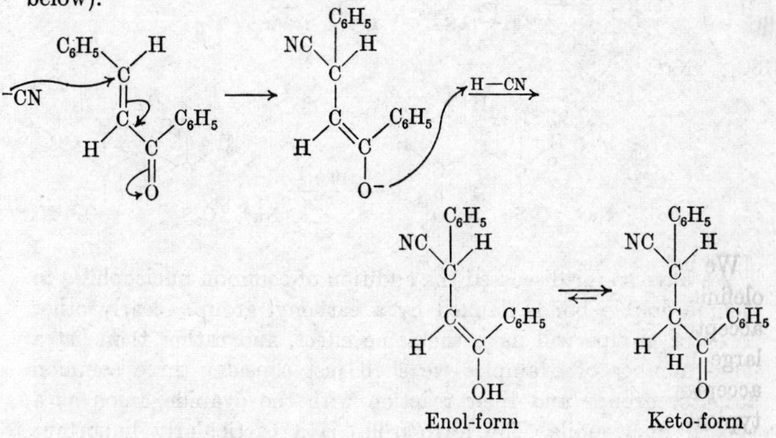

^-CN $\longrightarrow$ $\xrightarrow{H-CN}$ Enol-form $\rightleftharpoons$ Keto-form

Enol-form Keto-form

$$CH_3\ddot{N}H_2 + (CH_3)_2C{=}CH{-}C(CH_3){=}O \longrightarrow H_3C{-}\overset{+}{N}H_2{-}C(CH_3)_2{-}CH{=}C(CH_3){-}O^- \underset{\text{transfer}}{\overset{\text{Proton}}{\rightleftharpoons}}$$

$$\underset{\text{Enol-form}}{H_3C{-}NH{-}C(CH_3)_2{-}CH{=}C(CH_3){-}OH} \rightleftharpoons \underset{\text{Keto-form}}{H_3C{-}NH{-}C(CH_3)_2{-}CH_2{-}C(CH_3){=}O}$$

The addition of the bisulphite anion to aldehydes and methyl ketones was described in Part 1. It is interesting to find that the bisulphite anion adds to the β-carbon atom of an α,β-unsaturated aldehyde:

$$C_6H_5CH{=}CH{-}CH{=}O + HSO_3^-\ Na^+ \rightleftharpoons H{-}O_3S{-}CH(C_6H_5){-}CH{=}CH{-}O^-\ Na^+ \underset{\text{transfer}}{\overset{\text{Proton}}{\rightleftharpoons}}$$

$$Na^+\ {}^-O_3S{-}CH(C_6H_5){-}CH_2{-}CH{=}O \xrightarrow{Na^+HSO_3^-} Na^+\ {}^-O_3S{-}CH(C_6H_5){-}CH_2{-}CH(OH){-}SO_3^-\ Na^+$$

We have so far discussed the addition of common nucleophiles to olefinic double bonds flanked by a carbonyl group. Clearly other acceptor groups will have the same effect, and rather than list a large number of examples we shall just consider three common acceptor groups and their reaction with the cyanide anion as a typical nucleophile. The nitro-group is a particularly important acceptor and it makes an adjacent olefinic double bond very reactive towards nucleophiles:

$$\bar{N}C + C_3H_7CH{=}CH\overset{+}{N}(O^-){=}O \longrightarrow NC{-}C(C_3H_7)(H){-}CH{=}\overset{+}{N}(O^-)(O^-) \xrightarrow{H-CN} NC{-}C(C_3H_7)(H){-}CH{=}\overset{+}{N}(OH)(O^-) \rightleftharpoons C_3H_7CH(CN)CH_2NO_2$$

Aci-form Nitro-form

The sulphone group is likewise a powerful acceptor:

$$C_2H_5SO_2CH{=}CH_2 \xrightarrow[Na^+CN^-]{HCN} C_2H_5SO_2CH_2CH_2CN$$

and the cyano-group is a sufficiently powerful acceptor to activate a double bond to nucleophilic addition:

$$C_6H_5CH{=}C(C_6H_5)(CN) \xrightarrow[Na^+CN^-]{HCN} C_6H_5CH(CN){-}CH(CN)C_6H_5$$

We now come to consider nucleophilic carbon, i.e. compounds in which the nucleophilic centre is a carbon atom. The first molecules of this type that we considered were organometallic compounds, particularly Grignard reagents. In Chapter 11 of Part 1, we described the addition of Grignard reagents to carbonyl double bonds. In the light of our present discussion we might expect Grignard reagents to add to olefinic double bonds adjacent to electron-acceptors, such as a carbonyl bond. Such reactions do occur, but quite often they lead to a mixture of products resulting both from addition to the olefinic double bond and from addition to the carbonyl double bond (see below). If the acceptor group is of a kind that does not normally

$$C_6H_5CH{=}CHCOC_6H_5 + \overset{\delta-}{C_6H_5}{-}\overset{\delta+}{MgBr} \longrightarrow (C_6H_5)_2CHCH{=}C(C_6H_5)O^-MgBr^+ \xrightarrow{H_2O} (C_6H_5)_2CHCH_2COC_6H_5$$

$$+$$

$$C_6H_5CH{=}CHCOC_6H_5 + C_6H_5MgBr \dashrightarrow C_6H_5CH{=}CHC(C_6H_5)_2O^-MgBr^+ \dashrightarrow^{H_2O} C_6H_5CH{=}CHC(C_6H_5)_2OH$$

react with a Grignard reagent, addition may take place exclusively at the olefinic double bond.

$$\overset{\delta-}{C_2H_5}{-}\overset{\delta+}{MgBr} + (CH_3)_2C{=}CH{-}\overset{+}{N}(=O)O^- \longrightarrow C_2H_5C(CH_3)_2CH{=}\overset{+}{N}(O^-)O^-\ MgBr^+ \xrightarrow{H_2O} C_2H_5C(CH_3)_2CH_2NO_2$$

Chapter 9 describes how carbanions take part in substitution reactions and in addition reactions to carbonyl double bonds in a fashion analogous to that for other common nucleophiles. We should expect these carbanions also to add to olefinic double bonds that are activated by the presence of an acceptor group. This very important reaction is known as the Michael reaction. For example, diethyl

$$(EtO_2C)_2CH_2 + C_6H_5CH{=}CHCOC_6H_5 \xrightarrow{Na^+\ ^-OEt}$$

$$(EtO_2C)_2\bar{C}H + C_6H_5CH{=}CHC(=O)C_6H_5 \longrightarrow (EtO_2C)_2CH{-}CH(C_6H_5)CH_2COC_6H_5$$

malonate in the presence of sodium methoxide adds to α,β-unsaturated ketones. These Michael additions are reactions of great synthetic value since they provide yet another way of forming a carbon–carbon bond and lengthening carbon chains.

$$C_2H_5NO_2 + CH_2{=}CHCO_2Et \xrightarrow{Na^+\ ^-OEt}$$

$$\left[CH_3CH{=}\overset{+}{N}(\ddot{O}:^-)_2 \longleftrightarrow CH_3\ddot{C}H{-}N({=}\ddot{O})(\ddot{O}:^-) \right] + H_2C{=}CH{-}C({=}O)OEt \longrightarrow$$

$$CH_3CH(NO_2)CH_2CH_2CO_2Et$$

Rather than produce more examples let us summarize the facts just recorded. Hydrocarbon olefins add electrophiles but do not add nucleophiles. The introduction of electron-attracting groups at positions adjacent to an olefinic double bond retards electrophilic addition to that bond. Eventually, if there are sufficient powerful electron-attracting groups, electrophilic addition may be almost completely inhibited. On the other hand, the electron-attracting groups enhance the tendency of the double bond to add nucleophiles. Thus an olefinic double bond surrounded by fluorine atoms readily adds common nucleophiles. We should expect that electron acceptors would be even more effective than electron-attractors in activating a double bond to nucleophilic addition. This is exactly what we have found. We must now turn to consider nucleophilic addition to aromatic compounds.

Nucleophilic Substitution in Aromatic Compounds

In Part 1, when discussing the reactions of the benzene nucleus, we described its general inertness to chemical reactions and in this volume we have attributed the stability of the benzene nucleus to the delocalization of the π electrons. We have discussed at some length the so-called electrophilic substitution reactions of the benzene nucleus and described how these are really addition-with-elimination reactions. In Part 1 we concluded that the benzene nucleus which was attacked only by the most powerful electrophiles, would not, under any normal circumstances, be attacked by a nucleophile. Although hydrocarbon olefins will not add nucleophiles, at the beginning of this Chapter we described how surrounding the carbon–carbon double bond with sufficient electron-attracting groups activate it to nucleophilic addition. In the same way a sufficient number of electron-attracting groups activate the benzene nucleus to attack by nucleophiles; for instance, hexafluorobenzene reacts readily with sodium methoxide:

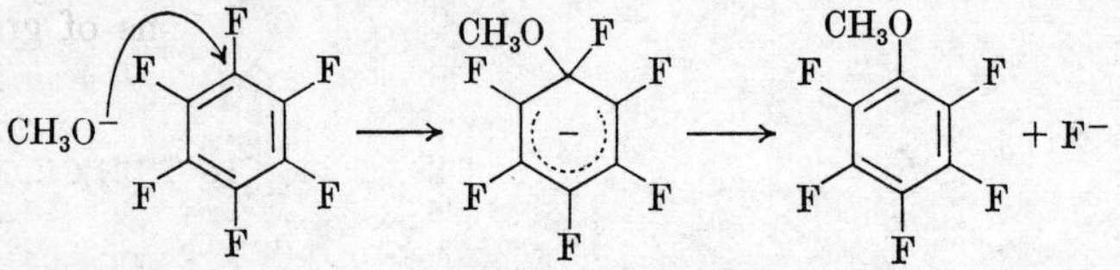

CH_3O^- F F F F F F ⟶ CH_3O F F F F F F ⟶ CH_3O F F F F F + F^-

The nucleophile adds to the benzene nucleus and we obtain an intermediate similar to the Wheland intermediate in electrophilic attack (but containing two more electrons). Just as the Wheland intermediate ejects a proton rather than adding an anion, so the negatively charged intermediate in this case ejects a halide anion

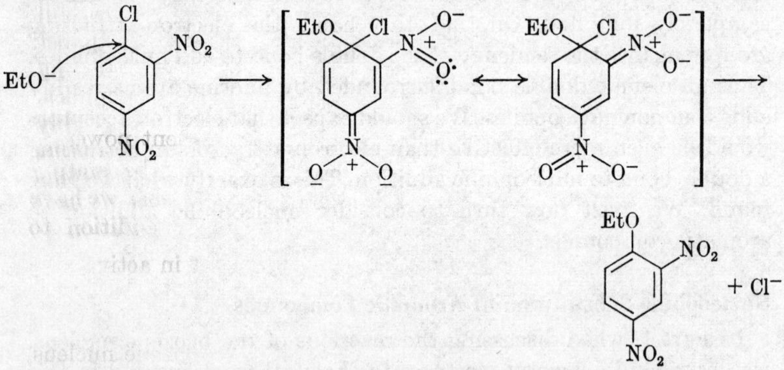

EtO^- Cl NO_2 NO_2 ⟶ [EtO Cl N^+ O^- O N O^- + O ⟷ EtO Cl N^+ O^- O^- N O + O^-] ⟶ EtO NO_2 NO_2 + Cl^-

rather than adding a proton. In both these reactions the delocalized aromatic ring is regenerated. Acceptor groups are more effective in facilitating nucleophilic addition to olefinic double bonds, and in the same way acceptor groups are more effective in facilitating nucleophilic substitution; the more acceptor groups there are in the molecule, the faster is the reaction (see Table 13.1).

It will also be apparent that the nucleophile will only attack positions *ortho* or *para* to the acceptor.

Table 13.1. Relative rates of nucleophilic addition-with-elimination (substitution)

$$ArCl + Na^+\ ^-OCH_3 \longrightarrow ArOCH_3 + Na^+Cl^-$$

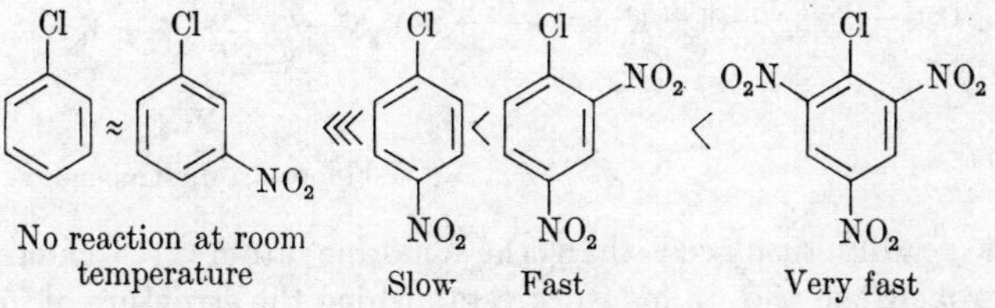

Chlorobenzene ≈ m-chloronitrobenzene	⋘ p-chloronitrobenzene	< 2,4-dinitrochlorobenzene	< 2,4,6-trinitrochlorobenzene
No reaction at room temperature	Slow	Fast	Very fast

In some cases the addend intermediate can be isolated (it should be noted, however, that the isolation of an intermediate of this kind does not *prove* that the reaction necessarily always goes through such an intermediate). Crystalline intermediates of this kind are known as Meisenheimer compounds.

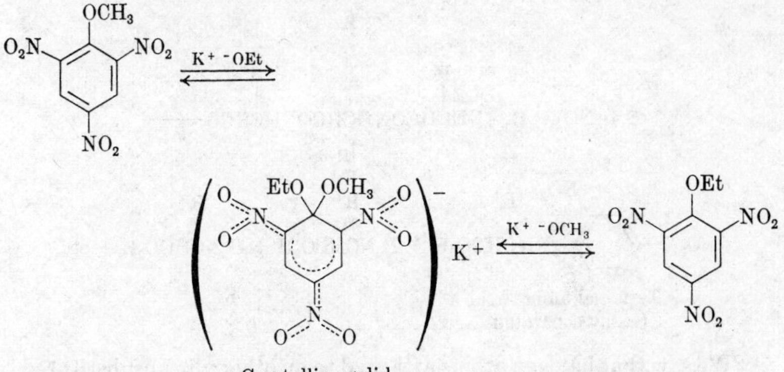

$$\text{2,4,6-}(O_2N)_3C_6H_2OCH_3 \underset{}{\overset{K^+\ ^-OEt}{\rightleftharpoons}} \left[\text{1-EtO-1-}CH_3O\text{-2,4,6-}(NO_2)_3C_6H_2\right]^- K^+ \underset{}{\overset{K^+\ ^-OCH_3}{\rightleftharpoons}} \text{2,4,6-}(O_2N)_3C_6H_2OEt$$

Crystalline solid

The displacement of halogens *ortho* or *para* to a nitro-group in an aromatic ring is a general reaction and we illustrate some common examples. The fluorine atom in 1-fluoro-2,4-dinitrobenzene is even

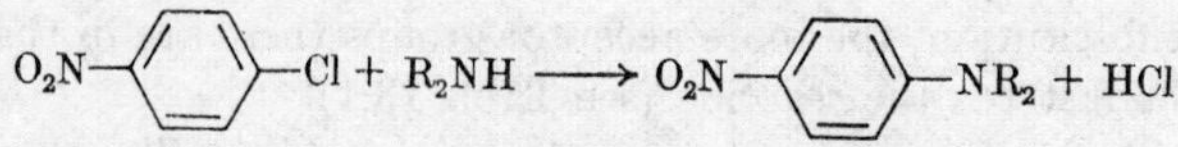

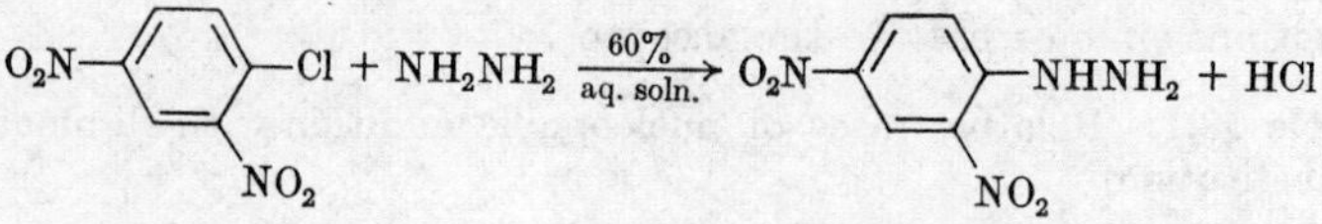

$$O_2N-C_6H_4-Cl + R_2NH \longrightarrow O_2N-C_6H_4-NR_2 + HCl$$

$$O_2N-C_6H_3(NO_2)-Cl + NH_2NH_2 \xrightarrow[\text{aq. soln.}]{60\%} O_2N-C_6H_3(NO_2)-NHNH_2 + HCl$$

2,4-Dinitrophenylhydrazine

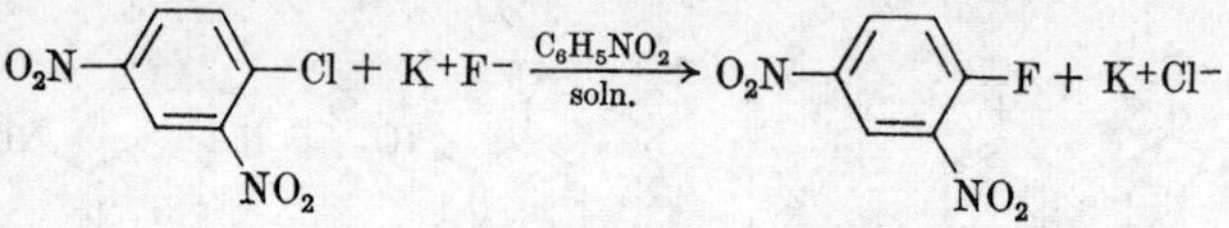

$$O_2N-C_6H_3(NO_2)-Cl + K^+F^- \xrightarrow[\text{soln.}]{C_6H_5NO_2} O_2N-C_6H_3(NO_2)-F + K^+Cl^-$$

1-Fluoro-2,4-dinitrobenzene

more readily displaced than the chlorine atom in 1-chloro-2,4-dinitrobenzene and, in his work establishing the structure of insulin (a protein molecule), Sanger used the fact that 1-fluoro-2,4-dinitrobenzene reacts extremely rapidly with amines; by this means he could always identify the terminal molecule at the end of a polypeptide chain.

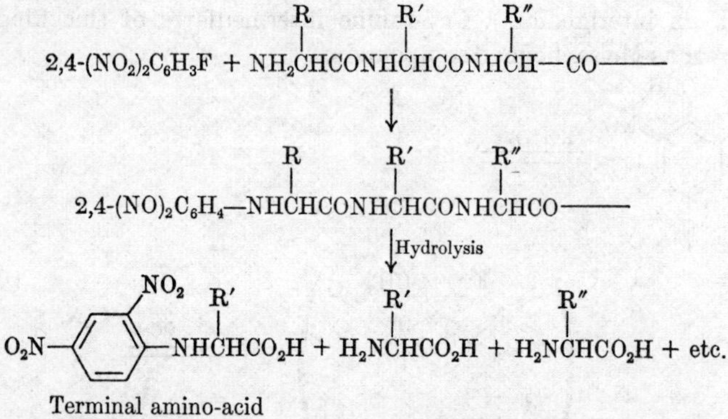

$$2,4\text{-}(NO_2)_2C_6H_3F + NH_2CH(R)CONHCH(R')CONHCH(R'')-CO-\!\!-\!\!-$$

$$\downarrow$$

$$2,4\text{-}(NO)_2C_6H_4-NHCH(R)CONHCH(R')CONHCH(R'')CO-\!\!-\!\!-$$

$$\downarrow \text{Hydrolysis}$$

$$O_2N-C_6H_3(NO_2)-NHCH(R')CO_2H + H_2NCH(R')CO_2H + H_2NCH(R'')CO_2H + \text{etc.}$$

Terminal amino-acid
(easily separated)

We saw that halogen atoms attached to an otherwise unsubstituted benzene nucleus are not replaced by nucleophilic substitution. We now see that halogen atoms *ortho* or *para* to powerful accepting groups, particularly nitro-groups, can be displaced, but that the reaction is not truly a substitution reaction, being rather an addition-with-elimination reaction; it is analogous to the so-called electrophilic aromatic substitution but involves nucleophiles instead of electrophiles. Nucleophilic addition-with-elimination reactions are much less important than electrophilic addition, but it is both interesting and important to see that the two reactions involve essentially the same mechanism. In electrophilic addition-with-elimination a hydrogen atom is usually the displaced species; certain other groups can be displaced but not a halogen atom. In nucleophilic addition-with-elimination it is usually a halogen atom that is displaced, though in special circumstances it may be a hydrogen atom. The difference is because a proton (H^+) is readily solvated and its departure from the cationic Wheland intermediate is facilitated. On the other hand, a hydride ion (H^-) is not readily solvated by most solvents, so that, although strong nucleophiles add to compounds such as *m*-dinitrobenzene, substitution often fails to take place because the anionic Wheland-type intermediate is unable to expel the hydride ion.

Problems

1. Predict the outcome of the following reactions:

(*a*) $CH_3CH{=}CHCO_2Et + HCN \xrightarrow[\text{excess}]{Na^+CN^-}$

(*b*) $C_6H_5CH{=}CHCOC_6H_5 + Na^+HSO_3^- \longrightarrow$

(*c*) $C_6H_5MgBr + C_6H_5CH{=}C(NO_2)C_6H_5 \longrightarrow$

(*d*) $O_2NCH_2CH_2CH_2NO_2 + 2CH_3COCH{=}CH_2 \xrightarrow{Na^+OH^-}$

(*e*) $NH_3 + CH_2{=}CHCN \xrightarrow[\text{excess}]{}$

(*f*) $CH_3CH{=}CHCO_2Et + CH_3CH(CO_2Et)_2 \xrightarrow{Na^+OEt^-}$

2. Which of the following aromatic compounds would react with sodium methoxide at normal temperatures?

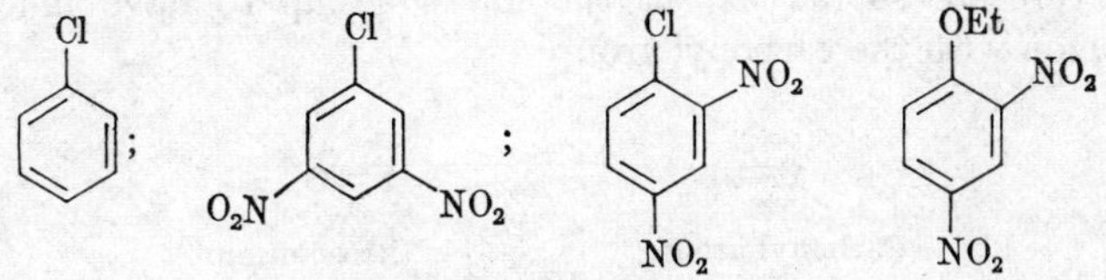

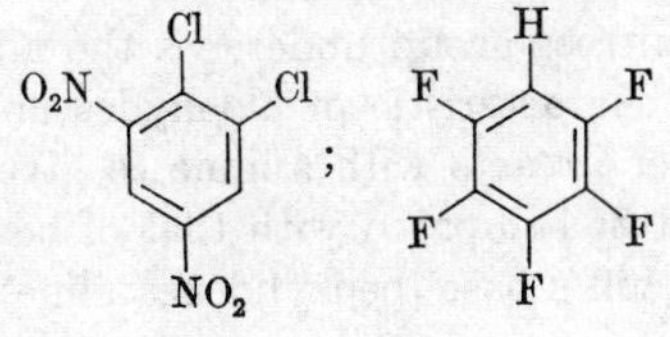

Cl (benzene ring); Cl, O_2N, NO_2 (benzene ring); Cl, NO_2, NO_2 (benzene ring); OEt, NO_2, NO_2 (benzene ring)

Cl, O_2N, Cl, NO_2 (benzene ring); H, F, F, F, F, F (benzene ring)

CHAPTER 14

The Nitroso- and the Hydroxyimino-group, and the Nitro-group

(a) Nitroso-group

The nitroso-group and the hydroxyimino-group bear the same relation to each other as the keto-group and the enol group:

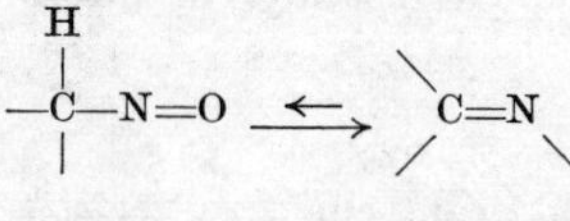

$$\underset{\text{Nitroso}}{-\overset{\displaystyle H}{\underset{|}{\overset{|}{C}}}-N{=}O} \rightleftarrows \underset{\text{Hydroxyimino}}{\gt C{=}N\diagdown OH}$$

The equilibrium lies so far on the side of the hydroxyimino-group that primary or secondary aliphatic nitroso-compounds exist only as transient intermediates. Aliphatic nitroso-compounds with the nitroso-group attached to a tertiary carbon are stable, the lower members existing as deep blue liquids at room temperature. However, greater interest lies in the chemistry of the aromatic nitroso-compounds, especially nitrosobenzene. On the basis of its electronic structure we should expect the nitroso-group to have much in common with the carbonyl group:

$$\underset{\text{Carbonyl group}}{\gt \overset{\delta+}{C}{=}\overset{\delta-}{\ddot{O}}:} \qquad \underset{\text{Nitroso-group}}{-\overset{\delta+}{\ddot{N}}{=}\overset{\delta-}{\ddot{O}}:}$$

We find indeed the nitroso-group undergoes the addition reaction with nucleophiles so characteristic of aldehydes and ketones. For example, nitrosobenzene reacts with aniline to give azobenzene, a reaction which should be compared with that of benzaldehyde and aniline to yield the Schiff's base (benzylideneaniline).

$$C_6H_5\ddot{N}H_2 + C_6H_5{-}N{=}O \longrightarrow C_6H_5{-}\underset{\displaystyle O^-}{\underset{|}{N}}{-}\overset{+}{N}H_2C_6H_5 \xrightarrow{-H_2O} C_6H_5{-}N{=}N{-}C_6H_5$$

Azobenzene

$$\text{Cf.}\quad C_6H_5\ddot{N}H_2 + C_6H_5CH{=}O \longrightarrow C_6H_5{=}\overset{\displaystyle H}{\overset{|}{\underset{\displaystyle O^-}{\underset{|}{C}}}}{-}\overset{+}{N}H_2C_6H_5 \xrightarrow{-H_2O}$$

$$C_6H_5{-}CH{=}NC_6H_5$$

Benzylidene aniline

In a similar way nitrosobenzene reacts with hydroxylamine yielding, among other products, a diazonium salt:

$$C_6H_5{-}N{=}O + NH_2OH \longrightarrow C_6H_5{-}N{=}N{-}OH$$

$$\rightleftharpoons C_6H_5\overset{+}{N}{\equiv}N + OH^-$$

$$\text{Cf.}\quad (C_6H_5)_2C{=}O + NH_2OH \longrightarrow (C_6H_5)_2C{=}NOH$$

With phenylhydroxylamine nitrosobenzene yields azoxybenzene:

$$C_6H_5{-}N{=}O + \underset{\text{Phenyl-hydroxylamine}}{C_6H_5NHOH} \longrightarrow \underset{\text{Azoxybenzene}}{C_6H_5{-}N{=}\underset{\displaystyle O^-}{\underset{|}{\overset{+}{N}}}{-}C_6H_5}$$

Nitrosobenzene reacts with diethyl malonate in the presence of a base, to yield an imine, this reaction being analogous to that of benzaldehyde with ethyl malonate in the Knoevenagel reaction (see p. 100):

$$C_6H_5{-}N{=}O + CH_2(CO_2Et)_2 \xrightarrow{Na^+\ ^-OEt} C_6H_5N{=}C(CO_2Et)_2$$

In all these reactions nucleophiles add to the nitrogen atom of the nitroso-group, just as they add to the carbon atom of the carbonyl group. A different type of addition is that in which nitric oxide adds to nitrosobenzene and ultimately yields the diazonium nitrate, the mechanism probably being as illustrated on page 164.

Nitric oxide has an unpaired electron and behaves as a stable free radical. In the present reaction sequence the first step is addition of

$$C_6H_5{-}N{=}O + 2NO \longrightarrow C_6H_5{-}\overset{+}{N}{\equiv}N \quad NO_3^-$$

$$Ar{-}N{=}O + NO\cdot \longrightarrow Ar{-}\underset{\substack{|\\ O\cdot}}{N}{-}N{=}O \xrightarrow{NO\cdot}$$

$$Ar{-}N(O){-}N{=}O \cdots N{=}O \longrightarrow \left[Ar{-}\ddot{N}{=}\ddot{N}{-}\ddot{O}{-}\overset{+}{N}(O^-){=}O \longleftrightarrow Ar\overset{+}{N}{\equiv}N \ :\ddot{O}{-}\overset{+}{N}(O^-){=}O \right]$$

this radical to the nitrogen–oxygen double bond of the nitroso-compound, thus generating a new radical which quickly reacts with a further molecule of nitric oxide. The addition compound then rearranges to give the diazonium nitrate. We have shown the reaction as involving free nitric oxide, but in solution the nitric oxide may be combined with a chlorine atom in the form of nitrosyl chloride or, more usually, in the form of N_2O_3. Thus, when an aromatic nitroso-compound is treated with an aqueous solution of nitrous acid, it is converted into the diazonium salt.

Nitroso-compounds are blue or blue-green in the liquid state or in solution. Solid nitroso-compounds are, however, colourless or yellow. This is due to the formation of dimers; both *cis*- and *trans*-dimers are known.

$$\begin{array}{c} R \quad O^- \\ \diagdown \; \diagup \\ \overset{+}{N} \\ \| \\ \underset{+}{N} \\ \diagup \; \diagdown \\ {}^-O \quad R \end{array} \qquad\qquad \begin{array}{c} R \quad O^- \\ \diagdown \; \diagup \\ \overset{+}{N} \\ \| \\ \underset{+}{N} \\ \diagup \; \diagdown \\ R \quad O^- \end{array}$$

trans *cis*

(b) Oximes

We have said that the hydroxyimino-group* is the enol form of the nitroso-compound. The first characteristic of these compounds that we must notice is that there are two geometric isomers analagous to the *cis*- and *trans*-isomers of the olefins.

* The HON= group is properly called hydroxyimino (an HN= group is an imino-group), but compounds containing a $\rangle C{=}NOH$ grouping are called oximes.

$$\begin{array}{c} R \quad R' \\ \backslash\;/ \\ C \\ \| \\ N \\ / \\ HO \end{array} \qquad \begin{array}{c} R \quad R' \\ \backslash\;/ \\ C \\ \| \\ N \\ \backslash \\ OH \end{array}$$

When an unsymmetrical ketone is treated with hydroxylamine, two isomeric oximes may be formed, though in a number of cases the proportion of one so exceeds that of the other that only one form is isolated.

Oximes are, as might be expected, amphoteric:

$$R_2C{=}N{-}OH \rightleftharpoons R_2C{=}N{-}O^- + H^+ \qquad K_a \approx 10^{-11}$$

$$R_2C{=}N{-}OH + H^+ \rightleftharpoons R_2C{=}N{-}\overset{+}{O}H_2 \qquad K_b \approx 10^{-12}$$

They undergo addition reactions, the most important of which is hydrogenation, yielding ultimately amines (cf. Chapter 18 of Part 1):

$$(H_3C)_2C{=}N{-}OH \xrightarrow{[H]} (H_3C)_2CHNH_2$$

Oximes also add halogens. However, instead of yielding an addition compound (the dihalogeno-compound), this reaction yields a halogeno-nitroso-compound:

$$(H_3C)_2C{=}N{-}O{-}H + Br{-}Br \longrightarrow (H_3C)_2C(Br){-}\overset{+}{N}{=}O{-}H \;\; Br^- \longrightarrow (CH_3)(H_3C)C(Br){-}N{=}O + HBr$$

Oximes also react with acid chlorides or acid anhydrides, to give *O*-acyl derivatives:

$$R_2C{=}N{-}OH \xrightarrow{(CH_3CO)_2O} R_2C{=}N{-}OCOCH_3$$

The way oximes behave when treated with strong acids under dehydrating conditions is of particular importance. A ketoxime (i.e.

an oxime of a ketone), when treated with a strong acid, undergoes a molecular rearrangement to yield an amide:

$$\underset{\text{R, R}'}{C}{=}N{-}OH + H^+ \rightleftharpoons R(R')C{=}N{-}\overset{+}{O}H_2 \longrightarrow \left(R^{+}\cdots C(R'){=}N\cdots OH_2 \right) \longrightarrow R'{-}C^+{=}N{-}R \xrightarrow{:OH_2}$$

$$R'C(\overset{+}{O}H_2){=}N{-}R \longrightarrow R'C({=}O){-}N(R)H + H^+$$

The reaction sequence should be carefully compared with that drawn for the pinacol–pinacolone change in Chapter 10. The driving force behind the pinacol–pinacolone change was the formation of an oxonium rather than a carbonium ion. In the present reaction the driving force is the formation of a carbonium ion rather than an iminium ion (i.e. nitrogen with six electrons in its outer shell is clearly of higher energy than a carbon atom with six electrons in its outer shell). This reaction is known as the Beckmann rearrangement. It is quite general and, besides sulphuric acid, other powerful dehydrating agents can be used. For example, when acetophenone oxime is treated with phosphorus pentachloride and the mixture is then drowned in water, the product is acetanilide. In this process the phosphorus pentachloride forms a phosphorochloridate ester which behaves as the leaving group. Note that we draw the migrating

$$H_3C(C_6H_5)C{=}N{-}OH \xrightarrow[2.\ H_2O]{1.\ PCl_5} H_3C{-}C({=}O){-}NHC_6H_5$$

Acetophenone oxime — Acetanilide

group as coming from the opposite side to the leaving group. This is clearly what we should expect when we compare this reaction with all our discussions of substitution reactions. If an unsymmetrical ketone forms two different oximes, then their rearrangement will give two different amides:

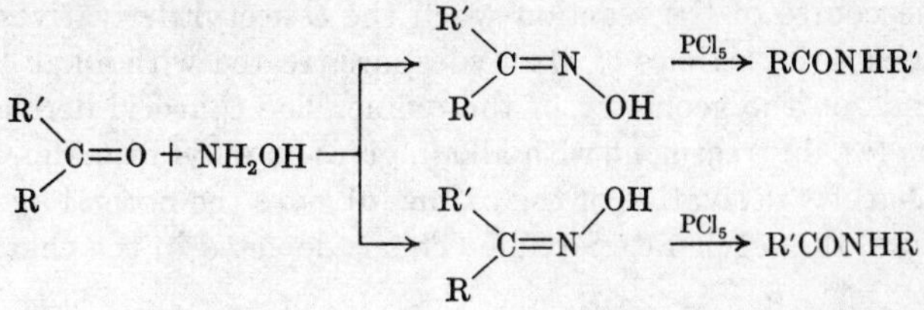

Proof that the reaction really is a *trans*(or *anti*)-process was provided by Meisenheimer in a very ingenious experiment. He made the two oximes (**2**) and (**3**) of a methyl naphthyl ketone (**1**). One of the oximes could be resolved into optical enantiomorphs, whereas the other could not. Examination of the structures shows that (**2**) can exist in two enantiomorphs. Rotation about the naphthyl carbonyl bond in (**2**) is restricted owing to steric hindrance between the hydroxyl group of the oxime and both the phenolic hydroxyl group at position 2 and the hydrogen atom at position 8 of the naphthalene ring. Thus the geometry of the two oximes was established. When these compounds were rearranged, oxime (**2**) gave the amide (**4**) in which the methyl group had migrated from the carbon atom of the oxime link to the nitrogen; whereas oxime (**3**) gave

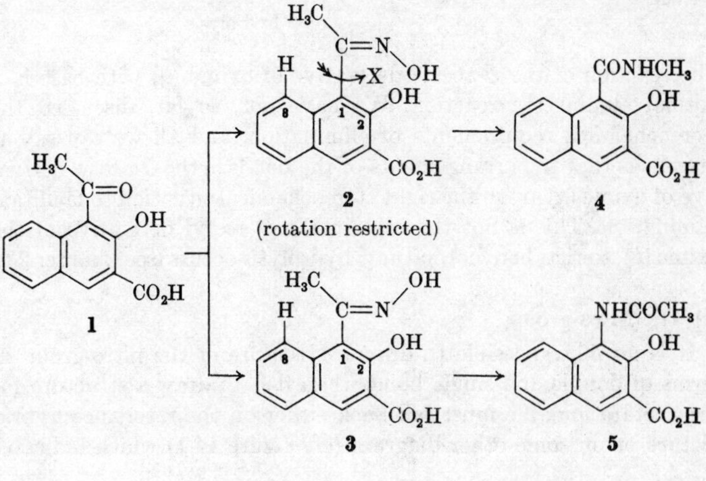

1 **2** (rotation restricted) **3** **4** **5**

amide (**5**) in which the naphthyl group had migrated from the carbon to the nitrogen.

The course of the reaction when the *O*-acetyl derivatives of the aldoximes (i.e. oximes of aldehydes) are treated with alkali likewise depends on the geometry of the oxime. The *O*-acetyl derivative of oxime (*a*), on treatment with alkali, gives an alkyl cyanide, whereas the *O*-acetyl derivative of the oxime (*b*) gives the normal hydrolysis product, the regenerated oxime. This is depicted in the chart.

$$R(H)C{=}O \longrightarrow \underset{(a)}{R(H)C{=}N{-}OH} \xrightarrow{(CH_3CO)_2O} R(H)C{=}N{-}O{-}C(=O)CH_3 \xrightarrow{\bar{O}H} R{-}C{\equiv}N + {}^{-}O_2C{-}CH_3 + H_2O$$

E_2 Elimination reaction

$$R(H)C{=}O \longrightarrow \underset{(b)}{R(H)C{=}N{-}OH} \xrightarrow{(CH_3CO)_2O} R(H)C{=}N{-}O{-}C(=O)CH_3 \xrightarrow{\bar{O}H} R(H)C{=}N{-}OH + {}^{-}O_2C{-}CH_3$$

Acyl hydrolysis

The reaction of the *O*-acetyl derivative of oxime (*a*) with base is a normal elimination reaction. We have not, so far, discussed the stereochemical requirements of elimination, and all we can say at present is that the arrangements of the bonds in the *O*-acetyl derivative of oxime (*a*) are in the right stereochemical situation to facilitate elimination. This is not the case in the *O*-acetyl derivative of the oxime (*b*), so that here normal acyl hydrolysis occurs (see Chapter 23).

(c) The Nitro-group

It is no more possible to draw a structure of the nitro-group in terms of double and single bonds than it is to draw a structure for the acetate ion. We must fall back either on the resonance hybrid picture or on some other diagram (cf. Figure 14.1) which indicates electron delocalization (cf. Chapter 7). Note that according to resonance theory the nitrogen atom in a nitro-group carries a positive charge. We should, therefore, expect the nitro-group to

$$\mathrm{R{-}\overset{+}{N}(=\ddot{O}\colon)({-}\ddot{O}\colon^-)} \longleftrightarrow \mathrm{R{-}\overset{+}{N}({-}\ddot{O}\colon^-)(=\ddot{O}\colon)}$$

Resonance structures for the nitro-group

behave as a very powerful electron-attractor. Even more important are the electron-accepting properties of the nitro-group discussed in Chapter 7. The importance of these accepting properties of the nitro-group were also apparent in the stability of the carbanions formed by treating aliphatic nitro-compounds with sodium ethoxide

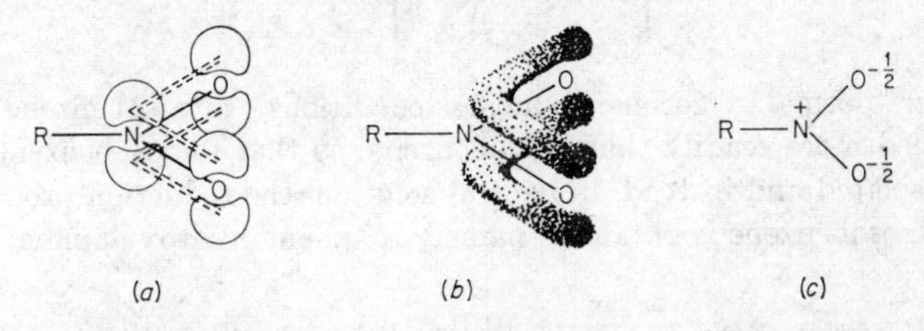

Figure 14.1. Orbital pictures for the nitro-group: (*a*) interacting *p* atomic orbitals, (*b*) π orbital 'electron clouds', and (*c*) charge distribution.

(see Chapter 9), in nucleophilic addition reactions (see Chapter 13) and in nucleophilic addition-with-elimination reactions of aromatic compounds (see Chapter 13). In Chapter 12 we discussed the tautomerism observed with the aliphatic nitro-compounds and drew attention to the fact that the *aci*-form is only present in very low concentrations under normal conditions.

$$\mathrm{C_6H_5{-}CH_2{-}\overset{+}{N}(=O)({-}O^-)} \rightleftarrows \mathrm{C_6H_5{-}CH{=}\overset{+}{N}(OH)({-}O^-)}$$

Nitro-form — Pale yellow liquid; *aci*-form — Colourless solid

The nitro-group does not undergo ordinary addition reactions, but the products of reduction and hydrogenation are extremely important. First and foremost comes the reduction of aromatic nitro-compounds to aromatic amines under acid conditions:

$$C_6H_5{-}NO_2 \xrightarrow[\text{Acid}]{[H]} C_6H_5NH_2$$

In neutral or alkaline conditions the reduction products can be very varied. The first step in the hydrogenation is the formation of the nitroso-compound:

$$C_6H_5{-}\overset{+}{N}(=O){-}O^- \xrightarrow{2[H]} \left[C_6H_5{-}N(O{-}H)(\ddot{O}{-}H) \right] \longrightarrow C_6H_5{-}N{=}O + H_2O$$

Under normal vigorous reducing conditions, nitrosobenzene is reduced more readily than nitrobenzene, so that nitrosobenzene is not isolated and instead, in neutral solution, the reduction product of nitrosobenzene is obtained, namely, *N*-phenylhydroxylamine:

$$C_6H_5{-}NO_2 \xrightarrow[\text{neutral soln.}]{\text{Zn dust}} (C_6H_5NO) \longrightarrow \underset{\textit{N}\text{-Phenyl-hydroxylamine}}{C_6H_5{-}NHOH}$$

In alkaline solution, with mild reducing agents, the nitroso-compound reacts with some of the phenylhydroxylamine—a reaction described above—to yield azoxybenzene:

$$2\,C_6H_5NO_2 \xrightarrow[\text{alkali}]{\text{Mild reduction}} C_6H_5NO + C_6H_5NHOH \longrightarrow \underset{\text{Azoxybenzene}}{C_6H_5N{=}\overset{+}{N}(O^-)C_6H_5}$$

Azoxybenzene can readily be reduced further to azobenzene and/or hydrazobenzene:

$$C_6H_5{-}N{=}\overset{+}{N}(O^-){-}C_6H_5 \xrightarrow{[H]} \underset{\text{Azobenzene}}{C_6H_5N{=}NC_6H_5} \xrightarrow{[H]} \underset{\text{Hydrazobenzene}}{C_6H_5NHNHC_6H_5}$$

Problem

Suggest intermediate steps for the following sequences:

(*a*) $CH_3CH_2COCH_3 \longrightarrow CH_3C(=NOH)\text{—}C(=NOH)CH_3$

(*b*) $C_6H_5COC_6H_5 \longrightarrow C_6H_5CONHC_6H_5$

(*c*) $C_6H_5NO \longrightarrow C_6H_5N{=}NC_6H_5$

CHAPTER 15

The Azo-, Diazo-, Diazonium, and Azide Groups

$$\underset{\text{Azo}}{R{-}N{=}N{-}R} \qquad \underset{\text{Diazo}}{R_2C{=}\overset{+}{N}{=}\overset{-}{N}} \qquad \underset{\text{Diazonium}}{R{-}\overset{+}{N}{\equiv}N} \qquad \underset{\text{Azide}}{R{-}N{=}\overset{+}{N}{=}\overset{-}{N}}$$

(a) Azo-compounds

The stability of an azo-, diazo-, diazonium, or azide group depends to a very large extent on the nature of the carbon residue to which it is attached. If the remainder of the molecule contains orbitals of π symmetry which can interact with the π orbitals of the nitrogen–nitrogen bond, then the molecule is very much more stable than if the remainder of the molecule is of a saturated aliphatic type. Thus azomethane, a gas, is readily decomposed by heat or light into two methyl radicals and a nitrogen molecule:

$$CH_3{-}N{=}N{-}CH_3 \longrightarrow CH_3\cdot + N_2 + CH_3\cdot$$

but azobenzene ($C_6H_5N{:}NC_6H_5$), is a very stable compound which melts at 68° and distils without decomposition.

Azo-compounds have a nitrogen–nitrogen double bond and thus, as expected, undergo addition reactions. The most important of these is hydrogenation to the symmetrical disubstituted hydrazine. This can usually be hydrogenated further to yield the corresponding amine by fission:

$$R{-}N{=}N{-}R \underset{HgO}{\overset{[H]}{\rightleftharpoons}} R{-}NH{-}NHR \xrightarrow{[H]} 2\,RNH_2$$

The conversion of the azo-compound into the symmetrical hydrazine can be reversed: the hydrazine can be oxidized to the azo-compound by mercuric oxide. This is an important reaction and is the way in which aliphatic azo-compounds are usually made.

Azodiisobutyronitrile is used as an initiator for radical polymerization and also as a 'blowing agent' in the production of plastic foam. It is prepared by just such a reaction:

$$(CH_3)_2C{=}O + NaCN + {}^{+}NH_3NH_2Cl^{-} \longrightarrow \left[(CH_3)_2\underset{\displaystyle OH}{\underset{|}{C}}{-}NH{-}\right]_2 \xrightarrow{^{-}CN}$$

$$\left[(CH_3)_2\underset{\displaystyle CN}{\underset{|}{C}}{-}NH{-}\right]_2 \xrightarrow[\text{or NaOCl}]{\text{HgO}} (CH_3)_2\underset{\displaystyle CN}{\underset{|}{C}}{-}N{=}N{-}\underset{\displaystyle CN}{\underset{|}{C}}(CH_3)_2$$

Azodiisobutyronitrile

As we should expect, azo-compounds can exist in geometrically isomeric forms. Azobenzene, for example, normally occurs in the *trans*-form, but irradiation with ultraviolet light converts this into the *cis*-form.

$$C_6H_5{-}N{=}N{-}C_6H_5 \xrightarrow{h\nu} C_6H_5{-}N{=}N{-}C_6H_5$$

trans m.p. 68.0°, $\mu = 0$ *cis* m.p. 71.4°, $\mu = 3.0$ D

Azo-compounds are all coloured substances; aliphatic azo-compounds usually are pale yellow and aromatic azo-compounds are darker, azobenzene for example being orange. Aromatic azo-compounds are of great technical importance in the dyestuffs industry. Azo-dyes form by far the largest single class of synthetic coloured materials used. They are principally formed by coupling diazonium salts with phenols. The commonest azo-dyes are red or yellow although azo-dyes are known and used for almost every shade.

(b) Diazo-compounds

In Part 1, when discussing the reactions of aliphatic amines with nitrous acid, we described how a diazonium salt was produced which usually decomposed rapidly to a carbonium ion and nitrogen. In certain cases, instead of losing nitrogen, the unstable aliphatic diazonium salt eliminated a proton to give a diazo-compound; for example, glycine ethyl ester (ethyl aminoacetate), when treated with nitrous acid, yields ethyl diazoacetate:

$$NH_2CH_2CO_2Et \xrightarrow{HNO_2} [N{\equiv}\overset{+}{N}{-}CH_2CO_2Et] \xrightarrow{-H^+} \overset{-}{N}{=}\overset{+}{N}{=}CHCO_2Et$$

Ethyl diazoacetate

It is not possible to write a classical structure for diazo-compounds without the inclusion of positive and negative charges. The canonical

$$\left[\rangle C{=}\overset{+}{N}{=}\overset{-}{\ddot{N}}: \longleftrightarrow \rangle \overset{-}{\ddot{C}}{-}\overset{+}{N}{\equiv}N: \right]$$

(*a*) (*b*)

form (*b*) shows the relation between a diazonium salt and a diazo-compound. The factor which determines whether or not treating an aliphatic amine with nitrous acid causes loss of nitrogen and formation of a carbonium ion, or loss of a proton and formation of a diazo-compound, is the nature of the substituent adjacent to the amino-group. If this is electron-attracting or electron-accepting, the diazonium salt becomes stabilized by loss of a proton; in the absence of an electron-acceptor or electron-attractor, the diazonium salt loses nitrogen to give the transient carbonium ion. As well as glycine ethyl ester, the following aliphatic amines yield diazo-compounds when treated with nitrous acid:

$$C_6H_5COCH_2NH_2 \xrightarrow{HNO_2} C_6H_5COCH{=}\overset{+}{N}{=}\overset{-}{N} \quad (C_6H_5CO \text{ an acceptor group})$$

$$CF_3CH_2NH_2 \xrightarrow{HNO_2} CF_3CH{=}\overset{+}{N}{=}\overset{-}{N} \quad (CF_3 \text{ an attractor group})$$

This does not mean to say that diazo-compounds lacking electron-acceptor or -attractor groups cannot be prepared. Such compounds

$$R'CONHCH_2R \xrightarrow{HNO_2} R'C(=O){-}N(N{=}O){-}CH_2R \xrightarrow{^-OH} R'C(OH)(O^-){-}N(N{=}O){-}CH_2R \longrightarrow$$

N-Nitroso-derivative

$$R'C(=O)OH + RCH_2{-}N{=}N{-}O^- \xrightarrow{H-OH} RCH{=}N{-}\ddot{N}{-}OH \longrightarrow RCH{=}\overset{+}{N}{=}\overset{-}{\ddot{N}}: + OH^-$$

are best prepared by treating the acyl derivative of the corresponding amine with nitrous acid and the resulting *N*-nitroso-compound with alkali, the reaction taking the course shown on p. 174.

Diazo-compounds can also be prepared by oxidation of the corresponding hydrazone in a reaction analogous to that described for azo-compounds:

$$R_2C{=}O \xrightarrow{NH_2NH_2} R_2C{=}NNH_2 \xrightarrow{HgO} R_2C{=}\overset{+}{N}{=}\overset{-}{N}$$

The most important diazo-compound is diazomethane CH_2N_2, which is conveniently prepared from *N*-methyl-*p*-toluenesulphonamide:

$$H_3C{-}C_6H_4{-}SO_2NHCH_3 \xrightarrow{HNO_2} H_3C{-}C_6H_4{-}SO_2N(NO)CH_3 \xrightarrow{KOH} CH_2N_2 + H_3C{-}C_6H_4{-}SO_3^-K^+$$

N-Methyl-*p*-toluene-sulphonamide

Diazomethane is a yellow, explosive, and poisonous gas and is normally handled for organic preparative purposes in ether solution. The reactions of aliphatic diazo-compounds are those that we should expect from inspection of the electronic distribution in the two canonical forms making up the resonance hybrid:

$$[RCH{=}\overset{+}{N}{=}\overset{-}{\ddot{N}}{:} \longleftrightarrow R\overset{-}{\ddot{C}}H{-}\overset{+}{N}{\equiv}N{:}]$$

Acids convert the diazo-compound into the diazonium salt and, as we have seen, aliphatic diazonium salts decompose to give nitrogen and a carbonium ion which may rearrange (see Chapter 10). In aqueous acid, alcohols are probable products, but in very concentrated acid solution derivatives formed by reaction of the carbonium ion with the anion of the acid are also obtained:

$$RCH{=}\overset{+}{N}{=}\overset{-}{N} \xrightarrow{HX} (RCH_2{-}\overset{+}{N}{\equiv}N\quad X^-) \longrightarrow (RCH_2^+ + N_2 + X^-)$$

$$(RCH_2^+ + N_2 + X^-) \xrightarrow{H_2O} RCH_2OH + H^+$$

$$(RCH_2^+ + N_2 + X^-) \xrightarrow{X^-} RCH_2X$$

The reaction of diazomethane with acids is extremely important since the methyl cation cannot rearrange and the product of the reaction is the methyl derivative of the acid or phenol, and it is used extensively to prepare methyl esters and ethers. The re-

$$\underset{\text{Phenol}}{C_6H_5OH} \xrightarrow{CH_2N_2} \underset{\text{Anisole}}{C_6H_5OCH_3}$$

$$\underset{\text{Phenylacetic acid}}{C_6H_5CH_2CO_2H} \xrightarrow{CH_2N_2} \underset{\text{Methyl phenylacetate}}{C_6H_5CH_2CO_2CH_3}$$

$$\underset{\text{Ethyl acetoacetate}}{(CH_3\underset{\underset{O}{\|}}{C}CH_2CO_2Et} \rightleftharpoons CH_3\underset{\underset{OH}{|}}{C}{=}CHCO_2Et) \xrightarrow[\text{(with the enol)}]{CH_2N_2} CH_3\underset{\underset{OCH_3}{|}}{C}{=}CHCO_2Et$$

action with enols to produce *O*-alkyl derivatives (i.e. vinyl ethers) is also important since it is the easiest way in which these compounds can be prepared.

In the absence of acids, diazomethane can behave as a nucleophile and add to carbonyl bonds, e.g. with carboxylic acid chlorides it yields diazo-ketones:

$$RC(=O)Cl + [\bar{\ddot{C}}H_2{-}\overset{+}{N}{\equiv}N: \longleftrightarrow CH_2{=}\overset{+}{N}{=}\overset{-}{\ddot{N}}:] \longrightarrow R\overset{Cl}{\underset{\bar{O}}{C}}{-}CH_2{-}\overset{+}{N}{\equiv}N \longrightarrow R\underset{\underset{O}{\|}}{C}{-}CH{=}\overset{+}{N}{=}\overset{-}{N} + HCl$$

If there is insufficient diazomethane, the diazo-ketone reacts with the hydrogen chloride formed to yield the corresponding chloro-ketone together with other decomposition products. In the presence of an excess of diazomethane, however, the hydrogen chloride reacts as fast as it is formed with diazomethane and good yields of diazo-ketone can be isolated. Diazo-ketones are important because in the presence of silver ions they rearrange, losing nitrogen to yield ketene derivatives:

$$\left[\begin{array}{c} R-C(=O)-CH=\overset{+}{N}=\overset{-}{N}: \end{array} \longleftrightarrow \begin{array}{c} R-C(-\overset{-}{O}:)=CH-\overset{+}{N}\equiv N: \end{array} \right] \longrightarrow O=C=C\begin{matrix} R \\ H \end{matrix} + N_2$$

A ketene

This rearrangement is known as the Wolff rearrangement. Notice that mechanistically it is very similar to the pinacol–pinacolone change and to the Beckman rearrangement already discussed. The product of this rearrangement is a ketene which has cumulative double bonds* and represents a grouping whose reactions we have not yet described (see Chapter 21). All that need concern us at the moment is that ketene reacts extremely rapidly with compounds containing a hydroxyl group, so that with water the ketene formed in the Wolff rearrangement yields the corresponding carboxylic acid and with ethanol the ethyl ester:

$$\begin{matrix} R \\ H \end{matrix}\!\!>C=C=O \begin{matrix} \xrightarrow{H_2O} RCH_2CO_2H \\ \xrightarrow{EtOH} RCH_2CO_2Et \end{matrix}$$

We thus have a sequence for converting an acid into the next acid of the homologous series:

$$RCO_2H \xrightarrow{SOCl_2} RCOCl \xrightarrow{CH_2N_2} RCOCHN_2 \xrightarrow[EtOH]{Ag^+} RCH_2CO_2C_2H_5$$

This reaction sequence is usually known as the Arndt–Eistert synthesis.

Since diazomethane adds to a carboxylic acid chloride we might also expect it to add to the carbonyl group in a ketone. It does, but the reaction products can be complicated by rearrangements that occur when the initial adduct loses nitrogen (as illustrated on p. 178). The formation of these products is clearly analogous to the Wolff rearrangement of the diazo-ketone.

* Cumulative double bonds are those which join at least three contiguous atoms, A=B=C. Conjugated double bonds are those which are separated by a single bond, A=B—C=D. The atoms may, but need not, be the same (see Part 1, Chapter 13).

$$RR'C{=}O + \bar{C}H_2{-}\overset{+}{N}{\equiv}\bar{N} \longrightarrow \left(RR'C(O^-){-}CH_2\overset{+}{N}_2 \right) \begin{cases} \xrightarrow{\text{R' migrates}} R(R'H_2C)C{=}O \\ \xrightarrow{\text{R migrates}} (RH_2C)R'C{=}O \\ \xrightarrow{\text{Ring closure}} RR'C{-}CH_2 \text{ (epoxide, ring through O)} \end{cases}$$

In Part 1 we described the addition of permanganate ions and of osmium tetroxide to carbon–carbon double bonds, so-called electrocyclic reactions (or four-centre reactions in Part 1). Diazoalkanes add to olefinic and acetylenic double bonds in a very similar electrocyclic reaction:

$$>C{=}C< + \overset{\text{VIII}}{Os}O_4 \longrightarrow \text{cyclic } >C{-}O{-}\overset{\text{VI}}{Os}O_2{-}O{-}C< \quad \text{(cf. Part I, page 62)}$$

$$>C{=}C< + R_2C{=}\overset{+}{N}{=}\bar{N} \longrightarrow \text{pyrazoline ring: } C{-}CR_2{-}N{=}N{-}C$$

$$-C{\equiv}C- + R_2C{=}\overset{+}{N}{=}\bar{N} \longrightarrow \text{ring: } C{=}C{-}N{=}N{-}CR_2$$

The synthesis of heterocyclic compounds is outside the scope of our present discussion but we shall conclude by giving just one specific example:

$$CH_2N_2 + HC{\equiv}CH \longrightarrow \text{Pyrazole (ring with N and NH)}$$

In Chapter 22 we shall discuss the reactive intermediates obtained when aliphatic diazo-compounds are heated or photolysed.

(c) **Aromatic Diazonium Salts**

In Part 1 some reactions of inorganic acids with organic compounds were described (Chapter 10), including the reaction of nitrous acid with primary amines and the resulting formation of diazonium salts, and it was explained also (Chapter 14) how, although aliphatic diazonium salts decomposed spontaneously to give carbonium ions, aromatic diazonium salts are moderately stable. The preparation of a diazonium salt, by treating an amine with nitrous acid, involves formation of the *N*-nitroso-amine, which rearranges immediately to the diazonium hydroxide:

$$\mathrm{Ar{-}NH_2} \xrightarrow[-\mathrm{H_2O}]{\mathrm{HNO_2}} \mathrm{Ar{-}NH{-}NO} \xrightarrow{\text{Fast}} \mathrm{Ar{-}N{=}N{-}OH}$$

$$\xrightarrow{\text{Fast}} \mathrm{Ar\overset{+}{N}{\equiv}N\ {}^{-}OH}$$

The kinetics of diazotization are very complicated because the nitrosating agent may be nitrous acid, dinitrogen trioxide N_2O_3, or, if the nitrous acid is prepared by treating an aqueous solution of sodium nitrite with dilute hydrochloric acid, nitrosyl chloride NOCl. Nitrosyl chloride and dinitrogen trioxide are more reactive nitrosating agents than nitrous acid itself. Diazotization is normally carried out in dilute aqueous solution between 0° and 5°. At higher temperatures the diazonium salt may start to decompose. Weakly basic amines such as 2,4-dinitroaniline require strongly acid solutions in which the much more reactive nitrosonium ion NO^+ is present.

The only other important method for preparing diazonium salts involves the reaction of nitrous acid with an aromatic nitroso-compound (cf. Chapter 14). Reactive aromatic nuclei (e.g. aromatic

$$\mathrm{C_6H_5OH} \xrightarrow{\mathrm{HNO_2}} \mathrm{HO{-}C_6H_4{-}NO} \rightleftharpoons \mathrm{O{=}C_6H_4{=}N{-}OH} \xrightarrow{2\mathrm{HNO_2}} \left[\, {}^{-}\ddot{O}{-}C_6H_4{-}\overset{+}{N}{\equiv}N: \;\longleftrightarrow\; \ddot{O}{=}C_6H_4{=}\overset{+}{N}{=}\ddot{N}:^{-} \,\right]$$

rings containing powerful donor groups, such as phenols and tertiary amines) react with nitrous acid to give *C*-nitroso-compounds; in the presence of an excess of nitrous acid these compounds react further to give good yields of the diazonium salt. Less reactive aromatic compounds can also be converted into diazonium salts through the nitroso-compounds, but the reaction sequence under these circumstances is more complicated.

Aromatic diazonium salts are true ionic salts and can sometimes be obtained crystalline. Simple mineral acid salts such as the diazonium chloride, nitrate, and sulphate are usually very soluble in water; the solid salts are liable to detonate, especially diazonium nitrates. Relatively stable diazonium salts can, however, be prepared. Important examples are the diazonium fluoroborates; these are considerably less soluble than the chlorides or nitrates, so that addition of fluoroboric acid or sodium fluoroborate to a solution of another diazonium salt results in the precipitation of the diazonium fluoroborate, which can be filtered off and stored as a dry solid material. Other stable diazonium salts include metallic salts such as the half zinc salt of *p*-dimethylaminobenzenediazonium chloride. This particular stabilized diazonium salt is used in the photocopying industry.

$$\left.\begin{array}{ll} \mathrm{Ar{-}\overset{+}{N}{\equiv}N} & \mathrm{Cl^-} \\ \mathrm{Ar{-}\overset{+}{N}{\equiv}N} & \mathrm{NO_3{}^-} \\ \mathrm{Ar{-}\overset{+}{N}{\equiv}N} & \mathrm{HSO_4{}^-} \end{array}\right\} \text{Very soluble in water; unstable in the solid state.}$$

$$\left.\begin{array}{ll} \mathrm{Ar{-}\overset{+}{N}{\equiv}N} & \mathrm{BF_4{}^-} \\ \mathrm{Ar{-}\overset{+}{N}{\equiv}N} & \mathrm{Cl^-,\ \tfrac{1}{2}\ ZnCl_2} \end{array}\right\} \text{Sparingly soluble in water; stable in the solid state}$$

The stability of a diazonium salt depends a great deal on the nature of the substituents in the benzene ring. Donor groups have a marked stabilizing effect:

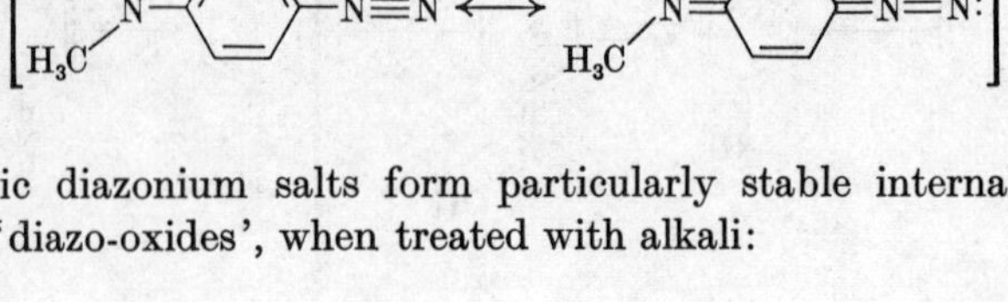

$$\left[(\mathrm{H_3C})_2\ddot{\mathrm{N}}{-}\mathrm{C_6H_4}{-}\overset{+}{\mathrm{N}}{\equiv}\ddot{\mathrm{N}} \longleftrightarrow (\mathrm{H_3C})_2\overset{+}{\mathrm{N}}{=}\mathrm{C_6H_4}{=}\overset{+}{\mathrm{N}}{=}\overset{-}{\ddot{\mathrm{N}}}{:}\right]$$

Phenolic diazonium salts form particularly stable internal salts, called 'diazo-oxides', when treated with alkali:

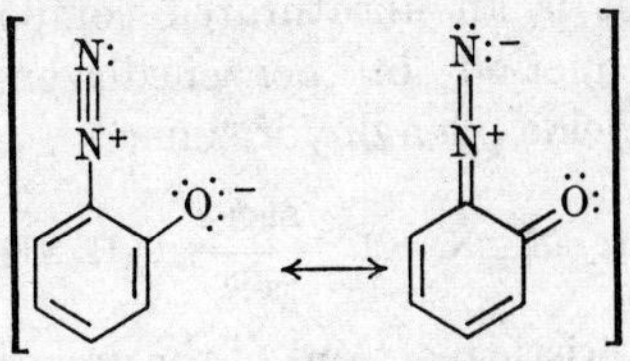

N: N:$^-$
N+ N+
Ö:$^-$ ⟷ Ö:

The behaviour of diazonium salts in alkaline media, as illustrated below, is very complicated. The coupling reaction so characteristic of

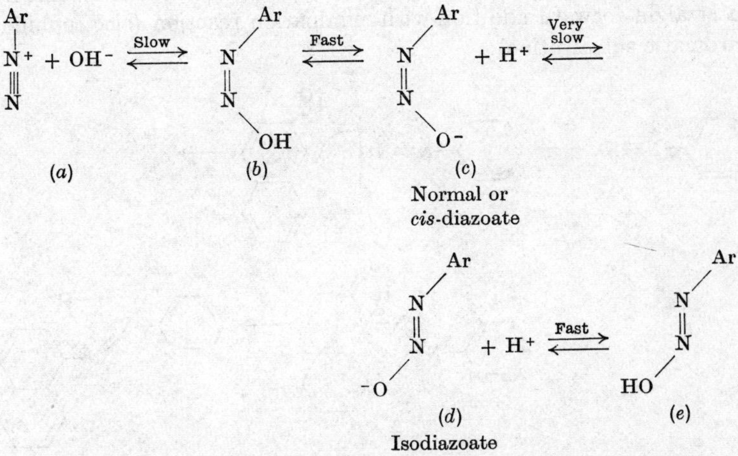

$$\mathrm{Ar{-}N^+{\equiv}N} + \mathrm{OH^-} \underset{}{\overset{\text{Slow}}{\rightleftharpoons}} \mathrm{Ar{-}N{=}N{-}OH} \overset{\text{Fast}}{\rightleftharpoons} \mathrm{Ar{-}N{=}N{-}O^-} + \mathrm{H^+} \overset{\text{Very slow}}{\rightleftharpoons}$$

(*a*) (*b*) (*c*)
Normal or
cis-diazoate

$$\mathrm{Ar{-}N{=}N{-}O^-} + \mathrm{H^+} \overset{\text{Fast}}{\rightleftharpoons} \mathrm{Ar{-}N{=}N{-}OH}$$

(*d*) (*e*)
Isodiazoate

aromatic diazonium salts (see below) is due to the diazonium cation (*a*). In the presence of strong alkali, the diazo-hydroxide (*b*) is formed which ionizes to yield the 'normal' or '*cis*-diazoate' ion. On acidification, the normal diazoate quite rapidly regenerates the diazonium cation (*a*). In some cases the sodium salt of the normal diazoate* is only moderately soluble and can be precipitated from an alkaline solution of the diazonium salt. The normal diazoate may sometimes be slowly converted into the 'iso-' or '*trans*-diazoate' (*d*), this being usually accelerated by heat. The 'isodiazoate', on acidification, yields the isodiazo-hydroxide (*e*), which does *not* dissociate to yield the diazonium salt (*a*), so that acidification of the 'isodiazoate' does not yield a solution which will couple immediately.

* These compounds are usually called 'diazotates' in the older literature.

A diazonium salt is an unsaturated compound and it can be hydrogenated. Reduction of benzenediazonium chloride with stannous chloride yields phenylhydrazine:

$$C_6H_5—\overset{+}{N}\equiv N\ \ Cl^- \xrightarrow[\text{-HCl}]{SnCl_2} C_6H_5NHNH_2$$

The most characteristic reactions of the arenediazonium cation are those in which it behaves as an electrophile. The coupling with phenoxide anions to produce brightly coloured azo-dyes is both the most dramatic and industrially the most important of them. This is a straight-forward addition-with-elimination reaction (electrophilic aromatic substitution):

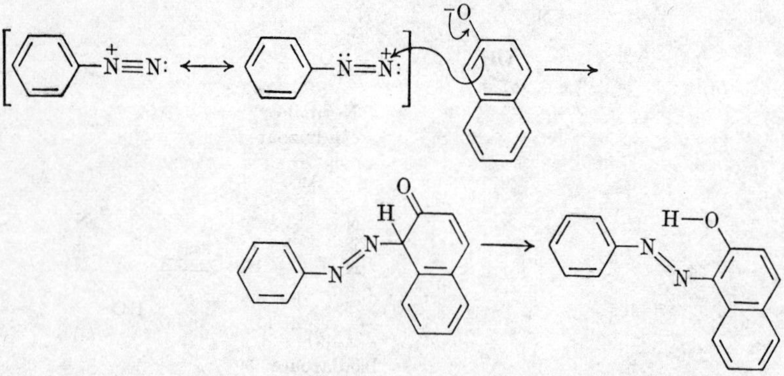

Bright red
benzeneazo-2-naphthol

Compared with the nitronium ion or the bromine cation, the diazonium cation is a very weak electrophile and, in general, arenediazonium salts couple only with phenoxide anions. Thus benzenediazonium chloride and 2-naphthol do not couple in acid solution; only on addition of alkali, with the resultant formation of the naphtholate anion, does coupling take place. In the presence of an excess of very strong alkali, however, the diazonium salt is converted into the diazoate and again no coupling occurs. The electrophilicity of a diazonium salt can be markedly affected by substituents in the benzene ring. Thus, *p*-methoxybenzenediazonium chloride will couple only with the phenoxide anion. Benzenediazonium chloride couples with trimethoxybenzene, while 2,4,6-trinitrobenzenediazonium chloride couples with the hydrocarbon, mesitylene (cf. Table 15.1).

Table 15.1. Effect of substituents on coupling of diazonium salts in alkaline solution

$Ar{-}N{\equiv}\overset{+}{N}$	Relative coupling rate with $C_6H_5O^-$	Couples with
$CH_3O{-}C_6H_4{-}\overset{+}{N}{\equiv}N$ (para)	0.1	$C_6H_5{-}O^-$
$C_6H_5{-}\overset{+}{N}{\equiv}N$	1	1,3,5-trimethoxybenzene (CH_3O, OCH_3, CH_3O)
$O_2N{-}C_6H_4{-}\overset{+}{N}{\equiv}N$ (para)	1,300	1,3-dimethoxybenzene (OCH_3, CH_3O)
2,4-dinitro: $O_2N{-}C_6H_3(NO_2){-}\overset{+}{N}{\equiv}N$	Too fast for measurement	$C_6H_5{-}OCH_3$
2,4,6-trinitro: $O_2N{-}C_6H_2(NO_2)_2{-}\overset{+}{N}{\equiv}N$	,, ,, ,, ,,	mesitylene (H_3C, CH_3, H_3C)

Diazonium salts, like other positively charged electrophiles, add to olefins, but the products can be complicated owing to polymerization and other side-reactions. Diazonium salts react, of course, with other nucleophiles, and benzenediazonium chloride couples readily with aniline:

$$C_6H_5{-}\overset{+}{N}{\equiv}N \quad H_2\ddot{N}{-}C_6H_5 \rightleftarrows C_6H_5{-}N{=}N{-}NH{-}C_6H_5 + H^+$$

Diazoaminobenzene

In the case of *N*,*N*-dimethylaniline the diazonium cation adds to the benzene ring (i.e. addition-with-elimination), and 'coupling' occurs at the *para*-position:

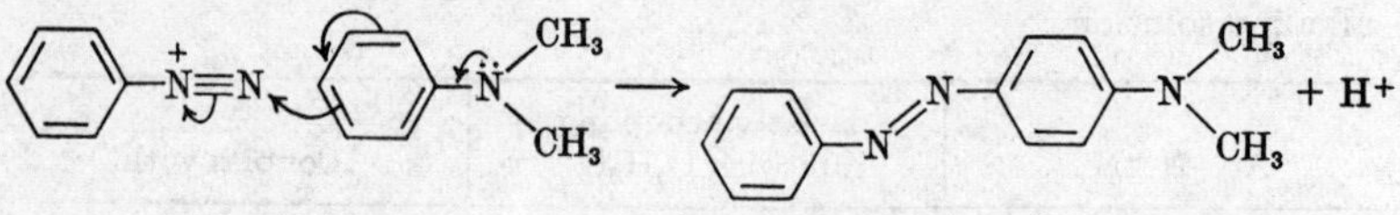

p-(*N*,*N*-Dimethylamino)azobenzene

We might expect, therefore, a limited amount of coupling to occur in the *para*-position of aniline itself. A little such reaction does occur. If diazoaminobenzene is treated with concentrated acid (notice that we have drawn the original coupling reaction as reversible), aniline and the diazonium salt are regenerated. On the other hand, treatment of aminoazobenzene $H_2N—C_6H_4—N{=}N—C_6H_5$ with acid does not result in regeneration of the diazonium salt, so that diazoaminobenzene can be rearranged in the presence of acid to aminoazobenzene:

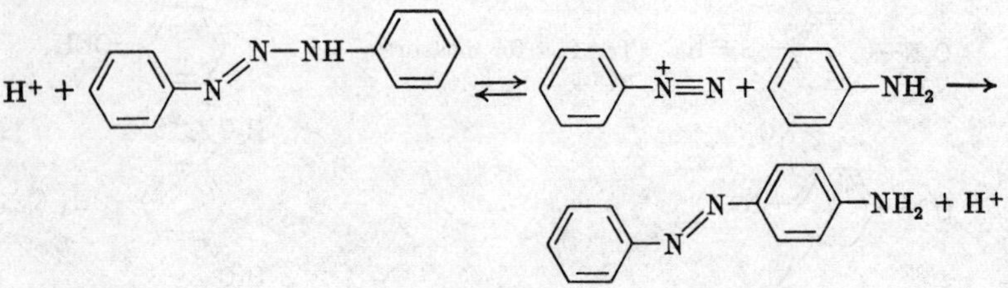

Coupling of the diazonium group with an amino-group can occur intramolecularly, and diazotization of *o*-phenylenediamine under normal conditions yields benzotriazole:

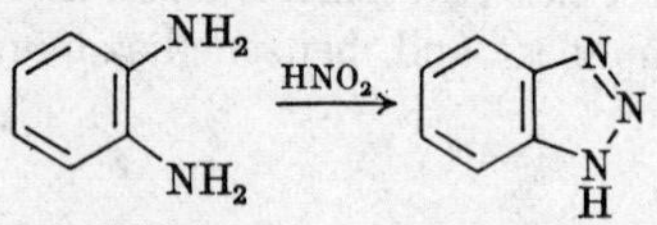

o-Phenylenediamine Benzotriazole

Benzenediazonium chloride couples with ammonia and with primary or secondary amines, but the products, which contain three nitrogen atoms in a row, are often very unstable. Benzenediazonium chloride couples also with hydroxylamines, the nature of the product depending on the pH of the medium; in neutral or mildly acid solution the product is phenyl azide:

$$C_6H_5\overset{+}{N}{\equiv}N + \ddot{N}H_2OH \longrightarrow C_6H_5{-}N{=}N{-}\overset{+}{N}H_2{-}OH \rightleftharpoons$$

$$C_6H_5{-}N{=}N{-}N(OH)H + H^+ \longrightarrow C_6H_5{-}N{=}\overset{+}{N}{=}\overset{-}{N} + H_2O$$

Phenyl azide

Another example of diazonium salts reacting with nucleophiles is the coupling with enolate anions:

$$C_6H_5{-}\overset{+}{N}{\equiv}N + \overset{-}{H}C(COCH_3)(CO_2Et) \longrightarrow$$

$$C_6H_5{-}N{=}N{-}HC(COCH_3)(CO_2Et) \rightleftharpoons C_6H_5NH{-}N{=}C(COCH_3)(CO_2Et)$$

Ethyl mesoxalate phenylhydrazone

Notice that the aliphatic azo-compound of the second of these reactions is in tautomeric equilibrium with a phenylhydrazone; thus, the same product could be prepared by the action of phenylhydrazine on ethyl mesoxalate, $C_6H_5NHNH_2 + O{=}C(CO_2Et)_2$. In some alkylanilines the diazonium group may 'bite the tail' of the molecule. A particularly simple example is the reaction that occurs when the diazonium salt derived from 2-methyl-5-nitroaniline is treated with alkali:

$$\text{5-nitro-2-methylaniline (}NH_2, CH_3, O_2N\text{)} \xrightarrow{HNO_2} \text{diazonium salt (}\overset{+}{N}\equiv N, CH_3, O_2N\text{)} \xrightarrow{OH^-} \text{6-Nitroindazole}$$

6-Nitroindazole

Here the methyl group is *para* to the powerful electron-accepting nitro-group and is therefore 'activated' (i.e. forms a carbanion with base): *o*-toluidine (*o*-methylaniline) does not yield indazole.

The remaining reactions of diazonium salts to be considered now are those in which nitrogen is lost from the molecule. Such a reaction can occur in two ways. First, straightforward decomposition of the diazonium salt to yield molecular nitrogen and the phenyl cation; or, secondly, donation of an electron to the diazonium salt, to yield a transient free radical which then breaks down to the phenyl radical and molecular nitrogen.

$$C_6H_5-\overset{+}{N}\equiv N \longrightarrow C_6H_5^+ + N_2\uparrow$$

$$C_6H_5-\overset{+}{N}\equiv N + e^- \longrightarrow C_6H_5\cdot + N_2\uparrow$$

When an aqueous solution of benzenediazonium chloride is heated, nitrogen is evolved and although a considerable amount of tar is formed the principal isolable material is phenol. This reaction proceeds by decomposition of the diazonium salt to the phenyl cation which then reacts with water. The whole process could be likened to an S_N1 hydrolysis of an alkyl halide.

$$Ar-\overset{+}{N}\equiv N \longrightarrow N_2 + Ar^+ \xrightarrow{H_2\ddot{O}} Ar-\overset{+}{O}H_2 \longrightarrow ArOH + H^+$$

In the presence of alcohol the corresponding phenyl ether is the predominant product (and not, as often stated, the hydrocarbon, although small amounts of this may be formed as by-product). In the presence of high concentrations of other ions such as chloride, some chlorobenzene is formed as well as phenol; but, as we shall see below, there are much better ways of converting diazonium salts into aryl chlorides.

We must now consider the reactions of diazonium salts in which nitrogen is lost, but which differ from those we have just discussed in that one-electron transfer is involved. These include a very important group of reactions known as Sandmeyer reactions. If benzenediazonium chloride is treated with cuprous chloride, nitrogen is evolved and chlorobenzene is produced. It would be very tempting to write down a reaction sequence as follows:

$$\mathrm{Ar\overset{+}{N}_2 + Cu^{I} \longrightarrow Ar\cdot + Cu^{II} + N_2}$$
$$\mathrm{Ar\cdot + Cu^{II} \longrightarrow Ar^+ + Cu^{I}}$$
$$\mathrm{Ar^+ + Cl^- \longrightarrow ArCl}$$

However, there is good reason to believe that the reaction is very much more complicated. The first step has been shown to involve formation of the anion $CuCl_2^-$. It appears that this anion then combines with the diazonium cation, and the resulting molecule breaks down as shown.

$$\mathrm{Cu^{I}Cl + Cl^- \rightleftharpoons (Cu^{I}Cl_2)^-}$$

$$\mathrm{Ar{-}\overset{+}{N}{\equiv}N + (Cu^{I}Cl_2)^- \longrightarrow Ar\!\!<\!\!\begin{smallmatrix} N{=}N \\ \\ Cl \end{smallmatrix}\!\!>\!\!Cu^{I}{-}Cl \longrightarrow Ar{-}Cl + N_2 + CuCl}$$

Similar reactions can be carried out with cuprous bromide and cuprous cyanide. With iodine, no cuprous salt is necessary and treatment of a diazonium solution with potassium iodide is sufficient to afford the aryl iodide. The oxidation potential of the iodide anion is of about the same order as that of cuprous cation. There is little doubt that this reaction again involves one-electron transfer, probably the ion analogous to $CuCl_2^-$ is I_3^-. Aryl azides can be formed in the same way, by using sodium azide in place of sodium iodide. Aryl fluorides cannot be prepared in this fashion, but they can be prepared by heating the solid diazonium fluoroborate (see above for a discussion of the formation of solid diazonium fluoroborates):

$$\mathrm{Ar{-}\overset{+}{N}{\equiv}N\ \ BF_4^- \xrightarrow{Heat} ArF + N_2 + BF_3}$$

Other reactions analogous to the Sandmeyer reaction are treatment of the diazonium salt with sodium cobaltinitrite which yields the nitro-compound, or with sodium arsenite which yields the aryl-arsonic acid. The former reaction is sometimes known as the Hodgson reaction, and the latter as the Bart reaction. The exact mechanism of these two reactions has not been completely worked out, but they probably involve one-electron transfer similar to that in the Sandmeyer reactions.

$$\mathrm{Ar\overset{+}{N}{\equiv}N \xrightarrow{3Na^+,\ Co(NO_2)_6^{3-}} ArNO_2} \quad \text{(Hodgson reaction)}$$

$$\mathrm{Ar\overset{+}{N}{\equiv}N \xrightarrow{Na_2HAsO_3} ArAsO(ONa)_2 + N_2 + H_2O} \quad \text{(Bart reaction)}$$

A final reaction we must consider is the replacement of the diazonium group by a hydrogen atom. This is most conveniently carried out by heating the diazonium salt with hypophosphorous acid. This reaction almost certainly involves the free aryl radical and can be represented as follows:

$$\mathrm{Ar\cdot + H{-}P(=O)(H){-}OH \longrightarrow ArH + \cdot P(=O)(H){-}OH}$$

$$\mathrm{Ar{-}\overset{+}{N}{\equiv}N + \cdot P(=O)(H){-}OH \longrightarrow Ar\cdot + N_2 + {}^{+}P(=O)(H){-}OH}$$

$$\mathrm{HO{-}P^{+}(=O)(H) + OH_2 \longrightarrow HO{-}P(=O)(H){-}OH + H^+}$$

Another way of carrying out this transformation is to prepare the solid diazonium fluoroborate and treat it with sodium borohydride; this clearly involves a completely different mechanism.

Some of the replacement reactions of aromatic diazonium salts are shown in the chart on p. 189.

Because they break down into free radicals, aromatic diazonium salts are sometimes used to prepare biaryl derivatives. Symmetrical biaryls are formed when finely divided copper is added to a heated, aqueous solution of a diazonium salt; for example, adding it to benzenediazonium chloride gives an approximately 25% yield of

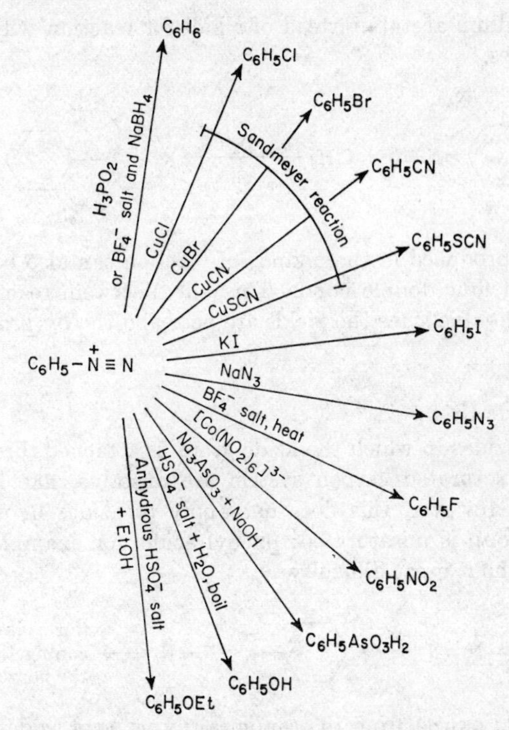

Replacement reactions of benzene diazonium salts.

biphenyl. Phenyl radicals are probably produced on the surface of the metal and combine to form the biphenyl derivative.

$$Br-C_6H_4-\overset{+}{N}\equiv N \xrightarrow{\text{Cu powder}} Br-C_6H_4-C_6H_4-Br$$

20–30%
4,4′-Dibromobiphenyl

Unsymmetrical biaryls are prepared by mixing a concentrated aqueous solution of a diazonium salt with some aromatic hydrocarbon, for example, benzene, and adding concentrated alkali to the solution while the reaction proceeds. This reaction is usually known as the Gomberg reaction. The yields are not high but the reaction is relatively simple to carry out. Slightly better yields can be obtained by using sodium acetate instead of alkali for reaction with diazonium chlorides.

$$o\text{-}O_2N\text{-}C_6H_4\text{-}\overset{+}{N}{\equiv}N + C_6H_6 \xrightarrow{Na^+\ ^-O_2CCH_3} o\text{-}O_2N\text{-}C_6H_4\text{-}C_6H_5 \quad 40\%$$

Radicals produced in these kinds of reaction can also be induced to add to olefinic double bonds. This, the Meerwein reaction, is of only limited value since the yields are poor and the by-products are many.

(d) The Azide Group

Organic azides in which the azide group is attached directly to a completely saturated carbon system are unstable, like hydrazoic acid itself. However, this does not apply to azides in which the attached group is unsaturated; phenyl azide, for example, can be handled without great difficulty:

$$\left[C_6H_5\text{-}\ddot{N}{=}\overset{+}{N}{=}\overset{-}{\ddot{N}}: \longleftrightarrow C_6H_5\text{-}\overset{-}{\ddot{N}}\text{-}\overset{+}{N}{\equiv}N: \longleftrightarrow \text{(forms involving conjugation with the benzene ring)} \right]$$

As we should expect from the canonical forms, aryl azides undergo electrocyclic reactions (four-centre additions) analogous to those of diazomethane and ozone. As well as adding to olefins, aryl azides add to acetylenes, imines, and even to enols.

$$>C{=}C< + Ar\text{-}N{=}\overset{+}{N}{=}N^- \longrightarrow \text{A 1,2,3-triazole}$$

A 1,2,3-triazole

$$H_3C\text{-}C(OH){=}CH\text{-}C({=}O)\text{-}OEt + C_6H_5N_3 \longrightarrow C_6H_5\text{-}N\text{(1,2,3-triazole, 5-}CH_3\text{, 4-}CO_2Et) + H_2O$$

Acyl azides in which the azide group can conjugate with the carbonyl group are also stable. They are readily prepared by treating an acid chloride with sodium azide:

$$Na^+N_3^- + R(Cl)C{=}O \longrightarrow N_3(R)(Cl)C{-}O^- \longrightarrow N_3(R)C{=}O + NaCl$$

Acyl azides are important because they rearrange when heated. The two terminal nitrogen atoms of the azide group form a very good leaving group and, on heating, nitrogen is evolved and the alkyl group migrates to fill the place from which the nitrogen atoms depart:

$$\left[R(C{=}O){-}\ddot{N}{=}\overset{+}{N}{=}\overset{-}{N}: \longleftrightarrow R(C{=}O){-}\overset{-}{N}{-}\overset{+}{N}{\equiv}N \longleftrightarrow R(C{=}N{-}\overset{+}{N}{\equiv}N){-}O^- \right] \longrightarrow \underset{\text{Isocyanate}}{O{=}C{=}N{-}R} + N_2$$

This, the Curtius rearrangement, should be compared with the pinacol–pinacolone rearrangement (Chapter 10, p. 114), the Beckmann rearrangement (Chapter 14, p. 166), and the Wolff rearrangement (this Chapter, p. 177). The reaction sequence is important because isocyanates react with water to evolve carbon dioxide and a primary amine, so that the Curtius rearrangement provides a route for converting a carboxyl group into an amino-group. Synthetically this

$$R{-}N{=}C{=}O + H_2O \longrightarrow R{-}NH_2 + CO_2$$

can be a very useful reaction. The reactions of isocyanates are discussed in more detail in Chapter 21.

Problems

1. What reaction would you expect between an ether solution of diazomethane and

(*a*) $CH_3OC_6H_4OH$; (*b*) *p*-$CH_3OC_6H_4CO_2H$; (*c*) *o*-$NO_2C_6H_4COCl$; (*d*) $C_6H_5CH{=}CHCOCH_3$?

2. Complete the following reaction sequences and indicate what feature is common to all of them:

(*a*) $$(CH_3)_2\text{—}\underset{\displaystyle OH}{\underset{|}{C}}\text{—}\underset{\displaystyle OH}{\underset{|}{C}}\text{—}(CH_3)_2 \xrightarrow{H_2SO_4}$$

(*b*) $$\begin{array}{l} CH_3 \qquad\qquad OH \\ \quad \diagdown \qquad\quad \diagup \\ \qquad C{=}N \\ \quad \diagup \\ C_6H_5 \end{array} \xrightarrow{PCl_5}$$

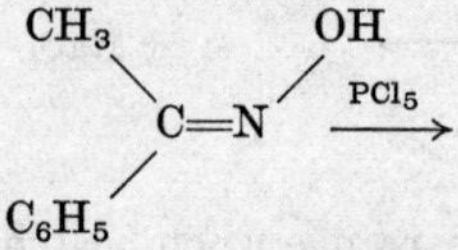

(*c*) $$C_6H_5COCHN_2 \xrightarrow[MeOH]{Ag}$$

(*d*) $$C_6H_5CH_2CON_3 \xrightarrow[\text{moist } C_6H_6]{\text{Heat in}}$$

CHAPTER 16

Synthesis (A): Aliphatic Compounds and the Use of Carbanions

In Chapter 9 we considered a wide variety of β-dicarbonyl compounds that formed stable carbanions. One important example was diethyl malonate $CH_2(CO_2Et)_2$. Diethyl malonate is readily prepared from sodium chloroacetate (which can be prepared from acetic acid) and sodium cyanide. The resulting sodium cyanoacetate is treated with sulphuric acid and ethanol, under which conditions hydrolysis of the cyano-group and esterification of the two carboxyl groups occurs in one process:

$$ClCH_2CO_2Na + NaCN \longrightarrow NC{-}CH_2CO_2Na \xrightarrow[H_2SO_4]{EtOH} CH_2(CO_2Et)_2$$

As stated in Chapter 9, diethyl malonate forms a stable carbanion when treated with a strong base:

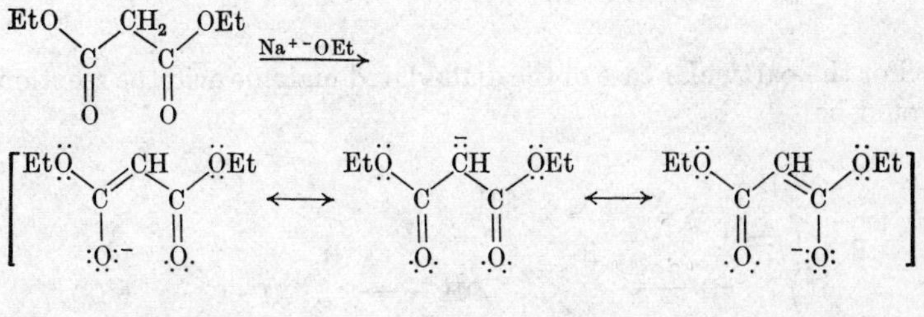

$$EtO{-}C(=O){-}CH_2{-}C(=O){-}OEt \xrightarrow{Na^{+}\,{}^{-}OEt}$$

$$\left[EtO{-}C(O^{-}){=}CH{-}C(=O){-}OEt \longleftrightarrow EtO{-}C(=O){-}\overset{..}{C}H^{-}{-}C(=O){-}OEt \longleftrightarrow EtO{-}C(=O){-}CH{=}C(O^{-}){-}OEt \right]$$

We also showed that these stable carbanions behave as nucleophiles, attacking alkyl halides:

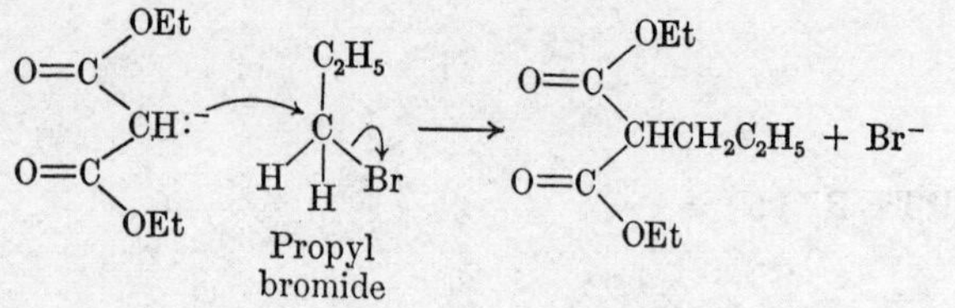

$$(O{=}C(OEt))_2CH{:}^- + \underset{\text{Propyl bromide}}{H_2C(C_2H_5){-}Br} \longrightarrow (O{=}C(OEt))_2CHCH_2C_2H_5 + Br^-$$

This reaction amounts to a method of lengthening the carbon chain which we can summarize as follows:

$$\text{RBr (or RI)} + CH_2(CO_2Et)_2 + \underset{\text{(1 mole)}}{Na^+ \ {}^-OEt} \xrightarrow[\text{soln.}]{\text{EtOH}} RCH(CO_2Et)_2$$

A second alkyl halide and another mole of sodium ethoxide can be added without isolation of the first alkylation product, and a disubstituted derivative of the malonic acid is then obtained:

$$\text{R'Br (or R'I)} + RCH(CO_2Et) + \underset{\text{(1 mole)}}{Na^+ \ {}^-OEt} \longrightarrow RR'C(CO_2Et)_2$$

To appreciate the full value of these reaction sequences in synthesis we must consider the properties of carboxylic acids with a carbonyl group in the β-position. Such acids, when heated, are readily decarboxylated probably in an intramolecular process as depicted below:

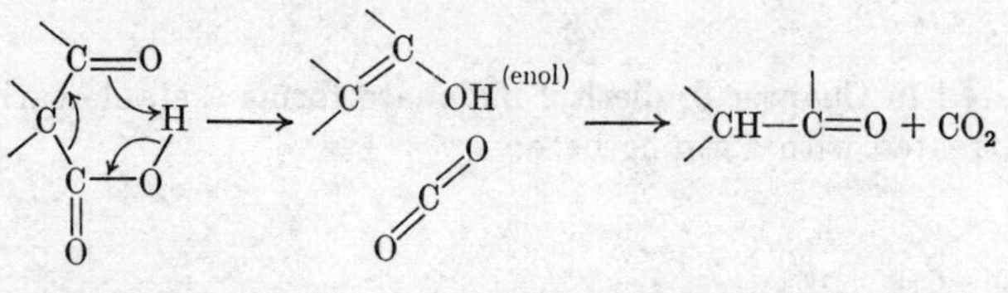

$$\text{>C(C{=}O)(C(=O)OH)} \longrightarrow \underset{\text{(enol)}}{\text{>C{=}C(OH)}} + O{=}C{=}O \longrightarrow \text{>CH{-}C{=}O} + CO_2$$

For the particular case of the dialkylated malonic acid the reaction would be:

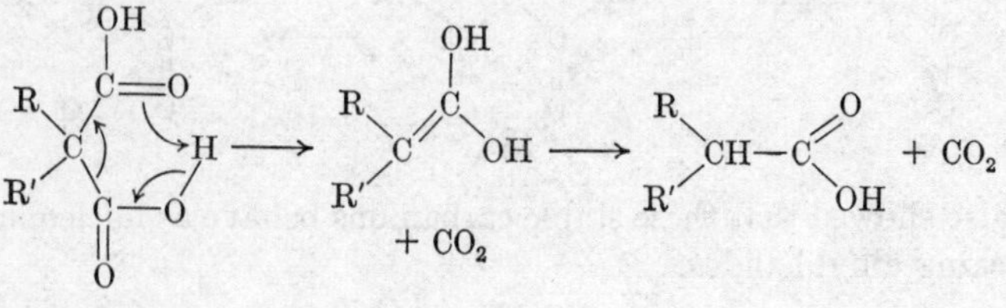

$$RR'C(COOH)(COOH) \longrightarrow RR'C{=}C(OH)OH + CO_2 \longrightarrow RR'CH{-}C(=O)OH + CO_2$$

This decarboxylation completes a reaction sequence of great synthetic utility. Diethyl malonate can be used for converting an alkyl halide into a carboxylic acid with two additional carbon atoms in the carbon chain, and for preparing branched-chain carboxylic acids.

$$RBr + CH_2(CO_2Et)_2 \xrightarrow{NaOEt \mid EtOH} RCH(CO_2Et)_2 \xrightarrow{Hydrol.} RCH(CO_2H)_2 \xrightarrow{Heat} RCH_2CO_2H$$

$$RCH(CO_2Et)_2 \xrightarrow{R'Br,\ NaOEt \mid EtOH} RR'C(CO_2Et)_2 \xrightarrow{Hydrol.} RR'C(CO_2H)_2 \xrightarrow{Heat} RR'CHCO_2H$$

In Chapter 9 we also gave a brief account of the reactions of diethyl malonate with dihalogenoalkanes and showed how cyclic compounds could be prepared. Although these reactions yield a mixture of products, Perkin was able to carry out the following transformations, using 1,3-dibromopropane, simply by changing the stoichiometry of the reaction. To prepare a cyclobutane derivative he treated 1 mole of 1,3-dibromopropane with one mole of diethyl malonate and two moles of sodium ethoxide:

$$H_2C(CH_2Br)_2 + CH_2(CO_2Et)_2 + 2\ NaOEt \longrightarrow \text{cyclo-}[H_2C{-}CH_2{-}C(CO_2Et)_2{-}CH_2]$$

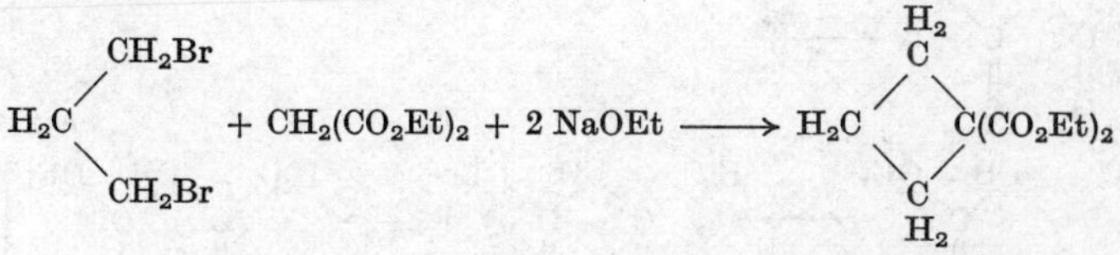

Et_2 cyclobutane-1,1-dicarboxylate

To prepare cyclopentane and cyclohexane derivatives he treated 1 mole of 1,3-dibromopropane with two moles of diethyl malonate and two moles of sodium ethoxide. These reactions involved further steps, as depicted below and on p. 196.

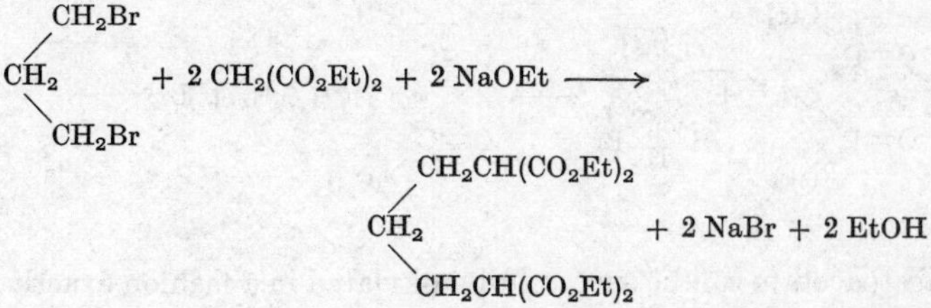

$$CH_2(CH_2Br)_2 + 2\ CH_2(CO_2Et)_2 + 2\ NaOEt \longrightarrow CH_2[CH_2CH(CO_2Et)_2]_2 + 2\ NaBr + 2\ EtOH$$

(A)

$$(A) + Br_2 + 2\,NaOEt \longrightarrow \begin{array}{l} CH_2{-}C(CO_2Et)_2 \\ CH_2 \quad\quad\quad | \\ CH_2{-}C(CO_2Et)_2 \end{array} + 2\,NaBr + 2\,EtOH$$

or

$$(A) + CH_2I_2 + 2\,NaOEt \longrightarrow \begin{array}{l} CH_2{-}C(CO_2Et)_2 \\ CH_2 \quad\quad\quad\quad CH_2 \\ CH_2{-}C(CO_2Et)_2 \end{array} + 2\,NaI + 2\,EtOH$$

The preparation of ethyl acetoacetate $CH_3COCH_2CO_2Et$ from two moles of ethyl acetate by treatment with sodium ethoxide in a Claisen condensation was described in Chapter 9. Ethyl acetoacetate can also be prepared from acetone *via* diketene. The details of the latter preparation (described in Chapter 21) are unimportant at the moment; what matters is that ethyl acetoacetate is a readily available compound. Like diethyl malonate, this compound also forms a stable carbanion:

$$H_3C{-}C(=O){-}CH_2{-}C(=O){-}OEt \xrightarrow{Na^+\,{}^-OEt}$$

$$\left[H_3C{-}C(O^-){=}CH{-}C(=O){-}\ddot{O}Et \longleftrightarrow H_3C{-}C(=O){-}\ddot{C}H^-{-}C(=O){-}\ddot{O}Et \longleftrightarrow H_3C{-}C(=O){-}CH{=}C(O^-){-}\ddot{O}Et \right]$$

Again like the anion from diethyl malonate, the anion from ethyl acetoacetate behaves as a nucleophile in a displacement reaction with an alkyl halide:

$$(CH_3)(O{=}C)\,CH{:}^-\,(C{=}O)(OEt) + C_2H_5CH_2Br \longrightarrow (CH_3)(O{=}C)\,CHCH_2C_2H_5\,(C{=}O)(OC_2H_5) + Br^-$$

Ethyl acetoacetate can be successfully alkylated in a fashion exactly analogous to the reactions already described for diethyl malonate:

$$RBr\ (\text{or } RI) + CH_3COCH_2CO_2Et + \underset{1\ \text{mole}}{NaOEt} \longrightarrow \underset{CH_3CO}{\overset{R}{}}\!\!\!>CHCO_2Et$$

Again as for diethyl malonate, dialkylation is possible:

$$\underset{(\text{or } R'I)}{R'Br} + CH_3CO\underset{R}{C}HCO_2Et + \overset{NaOEt}{\underset{(1\ \text{mole})}{\text{or } NaCPh_3}} \longrightarrow CH_3CO-\overset{R'}{\underset{R}{C}}-CO_2Et$$

Hydrolysis of the resulting alkylated ester with either dilute acid or dilute alkali yields the free β-keto-acid, which readily undergoes decarboxylation when heated:

$$CH_3CO\overset{R}{C}HCO_2Et \xrightarrow{\text{Dil. acid}} [H_3C-C(=O)-CH(R)-C(=O)-O-H] \longrightarrow [H_3C-C(OH)=CHR] + O=C=O \longrightarrow H_3C-CO-CH_2R + CO_2$$

We thus have a reaction sequence which can be used for making alkyl methyl ketones:

$$RBr + CH_3COCH_2CO_2Et$$

$$\downarrow \text{NaOEt} \mid \text{EtOH}$$

$$CH_3CO\overset{R}{C}HCO_2Et \xrightarrow[\text{hydrol.}]{\text{Dil. acid}} CH_3CO\overset{R}{C}HCO_2H \xrightarrow[-CO_2]{\text{Heat}} CH_3COCH_2R$$

$$\downarrow R'Br \mid \text{EtOH}$$

$$CH_3CO\overset{R}{\underset{R'}{C}}CO_2Et \xrightarrow[\text{hydrol.}]{\text{Dil. acid}} CH_3CO\overset{R}{\underset{R'}{C}}CO_2H \xrightarrow[-CO_2]{\text{Heat}} CH_3COCHRR'$$

In our previous description of the preparation of ethyl acetoacetate from ethyl acetate it was pointed out that the Claisen ester condensation is reversible. Alkylated derivatives of ethyl acetoacetate are, in effect, the product that we should get by cross-condensation between ethyl acetate and the esters derived from an alkanecarboxylic acid. Thus, if ethyl acetoacetate is treated with a strong base, in addition to ester hydrolysis (*a*), attack at the keto-group by the hydroxyl ions occurs and leads to carbon–carbon bond fission (*b*).

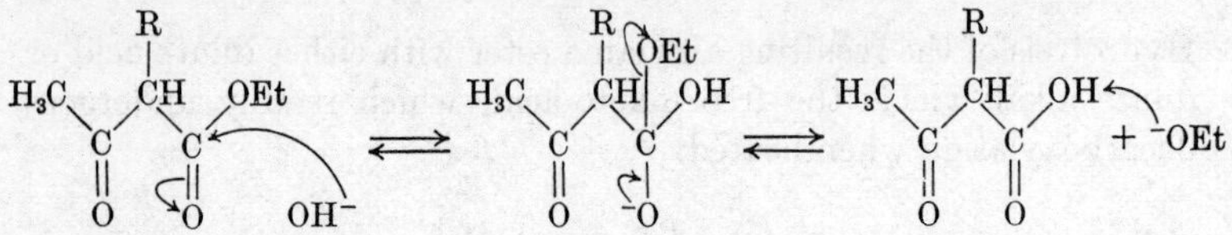

$$H_3C{-}C(=O){-}CH(R){-}C(=O){-}OEt + OH^- \rightleftharpoons H_3C{-}C(=O){-}CH(R){-}C(O^-)(OEt){-}OH \rightleftharpoons H_3C{-}C(=O){-}CH(R){-}C(=O){-}OH + {}^-OEt$$

(*a*) Normal alkaline hydrolysis of the ester group.

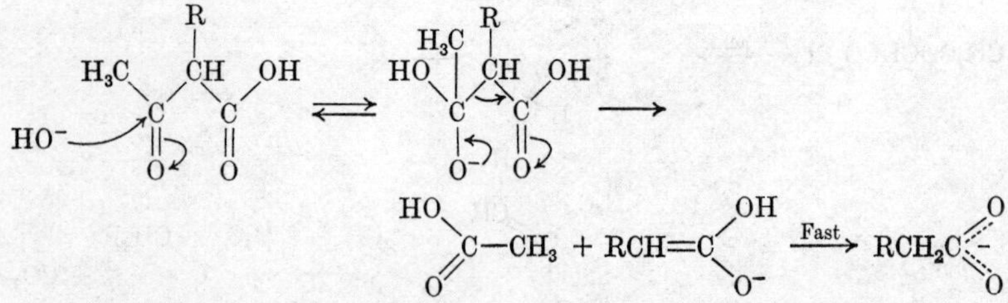

$$HO^- + H_3C{-}C(=O){-}CH(R){-}C(=O){-}OH \rightleftharpoons H_3C{-}C(HO)(O^-){-}CH(R){-}C(=O){-}OH \longrightarrow$$

$$HO{-}C(=O){-}CH_3 + RCH{=}C(OH){-}O^- \xrightarrow{\text{Fast}} RCH_2C(\cdots O)_2^-$$

(*b*) Carbon–carbon bond fission in the presence of strong alkali.

This means that by treating the ethyl alkylacetoacetate with a very strong base we can obtain acetic acid and a long-chain carboxylic acid. Ethyl acetoacetate could be used in place of diethyl malonate for the preparation of long-chain carboxylic acids, the products obtained by the two routes being identical. In practice, malonic ester is better for the preparation of long-chain acids. It is mainly for the preparation of ketones that ethyl acetoacetate is used in its simplest form. Some synthetic uses of ethyl acetoacetate in the preparation of diketones are illustrated below.

β-Diketones:

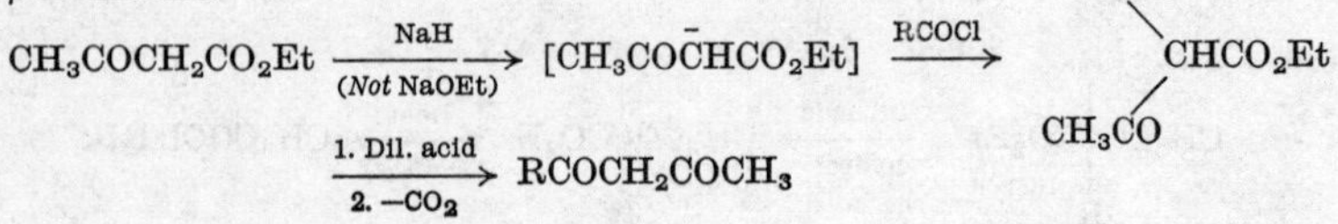

$$CH_3COCH_2CO_2Et \xrightarrow[\text{(}\textit{Not}\text{ NaOEt)}]{\text{NaH}} [CH_3CO\bar{C}HCO_2Et] \xrightarrow{\text{RCOCl}} (RCO)(CH_3CO)CHCO_2Et$$

$$\xrightarrow[\text{2. }-CO_2]{\text{1. Dil. acid}} RCOCH_2COCH_3$$

γ-Diketones:

$$CH_3COCH_2CO_2Et \xrightarrow[2.\ I_2]{1.\ NaOEt} \begin{array}{c} CH_3COCHCO_2Et \\ | \\ CH_3COCHCO_2Et \end{array} \xrightarrow[2.\ -CO_2]{1.\ Dil.\ acid} CH_3COCH_2CH_2COCH_3$$

δ-Diketones:

$$2\,CH_3COCH_2CO_2Et + RCHBr_2 \xrightarrow[2\ moles]{NaOEt} \begin{array}{l} CH_3COCHCO_2Et \\ \quad \diagdown CHR \\ \quad \diagup \\ CH_3COCHCO_2Et \end{array} \xrightarrow[2.\ -CO_2]{1\ Dil.\ acid} \begin{array}{l} CH_3COCH_2 \\ \quad \diagdown CHR \\ \quad \diagup \\ CH_3COCH_2 \end{array}$$

$$\downarrow NaOEt$$

(cyclohexenone ring bearing H_3C, CO_2Et, $-R$, O and CO_2Et substituents)

Various reactions involving the addition of carbanions to a carbonyl group were described in Chapter 9. Their synthetic utility is self-evident. Particularly important is the ring closure of esters of dibasic acids (the Dieckmann condensation) (see p. 99); an interesting example of this yielding a heterocyclic compound is illustrated:

$$EtN\begin{array}{l} \diagup CH_2CH_2CO_2Et \\ \diagdown CH_2CH_2CO_2Et \end{array} \longrightarrow EtN_1 \begin{array}{c} H_2C\text{——}CH_2 \\ {}_2H_2C\text{——}{}_3CH \\ \quad\quad \diagdown CO_2Et \end{array} {}_4C{=}O \longrightarrow EtN_1\bigcirc_4{=}O$$

1-Ethyl-
4-piperidone

The starting material, prepared from ethylamine and ethyl acrylate, undergoes normal Dieckmann condensation with sodium methoxide to yield the 3-ethoxycarbonyl derivative of the heterocyclic ketone. Hydrolysis of this ester yields a β-keto-acid which can be very readily decarboxylated to 1-ethyl-4-piperidone.

In Chapter 9 we described a number of reactions involving addition of a carbanion to a carbonyl compound that had no α-hydrogen atom. Particularly important examples of this reaction involve formaldehyde—in these a cross-Cannizzaro reaction may also occur. For example, an aqueous solution of acetaldehyde and formaldehyde, treated with a mild base such as calcium hydroxide or sodium carbonate, yields pentaerythritol. Students should take the opportunity of working out in detail all the steps involved in this reaction (see question 1). The overall stoichiometry of this reaction is:

$$CH_3CHO + 4\,CH_2O + H_2O \longrightarrow \underset{\text{Pentaerythritol}}{C(CH_2OH)_4} + HCO_2H$$

Another example of cross-condensation, involving a carbanion, is the Perkin reaction. The simplest example is the reaction of benzaldehyde with acetic anhydride in the presence of sodium or potassium acetate:

$$C_6H_5CHO + (CH_3CO)_2O \xrightarrow{CH_3CO_2^-} \underset{\text{Cinnamic acid}}{C_6H_5CH{=}CHCO_2H} + CH_3CO_2H$$

Again it is important that the individual reaction steps should be worked out.

The addition of carbanions to olefinic double bonds activated by an acceptor group (the so-called Michael reaction) is clearly of great synthetic importance. We give one further example, the synthesis of dimedone from diethyl malonate and mesityl oxide. Step 1 is the

$$(H_3C)_2C{=}CHCOCH_3 + CH_2(CO_2Et)_2 \xrightarrow[\text{Step 1}]{\text{NaOEt}} (H_3C)_2C(CH_2COCH_3){-}CH(CO_2Et)_2$$

$$\xrightarrow[\text{Step 2}]{\text{NaOEt}} \text{(cyclic: } (H_3C)_2C{<}\,CH_2{-}C{=}O,\ EtO_2C{-}HC{-}C(=O){-}CH_2\text{)} \xrightarrow[\text{Step 3}]{\text{KOH}} \text{Dimedone}$$

Dimedone

Michael addition. This is followed by a Dieckmann-type condensation, and the resulting cyclohexanedione ester is a β-keto-ester, which can be hydrolysed to the acid and then decarboxylated, see page 197.

A very important modification of the Michael addition is sometimes known as the Robinson ring-extension reaction. This involves the Michael addition of a cyclohexanone enolate anion to an alkyl vinyl ketone, followed by addition of the new enolate ion to the carbonyl group of the cyclohexanone (a Knoevenagel-type condensation).

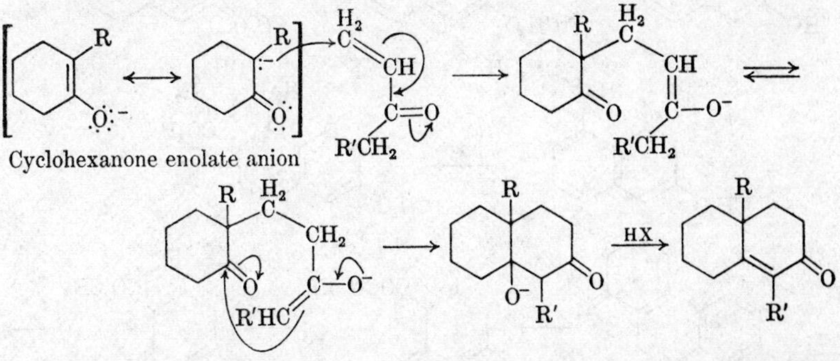

Cyclohexanone enolate anion

Because vinyl ketones are very unstable, a number of variants of this reaction have been developed, using ketones which undergo elimination to yield the vinyl ketone, *in situ*, when treated with base.

$$RCH_2COCH_2CH_2\overset{+}{N}Et_2CH_3 \xrightarrow{\text{Base}} RCH_2COCH{=}CH_2 \xleftarrow{\text{Base}} RCH_2COCH_2CH_2Cl$$

All the synthetic sequences described above involve carbanions. Only two types of reaction are involved: nucleophilic displacement of a halogen attached to a carbon atom by a carbanion; and addition of a carbanion to a carbon–oxygen or a carbon–carbon double bond activated by an electron-acceptor. The best way to become conversant with these synthetically valuable reactions is to devise reaction schemes for the preparation of moderately complex aliphatic compounds, such as are suggested at the end of this Chapter. In order to illustrate that these reactions are adequate for the synthesis of complex molecules we close now by describing Bachman's synthesis of equilenin.

Bachman's total synthesis of equilenin

The reaction sequence by which Bachman synthesized the naturally occurring steroid hormone equilenin is depicted on p. 202. Each step in this sequence is of a type discussed in the preceding Chapters. The student should work through this sequence step by step with the aid of the notes on pp. 203—204:

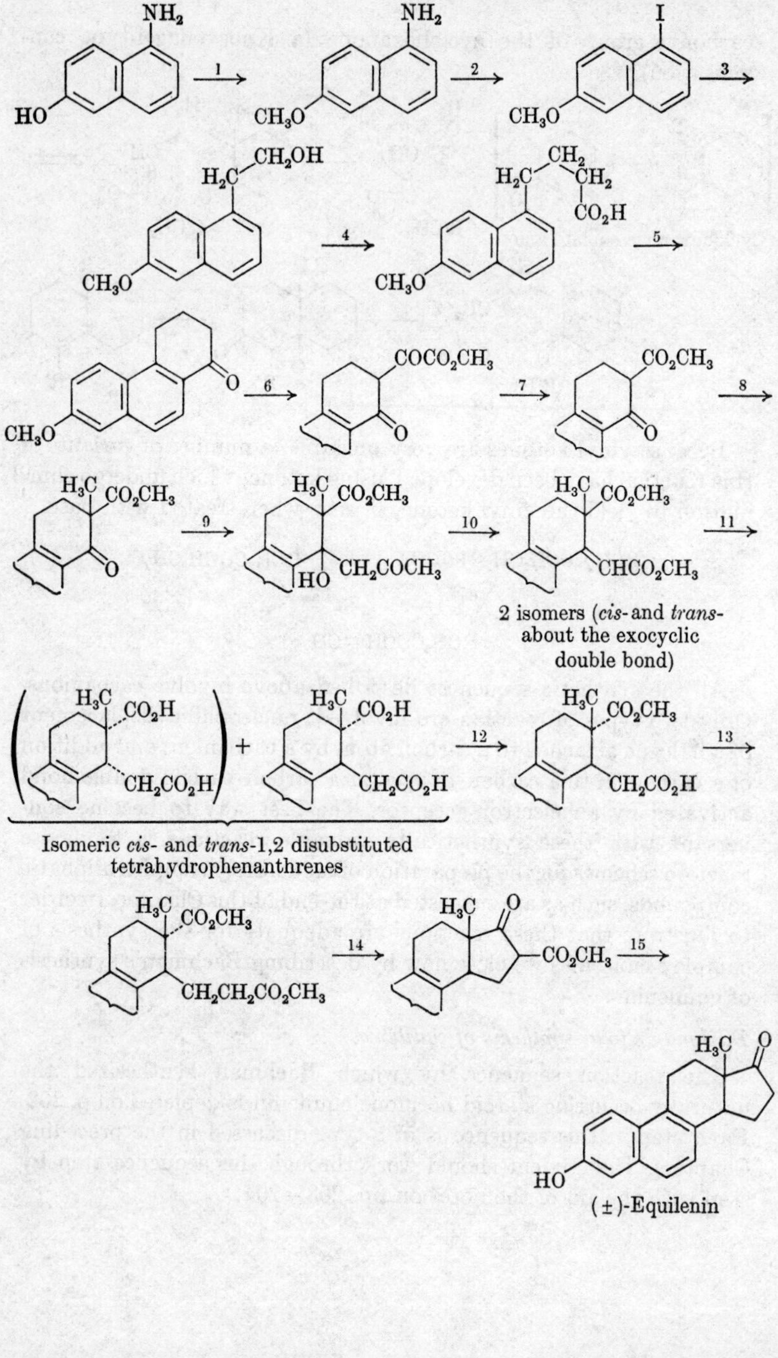

NH_2, HO $\xrightarrow{1}$ NH_2, CH_3O $\xrightarrow{2}$ I, CH_3O $\xrightarrow{3}$

H_2C, CH_2OH, CH_3O $\xrightarrow{4}$ H_2C, CH_2, CH_2, CO_2H, CH_3O $\xrightarrow{5}$

O, CH_3O $\xrightarrow{6}$ $COCO_2CH_3$, O $\xrightarrow{7}$ CO_2CH_3, O $\xrightarrow{8}$

H_3C, CO_2CH_3, O $\xrightarrow{9}$ H_3C, CO_2CH_3, HO, CH_2COCH_3 $\xrightarrow{10}$ H_3C, CO_2CH_3, $CHCO_2CH_3$ $\xrightarrow{11}$

2 isomers (*cis*- and *trans*- about the exocyclic double bond)

(H_3C, CO_2H, CH_2CO_2H) + H_3C, CO_2H, CH_2CO_2H $\xrightarrow{12}$ H_3C, CO_2CH_3, CH_2CO_2H $\xrightarrow{13}$

Isomeric *cis*- and *trans*-1,2 disubstituted tetrahydrophenanthrenes

H_3C, CO_2CH_3, $CH_2CH_2CO_2CH_3$ $\xrightarrow{14}$ H_3C, O, CO_2CH_3 $\xrightarrow{15}$

H_3C, O, HO

(±)-Equilenin

1. Conversion of the aminonaphthol into its methyl ether. This involves three reaction steps: acylation of the amino-group with acetic anhydride; treatment of the resulting amide with dimethyl sulphate in alkali, which results in the formation of the methyl ether; hydrolysis of the acetamido-group to generate 6-methoxy-1-naphthylamine.

2. This step involves diazotization of the naphthylamine and treatment of the diazonium salt with a solution of iodine and potassium iodide, a reaction analogous to the Sandmeyer reaction (Chapter 15).

3. This step involves conversion of the iodide into a Grignard reagent by treatment with magnesium in ether, followed by the reaction of this reagent with ethylene oxide to produce the substituted alcohol.

4. This step involves conversion of the alcohol into the alkyl bromide by treatment with phosphorus tribromide, followed by a normal malonic ester synthesis.

5. The carboxylic acid is converted into its acid chloride by treatment with thionyl chloride, and the subsequent ring closure is a Friedel–Crafts ketone synthesis. Stannic chloride was used as the Friedel–Crafts catalyst.

6. The ketone was treated with sodium ethoxide, and the resulting enolate anion condensed with dimethyl oxalate $(CO_2CH_3)_2$.

7. The pyruvic ester was pyrolysed, losing carbon monoxide.

8. The β-keto-ester, which is analogous to ethyl acetoacetate, was treated with sodium methoxide and methyl iodide, to yield the usual alkylation product.

9. This reaction involves treating the keto-ester with ethyl bromoacetate and zinc. This is the Reformatsky reaction, described in Part 1 (see Chapter 11).

10. The hydroxy-ester produced from the Reformatsky reaction readily loses water to yield two isomeric unsaturated esters.

11. The mixture of unsaturated esters was hydrogenated, to yield two saturated esters. The *trans*-dicarboxylic acid was separated for further reaction.

12. The *trans*-dicarboxylic acid was converted by diazomethane into the dimethyl ester (Chapter 15). This was then hydrolysed under mild conditions so that the tertiary carboxyl group, which is much more sterically hindered, was unaffected and the half-ester was isolated.

13. The Arndt–Eistert chain-lengthening process was employed (Chapter 15).

14. This step involves the Dieckmann ester condensation (see Chapter 9).

15. Hydrolysis of the keto-ester leaves a β-keto-carboxylic acid which is readily decarboxylated. In fact, the hydrolysis conditions had to be fairly vigorous because the methoxyl group, put on in step 1, had to be hydrolysed at the same time. Resolution of the racemic equilenin yielded dextrorotatory equilenin identical with the natural product. Equilenin found in the urine of pregnant mares is a mild estrogen, and probably the degradation product of a natural hormone. The important feature of the above sequence is that this comparatively complex natural product was synthesized from a readily obtainable compound by using reaction steps all of which have been described in the foregoing text.

Problems

1. Write the steps involved in the formation of pentaerythritol, $(HOCH_2)_4C$, from acetaldehyde and formaldehyde. Suggest possible by-products.

2. Suggest methods for synthesizing the following, all of which involve use of malonic ester or acetoacetic ester:

(*a*) H_5C_2 / H_3C \ $CHCHCH_2CO_2H$ (i.e. $(H_5C_2)(H_3C)CHCHCH_2CO_2H$)

(*b*) $n\text{-}C_5H_{11}COCH_3$

(*c*) cyclobutane with CO_2Et, CO_2Et on one carbon

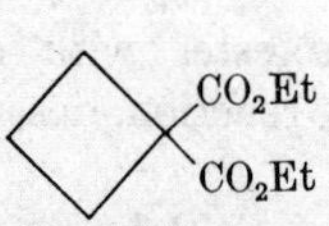

(*d*) cyclopentane with $COCH_3$, CO_2H on one carbon

(*e*) cyclohexane with CO_2Et and CO_2Et on 1,3-positions

(*f*) cyclopentanone with CO_2H on each carbon adjacent to C=O

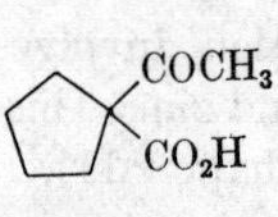

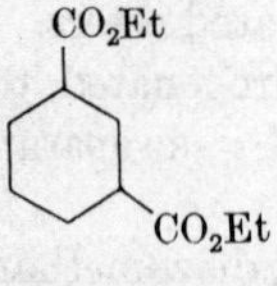

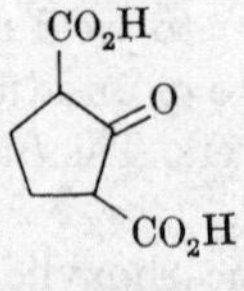

3. Suggest methods for synthesizing the following, all of which involve carbanions. (These are more difficult than those in question 2; use hints given below if necessary.)

(*a*) $CH_3CO[CH_2]_5CO_2H$

(*b*) $(C_6H_5)(CH_3)CH{-}CH(CO_2Et)(CH_2CO_2Et)$

(*c*) $(C_6H_5)(HO_2C)C{=}CHC_6H_5$

(*d*) $C_6H_5CH{=}CH{-}CH{=}CH{-}CHO$

(*e*) 5,5-dimethylcyclohexane-1,3-dione (H_3C, CH_3; O, O)

[Hints: (*a*) cyclohexane *via* enamine;
(*b*) diethyl succinate;
(*c*) and (*d*) benzaldehyde;
(*e*) mesityl oxide.]

CHAPTER 17

Synthesis (B): Aromatic Compounds

The most characteristic property of aromatic compounds is that they undergo addition-with-elimination reactions (so-called electrophilic substitution). It is very important to realize the tremendous variation in reactivity of aromatic molecules. This Chapter begins with the various types of addition-with-elimination reactions that aromatic compounds undergo, and particular attention will be paid to the variation in reactivity of different aromatic nuclei.

Addition-with-Elimination Reactions

(a) *Nitration*

Benzene, the archetype of aromatic compounds, is nitrated by a mixture of nitric acid and sulphuric acid, the active species being the nitronium ion NO_2^+ (see Part 1, Chapter 13). Toluene is appreciably more reactive, being nitrated approximately 25 times faster than benzene under the same conditions. As discussed in Chapter 8, the electron-repelling methyl group directs nitration to the *ortho*- and *para*-positions, as well as accelerating it. Nitrobenzene is nitrated approximately 10,000 times more slowly than benzene under the same conditions. The acceptor properties of the nitro-group greatly deactivate the benzene nucleus to attack by an electrophile, and the electron-attracting properties of the nitro-group direct substitution predominantly into the *meta*-position. Experimentally, this tremendous difference in rates means that nitration of benzene to yield nitrobenzene rather than *m*-dinitrobenzene can be achieved: the nitration of nitrobenzene is so much slower that very high yields of nitrobenzene uncontaminated with a dinitro-product can be obtained. On the other hand, phenol is nitrated approximately a thousand times more rapidly than benzene, and use of the nitric acid–sulphuric acid mixture would result in polynitration and tarry products; therefore phenol is normally nitrated in aqueous nitric acid; the mechanism of this reaction may be somewhat different from that discussed previously, but we shall not consider this in detail at the present time. As we should expect, aniline is still more reactive—so much so that its reaction with concentrated nitric acid has been used as a rocket propellent. To prevent polynitration and decomposition, aniline must always first be acylated, usually by acetic anhydride; the acetanilide can then be nitrated smoothly, under mild conditions, to yield *o*- and *p*-nitro-derivatives as we should expect from a donor substituent. As described in Chapters 7 and 8, the amino-group is an electron-donor and a π-electron-repeller: the ammonium cation, on the other hand, is not a donor and behaves as an electron-attractor. Thus, *N*,*N*-dimethyl-*p*-toluidine is nitrated in the 2-position (—NR_2 = 1) in solutions of low acidity and in the 3-position in solutions of high acidity.

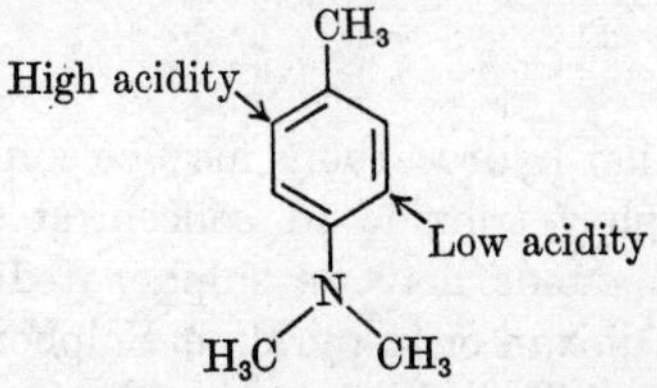

CH_3 High acidity Low acidity N H_3C CH_3

Positions nitrated depending on the acidity of the medium.

(*b*) *Halogenation*

Bromination or chlorination of benzene requires Lewis-acid-type catalysts (e.g. for bromination $Br^+FeBr_4^-$). Bromination of nitrobenzene is difficult to achieve. On the other hand, aniline reacts instantly with an aqueous solution of bromine, to precipitate the 2,4,6-tribromoaniline.

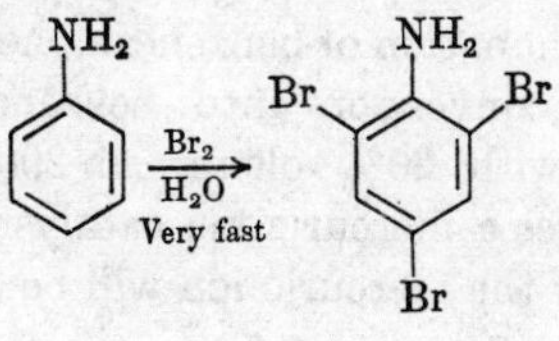

NH_2 $\xrightarrow[H_2O]{Br_2}$ Very fast NH_2 Br Br Br

Monobromination of aniline can be achieved by making acetanilide first and brominating the amide.

$$C_6H_5NH_2 \xrightarrow{(CH_3CO)_2O} C_6H_5NHCOCH_3 \xrightarrow[CCl_4]{Br_2} p\text{-}BrC_6H_4NHCOCH_3\ (80\%) + o\text{-}BrC_6H_4NHCOCH_3\ (20\%)$$

o-Bromoacetanilide is very much more soluble than the *p*-bromo-compound and they may be easily separated by recrystallization. *p*-Bromoaniline may then be formed by hydrolysis of the amide. Phenol likewise reacts extremely rapidly with bromine: treatment with bromine water (a solution of bromine in water) gives the tribromo-derivative extremely rapidly and this is precipitated as in the case of aniline. However, by carrying out the bromination at low temperatures in carbon disulphide solution, monobromination can be achieved to yield predominantly *p*-bromophenol.

(*c*) *Sulphonation*

Benzene and similar hydrocarbons may be sulphonated by oleum (a solution of sulphur trioxide in concentrated sulphuric acid). More reactive compounds may be sulphonated with solutions of sulphur trioxide in dioxan or in pyridine. Sulphonation differs from most other electrophilic addition-with-elimination reactions of aromatic compounds in that it is reversible. At temperatures between 100° and 200° an equilibrium is set up and the products of the reaction depend on the concentrations of the reactants.

Deliberate reversal of sulphonation (desulphonation) may usually be effected by heating the arenesulphonic acid with dilute sulphuric acid at 150–200°: this may be accomplished by carrying out the reaction in a sealed tube or by passing superheated steam through a solution of the sulphonic acid in dilute sulphuric acid.

Benzenesulphonic acid may be prepared by addition of benzene to 8% oleum at 0°. Sulphonation of benzenesulphonic acid is very much more difficult and conversion into benzene-*m*-disulphonic acid requires treatment with 20% oleum at 200°. Benzene-1,3,5-trisulphonic acid requires a mercuric ion catalyst with 15% oleum at 300°. The function of the mercuric ion will be discussed below.

Arenesulphonyl chlorides may sometimes be made directly by treating the aromatic compound with chlorosulphonic acid, $ClSO_3H$. For example, toluene gives a mixture of *o*- and *p*-toluenesulphonyl chloride. This is a reaction of some commercial importance as the *o*-toluenesulphonyl chloride is an intermediate in the manufacture of saccharin. The *p*-toluenesulphonyl chloride is also of value as it can be converted into chloramine-T which is used as a bactericide for purifying water.

CH_3 $ClSO_3H$ CH_3 SO_2Cl NH_3 CH_3 SO_2NH_2 1. $KMnO_4$ 2. $-H_2O$ O C NH SO_2

Saccharin

CH_3 SO_2Cl NH_3 CH_3 SO_2NH_2 NaOCl CH_3 SO_2NHCl

Chloramine-T

Another way of preparing the sulphonyl chlorides in one step is to treat the aromatic compound with a mixture of sulphuric acid and thionyl chloride:

$$ArH + H_2SO_4 + SOCl_2 \longrightarrow ArSO_2Cl + SO_2 + HCl$$

Sulphonation of phenols proceeds without difficulty under mild conditions. Sulphonation of aromatic amines is sometimes achieved by forming the amine sulphate and heating this solid at 180–200°:

CH_3 NH_3^+ HSO_4^- Heat 180° CH_3 NH_3^+ ^-O_3S

(*d*) *Mercuration*

Mercuration of aromatic compounds is a somewhat neglected reaction. From a mechanistic point of view it is complicated by the fact that two distinct reaction mechanisms are operative. In acid solution the reaction is a normal electrophilic addition of the mercuric cation, followed by elimination, analogous to nitration or halogenation. Under certain extreme conditions, however, mercuric acetate may decompose homolytically to yield radicals, but we shall not be concerned with reactions under these conditions. Benzene may be mercurated by treatment with mercuric acetate in glacial acetic acid at 100°:

$$C_6H_6 + Hg(OCOCH_3)_2 \longrightarrow C_6H_5HgOCOCH_3 + CH_3CO_2H$$

Phenol and aniline are mercurated very much more readily and polymercuration may occur. Benzene substituted by electron-acceptor groups undergoes electrophilic substitution only in strongly acidic media. Very often mercuration is an intermediate step in complex substitution reactions carried out in the presence of catalytic amounts of mercury. A particularly important process is the oxy-nitration of benzene in which benzene is treated with nitric acid and mercuric nitrate, the ultimate product being picric acid. Another example is the sulphonation of benzene-*m*-disulphonic acid, to yield the 1,3,5-trisulphonic acid referred to above. These reactions work because the mercury atom is easily replaced by other electrophiles (i.e. is more easily replaced than is a proton).

(*e*) *Nitrosation*

The nitrosonium ion is very much less reactive than the nitronium ion. Activated aromatic nuclei such as mesitylene (1,3,5-trimethylbenzene) and anisole (methyl phenyl ether) can be nitrosated by nitrosylsulphuric acid, $NO^+ \; HSO_4^-$. If the reaction mixture is simply poured into water, the unchanged nitrosylsulphuric acid reacts with the water to form nitrous acid and sulphuric acid and the former then reacts with the nitroso-compound to produce the diazonium salt (see Chapter 14). In order to obtain the nitroso-compound it is necessary to pour the reaction mixture into aqueous sulphamic acid (NH_2SO_3H), so that the nitrous acid is destroyed before it has a chance to react with the aromatic nitroso-compound. More reactive aromatic compounds, such as phenols and tertiary amines, react with aqueous nitrous acid (probably as the result of attack by either N_2O_3 or NOCl). With phenols the nitroso-compound formed only reacts slowly, to give the diazonium salt, because the nitrosophenol is in equilibrium with the quinone monoxime:

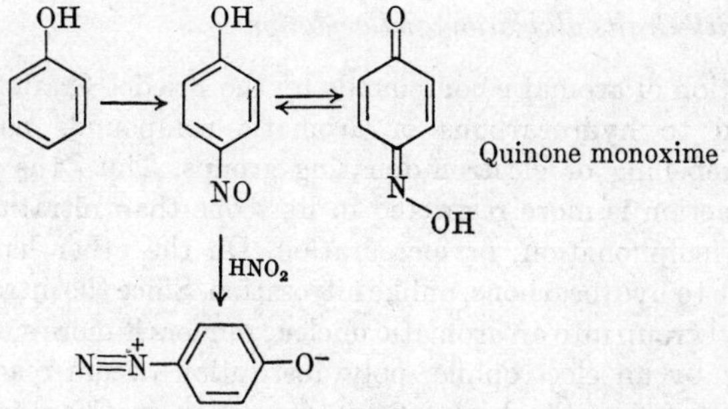

Nonetheless, it is impossible to prevent some formation of the diazonium salt (see Chapter 15). With aromatic tertiary amines such as *N*,*N*-dimethylaniline, nitrosation in acidic medium does not proceed to the diazonium salt because the protonated form of the nitroso-compound does not react further with nitrous acid:

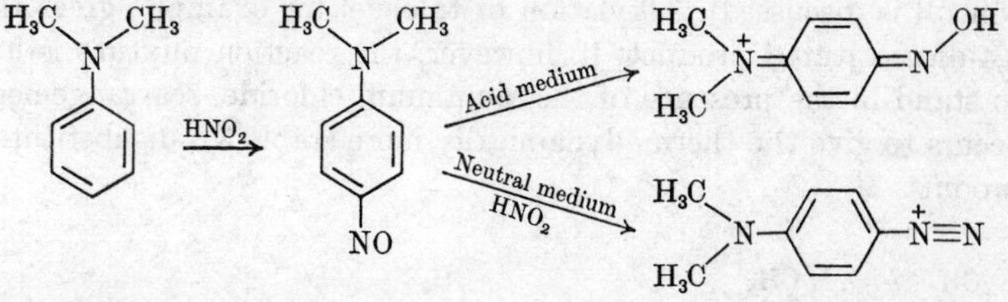

N,*N*-Dimethyl-*p*-nitrosoaniline can be obtained in almost quantitative yield by treatment of dimethylaniline with nitrous acid, followed by basification, the free nitroso-compound being released. Although the nitrosonium ion is far too unreactive to add to an aromatic ring substituted by electron-acceptors, aromatic nitroso-compounds can be prepared by treatment of aromatic compounds with nitrosylsulphuric acid in sulphuric acid solution in the presence of catalytic amounts of mercuric cations. As before, if the nitroso-compound is required, the mixture must be poured into sulphamic acid to prevent the formation of the diazonium salt:

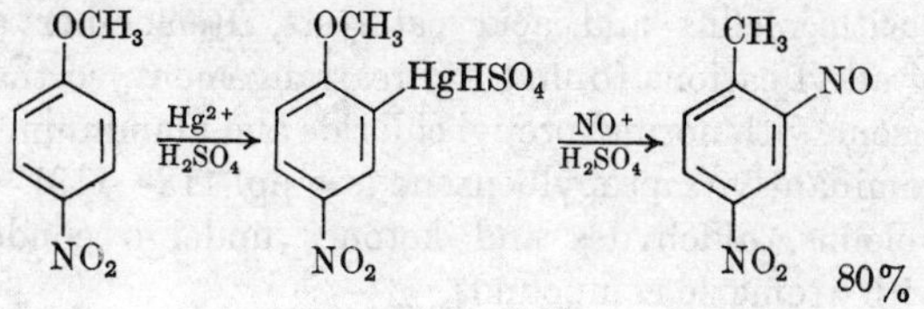

(*f*) *Friedel–Crafts alkylation and acylation*

Alkylation of aromatic compounds by the Friedel–Crafts reaction is limited to hydrocarbons or aromatic compounds containing electron-repelling or electron-donating groups. Thus, the Friedel–Crafts reaction is more restricted in its scope than nitration, halogenation, sulphonation, or mercuration. On the other hand, it is applicable to hydrocarbons, unlike nitrosation. Since the introduction of an alkyl group into an aromatic nucleus renders it more susceptible to attack by an electrophile, polysubstitution occurs readily. We should expect, on the basis of our discussion in Chapter 8, that polysubstitution would result in 1,2- or 1,4-disubstituted products. However, Friedel–Crafts alkylation is reversible, so that prolonged reaction will give the thermodynamically most stable product rather than that most rapidly formed (see Chapter 12, on tautomerism, where the difference between thermodynamic control and kinetic control is discussed). Alkylation of toluene, for example, gives the 2,4-disubstituted product; if, however, the reaction mixture is left to stand in the presence of the aluminium chloride, rearrangement occurs to give the thermodynamically more stable 3,5-disubstituted product.

CH_3 (toluene) $\xrightarrow[AlCl_3]{2RCl}$ CH_3, R, R (2,4-disubstituted) $\xrightarrow{\text{Prolonged reaction}}$ CH_3, R, R (3,5-disubstituted) + other isomers including $CH_3C_6H_2R_3$ and $CH_3C_6H_4R$

Alkylation with alkyl halides is usually carried out with aluminium chloride as catalyst, but ferric chloride, stannic chloride, boron trifluoride, and zinc chloride may each be used depending on the reactants. The mechanistic features of these reactions have been discussed in Chapter 10; particular attention should be paid to alkylation with olefins and acid catalysts. Remember also the tendency of alkyl cations to undergo rearrangement, so that treatment of benzene with normal propyl chloride and aluminium chloride yields predominantly *iso*propylbenzene (see pp. 118—119).

Besides olefins, aldehydes and ketones undergo condensation reactions with aromatic compounds.

$$CH_3CHO + H{-}OSO_3H \rightarrow H_3C\overset{+}{C}H(OH) \xrightarrow{m\text{-xylene}} HO{-}CH(CH_3){-}[\text{arenium ion}]{-}CH_3 \xrightarrow{-H_2O}$$

$$H_3C{-}C_6H_3(CH_3){-}\overset{+}{C}H(CH_3) \xrightarrow{m\text{-xylene}} H_3C{-}C_6H_3(CH_3){-}CH(CH_3){-}C_6H_3(CH_3){-}CH_3$$

An important application of this reaction is the condensation of chlorobenzene with chloral (trichloroacetaldehyde) in the presence of oleum:

$$2\,Cl{-}C_6H_5 + CCl_3CHO \xrightarrow[H_2SO_4]{10\%\,SO_3} Cl{-}C_6H_4{-}CH(CCl_3){-}C_6H_4{-}Cl$$

1,1,1-Trichloro-2,2-di-(*p*-chlorophenyl) ethane
[incorrectly called Dichloro-Diphenyl-Trichloroethane]

Closely related is a very important alkylation reaction called chloromethylation, in which the aromatic compound is treated with formaldehyde (as the polymer paraformaldehyde) and hydrogen chloride; zinc chloride is usually used as catalyst:

$$C_{10}H_8 + CH_2O + HCl \xrightarrow{ZnCl_2} C_{10}H_7CH_2Cl$$

The reaction probably involves the oxycarbonium ion ($CH_2{=}\overset{+}{\ddot{O}}H \leftrightarrow \overset{+}{C}H_2{-}\ddot{O}H$), but when water is eliminated after the first alkylation the resulting arylalkyl cation takes up a chloride anion:

$$ArH + H_2\overset{+}{C}{-}OH \longrightarrow \overset{+}{Ar}(H)(CH_2OH) \xrightarrow{-H_2O} ArCH_2^{+} \xrightarrow{Cl^-} ArCH_2Cl$$

Acylation with an acid chloride and aluminium chloride differs from the alkylation in that a molecular equivalent of aluminium chloride is required. This is because the ketone forms a complex with the aluminium chloride and the latter is not regenerated at the end of each reaction as it is in alkylation. Acylation may also be achieved with acyl cations (RCO^+), but these are less reactive than the acid chloride–aluminium chloride complex; so, although the acetyl cation (CH_3CO^+) attacks anisole (methyl phenyl ether) to yield 4-methoxyacetophenone it does not attack benzene. Acetophenone may readily be prepared by treating benzene with acetyl chloride or acetic anhydride and aluminium chloride.

A somewhat related reaction is the Fries reaction, in which the acetate ester of the phenol, treated with aluminium chloride, yields a mixture of the 2- and 4-hydroxyacetophenone:

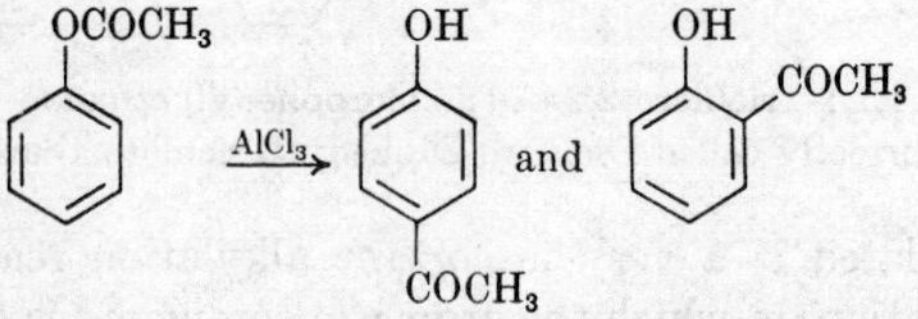

$OCOCH_3$ $\xrightarrow{AlCl_3}$ OH, $COCH_3$ and OH, $COCH_3$

Closely related to Friedel–Crafts acylation involving an acid chloride and aluminium chloride is the so-called Gattermann–Koch reaction, in which the aromatic compound is treated with carbon monoxide and HCl in the presence of aluminium chloride, to yield the corresponding aldehyde:

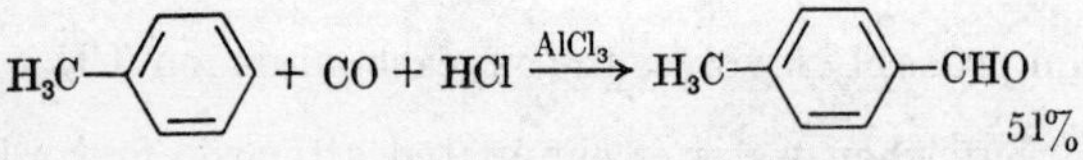

H_3C- + CO + HCl $\xrightarrow{AlCl_3}$ H_3C- $-CHO$ 51%

The yields are poor except with aromatic nuclei activated by either electron-repellers or, better, electron-donors. A variant involves using HCN in place of carbon monoxide and is known as the Hoesch synthesis. The latter reaction is restricted to phenols and very reactive compounds such as mesitylene (1,3,5-trimethylbenzene); alkyl cyanides may also be used in this case.

$$H_3C\text{-}C_6H_3(CH_3)_2 + HCN + HCl \longrightarrow (H)(\overset{+}{N}H_2\ Cl^-)C\text{-}C_6H_2(H_3C)(CH_3)(CH_3) \xrightarrow{H_2O} CHO\text{-}C_6H_2(H_3C)(CH_3)(CH_3)$$

(Probably $H\overset{+}{C}{=}NH\ Cl^-$)

$$HO\text{-}C_6H_4\text{-}OH + CH_3CN + HCl \xrightarrow{ZnCl_2} H_3C\text{-}C(=\overset{+}{N}H_2\ Cl^-)\text{-}C_6H_3(OH)(OH) \xrightarrow{H_2O} H_3C\text{-}C(=O)\text{-}C_6H_3(OH)(OH)$$

Introduction of Specific Substituents into the Benzene Ring and their Interconversion

(*a*) *The hydroxyl group*

The most important method for introducing the hydroxyl group into an aromatic nucleus is by fusion of the sodium salt of a sulphonic acid with sodium hydroxide.

$$C_6H_4(SO_3^-Na^+)_2 \xrightarrow[270^\circ]{NaOH} C_6H_4(OH)_2$$

Resorcinol

The other general route to phenols is from the amine to the diazonium salt which on decomposition in aqueous solution yields the phenol. Moderate yields of phenol can be obtained simply by heating the aqueous solution, but the product will be contaminated with azo-compounds, biphenyls, and tars. Better results are obtained by working in fairly concentrated sulphuric acid solution (40–50%) and at a high temperature, which results in rapid decomposition of the diazonium salt and reduces the tendency for the unconverted diazonium salt to couple with the phenol as it is formed. This route is, in general, of much less value than the sulphonic acid route, but it can be important when there are substituents which would render the compound sensitive to alkali, for example, nitro-groups.

$$m\text{-}NH_2C_6H_4NO_2 \xrightarrow{HNO_2} m\text{-}N_2^+C_6H_4NO_2 \xrightarrow[160^\circ]{H_2O/H_2SO_4(40\%)} m\text{-}HOC_6H_4NO_2 \quad (\sim 75\%)$$

In the laboratory, the replacement of halogen by a hydroxyl group is not an easy process unless there are acceptor groups in positions *ortho* and *para* to the halogen. Under these circumstances the halogen may be removed readily (see Chapter 13):

$$2,4\text{-}(NO_2)_2C_6H_3Cl \xrightarrow[Na_2CO_3]{Aq.} 2,4\text{-}(NO_2)_2C_6H_3OH \quad \left(\text{but } 3,5\text{-}(O_2N)_2C_6H_3Cl \xrightarrow[Na_2CO_3]{Aq.} \text{No reaction}\right)$$

Although the chlorine atom in chlorobenzene cannot be replaced by a hydroxyl group under normal laboratory conditions, this reaction was employed in the manufacture of phenol:

$$C_6H_5Cl \xrightarrow[300^\circ;\ \text{pressure}]{10\%\ NaOH} C_6H_5ONa \quad (+\ C_6H_5OC_6H_5)$$

(*b*) *The amino-group*

The most important way of introducing the amino-group into the aromatic nucleus is by the reduction of a nitro-compound. This reaction may be carried out chemically (e.g. by tin and hydrochloric acid) or catalytically (e.g. by hydrogen and Raney nickel). Partial reduction of a polynitro-compound can be important, and a particularly useful reaction is the partial reduction of *m*-dinitrobenzene to *m*-nitroaniline by yellow ammonium sulphide or hydrogen sulphide in alcoholic ammonia:

$$m\text{-}C_6H_4(NO_2)_2 \xrightarrow[\text{in EtOH}]{H_2S\ +\ NH_3} m\text{-}NH_2C_6H_4NO_2$$

With very reactive compounds such as *N*,*N*-dimethylaniline, nitrosation is a much more controlled reaction than nitration, and unsymmetrical *N*,*N*-dimethyl-*p*-phenylenediamine is best prepared by the nitrosation of dimethylaniline followed by reduction.

$$C_6H_5N(CH_3)_2 \xrightarrow[\text{in acid sol.}]{HNO_2} p\text{-}ON{\cdot}C_6H_4N(CH_3)_2 \xrightarrow[H_2O]{Zn + HCl} p\text{-}H_2N{\cdot}C_6H_4N(CH_3)_2$$

N,*N*-Dimethyl-*p*-phenylenediamine

The remaining important route for introducing an amino-group into the aromatic nucleus is from a carboxylic acid *via* the acid chloride, acyl azide, and a Curtius rearrangement (see Chapter 15):

$$ArCO_2H \longrightarrow ArCOCl \xrightarrow{NaN_3} ArCON_3 \longrightarrow ArNCO \xrightarrow{H_2O} ArNH_2$$

(*c*) *The carboxyl group*

Because of the stability of the aromatic nucleus, oxidation of alkyl side-chains results in the formation of arenecarboxylic acids. Benzoic acid may be prepared by the oxidation of toluene, and benzene-*p*-dicarboxylic acid (terephthalic acid) is prepared from *p*-xylene. This is an important reaction in the preparation of the polymer 'Terylene' (see Part 1, Chapter 17).

$$p\text{-}CH_3C_6H_4CH_3 \xrightarrow[\text{Industry: } O_2 + V_2O_5]{\text{Laboratory: } Na_2Cr_2O_7 + H_2SO_4\text{; or } KMnO_4} p\text{-}HO_2C{\cdot}C_6H_4{\cdot}CO_2H$$

Another way of converting methyl groups into carboxyl groups when attached to the aromatic nucleus is by radical chlorination, to yield the trichloromethyl group which is then readily hydrolysed by dilute aqueous alkali:

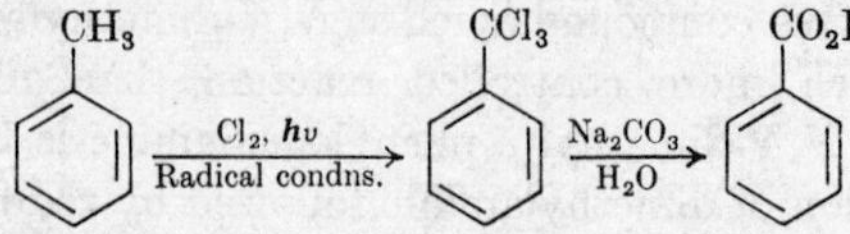

$$C_6H_5CH_3 \xrightarrow[\text{Radical condns.}]{Cl_2,\ h\nu} C_6H_5CCl_3 \xrightarrow[H_2O]{Na_2CO_3} C_6H_5CO_2H$$

A halogen substituent, particularly a bromine atom, may be converted into a carboxyl group, through the Grignard reagent in the usual way:

$$ArBr \xrightarrow[Et_2O]{Mg} ArMgBr \xrightarrow{CO_2} ArCO_2H$$

Similarly, the cyano-group may be hydrolysed to yield the corresponding carboxylic acid:

$$ArCN \xrightarrow[-H_2O]{H_2SO_4} ArCO_2H$$

The carboxyl group may be introduced directly into some phenols by the reaction of the phenoxide anion with carbon dioxide at fairly high pressures. This reaction, known as the Kolbe–Schmidt reaction, is used in the manufacture of salicylic acid:

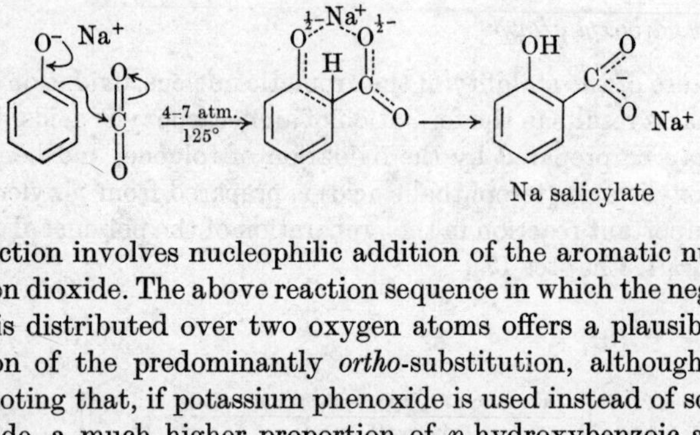

$$C_6H_5O^- Na^+ + O{=}C{=}O \xrightarrow[125^\circ]{4\text{–}7\ \text{atm.}} \text{[intermediate, } \tfrac{1}{2}-\ Na^+\ \tfrac{1}{2}-\text{]} \longrightarrow \text{Na salicylate}$$

Na salicylate

The reaction involves nucleophilic addition of the aromatic nucleus to carbon dioxide. The above reaction sequence in which the negative charge is distributed over two oxygen atoms offers a plausible explanation of the predominantly *ortho*-substitution, although it is worth noting that, if potassium phenoxide is used instead of sodium phenoxide, a much higher proportion of *p*-hydroxybenzoic acid is formed.

(*d*) *Aldehydes and ketones*

Ketones may be prepared by Friedel–Crafts acylation, as discussed above. Similarly, aldehydes may be prepared from reactive hydrocarbons by the Gattermann–Koch reaction or by the Hoesch synthesis, also described above. Benzaldehyde and its derivatives may be prepared from toluene by a variety of oxidizing agents. Much talked about is the reaction of toluene with chloryl chromide (CrO_2Cl_2), but more useful is the oxidation of the toluene by chromic anhydride in acetic anhydride solution:

$$\text{2-}BrC_6H_4CH_3 \xrightarrow[-AcOH-H_2SO_4]{CrO_3-Ac_2O} \text{2-}BrC_6H_4CH(OCOCH_3)_2 \xrightarrow{H^+-H_2O} \text{2-}BrC_6H_4CHO$$

Alternatively, the methylbenzene may be chlorinated under radical conditions to give the dichloromethyl group which can be hydrolysed to yield the aldehyde.

$$ArCH_3 \xrightarrow[\text{Radical condns.}]{Cl_2,\ h\nu} ArCHCl_2 \xrightarrow[H_2O]{Na_2CO_3} ArCHO$$

In alkylbenzenes the methylene group adjacent to the aromatic nucleus is more reactive than methylene groups further along the alkyl chain, so that careful oxidation can yield the alkyl aryl ketone.

$$C_6H_5CH_2CH_3 \xrightarrow[130°C,\ 5\ atm.]{O_2-Mn(OCOCH_3)_2} C_6H_5COCH_3$$

Conventional reactions of groups such as the cyano-group can be employed to prepare aromatic aldehydes and ketones. Thus an aryl cyanide may be reduced with stannous chloride and HCl, in the Stephen reaction, to yield an aldehyde:

$$ArCN \xrightarrow[-HCl]{SnCl_2} (ArCH{=}\overset{+}{N}H_2\quad HSnCl_6{}^-) \xrightarrow{H_2O} ArCHO$$

or an aryl cyanide will react with a Grignard reagent, to yield a ketone:

$$ArCN + RMgBr \longrightarrow (R)(Ar)C{=}NMgBr \xrightarrow{H_2O} (R)(Ar)C{=}O$$

(e) The cyano-group

The cyano-group is important as the stepping stone to a large number of other groups. One method of introducing it into an aromatic nucleus is the fusion of an arenesulphonic acid with potassium cyanide:

$$C_6H_5SO_3K + KCN \xrightarrow[\text{very high temp.}]{\text{Fuse}} C_6H_5CN + K_2SO_4$$

The yields by this procedure are usually moderate and, in general, the most important method for introducing the cyano-group is through the amine, the diazonium salt, and Sandmeyer reaction (see Chapter 15):

$$ArNH_2 \xrightarrow[H_2SO_4\text{–}H_2O]{HNO_2\text{–}} Ar\overset{+}{N}{\equiv}N \quad HSO_4^- \xrightarrow{CuCN} ArCN$$

(f) Halides

Chlorine and bromine atoms may be introduced into the aromatic nucleus by direct electrophilic addition-with-elimination reactions discussed above. The other important method of introducing these halogens, and the only satisfactory way of introducing isolated fluorine and iodine, is through diazonium salts. The Sandmeyer reaction can be used for chlorine and bromine. Fluorine requires the formation of the diazonium fluoroborate, and iodine requires a solution of potassium iodide. Both these reactions are discussed in Chapter 15. Two other much less important routes are available. The silver salt of a carboxylic acid, when treated with bromine in boiling carbon tetrachloride, yields the aryl halide (see Part 1, Chapter 18). The other minor route involves initial mercuration of the aromatic nucleus and replacement of the mercury derivative by the halogen:

$$C_6H_6 + Hg(O_2CCH_3)_2 \xrightarrow{100°} C_6H_5HgO_2CCH_3 + HO_2CCH_3$$

Phenylmercuric acetate

$$C_6H_5HgO_2CCH_3 \xrightarrow{Br_2} C_6H_5Br + BrHgO_2CCH_3$$

Directive Effects and the Correct Orientation of Substituents

As well as the various types of reaction by which particular groupings may be introduced into an aromatic nucleus, we have another major problem to consider, namely, the introduction of a specific group at a specific position in the nucleus.

Suppose that we wish to obtain *m*-nitrotoluene; nitration of toluene yields predominantly *p*-nitrotoluene, together with some of the *ortho*-isomer owing to the directive effect of the methyl group. The Friedel–Crafts reaction fails with nitrobenzene because, as described above, Friedel–Crafts alkylation is restricted to benzene and alkylbenzenes. The best procedure is outlined in the annexed reaction sequence. The first step (A) involves nitration of toluene,

$$C_6H_5CH_3 \xrightarrow[A]{HNO_3} p\text{-}CH_3C_6H_4NO_2 \xrightarrow[B]{Sn-HCl} p\text{-}CH_3C_6H_4NH_2 \xrightarrow[C]{(CH_3CO)_2O} p\text{-}CH_3C_6H_4NHCOCH_3 \xrightarrow[D]{HNO_3}$$

$$CH_3C_6H_3(NO_2)NHCOCH_3 \xrightarrow[E]{H^+-H_2O} CH_3C_6H_3(NO_2)NH_2 \xrightarrow[F]{HNO_2-H_2SO_4} CH_3C_6H_3(NO_2)N_2^+ \xrightarrow[G]{H_3PO_2} m\text{-}CH_3C_6H_4NO_2$$

which gives predominantly the *para*-isomer; 20% of the *ortho*-compound is formed but separation is easy. Step (B) is the reduction to the amine. This introduces a donor group which directs a further entering substituent *ortho–para* to itself, and the directive effect of a donor group is always more powerful than that of an electron-repeller such as the methyl group. However, before we can nitrate this *p*-toluidine we must protect the amino-group and so reduce the compound's reactivity so that we need reaction (C). The resulting *p*-methylacetanilide is then nitrated (step D), to give the 2-nitro-4-methylacetanilide. Hydrolysis (E) yields the free amine, which may then be diazotized (F) and treated with hypophosphorous acid (G) to yield the desired compound. The important feature in this reaction sequence is that we introduce an amino-group to provide correct orientation of our nitration; we then remove this group by diazotization and treatment with hypophosphorous acid.

Another example would be the preparation of 1,3,5-tribromobenzene. If we brominated benzene, the first bromine atom introduced behaves as a donor group, and substitution occurs *ortho–para* to it, yielding *o*- and *p*-dibromobenzene. However, bromination of aniline, as we know, proceeds extremely readily, to give 2,4,6-tribromoaniline, and diazotization of the amino-group followed by treatment with hypophosphorous acid or boiling with alcohol will give the desired compound (this diazonium salt is exceptional in that boiling with ethanol gives good yields of the hydrocarbon).

NH_2 $\xrightarrow{Br_2—H_2O}$ (Br, Br, Br; NH_2) $\xrightarrow{HNO_2—H_2SO_4}$ (Br, Br, Br; N_2^+) $\xrightarrow{H_3PO_2}$ (Br, Br, Br)

Another simple example is the preparation of *p*-dinitrobenzene. This can be most readily prepared from aniline, by conversion into *p*-nitroaniline. The amino-group may then be diazotized and treated

NH_2 $\xrightarrow{(CH_3CO)_2O}$ $NHCOCH_3$ $\xrightarrow[H_2SO_4;\ 0°]{HNO_3—}$ $NHCOCH_3$ / NO_2 $\xrightarrow{H^+\ —H_2O}$ NH_2 / NO_2

NH_2 / NO_2 $\xrightarrow{HNO_2}$ N_2^+ / NO_2 $\xrightarrow{Na^+{}_3[Co(NO_2)_6]^{3-}}$ NO_2 / NO_2

NH_2 / NO_2 $\xrightarrow{CF_3CO_3H\ —CHCl_3}$ NO_2 / NO_2

with sodium cobaltinitrite according to the Hodgson reaction (see Chapter 15) or oxidized with trifluoroperacetic acid in a reaction that we have not discussed above. This oxidation is a fairly general reaction, namely, the oxidation of an amine, particularly arylamines, with trifluoroperacetic acid in chloroform solution, to the corresponding nitro-compound.

Carboxyl is an acceptor group and is *meta*-directing. However, *p*-aminobenzoic acid can readily be prepared from toluene by the sequence shown.

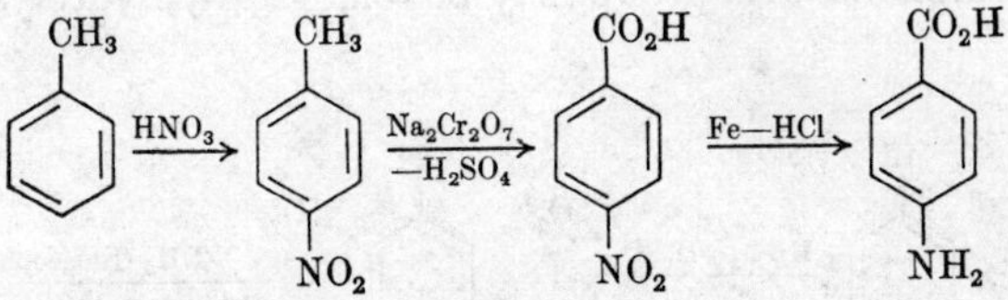

CH_3 $\xrightarrow{HNO_3}$ CH_3 / NO_2 $\xrightarrow[-H_2SO_4]{Na_2Cr_2O_7}$ CO_2H / NO_2 $\xrightarrow{Fe-HCl}$ CO_2H / NH_2

The nitro-group may be introduced to direct future substitution *meta* to it, and may then be removed by reduction to the amino-group followed by diazotization, etc.; or the nitro-group may be converted into an amino-group which directs *ortho–para* and again may be removed by diazotization. These simple techniques should make it possible to introduce a substituent into any desired position. However, another substituent which is also frequently used in this fashion is the sulphonic acid group. As we have described, the sulphonic acid group can be removed by treatment with dilute sulphuric acid at 160–200°. Thus *o*-nitroaniline may be prepared by the sequence illustrated:

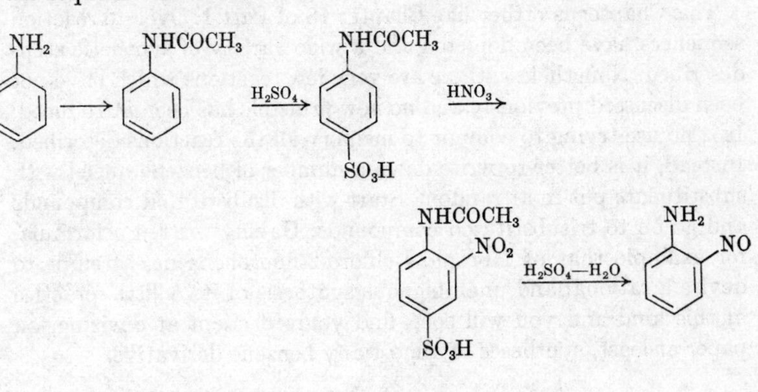

NH_2 $\longrightarrow$ $NHCOCH_3$ $\xrightarrow{H_2SO_4}$ $NHCOCH_3$ / SO_3H $\xrightarrow{HNO_3}$

$NHCOCH_3$ / NO_2 / SO_3H $\xrightarrow{H_2SO_4-H_2O}$ NH_2 / NO_2

Another example of the same technique is the preparation of 2,6-dinitroaniline. In this case we start with chlorobenzene which is sulphonated predominantly in the *para*-position. The sulphonic acid group and the chlorine atom both direct nitration into positions *meta* to the sulphonic acid group. We then have a chlorine atom attached to a benzene ring, with nitro-groups in the two positions *ortho* to it, and also an acceptor sulphonic acid group in the *para*-position. The chlorine atom therefore reacts readily with ammonia, to give the corresponding amine. Acidification gives the amino-sulphonic acid, and the sulphonic acid group may be removed by hydrolysis.

Cl $\xrightarrow{H_2SO_4}$ Cl, SO_3H $\xrightarrow{HNO_3—H_2SO_4}$ Cl, O_2N, NO_2, SO_3H $\xrightarrow{NH_3 \text{ (Nucleophilic addition-with-elimination)}}$

NH_2, O_2N, NO_2, SO_3H $\xrightarrow[—H_2O]{H_2SO_4—HCl}$ NH_2, O_2N, NO_2

Notes to the student

This Chapter is rather like Chapter 18 of Part 1. Over 40 reaction sequences have been depicted and a wide variety of synthetic steps described. Nonetheless, there are very few reactions which have not been discussed previously and no new principle has been introduced. It is no use trying to commit to memory all the reactions described. Instead, it is better to write down a number of benzene nuclei with substituents put in at random. Start with disubstituted compounds and go on to trisubstituted compounds. Having written a formula, for example that of 1-bromo-4-chloro-2-fluorobenzene, attempt to devise a rational and unambiguous synthesis of it. A little practice of this kind and you will soon find yourself adept at devising, on paper at least, syntheses of almost any benzene derivative.

Problems

1. Devise methods for the synthesis of all the dinitrotoluenes and all the tetrachlorobenzenes.

2. Starting from readily accessible benzene derivatives, how would you prepare:

(*a*) $(CH_3)_3\overset{+}{N}$, Br (*b*) OH, $(CH_3)_3C$, $C(CH_3)_3$, $C(CH_3)_3$ (*c*) $COCH_3$, OH, OH

3. How may the following conversions be effected:

(*a*) CO, O, CO ⟶ NH_2, CO_2H

(*b*) OH ⟶ CH_3O, $CH{=}CHCO_2H$

(*c*) ⟶ C_6H_5

(*d*) CO, O, CO ⟶ O, CH_3, O

CHAPTER 18

Molecules with more than One Functional Group

We must now consider what happens when there is more than one functional group in a molecule. Let us begin with an extreme example, in which an acidic group, such as carboxyl, and a basic group, such as amino, are both present in the same molecule. Such compounds are called amino acids. The simplest amino acid is aminoacetic acid, known as glycine. At first sight we might write this molecule as $NH_2CH_2CO_2H$. In fact, no such species exists. The carboxyl group is acidic and the amino group is basic, so in neutral solution glycine exists with a positive charge and a proton on the nitrogen atom (giving an ammonium cation) and the carboxylate group as an anion. This type of molecule is sometimes known as an internal salt or, more usually, by the German term *Zwitterion*. Although the zwitterion carries charges it does not migrate, either to the anode or to the cathode, when its solution is electrolysed. Addition of alkali removes a proton from the nitrogen atom and the molecule thus becomes negatively charged and migrates to the anode. On the other hand, addition of acid adds a proton to the carboxylate part of the molecule which, now positively charged, migrates to the cathode.

$$H_2N{-}CH_2{-}C\begin{matrix}O\\ \cdots\\ O\end{matrix}^{-} \underset{OH^-}{\overset{H^+}{\rightleftharpoons}} H_3\overset{+}{N}{-}CH_2{-}C\begin{matrix}O\\ \cdots\\ O\end{matrix}^{-} \underset{OH^-}{\overset{H^+}{\rightleftharpoons}} H_3\overset{+}{N}{-}CH_2{-}C\begin{matrix}{=}O\\ \\ OH\end{matrix}$$

Migrates to anode — Zwitterion does not migrate — Migrates to cathode

In an acidic solution the amino acid migrates to the cathode, but, if a base is gradually added to the solution, a stage is eventually reached where no migration occurs. Addition of further base will result in the amino acid migrating to the anode. The pH at which no migration occurs is called the *isoelectric point*.

At first sight, dibasic acids are a very different proposition to amino acids, but once one acidic group of a dibasic acid is dissociated we have a basic group and an acidic group together in the same molecule. We can write the dissociation of a dibasic acid as follows:

$$HO_2C(CH_2)_nCO_2H \underset{k_{-1}}{\overset{k_1}{\rightleftharpoons}} HO_2C(CH_2)_nCO_2^- + H^+ \qquad K_1 = \frac{k_1}{k_{-1}}$$

$$HO_2C(CH_2)_nCO_2^- \underset{k_{-2}}{\overset{k_2}{\rightleftharpoons}} {}^-O_2C(CH_2)_nCO_2^- + H^+ \qquad K_2 = \frac{k_2}{k_{-2}}$$

There are two ionizable protons in the undissociated acid, and after complete dissociation there are two sites in the dianion for their return. If the carboxyl groups were completely insulated from one another we should have $k_1 = 2k_2$ and $k_{-2} = 2k_{-1}$, hence $K_1 = 4K_2$. Thus, on purely statistical grounds, we should expect the first dissociation constant of a dicarboxylic acid to be four times greater than the second. Experimentally it is found that the ratio K_1/K_2 is always greater than four. The further the carboxyl groups are separated, i.e. in our generalized formula the greater n is, the nearer the ratio K_1/K_2 approaches four.

Table 18.1. Linear dicarboxylic acids

Acid	Common name	n	10^5K_1	10^6K_2	K_1/K_2
HO_2CCO_2H	Oxalic	0	3470	3.98	8700
$HO_2C(CH_2)CO_2H$	Malonic	1	177	4.37	405
$HO_2C(CH_2)_2CO_2H$	Succinic	2	7.36	4.50	16.4
$HO_2C(CH_2)_3CO_2H$	Glutaric	3	4.60	5.34	8.6
$HO_2C(CH_2)_4CO_2H$	Adipic	4	3.90	5.29	7.4
$HO_2C(CH_2)_5CO_2H$	Pimelic	5	3.33	4.87	6.8
$HO_2C(CH_2)_6CO_2H$	Suberic	6	3.07	4.71	6.5
$HO_2C(CH_2)_7CO_2H$	Azelaic	7	2.82	4.64	6.1
$HO_2C(CH_2)_8CO_2H$	Sebacic	8	2.75	4.60	6.0

We must now refer back to Chapters 6 and 7, especially Chapter 6. The reason why the ratio K_1/K_2 is so much greater than four in oxalic, malonic, and succinic acid is that in the completely dissociated anions of these acids we have two negatively charged groups close together and we are concentrating charge. The additional free energy ($\Delta G^0 = -\boldsymbol{R}T \ln K_1 + \boldsymbol{R}T \ln 4K_2$) must be due to the interaction of the negative charges of the carboxyl groups. Notice that as the carboxyl groups become further apart so this additional free energy becomes less, just as we saw in Table 1, Chapter 7, for ω-chloro carboxylic acids. Very similar effects have been observed in the basic dissociation constants of the linear diamines, though these have been studied in far less detail.

So far we have considered molecules with two extreme types of functional group, i.e. acidic groups and basic groups in the same molecule. We must now consider what happens when a molecule contains two groups which are likely to interact to form a new compound. For example, if an alcoholic hydroxyl group and a carboxyl group are present in the same molecule, we must consider whether the molecule will 'bite its own tail' and form a cyclic ester. Experiment shows that such reactions occur extremely readily if the steric situation is favourable. Thus γ- and δ-hydroxy acids exist as their salts, but on acidification they readily give internal esters known as lactones.

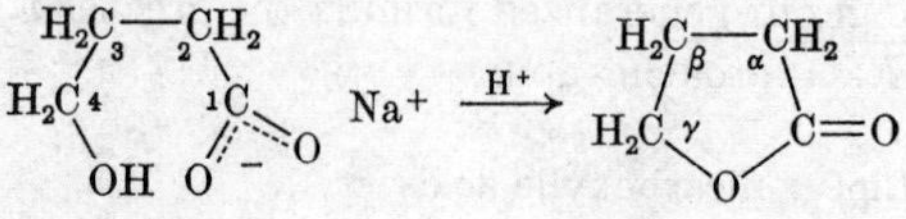

H_2C_4—$_3$—$_2$—$_1C(=O)O^-$ with OH; Na^+ $\xrightarrow{H^+}$ $H_2C(\gamma)$, $H_2C(\beta)$—$CH_2(\alpha)$, C=O, O

Sodium 4-hydroxybutyrate → γ-Butyrolactone

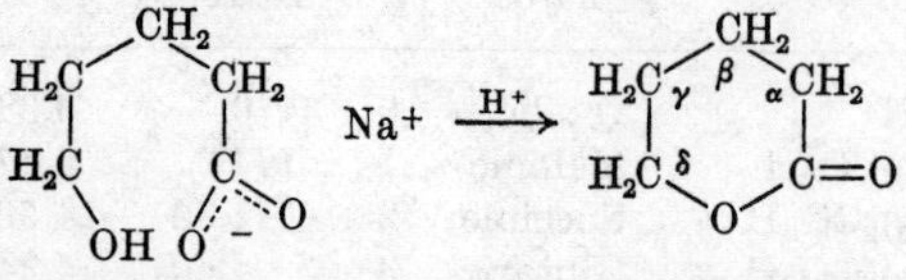

Na^+ $\xrightarrow{H^+}$ (γ, β, α, δ)

Sodium 5-hydroxyvalerate → δ-Valerolactone

In a similar way, 4- and 5-amino carboxylic acids will yield lactams when heated:

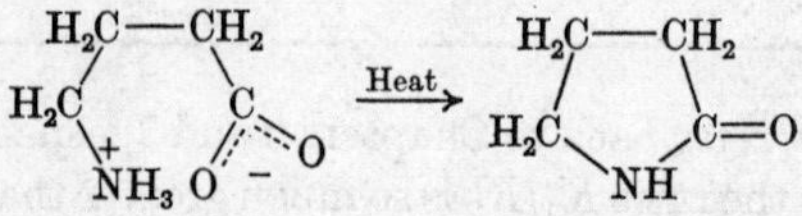

$\xrightarrow{\text{Heat}}$

4-aminobutyric acid → γ-Butyrolactam (Pyrrolidone)

It has been found possible to prepare lactones and lactams with ring sizes other than 5 or 6. For example, β-propiolactone, a 4-membered ring system can be prepared. There is, however, a very big difference. Free 4-hydroxybutyric acid does not exist because the lactone is formed so readily. On the other hand β-propiolactone reacts with water to yield 3-hydroxypropionic acid. Similarly, lactones and lactams with ring systems containing more than six members can be prepared, but these compounds, though not showing the great reactivity of β-propiolactone, are much more reactive than their five- and six-membered-ring analogues. The full reason for the exceptional stability of the five- and six-membered ring systems is outside the scope of the present Chapter and we must content ourselves by saying that the problem is a steric one. A particularly important example of the comparative instability of a lactam with a ring system containing more than six members is caprolactam (hexanolactam). This compound is prepared on a large scale from cyclohexane *via* cyclohexanone oxime which undergoes Beckmann rearrangement with sulphuric acid to yield caprolactam. Heated in the presence of a base, the caprolactam ring opens and a linear polyamide is formed. This material is known as Nylon-6 and for some purposes, its properties are superior to those of Nylon-66 described in Chapter 17 of Part 1. An analogous reaction with δ-valerolactam does not occur, although butyrolactam can be polymerised.

$$\text{Cyclohexanone oxime} \xrightarrow[\text{Beckmann rearrangement}]{H_2SO_4} \text{Caprolactam} \xrightarrow[\text{heat}]{\text{Base;}} \left(NHCO(CH_2)_5\right)_n \ \text{'Nylon 6'}$$

Clearly, an α-hydroxy acid cannot 'bite its own tail', but the possibility that the carboxyl group of one molecule may condense with the alcoholic group of another remains. These reactions in which one molecule of an acid 'bites the tail' of another molecule occur particularly readily with the α-hydroxy and α-amino carboxylic acids. Thus lactic acid, heated in the presence of an acid catalyst, yields a dimeric ester called a lactide:

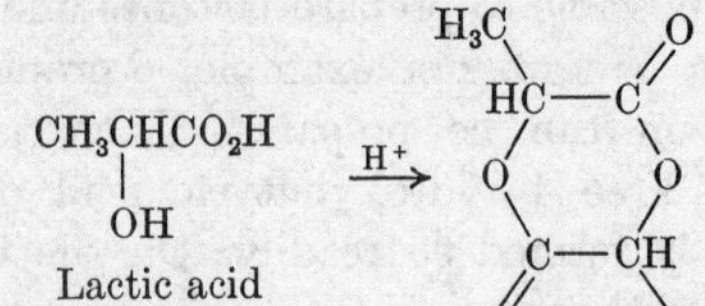

$$CH_3CH(OH)CO_2H \xrightarrow{H^+} \text{Lactide}$$

Lactic acid (2-hydroxypropionic acid)

Similarly, two molecules of glycine ethyl ester condense with the elimination of two molecules of ethanol to form a similar cyclic compound:

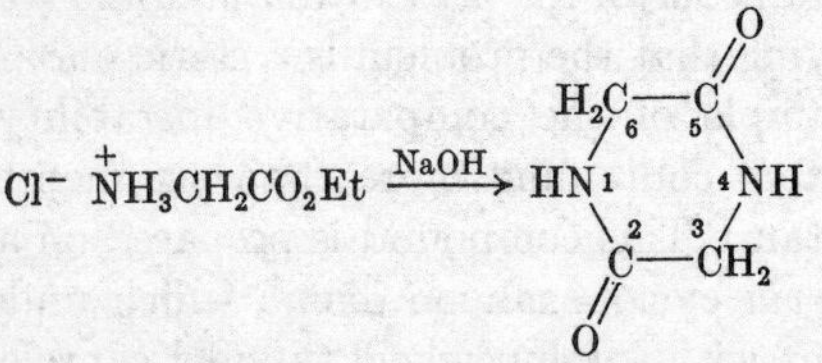

$$Cl^- \overset{+}{N}H_3CH_2CO_2Et \xrightarrow{NaOH} \text{Piperazine-2, 5-dione (diketopiperazine)}$$

Acetic anhydride could be regarded as the acetyl derivative of acetic acid and, similarly, dibasic acids form cyclic anhydrides. Again the facility with which such anhydrides are formed depends on the ring size. Heating succinic acid, glutaric acid, or phthalic acid alone or in the presence of acetic anhydride results in the formation of the cyclic anhydride of the dibasic acid.

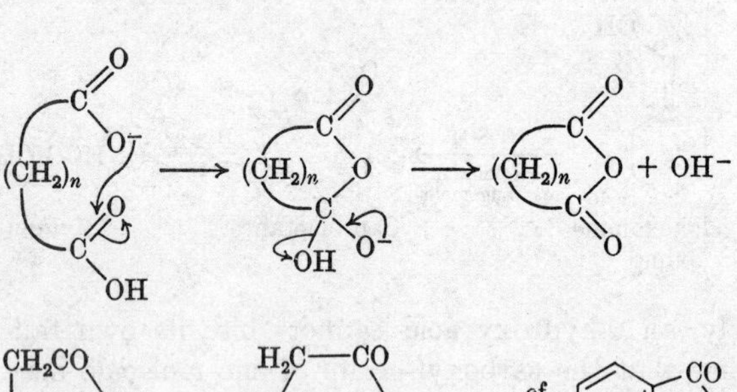

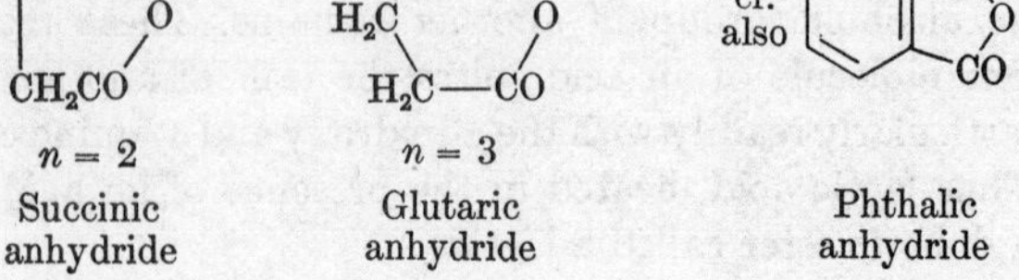

$$(CH_2)_n(CO_2^-)(CO_2H) \longrightarrow \text{tetrahedral intermediate} \longrightarrow (CH_2)_n(CO)_2O + OH^-$$

$n = 2$	$n = 3$	cf. also
Succinic anhydride	Glutaric anhydride	Phthalic anhydride

In a very similar way these acids will form mono- and di-amides and also cyclic imides. The imide can usually be formed simply by heating the diamide.

```
CH2CO                 H2C—CO                         CO
|    \               /      \          cf.   (C6H4)   \
|     NH          H2C        NH        also            NH
|    /               \      /                         /
CH2CO                 H2C—CO                         CO
n = 2                 n = 3
Succinimide           Glutarimide                 Phthalimide
```

We should perhaps remember at this point the importance of the *N*-bromo-derivative of succinimide in free radical reactions (Chapter 11) and the use of phthalimide in Gabriel's synthesis of primary aliphatic amines (Part 1, Chapter 18). Adipic acid, on the other hand, does not form a seven-membered cyclic anhydride or amide, although under special circumstances a linear polymeric anhydride can be prepared.

Other examples of cyclic addition to carbonyl groups have been discussed previously. For example, the Dieckmann reaction in which the diesters of adipic and pimelic acid cyclise in the presence of a strong base (Chapter 9, p. 99):

```
          O
H2C—C//
 /    \
H2C    OEt      Na+ -OEt
 \          OEt ———————→
H2C—CH2—C
          \\O

[ H2C—C(=O:)(ÖEt)                    H2C—C(:O:)(ÖEt)                   ]
[ H2C                ÖEt   ←——→     H2C               ÖEt              ] ——→
[ H2C—CH=C(—O:−)                     H2C—C̈H—C(=O:)                     ]

H2C—C(O−)(OEt)                          H2C—C=O
H2C     |          ——→                  H2C    |             + −OEt
H2C—CHCO2Et                             H2C—CHCO2Et
```

If, instead of using the esters, the calcium salts of adipic or pimelic acid are heated, cyclic ketones are formed directly. On the other hand, heating the calcium salts of succinic or glutaric acid yields the corresponding acid anhydrides. The observation that succinic and glutaric acid usually give cyclic anhydrides, whereas adipic and pimelic acid usually yield ketones, is sometimes formulated as a rule called the Blanc rule. It has been used to determine

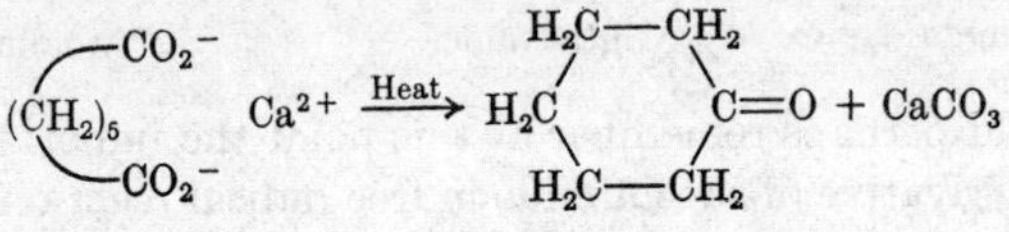

```
                                       H2C——CH2
      ┌—CO2⁻                          /        \
(CH2)5       Ca2+  ——Heat——>  H2C            C=O + CaCO3
      └—CO2⁻                          \        /
                                       H2C——CH2

   Calcium                        Cyclohexanone
   pimelate
```

the ring sizes in compounds of unknown constitution. Thus a five-membered ring on oxidation gives a glutaric acid derivative which cyclises to an anhydride, whereas a six-membered ring on oxidation gives an adipic acid derivative which cyclises to a ketone. The rule, however, breaks down for certain substituted compounds, and in one very famous case it led to the wrong structure for an important group of natural products called the steroids.

Another important example of 'tail-biting' occurs with carbohydrate molecules. For example, glucose, $C_6H_{12}O_6$, shows reactions of an aldehyde. Since it can also be shown to be a linear molecule its structure must be *A*. However, we should expect molecule *A* which contains both the free aldehyde group and the hydroxyl groups to form a cyclic hemiacetal; both the six-membered ring *B* and five-membered ring *C* are possible. In solution the open-chain

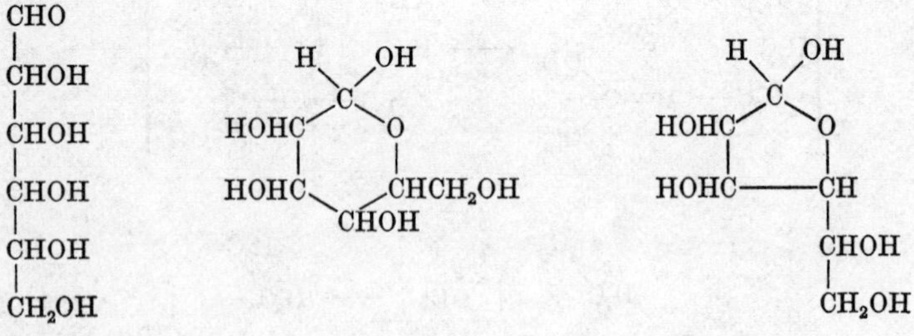

```
CHO
|                  H     OH                          H     OH
CHOH                \   /                             \   /
|                     C                                 C
CHOH           HOHC/     \O                     HOHC/     \O
|                |         |                      |         |
CHOH           HOHC\     /CHCH2OH              HOHC————————CH
|                   CHOH                                    |
CHOH                                                      CHOH
|                                                           |
CH2OH                                                     CH2OH
 (A)                 (B)                                   (C)
```

and both ring structures occur in equilibrium, although the six-membered ring system (B) greatly predominates. In the solid state glucose exists in the six-membered hemiacetal structure.

Most of the examples of 'tail-biting' we have described involve cyclic additions to a carbonyl group. Clearly, ring-forming reactions involving a substitution process are equally possible and under suitable circumstances, a halogeno alcohol forms a cyclic ether. An extreme example is the formation of epoxides from chlorohydrins on treatment with alkali:

$$\underset{\substack{\text{Ethylene chlorohydrin}\\\text{(2-Chloroethanol)}}}{ClCH_2CH_2OH} \xrightarrow{NaOH} \underset{\substack{\text{Ethylene oxide}\\\text{(Oxirane)}}}{H_2C\overset{O}{—}CH_2}$$

Another rather specialized example of cyclic substitution is that undergone by the salts of 1,4-diaminobutane (common name, putrescine) and 1,5-diaminopentane (common name, cadaverine) which, when heated, yield cyclic amines:

$$\underset{\substack{\text{1,4-Diaminobutane}\\\text{(putrescine)}}}{H_2N\text{—}CH_2CH_2CH_2CH_2\text{—}NH_3^+\ Cl^-} \longrightarrow \underset{\substack{\text{Pyrrolidinium}\\\text{chloride}}}{\text{(CH}_2)_4NH_2^+\ Cl^-} + NH_3$$

We now have to consider a quite different consequence of having two functional groups in the same molecule. We will begin with a very simple example. The way in which glucose forms a cyclic hemiacetal with itself is referred to above, and in Chapter 7 of Part 1

$$(H_3C)_2C{=}O + 2\,CH_3OH \xrightarrow{H^+} (H_3C)_2C(OCH_3)_2 + H_2O$$

$$\begin{array}{c}CH_2OH\\|\\CH_2OH\end{array} + (H_3C)_2C{=}O \longrightarrow \begin{array}{c}H_2C\text{—}O\\|\\H_2C\text{—}O\end{array}\!\!>C(CH_3)_2$$

the reaction of acetone with methanol in the presence of HCl (see p. 233) to yield a ketal was described. Similarly, acetone reacts with ethylene glycol, to form a cyclic ketal. Such cyclic ketals can also be formed with a 1,3- as well as a 1,2-diol, but not with dihydric alcohols in which the hydroxyl groups are more widely separated. Clearly there is nothing special about these reactions. They are simply due to the fact that the two hydroxyl groups are close enough to each other for a cyclic ketal to be sterically possible. Nevertheless, in certain circumstances, this ability to form cyclic compounds can be extremely important. For example, 1,2-glycols react with periodic acid to yield a cyclic ester, which is unstable and decomposes to give iodate and two carbonyl compounds that result from carbon–carbon bond fission. This is a general reaction of

1,2-glycol

Possibly hydrated

extreme importance, both preparatively and analytically. A 1,2-glycol is split *quantitatively* with acidic sodium periodate solution to yield two carbonyl compounds. Since the reaction is quantitative, it may be used to detect not only the presence of a 1,2-glycol in the molecule but, if more than one such grouping is present, the number of such groups:

$$RCH(OH)-CH(OH)R' \xrightarrow[H^+]{NaIO_4} RCHO + R'CHO + NaIO_3 \quad \text{Quantitatively}$$

It is worth briefly noting what happens with a 1,2,3-trihydric alcohol:

$$\begin{array}{l} R{-}CHOH \\ \quad | \\ \quad CHOH \\ \quad | \\ \quad CH_2OH \end{array} \xrightarrow[\text{2 moles NaIO}_4]{\text{Takes up exactly}} RCHO + HCO_2H + HCHO$$

The formation of formic acid, which can be determined quantitatively, indicates the presence of a CHOH group flanked by two other hydroxyl groups and the presence of formaldehyde, which can also be determined quantitatively, indicates the presence of a CH_2OH group adjacent to another hydroxyl group. The value of these reactions in establishing structures is obvious (see Chapter 26).

Lead tetra-acetate behaves in almost the same way with 1,2-glycols. It was originally thought that this also involved the formation of a cyclic compound involving the lead, but present evidence based on kinetic studies suggests that the base-catalysed reaction proceeds as illustrated:

$$\text{R-CHOH-CHOH-R}' \xrightarrow{\text{Base}} \text{R-CHO}^- \text{-CHOH-R}' + Pb(CH_3CO_2)_2(CO_2CH_3)_2 \xrightarrow{-\,^-CO_2CH_3}$$

$$\text{R-CHOPb}(CO_2CH_3)_3\text{-CHOH-R}' \xrightarrow{+OH^-} \text{R(H)C(-O-Pb}(CO_2CH_3)_3)\text{-C(H)(R}')\text{-O-H}\;\; ^-OH \longrightarrow \begin{array}{c} RCHO \\ R'CHO \end{array} + \begin{array}{c} \bar{Pb}(CO_2CH_3)_3 \\ H_2O \end{array}$$

A wide variety of further reactions undergone by molecules with more than one functional group could be considered. For example, the reactions of α-diketones with α-diamines:

$$\text{R-C(=O)-C(=O)-R}' + \text{o-}C_6H_4(NH_2)_2 \longrightarrow \text{2-R-3-R}'\text{-quinoxaline}$$

However, such reactions involve no new principles. Probably the most important conclusion to be drawn from the reactions discussed in this Chapter is that the presence of more than one functional group in a molecule may result in reactions that cannot occur with monofunctional compounds, but these reactions are always straightforward extensions of the reactions undergone by the functional groups concerned. For instance, dicarboxylic acids behave in exactly the way that we should predict from our knowledge of the reactions of a single carboxyl group and of the electronic effects in organic molecules discussed in Chapter 7.

Problems

1. (*a*) The following are the observed acid dissociation constants $K_a = [H^+][B]/[BH^+]$ of three aliphatic diamines:

	$K_1(2H^+)$	$K_2(H^+)$
Ethane-1,2-diamine	1×10^{-7}	8×10^{-11}
Propane-1,3-diamine	2.9×10^{-9}	3.0×10^{-11}
Butane-1,4-diamine	4.5×10^{-10}	1.6×10^{-11}

What would the ratio K_1/K_2 be if the only effect was a statistical one? How can the fact that the ratio is larger be explained in qualitative terms?

(*b*) Glycine (aminoacetic acid) has two dissociation constants $pK_1 = 2.4$ and $pK_2 = 9.8$. Explain this phenomenon, and comment on the magnitude of pK_2 ($CH_3\overset{+}{N}H_3 \rightleftharpoons CH_3NH_2 + H^+aq.$ $pK = 10.6$).

2. Explain the following observations:

(*a*) A sodium salt, $C_4H_7O_3Na$, when treated with dilute acid yields a neutral substance, $C_4H_6O_2$. Treatment with alkali reconverted the neutral substance into the salt.

(*b*) An acidic compound, $C_6H_{10}O_4$, on distillation with acetic anhydride, gave a neutral compound, C_5H_8O, and carbon dioxide.

(*c*) Three linear compounds, A, B, and C, all have the molecular formula $C_5H_{12}O_3$. A gave no reaction with periodate, B reacted with one mole of periodate, and C reacted with two moles of periodate during which reaction one mole of formic acid was formed.

CHAPTER 19

Quinones and Related Compounds

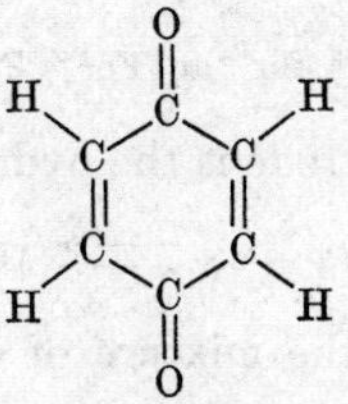

Quinone
(p-benzoquinone),
yellow crystals,
m.p. 116°

As the formula shows, quinone is an unsaturated cyclic diketone; it has all the reactions expected of such a compound. Before we discuss these, however, we must note the following reaction scheme:

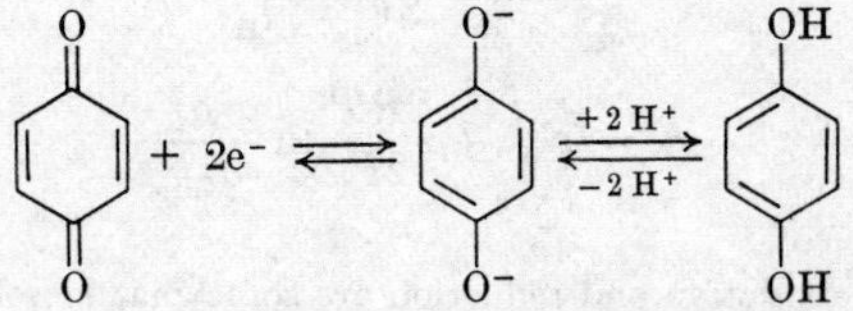

$$\text{quinone} + 2e^- \rightleftharpoons {}^-O{-}C_6H_4{-}O^- \underset{-2\,H^+}{\overset{+2\,H^+}{\rightleftharpoons}} HO{-}C_6H_4{-}OH$$

This shows how it is possible to add two electrons to a quinone, converting it from an unsaturated diketone into the dianion of a dihydric phenol. The reaction is very similiar to that of inorganic metal ions. For example, the stannic ion can accept two electrons to become a stannous ion:

$$Sn^{4+} + 2e^- \rightleftharpoons Sn^{2+} \qquad \ldots(a)$$

In this sense the chemistry of quinones resembles that of inorganic electrolytes. Notice that this is a true oxidation–reduction process.* Oxidation is the loss of electrons and reduction the gain of electrons. In an equilibrium of this kind, the oxidizing (or reducing) power of the solution can be measured by inserting an electrode unattacked by the solution, e.g. a platinum electrode. If the system is reducing, there will be a tendency for electrons to pass from the solution to the electrode. This kind of electrode system in which both oxidized and reduced species are present, each being converted into the other by a minute change of potential from the equilibrium value, is called a redox electrode. The combination of two such redox electrodes forms the familiar oxidation–reduction cell:

$$\mathrm{Pt}|\mathrm{Sn^{2+}\ \ Sn^{4+}aq}\|\mathrm{Fe^{2+}\ \ Fe^{3+}aq}|\mathrm{Pt}$$

An important redox electrode is the hydrogen electrode:

$$\mathrm{H^{+}aq + e^{-} \rightleftharpoons \tfrac{1}{2}H_2\ gas} \qquad \ldots(b)$$

Similarly a solution of the mixture of quinone and quinol (*p*-dihydroxybenzene) can form another such system:

$$\underset{\text{Quinol}}{\mathrm{C_6H_4(OH)_2}} \rightleftharpoons \underset{\text{Quinone}}{\mathrm{C_6H_4O_2}} + \mathrm{2H^{+} + 2e^{-}} \qquad \ldots(c)$$

The electrode potentials of such systems would be as follows:

$$E = E_c{}^a - \frac{RT}{2F}\ln\frac{a_{\mathrm{Sn}^{2+}}}{a_{\mathrm{Sn}^{4+}}} \qquad \ldots(a)$$

$$E = E_c{}^b - \frac{RT}{2F}\ln\frac{p_{\mathrm{H}_2}^{1/2}}{a_{\mathrm{H}^+}} \qquad \ldots(b)$$

$$E = E_0{}^c - \frac{2RT}{2F}\ln\frac{a_{\mathrm{Q}}a_{\mathrm{H}^+}^2}{a_{\mathrm{QH}_2}} \qquad \ldots(c)$$

* The terms oxidation and reduction are somewhat loosely used in organic chemistry. Strictly, the terms should be used only for reactions involving the loss or gain of electrons. However, the conversion of an aldehyde into an alcohol is often spoken of as a reduction. It would be much better described as a hydrogenation (analogous to the hydrogenation of ethylene.) The only justification for referring to the reaction as a reduction is that using *some* reagents the first step is a reduction:

$$\mathrm{RCHO + 2e^{-} \longrightarrow (RCHO)^{2-} \xrightarrow{2H^{+}} RCH_2OH}$$

where a_X = the activity of species X, p_{H_2} = partial pressure of hydrogen, F = Faraday.

There is no method for evaluating the absolute potential of an electrode, so it is customary to refer all electrodes to the hydrogen electrode under standard conditions (H_2 1 atmosphere, H^+ unit activity, $E_0{}^b = 0$). The standard electrode potentials of quinones (based on the hydrogen electrode) and of their substituted derivatives are of considerable interest:

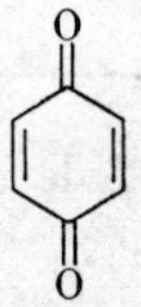

(A)
$E_0 = 0.699$ volt (H_2O)
$= 0.715$ volt (EtOH)

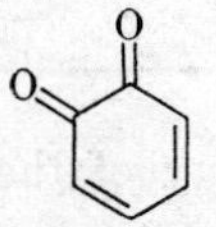

(B)
$E_0 = 0.792$ volt (H_2O)

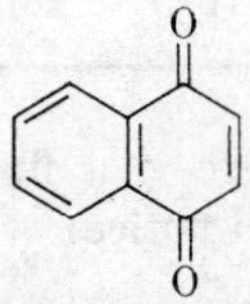

$E_0 = 0.470$ volt (H_2O)
$= 0.484$ volt (EtOH)

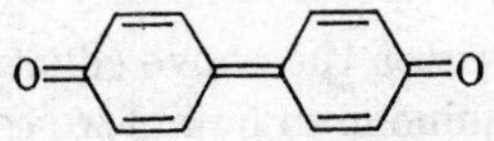

$E_0 = 0.95$ volt (EtOH)

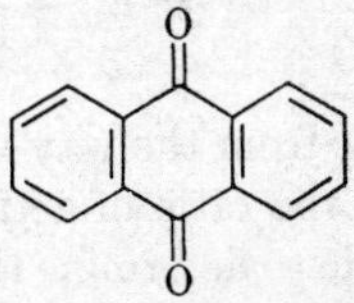

$E_0 = 0.154$ volt (EtOH)

Notice that *o*-benzoquinone (B) has a higher electrode potential than *p*-benzoquinone (A), suggesting that the *o*-quinonoid system is of higher energy and less stable than the *p*-quinonoid system. Notice also that fused rings stabilize the quinone, making it a less powerful oxidizing agent. Fieser has made some very interesting studies on the effect of substituents on the electrode potential of quinones. In Table 19.1 the values are expressed as the difference between the electrode potential of the substituted naphthoquinone and that of unsubstituted naphthoquinone. Groups that repel or donate electrons lower the electrode potential, i.e. the quinone is a less powerful oxidizing agent. This is as we should expect; it accepts electrons less readily. On the other hand, electron-attractors or -acceptors increase the ability of the quinone to accept electrons and so increase its oxidizing properties.

Table 19.1

X	ΔE_0 (millivolts)	X	ΔE_0 (millivolts)
$NHCH_3$	−253	C_6H_5	−32
NH_2	−210	$OCOCH_3$	−9
OH	−128	Cl	+24
OCH_3	−131	SO_3Na	+69
CH_3	−76	$SO_2C_6H_4CH_3$	+121

There is a further property of the quinone electrode that we should notice:

$$E = E_0 - \frac{RT}{2F} \ln \frac{a_Q a_{H^+}^2}{a_{QH}}$$

$$= E_0 - \frac{RT}{2F} \ln \frac{a_Q}{a_{QH_2}} - \frac{RT}{F} \ln a_{H^+}$$

It is clear from the way we have written the above equation that if we knew the ratio of activities of quinone to quinol we could use the electrode to determine the pH. As it happens, quinone and quinol form a 1:1 molecular complex (a charge-transfer complex) known as 'quinhydrone'. Quinhydrone is much less soluble than either quinone or quinol, so that it is easily prepared, and it can be purified by recrystallization. This makes it possible to take exactly equal concentrations of quinone and quinol and these can be added to a solution whose pH is to be determined. Under these circumstances the second term in the equation vanishes since, in dilute solution, $C_Q/C_{QH_2} = 1 \approx a_Q/a_{QH_2}$. The quinhydrone electrode used to be used extensively as a means of determining pH. Nowadays its place has been taken completely by the glass electrode.

Apart from its properties as an oxidizing agent, quinone behaves like an unsaturated ketone. Thus, it reacts with hydroxylamine to produce, first, the quinone monoxime and then the dioxime. Notice that the monoxime is identical with the compound obtained by treating phenol with nitrous acid. We have referred to this type of tautomerism in previous Chapters.

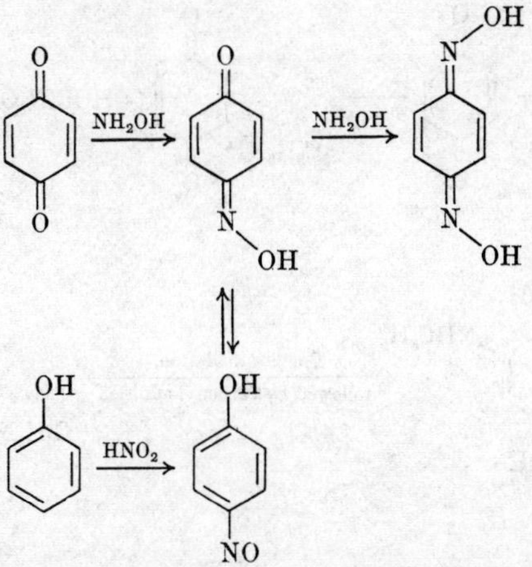

Phenylhydrazine behaves as a reducing agent and is oxidized by quinone, but substituted phenylhydrazines will react with the carbonyl groups in the normal manner. In general, strong nucleophiles tend to add in a conjugate fashion, i.e. to the olefinic carbon atom. For example, aniline reacts as follows:

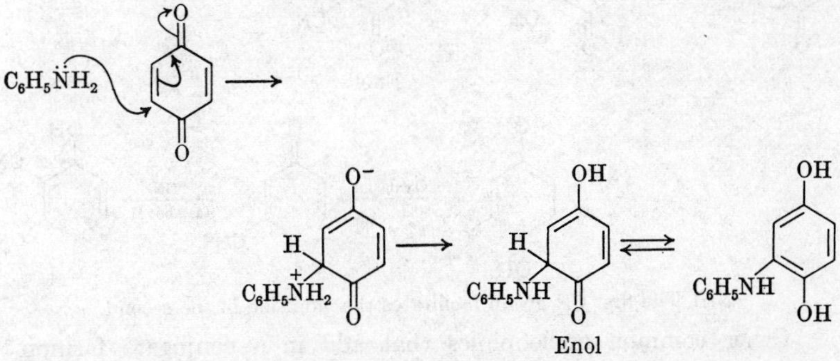

Enol

Now we have seen from the table of standard electrode potentials that an electron-donor group lowers the electrode potential, so that in the presence of an excess of quinone the anilinoquinol will be oxidized. The resulting anilinoquinone will then react further with more aniline:

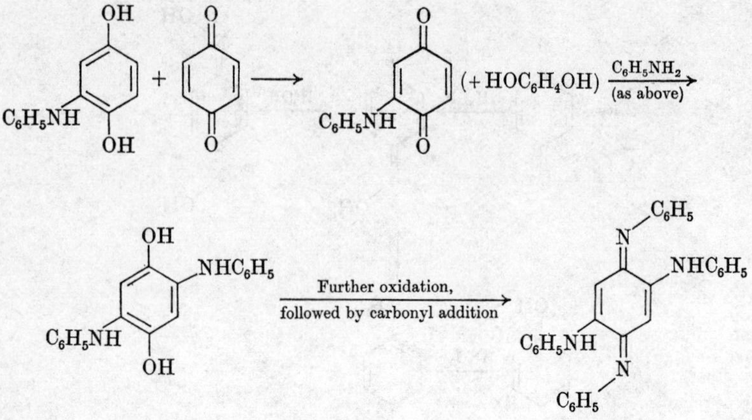

$(+ HOC_6H_4OH)$ $\xrightarrow[\text{(as above)}]{C_6H_5NH_2}$

Further oxidation, followed by carbonyl addition

Azophenine

This conjugate addition of nucleophiles to the olefinic carbon atom of quinones is a very general reaction. A good example is the addition of cyanide, as illustrated.

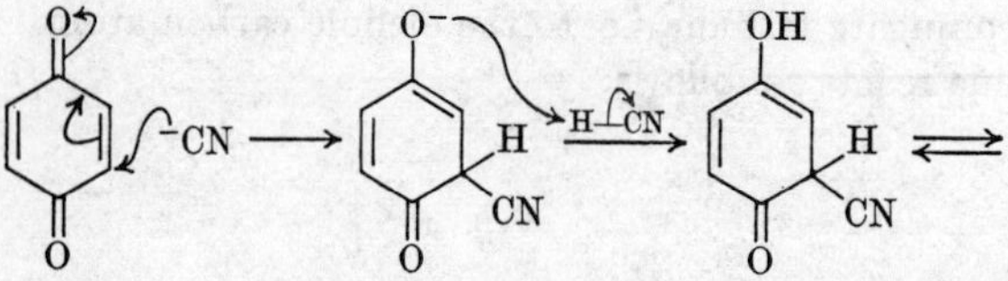

Enol

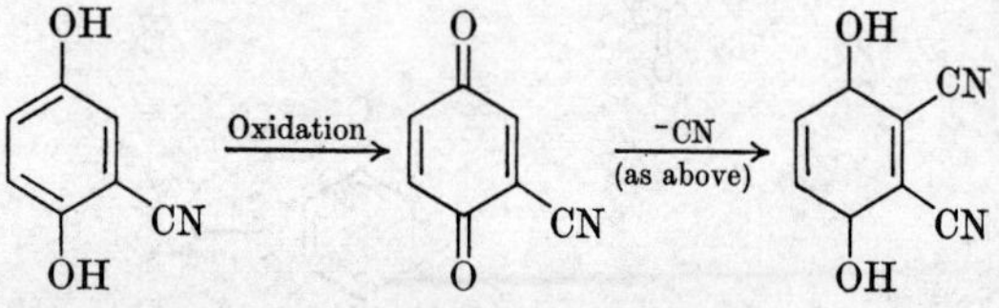

Oxidation

^-CN (as above)

N.B. The first CN group facilitates the addition of the second.

Other common nucleophiles that add in a conjugate fashion include the bisulphite anion, the diethyl malonate and ethyl cyanoacetate enolate anions, Grignard reagents (both 1,2-addition to the carbonyl and 1,4-conjugate addition occur in this case), and thiol anions. We do not depict these reactions since they are straightforward and exactly analogous to the additions of aniline and cyanide shown above.

Electrophiles as well as nucleophiles add to an isolated carbonyl double bond, so acids will add electrophilically to quinones, usually in a conjugate fashion. For example, the addition of hydrogen chloride yields 2-chloroquinol:

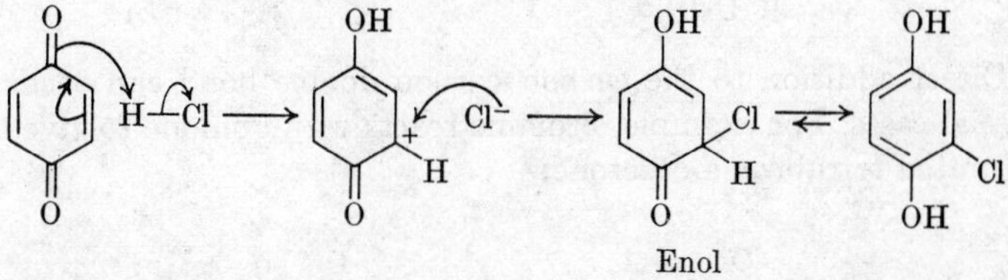

Enol

A similar reaction occurs with acetic anhydride and a strong acid. We do not depict all the individual steps of this process as they are clearly analogous to the addition of hydrogen chloride:

$$+ (CH_3CO)_2O + H^+ \longrightarrow$$

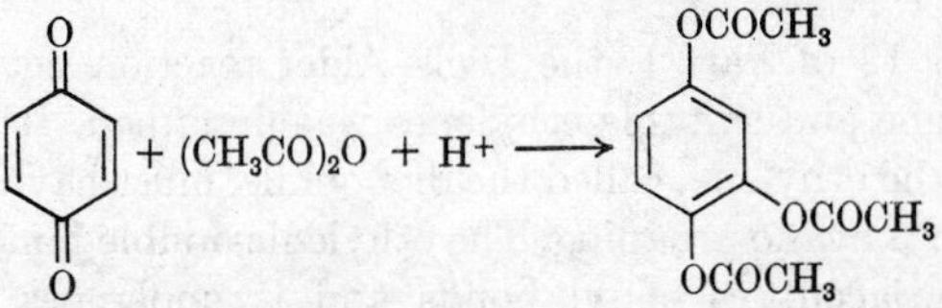

The reduction of a quinone is a two-stage process and in many cases the presence of the semiquinone radical ion can be established.

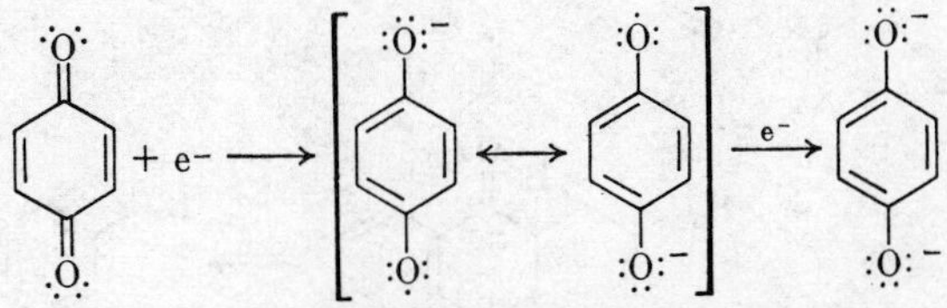

Semiquinone radical ion

In some reactions this one-electron transfer can be important; for example, when quinone is treated with a diazonium salt, nitrogen is evolved. The overall mechanism of the reaction is possibly as follows:

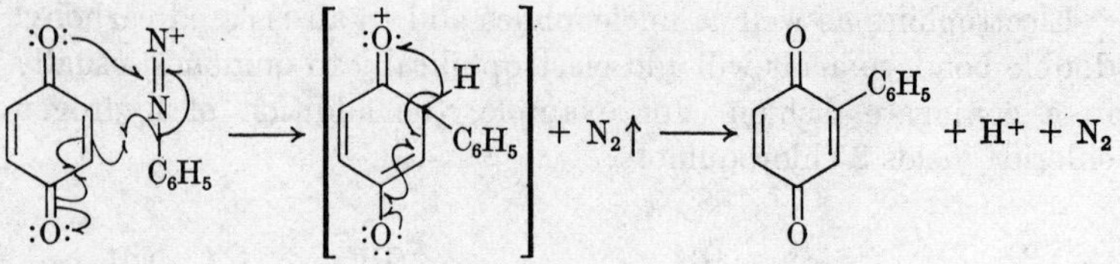

Direct addition to the carbon–carbon double bond can occur in special cases. For example, bromine reacts with quinone to give the saturated tetrabromo-diketone:

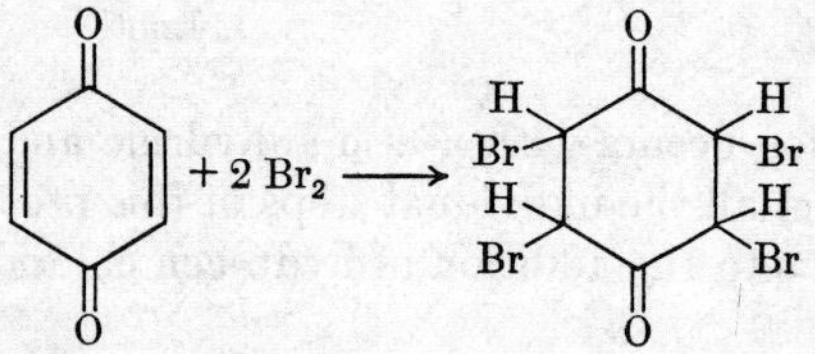

$$\text{quinone} + 2\,Br_2 \longrightarrow \text{tetrabromo-diketone}$$

In Chapter 12 of Part 1, the Diels–Alder reaction between a conjugated diene and suitable ethylenes was described. It was emphasized that the ethylene, called the dienophile, must have an adjacent carbonyl or a cyano grouping. The ethylenic double bond in quinone has two adjacent carbonyl bonds and it undergoes Diels–Alder reactions very readily, acting as the dienophile.

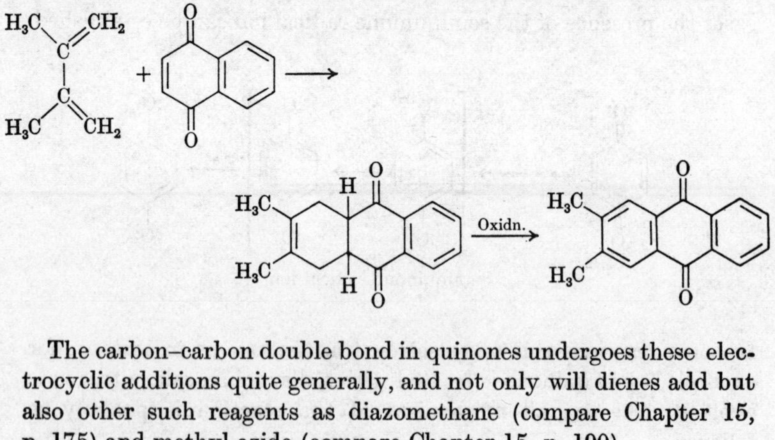

The carbon–carbon double bond in quinones undergoes these electrocyclic additions quite generally, and not only will dienes add but also other such reagents as diazomethane (compare Chapter 15, p. 175) and methyl azide (compare Chapter 15, p. 190).

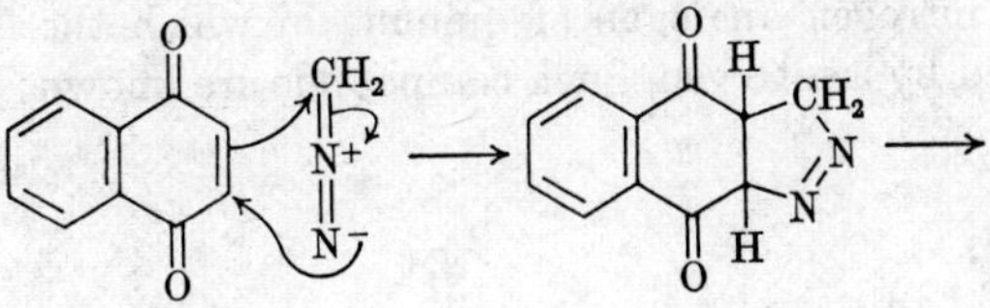

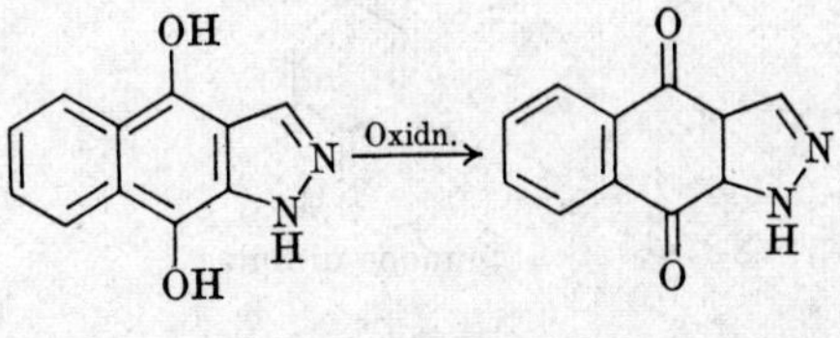

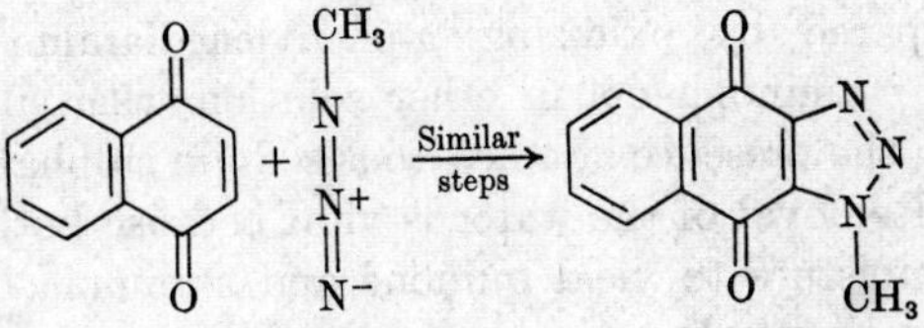

Oxidn.

Similar steps

As we have seen, *o*-benzoquinone has a much higher electrode potential than *p*-benzoquinone. It is also very much more reactive and, although it can be prepared crystalline, it decomposes within half an hour of its preparation. In solution it undergoes a Diels–Alder reaction with itself (i.e. one molecule behaves as a diene and one as a dienophile):

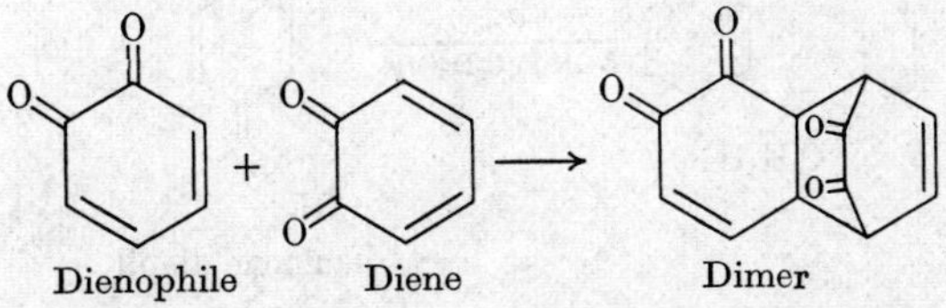

Dienophile Diene Dimer

With cyclopentadiene or cyclohexadiene, *o*-benzoquinone behaves as a dienophile although the alternative adduct is formed on heating:

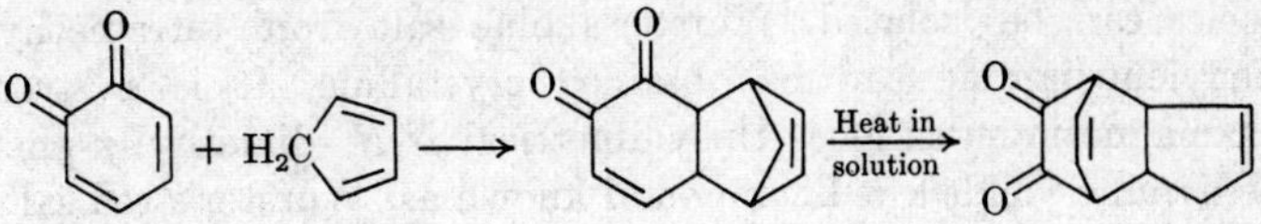

Heat in solution

We can visualize a nitrogen analogue of quinone in which the oxygen atom is replaced by a nitrogen. Such compounds are known as quinone imines:

NH / NH

Quinone di-imine

NH / O

Quinone imine

Quinone di-imine and monoimine are both extremely unstable. They can be prepared by oxidizing *p*-phenylenediamine (*p*-diaminobenzene) or *p*-aminophenol in ether solution with silver oxide or lead oxide in the presence of anhydrous sodium sulphate (to remove the water). Removal of the water is vital because both compounds react with moisture to yield quinone and ammonia. Even in dry ether solution they polymerize and are rapidly decomposed by light. *N*-Substituted derivatives are very much more stable and quinone dianil can be prepared from *N*,*N'*-diphenyl-*p*-phenylenediamine by direct oxidation with oxygen in alcoholic potassium hydroxide:

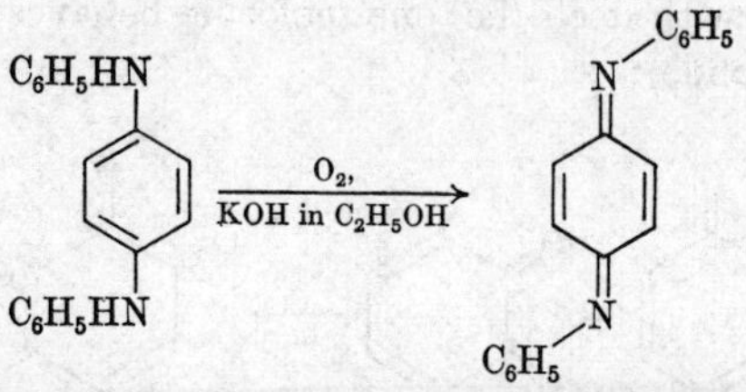

C_6H_5HN / C_6H_5HN $\xrightarrow[\text{KOH in } C_2H_5OH]{O_2,}$ $N-C_6H_5$ / $N-C_6H_5$

Quinone dianil

In some cases, very interesting intermediate compounds are formed. For example, if *N*,*N'*-dimethyl- or *N*,*N*,*N'*,*N'*-tetramethyl-*p*-phenylenediamine is oxidized, brightly coloured intermediate species can be isolated. Wurster's blue salt from tetramethyl-*p*-phenylenediamine can be obtained crystalline. It is a strongly paramagnetic substance. Unsymmetrical *N*,*N'*-dimethyl-*p*-phenylenediamine yields a red compound known as 'Wurster's red salt'.

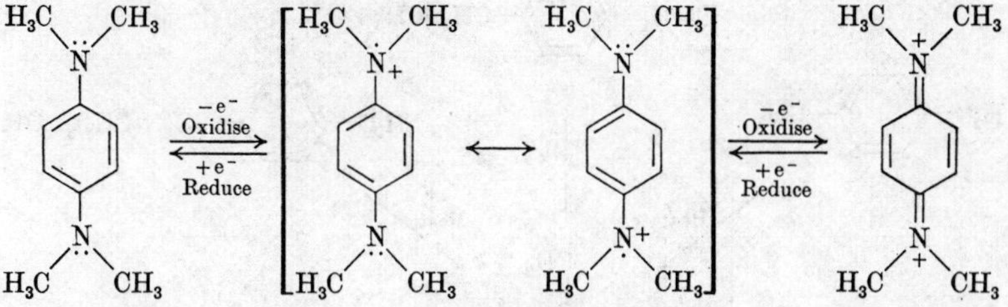

Colourless ⇌ (−e⁻ Oxidise / +e⁻ Reduce) [Blue 'Wurster's blue salt'] ⇌ (−e⁻ Oxidise / +e⁻ Reduce) Colourless

Blue
'Wurster's blue salt'

Since a quinone di-imine can exist even if it is very reactive, it is logical for us to ask, can the nitrogen atom be replaced by a carbon atom? The resulting compounds are known as quinomethanes and quinodimethanes:

Quinomethane Quinodimethane

Quinodimethane is sometimes known as *p*-xylylene and it has been prepared by pyrolysis of *p*-xylene, by reaction of magnesium with a chloro ether, and by thermal decomposition of a quaternary salt.

Quinodimethane exists only in the vapour phase or in solution at low temperatures. The moment the solution warms up, polymerization takes place. The annexed diagram shows some of the polymers formed, including the very interesting dimer, the trimer, and the linear polymer. Quinodimethane reacts with almost any reagent; its reactions with oxygen, sulphur dioxide, chlorine, and hydrogen chloride are illustrated on p. 248.

Quinodimethane is so reactive that it is only known as a reaction intermediate. Just as quinone di-imine derivatives with substituents on the nitrogen atoms are relatively stable, so the presence of substituents on the terminal carbon atoms stabilize quinodimethanes. Thus, the tetraphenyl derivative (*C*) can be isolated and is comparatively stable (see p. 248).

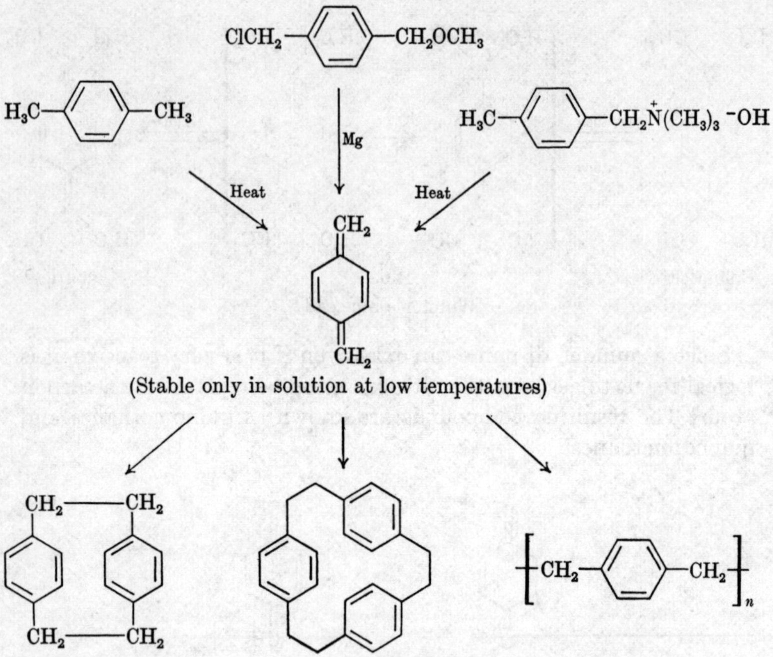

$ClCH_2$–C₆H₄–CH_2OCH_3 $\xrightarrow{Mg}$

H_3C–C₆H₄–CH_3 $\xrightarrow{Heat}$

H_3C–C₆H₄–$CH_2\overset{+}{N}(CH_3)_3\ {}^{-}OH$ $\xrightarrow{Heat}$

CH_2=C₆H₄=CH_2

(Stable only in solution at low temperatures)

CH_2—CH_2 / CH_2—CH_2

$\left[CH_2 - C_6H_4 - CH_2 \right]_n$

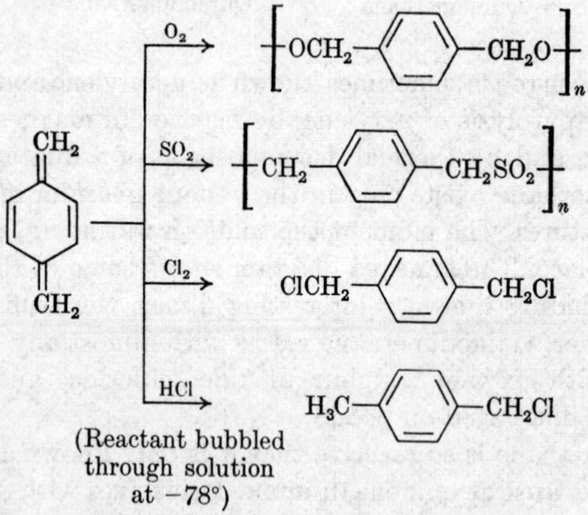

$\xrightarrow{O_2}$ $\left[OCH_2 - C_6H_4 - CH_2O \right]_n$

$\xrightarrow{SO_2}$ $\left[CH_2 - C_6H_4 - CH_2SO_2 \right]_n$

$\xrightarrow{Cl_2}$ $ClCH_2 - C_6H_4 - CH_2Cl$

$\xrightarrow{HCl}$ $H_3C - C_6H_4 - CH_2Cl$

(Reactant bubbled through solution at −78°)

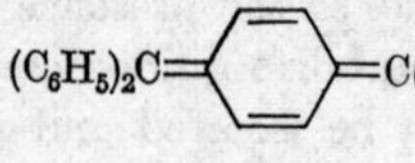

$(C_6H_5)_2C{=}C_6H_4{=}C(C_6H_5)_2$

(*C*)

$(C_6H_5)_2C{=}\left[C_6H_4\right]_n{=}C(C_6H_5)_2$

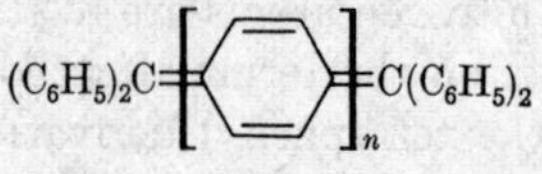

(*D*)

It would be clearly possible to link together a number of cyclohexadiene rings as shown in formula (*D*). We have seen, however, that the corresponding quinone (where $n = 2$) has a higher electrode potential (i.e. is less stable) than quinone itself, so we might expect that the larger n becomes the more unstable the quinodimethane will become. On the other hand, the tendency of the molecule to rearrange to yield a diradical (*E*) will increase since more stable 'benzenoid' rings will be formed. The diradical has unpaired elec-

$$(C_6H_5)_2\dot{C}\uparrow-\left[C_6H_4\right]_n-\dot{C}\uparrow(C_6H_5)_2$$

(*E*)

trons with parallel spins centred on the benzylic carbon atoms (a triplet). Notice that you cannot have resonance between the triplet (*E*) and the quinodimethane structure (*D*) (cf. Chapter 4; resonance is not possible between structures where the numbers of pairs of electrons with opposite spins differ). The molecule with $n = 2$ is known as Tschitschibabin's hydrocarbon and at normal temperatures it is almost entirely in the quinodimethane form. But for $n = 3$ and $n = 4$ there is good evidence that there is an appreciable percentage of the compound present in the triplet state at room temperature.

Simple quinomethanes are unknown except as reaction intermediates, but derivatives of both quinomethane and quinone imines are very important as very highly coloured materials used as dyes. Fuchsone and the Homolka-type base are yellow compounds

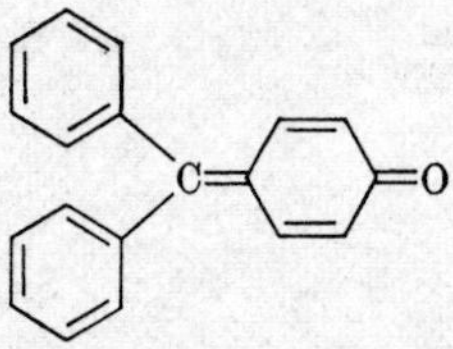

Fuchsone

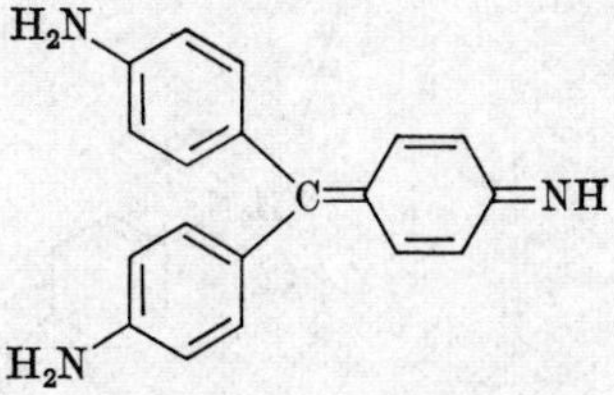

Homolka-type base

like quinone itself. Treatment of the Homolka base, shown above, with cida yields an intensely coloured salt. This salt is the dye

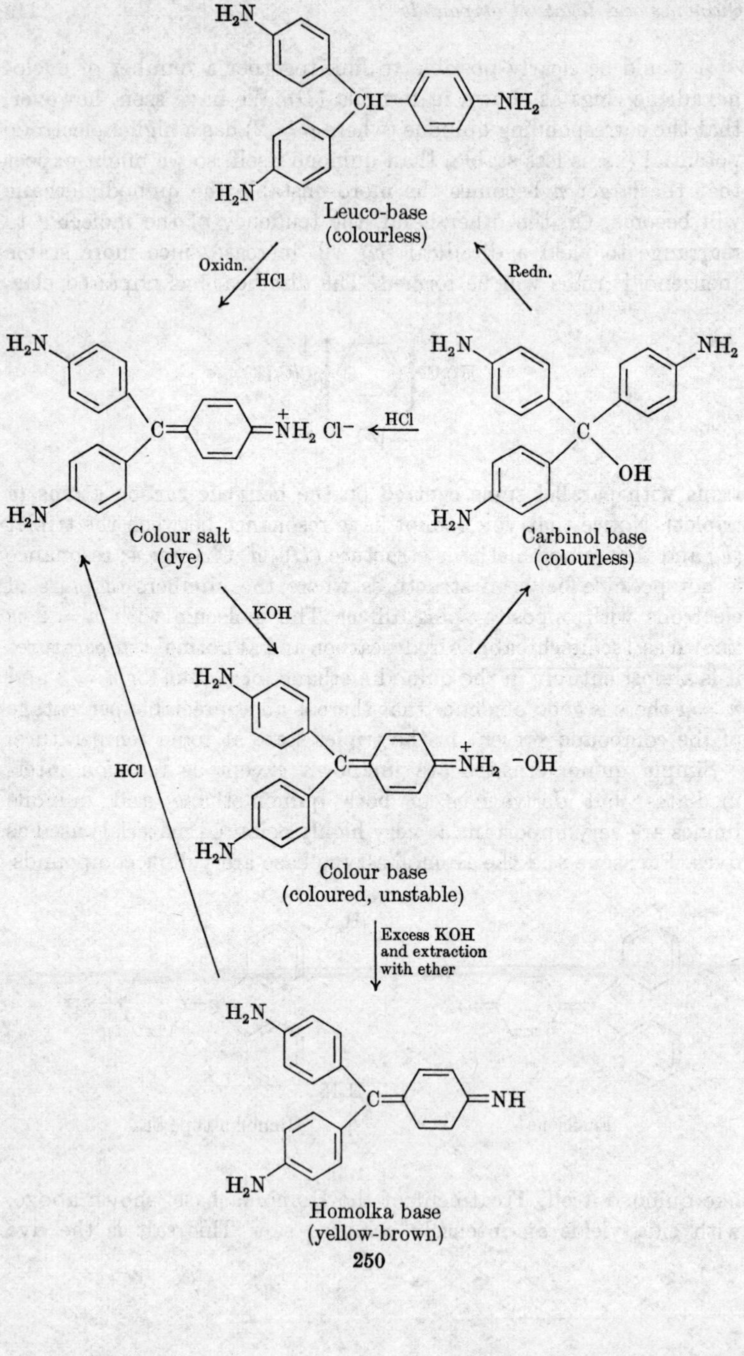

H_2N NH_2 H_2N CH

Leuco-base
(colourless)

Oxidn. HCl

Redn.

H_2N C= $\overset{+}{N}H_2$ Cl^- H_2N

HCl

H_2N NH_2 C OH H_2N

Colour salt
(dye)

Carbinol base
(colourless)

KOH

H_2N C= $\overset{+}{N}H_2$ ^{-}OH H_2N

HCl

Colour base
(coloured, unstable)

Excess KOH
and extraction
with ether

H_2N C= NH H_2N

Homolka base
(yellow-brown)

'Pararosaniline', which is intense red with a slightly bluish lustre. The relationship between the dye, its colourless reduced form, and the Homolka base are illustrated. Notice that the dye and the colour base have been drawn as a single canonical form but, in fact, there are three canonical forms with positive changes on different nitrogen atoms (each with three more different Kekulé forms):

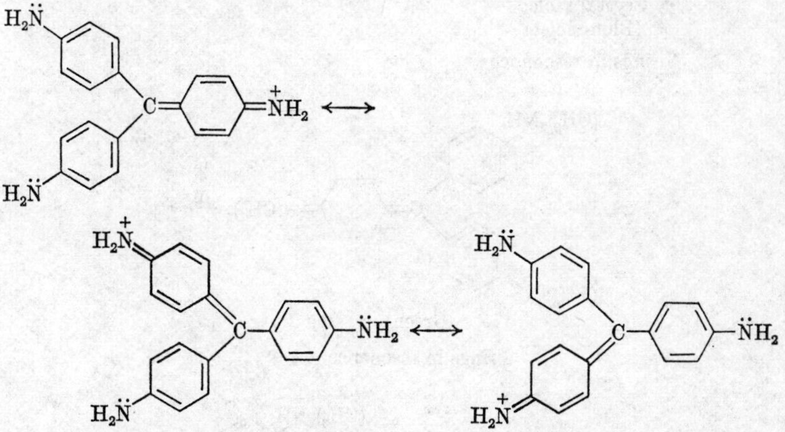

The number of canonical forms involving quinonoid structures is definitely associated with the light-absorbing properties of the molecule. This is well illustrated in the effect that increasing acid concentration has on the colour of these dyes. For example, Crystal Violet, a blue-violet dye, on treatment with acid, is converted first into a green dye salt and finally into a yellow-brown triply charged cation (see p. 252).

Although the most important dyestuffs of this class are of the quinone imine type, there are, as we have indicated, also quinomethane derivatives; a common example is phenolphthalein (p. 253).

The rosaniline-type dyes were originally prepared by oxidation of mixtures of aniline and toluidines. For example, two moles of aniline and 1 mole of *p*-toluidine, on oxidation, gave Pararosaniline itself. New methods have been devised and mixtures of aniline and formaldehyde are now used to synthesize Pararosaniline. Crystal Violet can be prepared from dimethylaniline and phosgene by a

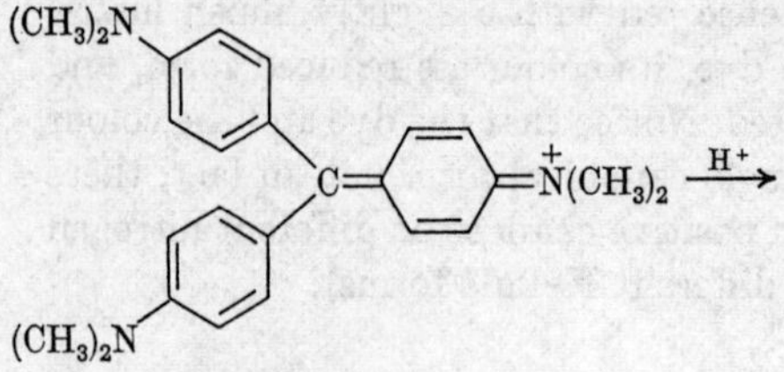

$(CH_3)_2N$

$(CH_3)_2N$

$C{=}$ $=\overset{+}{N}(CH_3)_2$ $\xrightarrow{H^+}$

Crystal violet
(Blue-violet)
3 rings in resonance

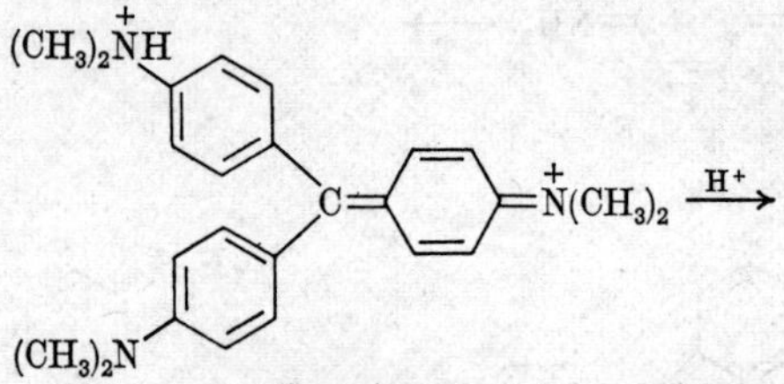

$(CH_3)_2\overset{+}{N}H$

$(CH_3)_2N$

$C{=}$ $=\overset{+}{N}(CH_3)_2$ $\xrightarrow{H^+}$

(Green)
2 rings in resonance

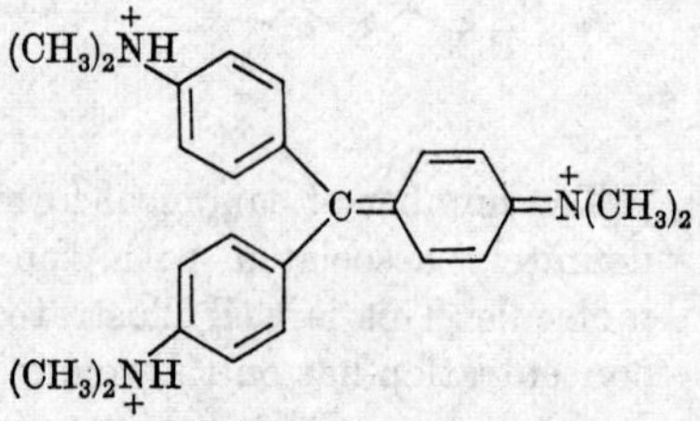

$(CH_3)_2\overset{+}{N}H$

$(CH_3)_2\overset{+}{N}H$

$C{=}$ $=\overset{+}{N}(CH_3)_2$

(Yellow-brown)
one quinone imine ring

two-step process. Phenolphthalein, on the other hand, is prepared by fusing phthalic anhydride directly with phenol, usually in the presence of a strong acid catalyst. In spite of the intense colours of these substances they are not very satisfactory as commercial dyestuffs, as they are neither light-fast nor washing-fast. They are mainly used for fabric which is intended for the most expensive evening gowns, which never see direct sunlight and are never washed.

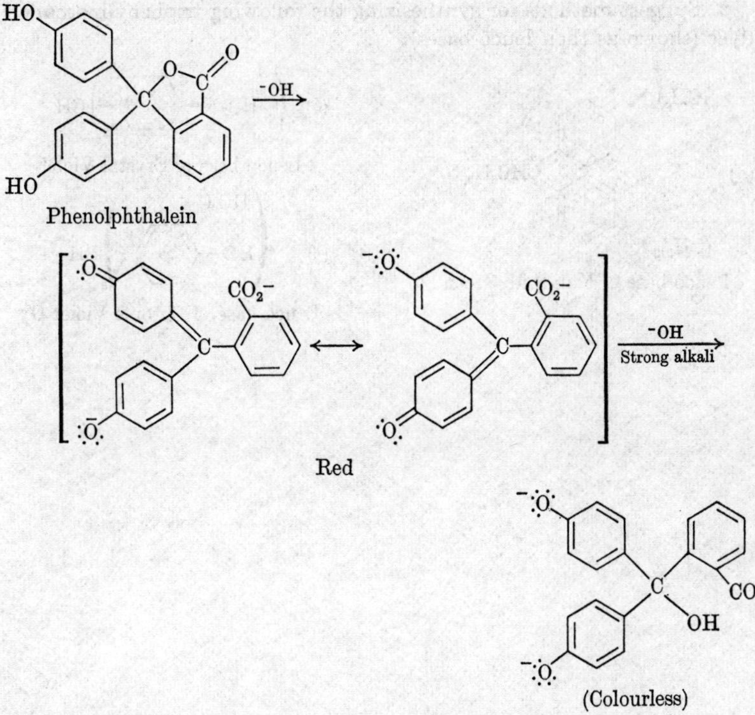

HO HO O C O C $\xrightarrow{^-OH}$

Phenolphthalein

:Ö: CO_2^- C :Ö:⁻ $\longleftrightarrow$ ⁻:Ö: CO_2^- C :Ö: $\xrightarrow[\text{Strong alkali}]{^-OH}$

Red

⁻:Ö: C CO_2^- OH ⁻:Ö:

(Colourless)

Problems

1. Starting from *p*-benzoquinone, how would you synthesize:

(*a*) H_3C H_3C O O (*b*) O OCH_3 CH_3O O (*c*) O N OCH_3 CH_3O NO_2

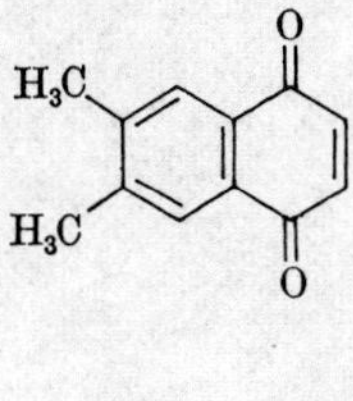

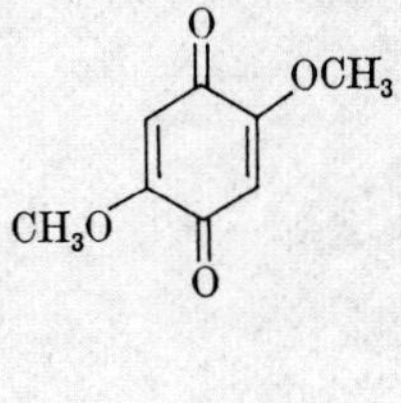

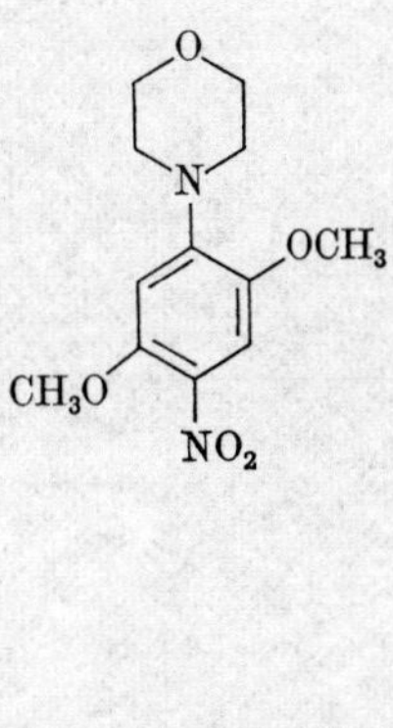

2. Suggest methods for synthesizing the following triphenylmethane dyes (shown as their leuco bases):

(*a*) $[(CH_3)_2N-C_6H_4]_2CHC_6H_5$

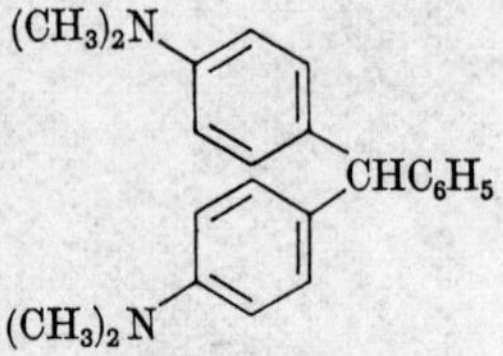

Leuco base of Malachite Green

(*b*) $((CH_3)_2N-C_6H_4)_3CH$

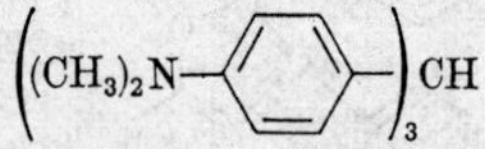

Leuco base of Crystal Violet

(*c*) $(HO_2C(HO)C_6H_3)_3CH$

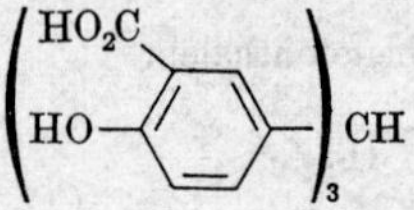

Leuco base of Chrome Violet Gy

CHAPTER 20

Polycyclic Aromatic Compounds

Napthalene

The simplest polycyclic aromatic compound is naphthalene, in which two benzene rings are *ortho*-fused together. A consideration of some of the chemical properties of naphthalene will serve as an introduction to the properties of polycyclic aromatic compounds in general. Notice that, whereas in benzene all the carbon atoms are equivalent, in naphthalene there are three different types of carbon atom. The numbering is shown in the formula. Substituents attached

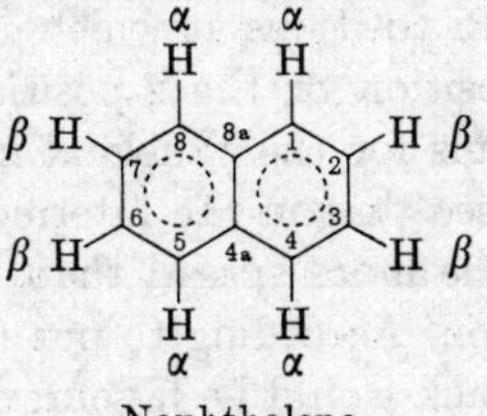

Naphthalene

to carbon atom 1 are sometimes given the prefix α, and those attached to carbon atom 2 the prefix β.

We can draw three Kekulé structures for naphthalene to represent the principal canonical forms of the resonance hybrid:

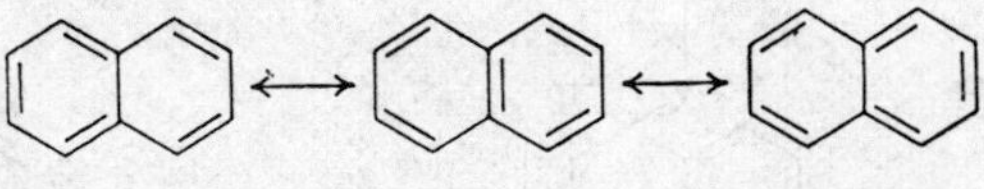

In two of these the bond between carbon atoms 1 and 2 is drawn as a double bond, whereas the bond between carbon atoms 2 and 3 is only a double bond in one of them. If the three Kekulé structures are assumed to contribute equally to the ground state of the naphthalene molecule, then the bond between carbon atoms 1 and 2 must have more double bond character than the bond between carbon atoms 2 and 3. This argument implies that, unlike benzene, where all the bonds are exactly equivalent, in naphthalene the mobile π-electrons are more concentrated in some bonds than in others. It is very interesting to find that accurate X-ray determination of the bond lengths in naphthalene show that the bonds are not of equal length.

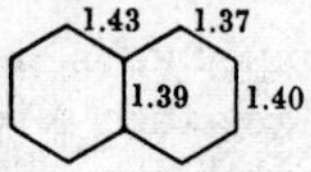

1.43 1.37 1.39 1.40

(Cf. carbon bond lengths
in ethylene 1.34 Å
in benzene 1.40 Å
in ethane 1.54 Å)

Bond lengths in naphthalene

As we should expect, naphthalene undergoes the same addition-with-elimination reactions (electrophilic substitution reactions) as benzene. There are two positions at which such a reaction can take place, either the 1-position or the 2-position. If we draw all the possible canonical forms for the Wheland intermediates for attack at both positions we see that in the intermediate for attack at the 1-position the charge is more spread than in the intermediate for attack at the 2-position. According to our hypothesis (see Chapter 8) we predict that attack would be favoured at the 1-position. This

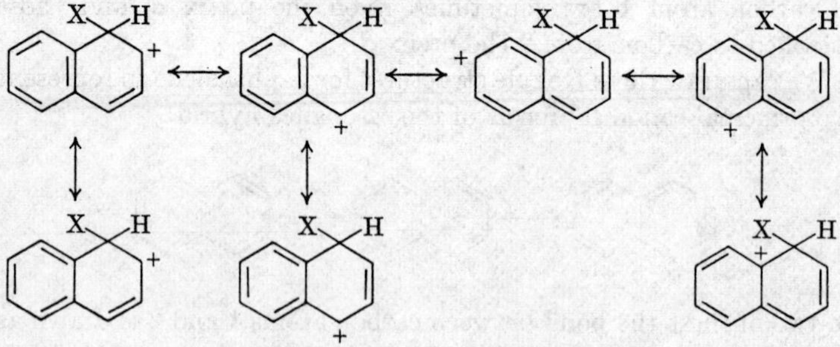

Attack at the 1-position (7 canonical forms)
(4 retaining a 'benzene' ring)

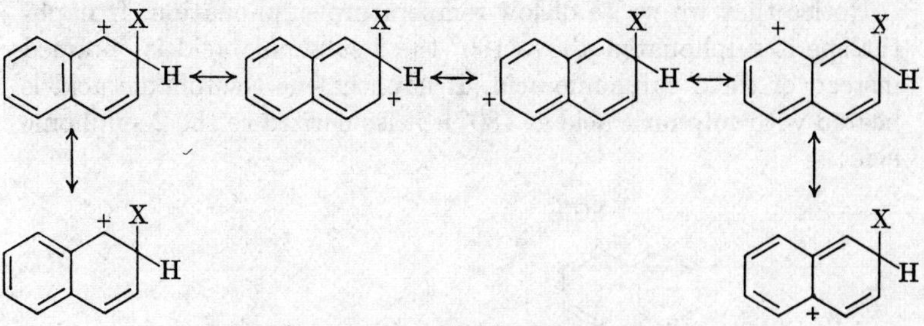

Attack at the 2-position (6 canonical forms)
(2 retaining a 'benzene' ring)

argument is based on the important assumption that all the ionic structures drawn have equal weight in the resonance hybrid. (Another popular argument, hard to justify, assumes that structures containing intact 'benzene rings' are of lower energy than those involving quinonoid structures. On this basis the Wheland intermediate for attack at the 1-position has four lower-energy structures, whereas the Wheland intermediate for attack at the 2-position has only two.) Simple molecular-orbital theory also predicts that the Wheland intermediate involving attack at the 1-position is the more stable. Experimentally, it is found that nitration, halogenation, and sulphonation (at low temperatures) all occur preferentially in the 1-position.

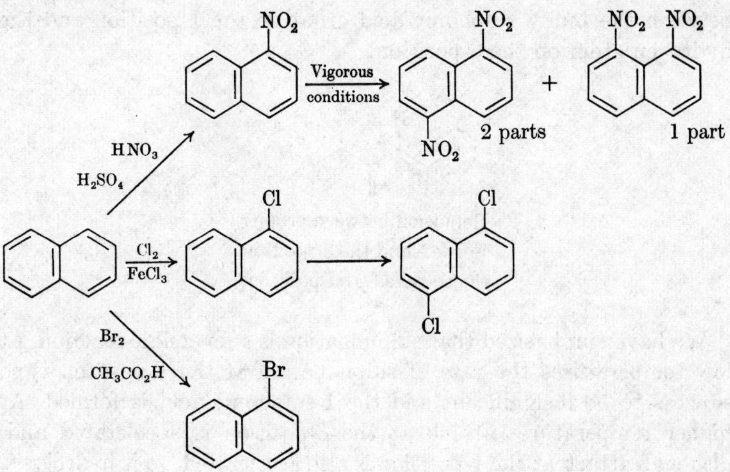

Notice that we wrote of low-temperature sulphonation. If naphthalene is sulphonated above 160° the 2-sulphonic acid is obtained instead of the 1-sulphonic acid. If naphthalene-1-sulphonic acid is heated with sulphuric acid at 180° it is isomerized to the 2-sulphonic acid:

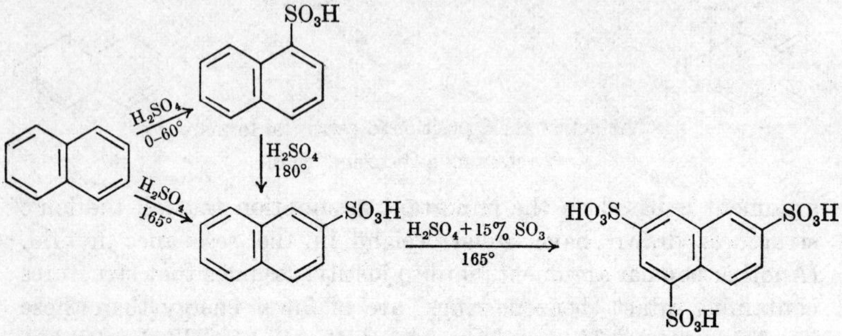

This is yet another example of the importance of distinguishing between the compound which is formed most rapidly and that which is most stable, i.e. kinetic control of a reaction *versus* thermodynamic control of a reaction (see Chapter 12, p. 140, and Chapter 17, p. 212). The 1-sulphonic acid is formed more rapidly for the reasons we have already discussed. The 2-sulphonic acid is thermodynamically more stable, probably because of steric interaction between the bulky sulphonic acid group in the 1-position and the hydrogen atom on the 8-position.

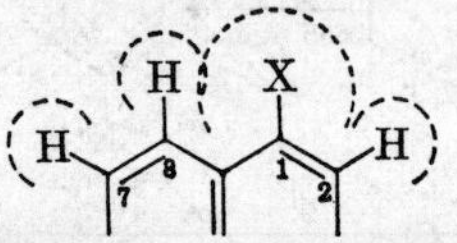

Repulsion between groups in the 1 and the 8-position, the so-called *peri*-positions.

We have emphasized that sulphonation is a reversible reaction. At low temperatures the rate of sulphonation at the β-position is so slow as to be insignificant and the 1-sulphonic acid is formed. At higher temperatures attack at the β-position is accelerated and, although attack at the 1-position is also accelerated, so is hydrolysis.

At temperatures above 170° an equilibrium situation is produced so that the thermodynamically more stable product is the one isolated.

The importance of steric repulsion in the 1-position also manifests itself in reactions that are kinetically controlled. Friedel–Crafts acylation of naphthalene in carbon disulphide solution at low temperatures, by acetyl chloride and aluminium chloride, yields a mixture of both ketones in which methyl 1-naphthyl ketone predominates:

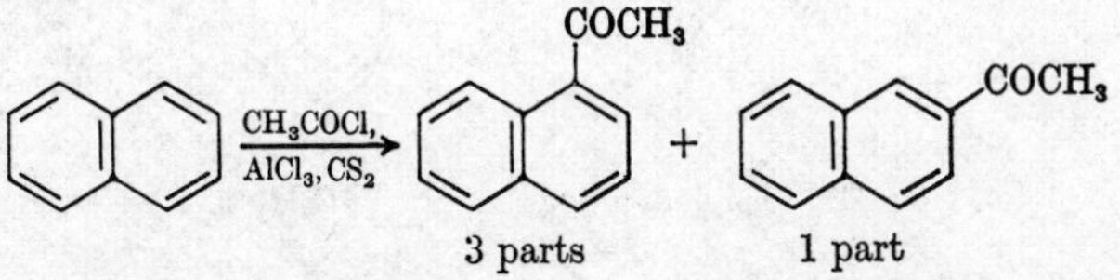

$COCH_3$ $COCH_3$

$\xrightarrow[AlCl_3,\ CS_2]{CH_3COCl,}$ +

3 parts 1 part

If, instead of carbon disulphide, nitrobenzene is used as solvent, then methyl 2-naphthyl ketone is almost the only product. The complex between acetyl chloride and aluminium chloride is bulky in any case, but there is good reason to believe that, in nitrobenzene, the solvent molecule is also associated with this complex. Probably this is the cause of the change in isomer distribution.

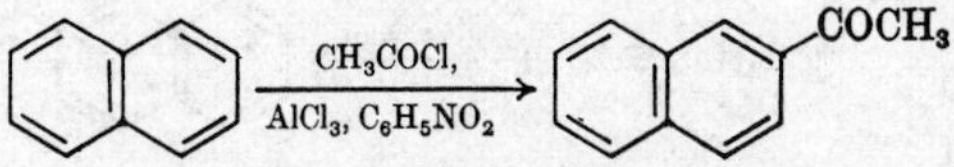

$\xrightarrow[AlCl_3,\ C_6H_5NO_2]{CH_3COCl,}$ $COCH_3$

So far the directive effects of substituents in the naphthalene nucleus have been in accord with the discussion in Chapter 8. The results of polynitration, polychlorination, and high-temperature polysulphonation are depicted above. In nitration, for example, the electron-attracting and -accepting properties of the nitro-group deactivate further substitution in the ring to which the nitro-group itself is attached and subsequent attack occurs exclusively in the other ring. In chlorination, although the chlorine atom is an electron-donor, we know that its electron-attracting properties make it a deactivating group, so that here also substitution occurs in the other ring. However, this apparent simplicity is not preserved when we come to consider the donor groups. 1-Naphthol (α-naphthol) couples with benzenediazonium chloride in alkaline solution to yield a mixture of the 2- and the 4-azo-dye, exactly as we should expect:

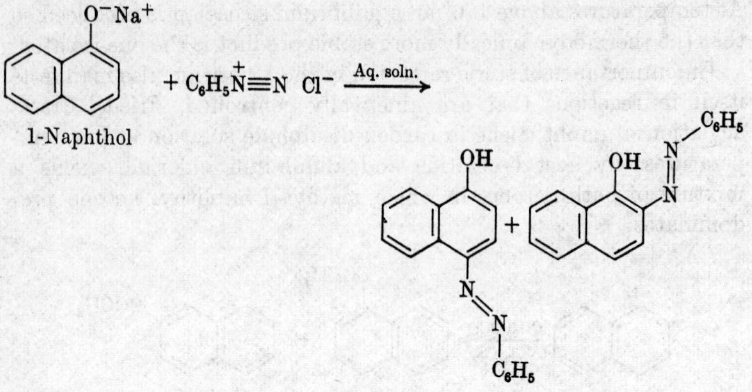

O^-Na^+

$+ C_6H_5\overset{+}{N}{\equiv}N\ Cl^- \xrightarrow{\text{Aq. soln.}}$

1-Naphthol

OH, $N{=}N{-}C_6H_5$ + OH, $N{=}N{-}C_6H_5$

With 2-naphthol, however, coupling occurs exclusively in the 1-position and no coupling occurs at all in the *ortho*-position (the 3-position).

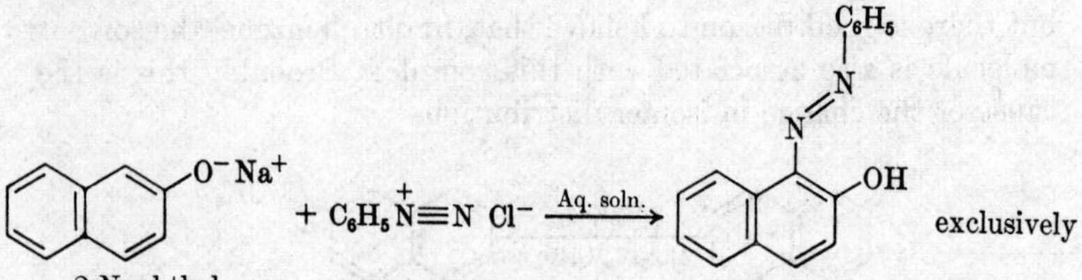

O^-Na^+ $+ C_6H_5\overset{+}{N}{\equiv}N\ Cl^- \xrightarrow{\text{Aq. soln.}}$ $C_6H_5{-}N{=}N$, OH exclusively

2-Naphthol

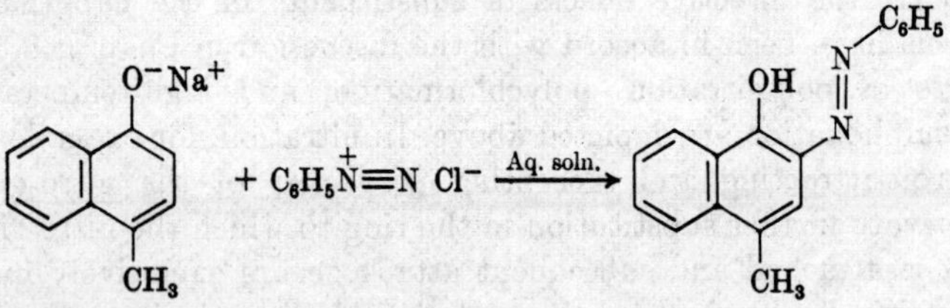

O^-Na^+, CH_3 $+ C_6H_5\overset{+}{N}{\equiv}N\ Cl^- \xrightarrow{\text{Aq. soln.}}$ OH, $N{=}N{-}C_6H_5$, CH_3

4-Methyl-1-naphthol

CH_3, O^-Na^+ $+ C_6H_5\overset{+}{N}{\equiv}N\ Cl^- \xrightarrow{\text{Aq. soln.}}$ No reaction

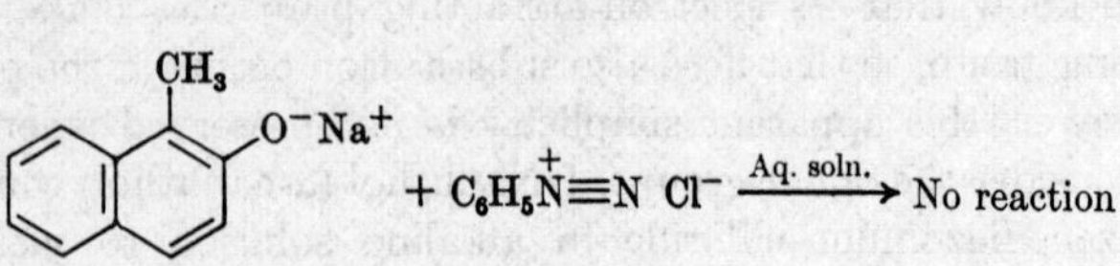

1-Methyl-2-naphthol

This exclusive attack at the 1-position in 2-naphthol becomes even more striking when the reactions of 4-methyl-1-naphthol and 1-methyl-2-naphthol are compared. The former compound couples readily in the 2-position; the latter does not react at all with benzene-diazonium chloride in alkaline solution. If the donor group is in the 1-position we can draw two canonical forms for the Wheland intermediate, in which the positive charge is centred on the donor group for attack both at the 2-position and at the 4-position:

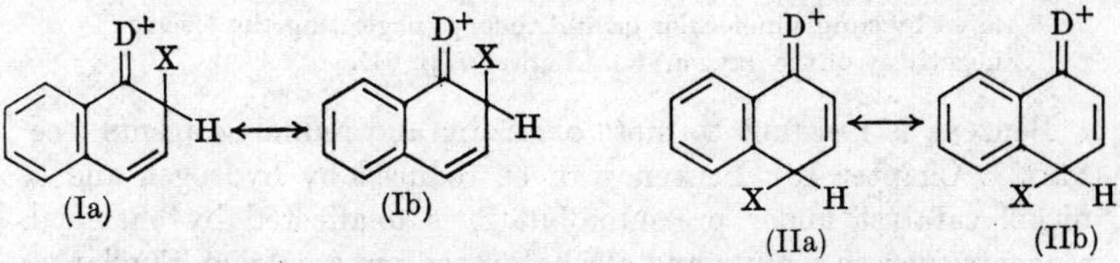

The canonical forms for Wheland intermediates in which the positive charge is centred on the donor group. (Cf. Chapter 8, p. 82.)

When the donor group is in the 2-position and attack occurs at the 3-position, there is only one structure we can draw for the Wheland intermediate in which the charge is centred on the donor group; on the other hand, two such structures can be drawn for the Wheland intermediate involving attack at the 1-position:

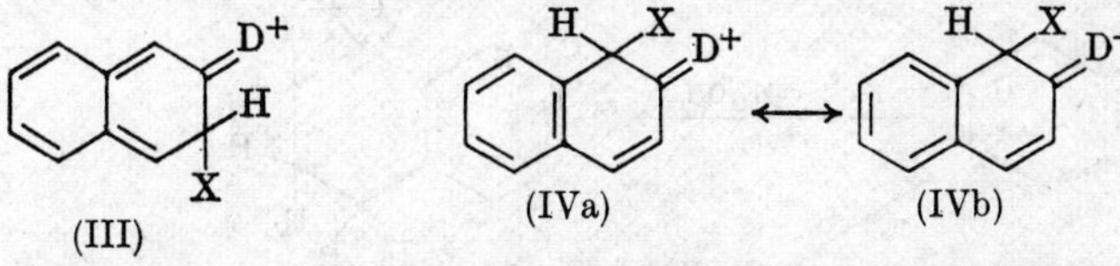

The canonical forms for Wheland intermediates in which the positive charge is centred on the donor group. (cf. Chapter 8, p. 82).

If we invoke the argument that structures with 'benzene rings' are of lower energy than those containing only quinonoid structures, then the difference between structure (III) (the donor group in the 2-position and attack at the 3-position) and the other structures (Ia and Ib, IIa and IIb, IVa and IVb) becomes even more striking. Although we have cautioned against paying too much attention to electron distribution in the ground state, the annexed charge distribution calculated by using simple molecular-orbital theory for the 1-and 2-naphtholate anions is not without interest:

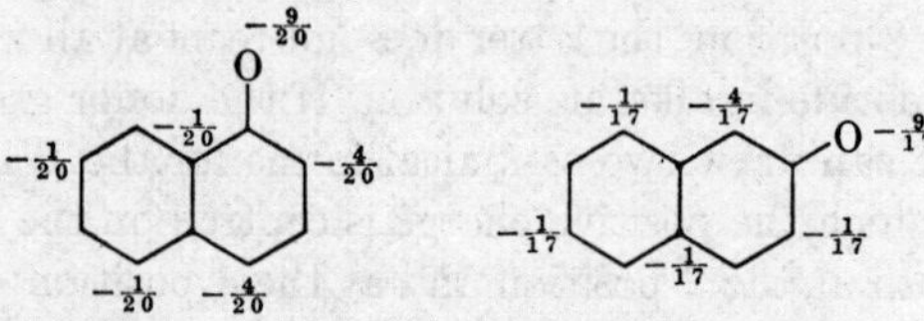
$-\frac{9}{20}$
O
$-\frac{1}{20}$
$-\frac{1}{20}$
$-\frac{4}{20}$
$-\frac{1}{20}$
$-\frac{4}{20}$
$-\frac{1}{17}$
$-\frac{4}{17}$
O
$-\frac{9}{17}$
$-\frac{1}{17}$
$-\frac{1}{17}$
$-\frac{1}{17}$

Charge distribution in the 1- and 2-naptholate anions as calculated by simple molecular-orbital theory (neglecting the electronegativity of the oxygen) (cf. Chapter 7, p. 64).

Benzene is resistant to most oxidizing and reducing agents (see Part 1, Chapter 13). Benzene can be reduced by hydrogen and a nickel catalyst under pressure, but it is unaffected by chemical reagents such as sodium and alcohol or sodium amalgam. Similarly, although benzene reacts with ozone and under forcing conditions

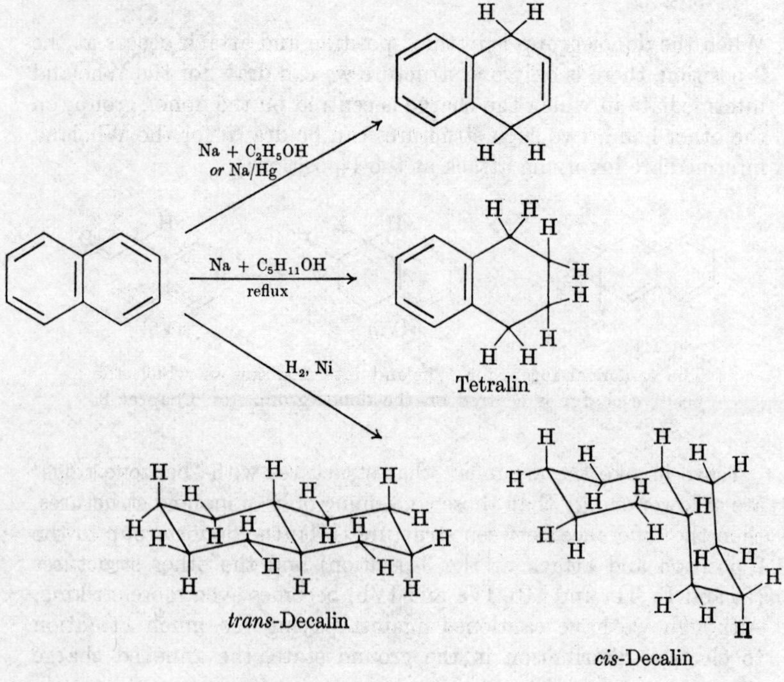
Na + C_2H_5OH
or Na/Hg
Na + $C_5H_{11}OH$
reflux
H_2, Ni
Tetralin
trans-Decalin
cis-Decalin

with very powerful chemical oxidizing agents, benzoquinone cannot readily be prepared by the oxidation of benzene itself. Naphthalene behaves more like an unsaturated compound. It can be reduced to a dihydro-derivative by sodium and alcohol or sodium amalgam, and to a tetrahydro-derivative (called tetralin) by sodium in refluxing amyl alcohol. Reduction with hydrogen and a nickel catalyst yields *cis*- and *trans*-decalin.

Treatment of naphthalene with chromic oxide in acetic acid gives a small yield of 1,4-naphthoquinone. 2-Methylnaphthalene, treated with the same reagents, gives a good yield of the corresponding naphthoquinone. This result should be compared with the formation of benzoic acid by treating toluene with the same reagents.

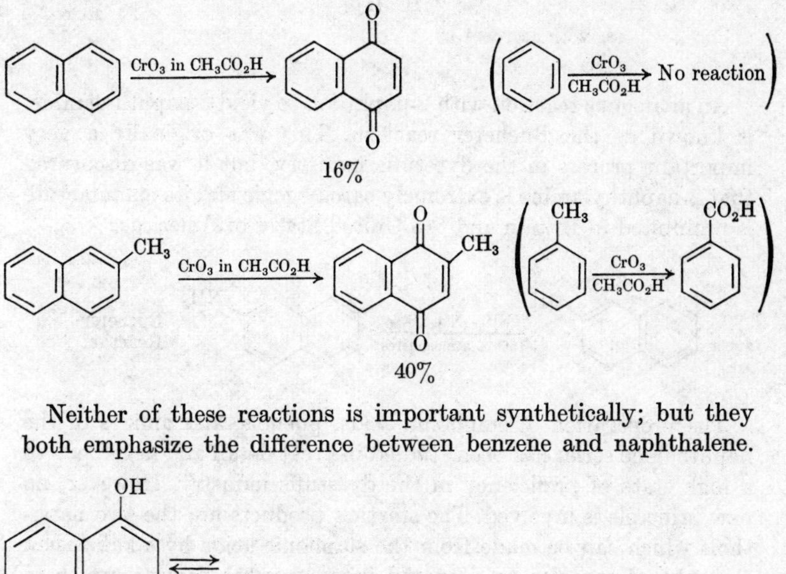

CrO_3 in CH_3CO_2H — 16% (CrO_3 / CH_3CO_2H → No reaction)

CH_3 — CrO_3 in CH_3CO_2H — CH_3 — 40% (CH_3 — CrO_3 / CH_3CO_2H → CO_2H)

Neither of these reactions is important synthetically; but they both emphasize the difference between benzene and naphthalene.

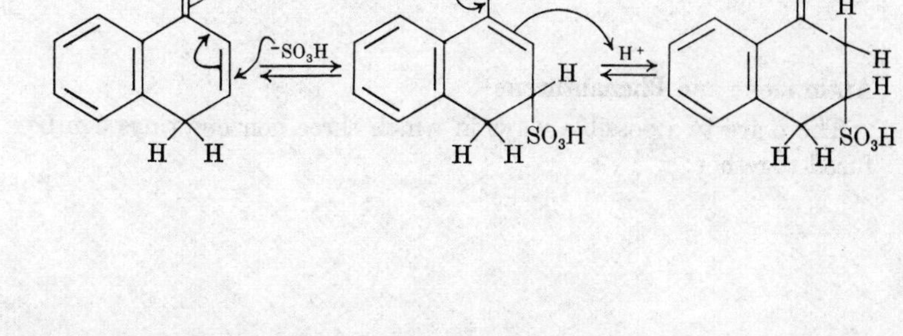

OH ⇌

O — H H ⇌ ($^-SO_3H$) — ^-O — H — SO_3H — H H ⇌ (H^+) — O — H — H — H — SO_3H — H H

The naphthols also exhibit properties differing from those of phenol. Phenol shows no tendency to tautomerize to cyclohexadienone, but 1-naphthol reacts with sodium bisulphite to yield 1-oxotetralin-2-sulphonate; presumably the naphthol tautomerizes to the keto-form (see p. 263). The resulting sulphonic acid reacts with ammonia to give 1-naphthylamine:

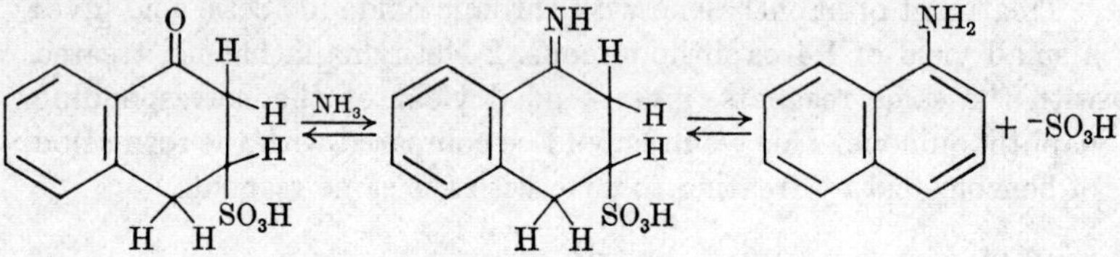

$\xrightarrow{NH_3}$ $\rightleftharpoons$ $+\ {}^{-}SO_3H$

(as with acetone)

An analogous reaction with 2-naphthol, to yield 2-naphthylamine, is known as the Bucherer reaction. This was originally a very important process in the dyestuffs industry, but it was discovered that 2-naphthylamine is extremely carcinogenic and its manufacture is prohibited in Britain and the United States of America.

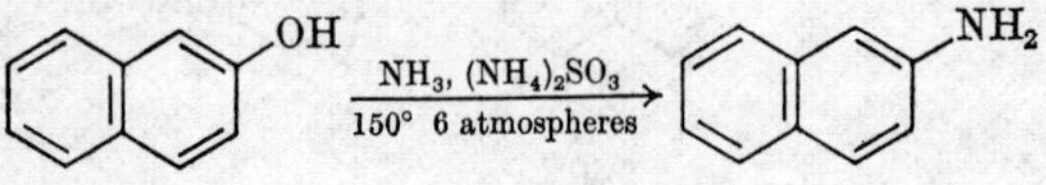

$\xrightarrow[150^\circ\ \ 6\ \text{atmospheres}]{NH_3,\ (NH_4)_2SO_3}$ Bucherer Reaction

The preparation of sulphonic acids, phenols, and amines of the naphthalene series has been studied in great detail and developed to a high state of proficiency in the dyestuffs industry. However, no new principle is involved. The starting products are the two naphthols which can be made from the sulphonic acids by alkali fusion (1-naphthol can also be prepared from 1-naphthylamine, which in turn is prepared from 1-nitronaphthalene, itself prepared by nitration of naphthalene).

Anthracene and Phenanthrene

There are two possible ways in which three benzene rings can be fused together:

Anthracene

Phenanthrene

The linear molecule is called anthracene for which four Kekulé structures can be drawn:

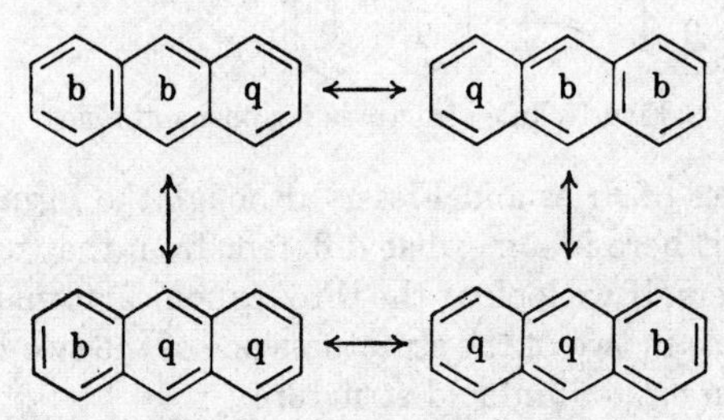

Four Kekulé structures for anthracene

Although these structures are not equivalent to each other, in valence-bond calculations they are presumed to have the same energy. It follows, therefore, that the ground state must have an equal weighting of all four structures. As in the case of naphthalene we get numbers indicating the percentage double-bond character of each bond:

Double-bond character

Observed bond lengths in Å

The angular molecule is called phenanthrene, for which five Kekulé structures can be drawn (p. 266). In four of these the 9,10-bond is drawn as a double bond. We should, therefore, expect a very high degree of double-bond character, for this part of the molecule.

The relative reactivity of polycyclic aromatic compounds is acounted for by an empirical theory that is associated particularly

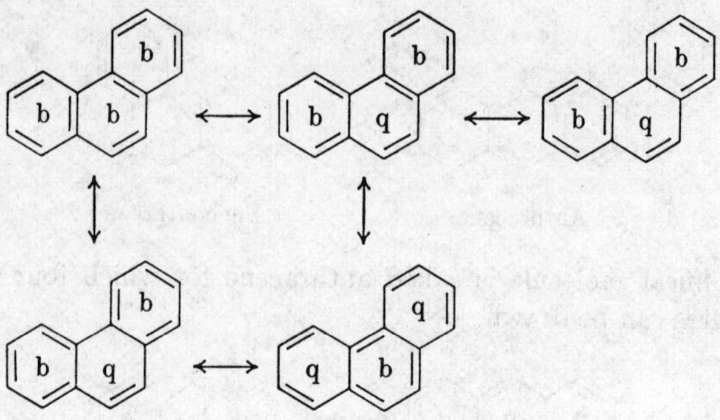

Five Kekulé structures for phenanthrene

with the names of Fries and Fieser, although the form in which we shall present it here is somewhat different from that used by either of these workers. If we look at the three canonical forms of naphthalene we see that in two of the structures one of the two benzene rings is drawn as an *ortho*-quininoid structure:

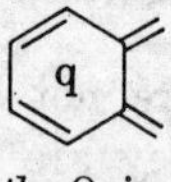

ortho-Quinoid
structure

In the complete resonance hybrid for naphthalene we have exactly half as many quinonoid structures as we have benzenoid structures. Similarly, in phenanthrene in the total resonance hybrid there are half as many quinonoid structures as benzenoid structures. In anthracene, however, there are equal numbers of quinonoid and benzenoid structures. Now, according to the theory, the more quinonoid structures there are in the complete resonance hybrid the more reactive the molecule will be. It would be incorrect to associate the number of quinonoid structures with lack of stability. The resonance (or π-electron) energy of anthracene is approximately three times that of benzene and it should not be thought that anthracene is unstable. Anthracene is reactive (quite a different matter) because its lowest unoccupied orbital is of comparatively low energy (and likewise, because of the pairing properties of aromatic molecular orbitals, its highest occupied orbital is of comparatively high energy). It can be shown that for linear polycyclic aromatic compounds (which happen to be those in which quinonoid structures predominate) the energy difference between the highest occupied orbital and the lowest unoccupied orbital will decrease as the number of rings is increased. This provides some quantum-mechanical justification of the Fries–Fieser theory. The fusion of an angular ring will not in general have the same effect on the orbital spacing.

The 9,10-positions in anthracene are exceedingly reactive and anthracene reacts with maleic anhydride to form a Diels–Alder adduct and, in the presence of light, it reacts with oxygen to give a photo-oxide. When irradiated in the absence of oxygen, it dimerizes.

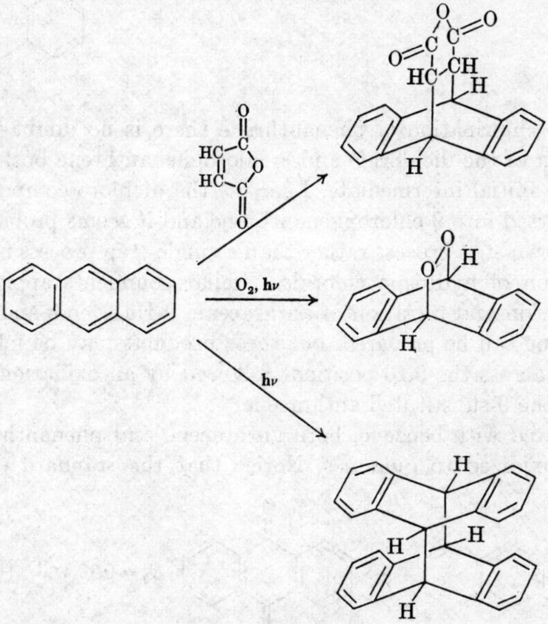

Phenanthrene undergoes none of these reactions. It is inert to maleic anhydride (under normal conditions) and does not react with oxygen in the presence of light, nor does it form a photo-dimer. It does, however, undergo addition reactions extremely readily in the 9(10)-position. The ease with which anthracene and phenanthrene undergo addition reactions makes it difficult to distinguish between the addition-with-elimination reactions like those of benzene and complete two-step addition followed by elimination of a complete molecule.

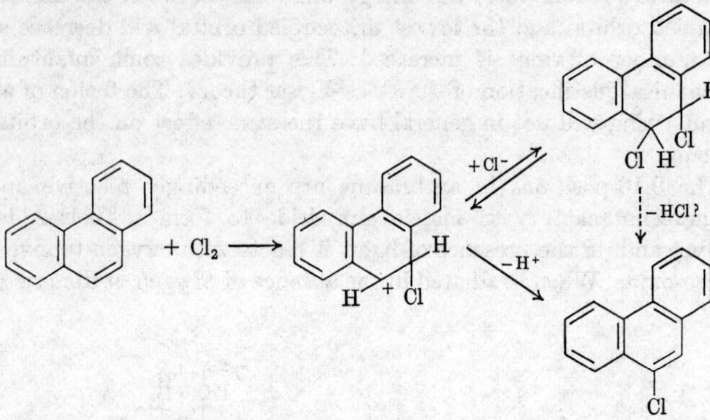

$+ Cl_2 \longrightarrow$ H + Cl H; $+Cl^-$ ⇌ Cl H H Cl; $-HCl?$; $-H^+$ → Cl

In the chlorination of phenanthrene there is no doubt that the formation of the dichloride and 9-chlorophenanthrene both involve the same initial intermediate. Likewise the dichloro-compound can be converted into 9-chlorophenanthrene and it seems probable that this is a two-step process rather than a single-step process involving elimination of hydrogen chloride. Similar comments apply to the halogenation and nitration of anthracene. 9-Halogeno- and 9-nitro-anthracene can be prepared, but these products may be formed by addition across the 9,10-positions followed by an oxidation process to yield the 9-substituted anthracene.

Compared with benzene, both anthracene and phenanthrene are readily oxidized to quinones. Notice that the standard electrode

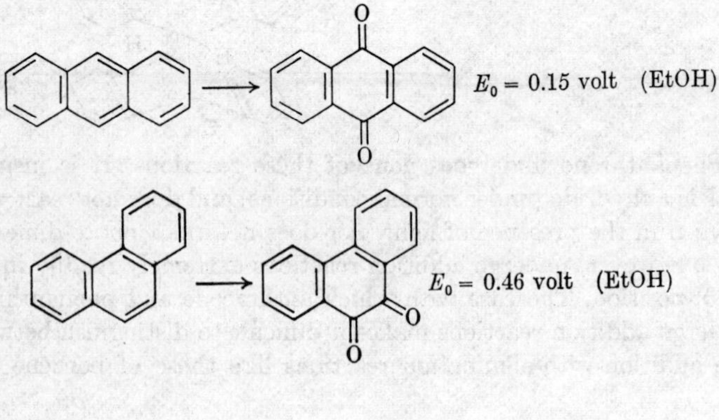

$\longrightarrow$ O O $E_0 = 0.15$ volt (EtOH)

$\longrightarrow$ O O $E_0 = 0.46$ volt (EtOH)

potential of 9,10-phenanthraquinone is much lower than that of benzoquinone and even slightly lower than that of 1,4-naphthoquinone (although it is an *ortho*-quinone).

Polycyclic Compounds

It is convenient to divide the higher polycyclic aromatic compounds into three classes: linear, angular, and condensed. The most striking differences arise between the linear compounds on the one hand and the angular and condensed compounds on the other. These are already illustrated by the difference in chemical reactivity of anthracene and phenanthrene. As we have indicated, the linear polycyclic aromatic compounds become increasingly reactive as additional rings are added; they also become more intensely coloured.

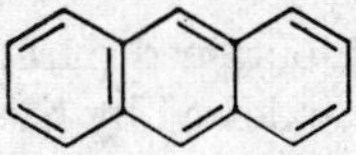

Anthracene, colourless with a blue fluorescence

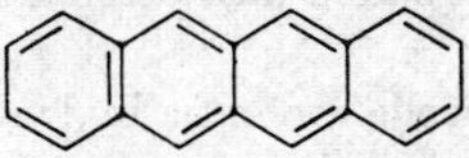

Naphthacene (tetracene), orange

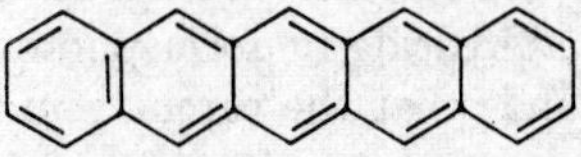

Pentacene, blue

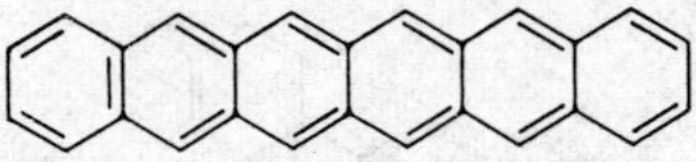

Hexacene, green

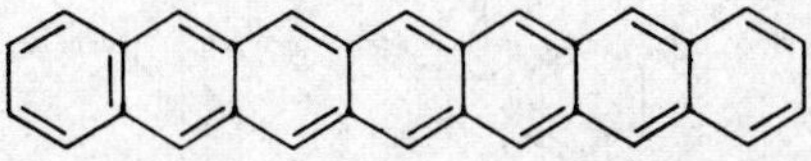

Heptacene, greenish-black

Heptacene is so reactive that it has not yet been obtained in a pure state. Hexacene has been obtained pure by sublimation *in vacuo*; it can be recrystallized provided the solution is kept in an inert atmosphere; it reacts instantly with maleic anhydride and with oxygen. The marked stabilizing influence of angular benzo-groups on a linear polycyclic aromatic compound is well illustrated by the properties of the dibenzo- and the tetrabenzo-heptacene illustrated on p. 270.

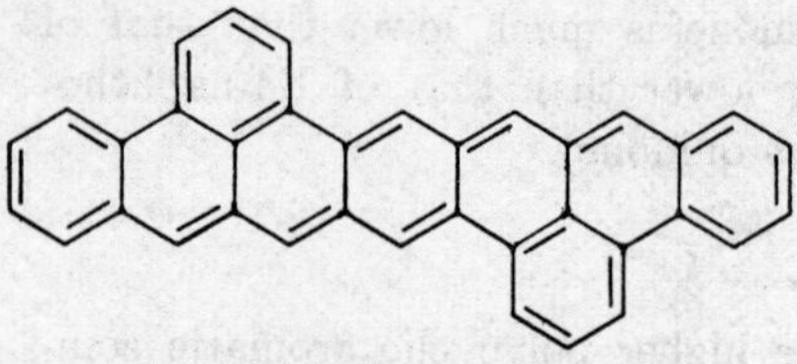

5, 6: 14,15-Dibenzoheptacene
(violet-red).

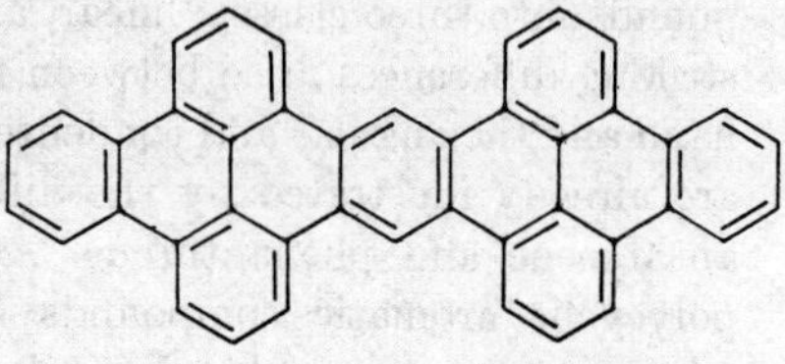

5,6:8,9:14,15:17,18-Tetrabenzoheptacene
(orange)

Heptacene is so reactive that it has not yet been obtained pure; dibenzoheptacene is stable in solution but is readily oxidized by the air; tetrabenzoheptacene is a stable orange compound unaffected by air.

After phenanthrene the next angular polycyclic aromatic hydrocarbons, chrysene and dibenzanthracene, are colourless compounds which do not react readily with maleic anhydride or form photo-oxides. It is among this series of compounds that the carcinogenic

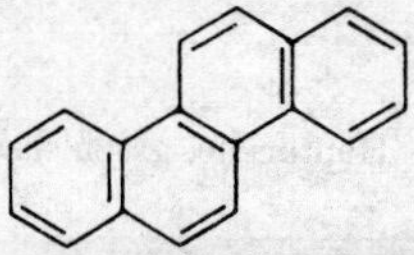

Chrysene

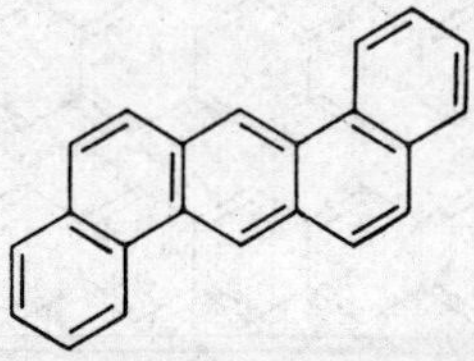

1,2:5,6-Dibenzoanthracene

hydrocarbons are found. Simple examples of polycyclic aromatic hydrocarbons showing carcinogenic activity are 1,2:5,6-dibenz-anthracene shown above, 2,3-benzopyrene, and 3,4:9,10-dibenzo-pyrene. Examples of condensed ring hydrocarbons include pyrene, perylene, and coronene. These compounds are less reactive and more 'aromatic' than the angular compounds. None of them shows carcinogenic activity.

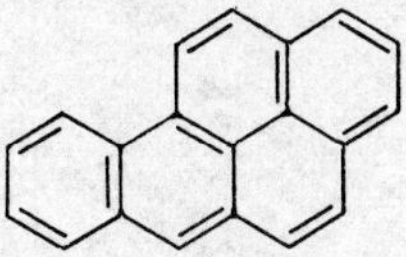

2,3-Benzopyrene

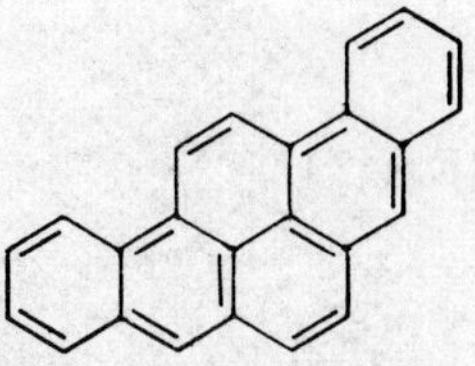

3,4:9,10-Dibenzopyrene

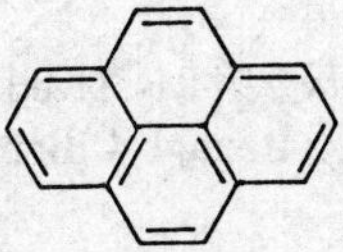

Pyrene

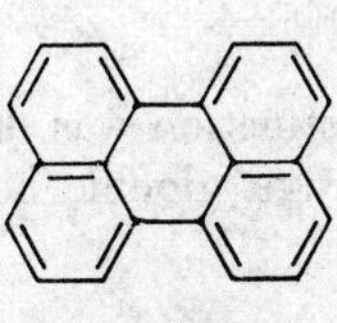

Perylene

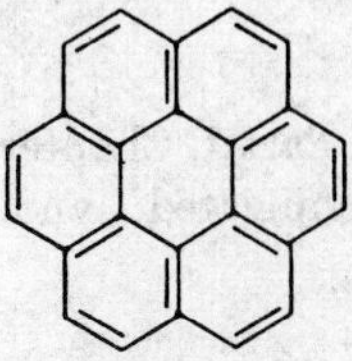

Coronene

Problems

1. Suggest syntheses, starting with naphthalene, of the following naphthalenesulphonic acids, all of which have been employed as intermediates in the dyestuff industry:

(*a*) 4-Aminonaphthalene-1-sulphonic acid (Naphthionic acid)

(*b*) 5-Aminonaphthalene-1-sulphonic acid (Laurent acid)

(*c*) 5-Aminonaphthalene-2-sulphonic acid (Cleve acid)

(*d*) NH_2 OH; HO_3S SO_3H (H-acid)

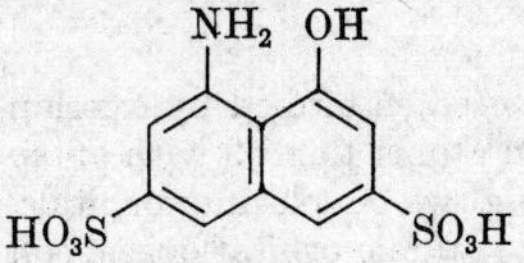

(*e*) HO_3S NH_2; OH (J-acid)

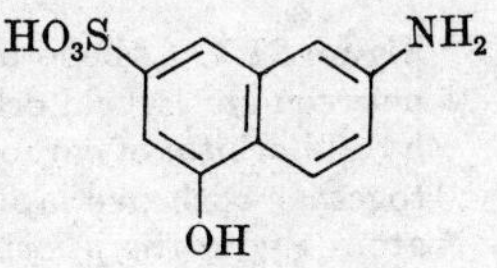

(*f*) NH_2 OH; SO_3H (1,2,4-acid)

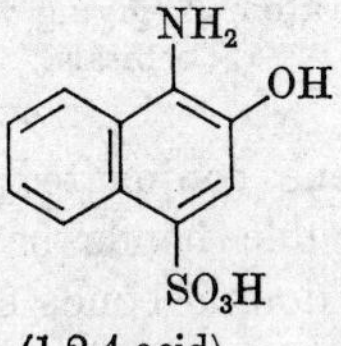

2. How might the following conversions be effected:

(*a*) 2,5-Dimethylphenylacetic acid into 1,4-dimethylphenanthrene-10-carboxylic acid

(*b*) 1-Methylnapthalene into 3-methyl-1,2-benzanthracene.

CHAPTER 21

Cumulative Double Bonds

In Part 1, Chapter 13, a distinction was drawn between 'isolated', 'conjugated', and 'cumulative' double bonds: e.g. hexa-1,4-diene

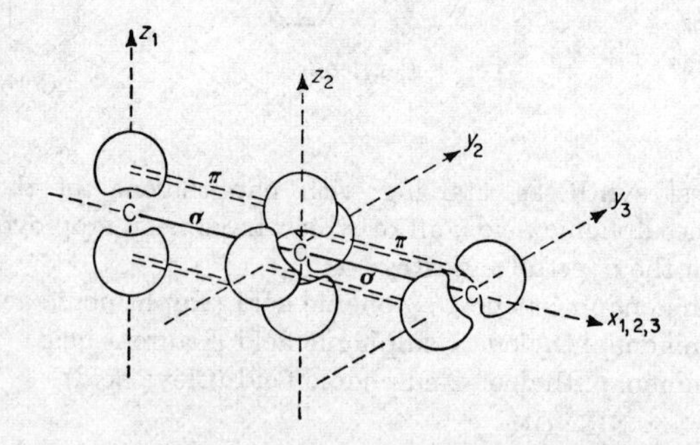

Figure 21.1. Allene bonding represented as formed by overlap between sp^2 hybrid orbitals of carbon atoms 1 and 3 with an sp hybrid orbital of carbon atom 2, giving two σ molecular orbitals, together with overlap between the p_z atomic orbital on carbon atom 1 with the p_z orbital of carbon 2 and independent overlap between the p_y orbital of carbon 2 with the corresponding p_y orbital of carbon atom 3, giving two orthogonal π molecular orbitals.

isolated; hexa-2,4-diene conjugated; and hexa-1,2-diene cumulative). Conjugated double bonds have received considerable discussion, and isolated double bonds show no special characteristics due solely to there being more than one of them. The first feature to notice about cumulative double bonds is that the π-molecular orbitals of the two double bonds are orthogonal, i.e. are independent of one another, just as the $2p_x$ and $2p_y$ orbitals of an atom are independent of each other. Molecules with cumulative double bonds are known as allenes and the simplest ($CH_2{=}C{=}CH_2$ propa-1,2-diene) as allene itself.

The 'electron clouds' are shown diagrammatically in Figure 21.2. Whereas the four bonds emanating from an ethylenic double bond all lie in one plane, the four bonds emanating from an allene group

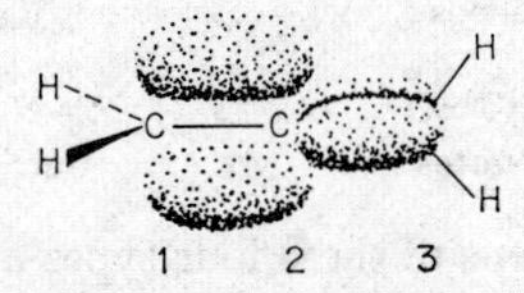

Figure 21.2. 'π-Electron density clouds' in the allene molecule. The hydrogen atoms attached to carbon atom 1 project at right angles to the plane of the paper. The two hydrogen atoms attached to carbon atom 3 lie in the plane of the paper.

lie in two planes at right angles to each other. If the two terminal carbon atoms of the allene bond carry the same pair of non-identical substituents (or different pairs) the whole molecule becomes chiral (see Chapter 23) and optical enantiomers are possible. The first

Enantiomeric allenes

compound of this type to be resolved was 1,3-di-α-naphthyl-1,3-diphenylallene (**1**).

$$(\alpha\text{-}C_{10}H_7)(C_6H_5)C{=}C{=}C(C_6H_5)(\alpha\text{-}C_{10}H_7)$$

(**1**)

Dextrorotatory isomer has $[\alpha]_D^{17} + 437°$

Cumulative double bonds are not restricted to carbon–carbon–carbon linkages. There are carbon–carbon–oxygen linkages, in the ketenes; nitrogen–carbon–oxygen linkages, in the isocyanates; and nitrogen–carbon–nitrogen linkages, in the carbodi-imides. We have already considered carbon–nitrogen–nitrogen and nitrogen–

nitrogen–nitrogen linkages of this kind under the heading of diazo compounds and azides, Chapter 15):

$$\begin{matrix} R & & R \\ & C{=}C{=}C & \\ R & & R \end{matrix} \qquad \begin{matrix} R & \\ & C{=}C{=}O \\ R & \end{matrix}$$

Allenes Ketenes

$$R{-}N{=}C{=}O \qquad R{-}N{=}C{=}N{-}R$$

Isocyanates Carbodi-imides

The chemical reactions of these four types of group have much in common, but since all such compounds are reactive and require special methods of preparation we shall consider them separately, and contrary to our usual practice, shall begin with their preparations.

(a) Allenes

Numerous methods of preparing allenes have been reported but many of these are restricted to special compounds and will not be discussed. Elimination reactions described in Part I for the preparation of olefins are in certain instances applicable for allenes, e.g. elimination from halogeno- or hydroxy-propenes.

$$\underset{(X = \text{Halogen})}{RR'C{=}C(X){-}CHR''R'''} \xrightarrow{-HX} RR'C{=}C{=}CR''R''' \xleftarrow{-HY} \underset{(Y = \text{Halogen or hydroxyl})}{RR'C(Y){-}CH{=}CR''R'''}$$

Success of reactions of this kind depends on the nature of the substituents R, R′, R″, R‴. In suitable cases the conversion of a substituted propene to an allene may be carried out by addition of bromine, followed by elimination of one molecule of hydrogen bromide to yield the α-halogenopropene, followed in turn by elimination of a second molecule of hydrogen bromide. In a similar way a substituted 1,2,3-trihalogenopropane may be converted into an allene by elimination of the hydrogen halide, followed by elimination of two halogen atoms by treatment with zinc:

$$CH_2BrCHBrCH_2Br \xrightarrow{KOH} CH_2{=}CBrCH_2Br \xrightarrow{Zn} CH_2{=}C{=}CH_2$$

Originally, yields of allene by this process (the Gustavson method) were rather poor, but more recent Russian work, in which butyl acetate is used as solvent, has improved the reaction so that yields of over 95% may be obtained. The remaining important general method for making allenes involves the rearrangement of acetylenes. The reversible rearrangement of allenes into acetylenes is discussed below.

Allenes undergo the usual addition reactions of unsaturated hydrocarbons. Remember that the π orbitals are orthogonal, i.e. they do not interact. Addition of hydrogen bromide to allene takes the expected course, to yield 2-bromopropene. In addition, however, cyclic compounds are formed.

$$CH_2{=}C{=}CH_2 + H{-}Br \xrightarrow{-80^\circ} CH_2{=}\overset{+}{C}{-}CH_3 \xrightarrow{Br^-} CH_3CBr{=}CH_2\ (13\%) \xrightarrow{HBr} CH_3CBr_2CH_3\ (39\%)$$

or

$$CH_2{=}\overset{+}{C}{-}CH_3 \xrightarrow{CH_2=C=CH_2} CH_2{=}C(CH_3){-}CH_2{-}\overset{+}{C}{=}CH_2 \xrightarrow{Br^-} \text{1-bromo-1-methyl-3-methylenecyclobutane (H}_2\text{C–CBr(CH}_3\text{)–CH}_2\text{–C(=CH}_2\text{))} \xrightarrow{HBr} \text{1,3-dibromo-1,3-dimethylcyclobutane}$$

52% (mainly *trans*)

Addition of HCl to buta-1,2-diene, as well as yielding the expected 2-chlorobut-2-ene, gives an appreciable yield of the acetylene, but-2-yne. The initial addition step takes place as expected, but the resulting carbonium ion can be stabilized not only by reaction with a chloride anion but also by ejection of a proton:

$$CH_3CH{=}C{=}CH_2 + H{-}Cl \xrightarrow{-78^\circ} CH_3CH{=}\overset{+}{C}{-}CH_3$$

$$CH_3CH{=}\overset{+}{C}{-}CH_3 \xrightarrow{+Cl^-} CH_3CH{=}CClCH_3\quad 60\%$$

$$CH_3CH{=}\overset{+}{C}{-}CH_3 \xrightarrow{-H^+} CH_3C{\equiv}CCH_3\quad 40\%$$

In the latter process the allene has been rearranged to the acetylene.

Base-catalysed rearrangements of allenes to acetylenes are more common than the acid-catalysed reactions. Allene itself reacts with sodium in ether to give the corresponding sodium acetylide:

$$CH_2{=}C{=}CH_2 \xrightarrow[(C_2H_5)_2O]{Na} CH_3C{\equiv}CNa$$

At high temperatures in a basic medium an equilibrium can be set up between an allene and an acetylene. For example, the interconversion of pent-1-yne, penta-1,2-diene, and pent-2-yne has been studied and at 175° (using 4N-KOH in ethanol) the following equilibrium is established:

$$\underset{1.3\%}{CH_3CH_2CH_2C{\equiv}CH} \rightleftharpoons \underset{3.5\%}{CH_3CH_2CH{=}C{=}CH_2} \rightleftharpoons \underset{95.2\%}{CH_3CH_2C{\equiv}CCH_3}$$

This equilibrium can be depicted as follows:

$$CH_3CH_2CH_2C{\equiv}C{:}^-$$
$$\downarrow +H^+ \quad \uparrow -H^+$$
$$CH_3CH_2CH_2C{\equiv}CH$$
$$\downarrow -H^+ \quad \uparrow +H^+$$
$$[CH_3CH_2\overset{..}{\bar{C}}H{-}C{\equiv}CH \longleftrightarrow CH_3CH_2CH{=}C{=}\overset{..}{\bar{C}}H]$$
$$\downarrow +H^+ \quad \uparrow -H^+$$
$$CH_3CH_2CH{=}C{=}CH_2$$
$$\downarrow -H^+ \quad \uparrow +H^+$$
$$[CH_3CH_2\overset{..}{\bar{C}}{=}C{=}CH_2 \longleftrightarrow CH_3CH_2C{\equiv}C{-}\overset{..}{\bar{C}}H_2]$$
$$\downarrow +H^+ \quad \uparrow -H^+$$
$$CH_3CH_2C{\equiv}CCH_3$$

An important example of the rearrangement of an allene is involved in the industrial synthesis of chloroprene. Vinylacetylene is treated with hydrogen chloride in the presence of cuprous chloride. The first product of this reaction results from 1,4-addition to the vinylacetylene and is 1-chloro-2,3-butadiene. This rearranges under the influence of the copper salt to yield chloroprene.

$$CH{\equiv}CCH{=}CH_2 \xrightarrow{HCl} CH_2{=}C{=}CHCH_2Cl \xrightarrow{CuCl} CH_2{=}CCl{-}CH{=}CH_2$$

'Chloroprene' used for synthetic rubber

Molecules containing more than two cumulative double bonds joined together are known as 'cumulenes'. Tetraphenylbutatriene can be prepared from benzophenone and dilithium acetylide. The resulting butynediol is reduced with stannous chloride and HCl, or with phosphorus tribromide in pyridine:

$$(C_6H_5)_2C{=}O + LiC{\equiv}CLi \longrightarrow (C_6H_5)_2\underset{OH}{\underset{|}{C}}{-}C{\equiv}C{-}\underset{OH}{\underset{|}{C}}(C_6H_5)_2 \longrightarrow$$

$$(C_6H_5)_2C{=}C{=}C{=}C(C_6H_5)_2$$

1,1,4,4-Tetraphenylbuta-1,2,3-triene

Hexapentaenes are made similarly and are moderately stable compounds:

$$ClCH_2C{\equiv}CCH_2Cl \xrightarrow{4NaNH_2} NaC{\equiv}C{-}C{\equiv}CNa \xrightarrow{R_2CO} R_2\underset{OH}{\underset{|}{C}}{-}C{\equiv}C{-}C{\equiv}C{-}\underset{OH}{\underset{|}{C}}R_2$$

$$\xrightarrow[\text{or } PCl_3 + C_5H_5N]{SnCl_2 + HCl} R_2C{=}C{=}C{=}C{=}C{=}CR_2$$

A hexapentaene

Butatrienes are yellow; hexapentaenes are red and crystalline; octaheptaenes and decanonaenes have been prepared in solution but appear too unstable to be obtained in a solid state.

(b) Ketenes

Ketene itself is prepared by the pyrolysis of acetone:

$$CH_3COCH_3 \xrightarrow{700°} CH_2{=}C{=}O$$

The mechanism is not known with complete certainty but probably involves initial breakdown to a methyl and an acetyl radical; at this temperature the latter would break down further to give a second methyl radical and carbon monoxide. The methyl radicals attack acetone molecules and the resulting acetonyl radical breaks down intramolecularly to yield ketene and another methyl radical, so maintaining the chain:

$$CH_3COCH_3 \xrightarrow{\Delta H} CH_3\cdot + CH_3CO\cdot \longrightarrow 2CH_3\cdot + CO$$

$$CH_3\cdot + CH_3COCH_3 \longrightarrow CH_4 + \dot{C}H_2COCH_3$$

$$H_2\dot{C}-C(=O)-CH_3 \longrightarrow CH_2{=}C{=}O + CH_3\cdot$$

Ketene is stable in the gas phase but the moment it liquefies it forms a dimer. Ketene dimer is a lactone and reacts with ethanol to form ethyl acetoacetate.

$$[CH_2{=}C{=}\ddot{O}: \longleftrightarrow CH_2{=}\overset{+}{C}-\ddot{\overset{-}{O}}:]$$

$$CH_2{=}C{=}O + CH_2{=}C{=}O \longrightarrow CH_2{=}C(-O^-)\ \ CH_2{=}\overset{+}{C}-O \longrightarrow H_2C-C{=}O,\ CH_2{=}C-O \text{ (four-membered ring)}$$

$$H-\ddot{O}-Et + H_2C-C{=}O,\ CH_2{=}C-O \longrightarrow H_2C{=}C(-O^-)-CH_2-C(=O)-\overset{+}{O}(H)Et \longrightarrow H_3C-C(=O)-CH_2-C(=O)-OEt$$

Because ketene is very low-boiling and polymerizes in the liquid state, reactions are normally carried out by bringing the gas into contact with the other reactant either in solution or in the liquid phase. The carbonyl double bond is more polarized than the ethylenic double bond; thus most reactions are initiated by addition to one or other end of the carbonyl bond. Electrophiles add to the oxygen atom; thus hydrogen chloride adds to ketene to yield acetyl chloride;

$$CH_2{=}C{=}O + H-Cl \longrightarrow CH_2{=}C^+(OH)\ \ Cl^- \longrightarrow \underset{\text{Enol}}{CH_2{=}C(HO)Cl} \rightleftarrows CH_3-C(=O)Cl$$

$$CH_2{=}C{=}O + H-OCOCH_3 \longrightarrow CH_2{=}C^+(OH)\ \ \bar{O}COCH_3 \longrightarrow \underset{\text{Enol}}{CH_2{=}C(HO)OCOCH_3} \rightleftarrows CH_3-C(=O)-O-C(=O)-CH_3$$

and acetic acid adds to give acetic anhydride. The latter reaction is used in the manufacture of acetic anhydride. Notice that the final product of both these reactions is that to be expected from addition across the ethylenic bond and not across the carbonyl bond.

Nucleophiles likewise add across the carbonyl bond attacking the central carbon atom in this case. Although the addition occurs

$$CH_2{=}C{=}O + \bar{C}N \longrightarrow \left[H_2C{=}C(CN){-}\ddot{\underset{..}{O}}{:}^- \longleftrightarrow H_2\ddot{C}^-{-}C(CN){=}O \right] \xrightarrow{HCN} H_3C{-}C(CN){=}O$$

$$CH_2{=}C{=}O + H\ddot{O}{-}Et \longrightarrow H_2C{=}C(O^-){-}O^+(H)Et\ ^{*} \longrightarrow H_3C{-}C(OEt){=}O$$

$$CH_2{=}C{=}O + {:}NH_2R \longrightarrow H_2C{=}C(O^-){-}N^+(H)(H)R\ ^{*} \longrightarrow H_3C{-}C(NHR){=}O$$

across the carbonyl bond the ultimate product is again that which we should obtain if the addition were across the ethylenic bond.

Ketene is very reactive and can be used to give the acetyl derivatives of a wide variety of compounds. In particular, it reacts with enols to give enol acetates (see Chapter 12).

* We have depicted this reaction as an intramolecular proton transfer, since this correctly shows the overall course of the reaction. It is more likely that

$$\ddot{S}{-}H \quad H_2C{=}C(O^-){-}O^+(H)Et \quad \dot{S}{-}H$$

protons are transferred to and from the solvent, but to depict this would be cumbersome without adding significantly to our interpretation of the reaction.

We have limited our discussion so far to the reactions of ketene itself. Ketene is an important chemical made on a large scale and used industrially, particularly for the manufacture of acetic anhydride. However, the ketene grouping is a general one and both aldoketenes (RCH=C=O) and ketoketenes ($R_2C{=}C{=}O$) are known. For example, diphenylketene can be prepared from the corresponding acid chloride by treatment with base:

$$(C_6H_5)_2CHCOCl \xrightarrow{Et_3N} (C_6H_5)_2C{=}C{=}O$$

This is a fairly general method of preparation and is a normal elimination reaction.

(c) Isocyanates

Isocyanates can be prepared by heating together the appropriate dialkyl sulphate and potassium cyanate in the presence of dry sodium carbonate. This amounts to a direct displacement reaction.

$$R_2SO_4 + K^+NCO^- \xrightarrow{Na_2CO_3} R{-}N{=}C{=}O + ROSO_3^-K^+$$

$$[:\ddot{\bar{O}}{-}C{\equiv}N: \longleftrightarrow \ddot{O}{=}C{=}\ddot{\bar{N}}:] \quad CH_3{-}CH_2{-}OSO_3C_2H_5 \longrightarrow O{=}C{=}NC_2H_5 + \bar{O}SO_3C_2H_5$$

Another important preparation involves treating a primary amine with phosgene (carbonyl chloride). The intermediate carbamoyl chloride loses hydrogen chloride to yield an isocyanate:

$$C_6H_5NH_2 + COCl_2 \longrightarrow \underset{\text{N-Phenylcarbamoyl chloride}}{C_6H_5NHCOCl} \xrightarrow{Heat} \underset{\text{Phenyl isocyanate}}{C_6H_5{-}N{=}C{=}O} + HCl$$

Most of the common isocyanates are volatile liquids with a powerful and extremely unpleasant smell. If pure they can be kept for months without change, but traces of impurities cause them to polymerize:

$$3\,R{-}N{=}C{=}O \longrightarrow \text{(RNCO)}_3 \text{ (cyclic trimer: six-membered ring of alternating N–R and C=O)}$$

The chemical reactions of isocyanates are extremely similar to those of ketenes. Particularly important are the reactions with nucleophilic substances.

$$C_6H_5N{=}C{=}O + H_2\ddot{O} \longrightarrow C_6H_5N{=}C(O^-){-}O^+H_2 \longrightarrow C_6H_5NH{-}C({=}O){-}OH \longrightarrow C_6H_5NH_2 + CO_2$$

$$C_6H_5N{=}C{=}O + H\ddot{O}Et \longrightarrow C_6H_5N{=}C(O^-){-}O^+(H)Et \longrightarrow C_6H_5NH{-}C({=}O){-}OEt$$

O-Ethyl-*N*-phenylurethane

$$C_6H_5N{=}C{=}O + H_2\ddot{N}C_6H_5 \longrightarrow C_6H_5N{=}C(O^-){-}N^+H_2C_6H_5 \longrightarrow C_6H_5NH{-}C({=}O){-}NHC_6H_5$$

N,*N'*-Diphenylurea

Phenyl isocyanate is often used to prepare crystalline derivatives of alcohols, phenols, and amines, the resulting phenylurethanes or substituted phenylureas being crystalline and easily purified. Aryl di-isocyanates (e.g. *p*-phenylene di-isocyanate) are used in the industrial preparation of polyurethane elastomers. The basic reaction involves the di-isocyanate reacting with diols (or diamines) to produce long chains containing urethane links.

An important degradative reaction involves treating an amide with sodium hypobromite (i.e. bromine in aqueous alkali). This reaction is known as the Hofmann degradation. The overall reaction is a conversion of an amide into an amine. The steps of the reaction involve *N*-bromination of the amide, followed by elimination of HBr to yield the isocyanate; as shown on p. 282, the isocyanate then reacts with water to yield the amine. This reaction should be

* See footnote on page 279.

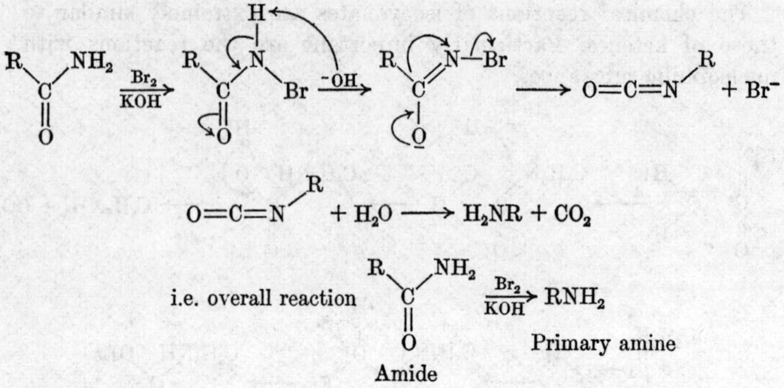

$$RC(=O)NH_2 \xrightarrow[KOH]{Br_2} RC(=O)NHBr \xrightarrow{^-OH} RC(O^-)=N{-}Br \longrightarrow O{=}C{=}N{-}R + Br^-$$

$$O{=}C{=}N{-}R + H_2O \longrightarrow H_2NR + CO_2$$

i.e. overall reaction

$$\underset{\text{Amide}}{RC(=O)NH_2} \xrightarrow[KOH]{Br_2} \underset{\text{Primary amine}}{RNH_2}$$

compared very carefully with the Curtius rearrangement described at the end of Chapter 15.

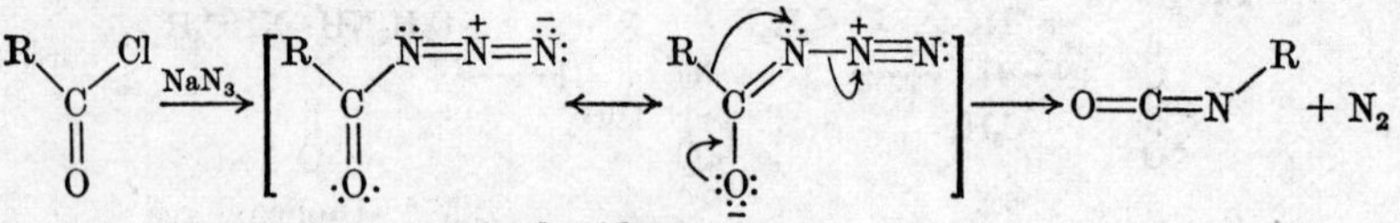

$$RC(=O)Cl \xrightarrow{NaN_3} \left[RC(=O)\ddot{N}{-}\overset{+}{N}{=}\overset{-}{N}: \longleftrightarrow RC(O^-)=\ddot{N}{-}\overset{+}{N}{\equiv}N: \right] \longrightarrow O{=}C{=}N{-}R + N_2$$

Acyl azide

Curtius rearrangement (Chapter 15, p. 191)

(d) Carbodi-imides

A common preparation of carbodi-imides involves oxidation of the corresponding thiourea:

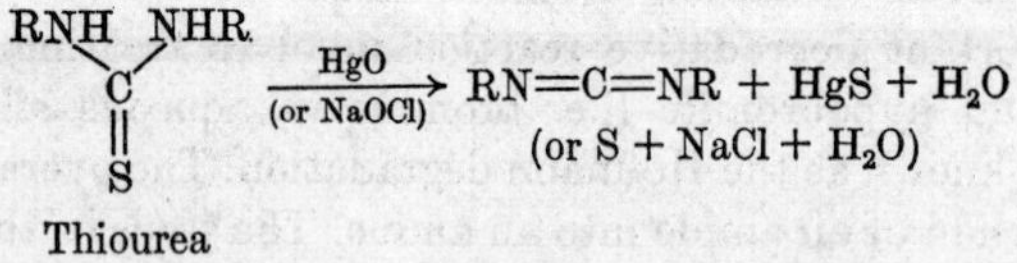

$$\underset{\text{Thiourea}}{(RNH)(NHR)C{=}S} \xrightarrow[\text{(or NaOCl)}]{HgO} RN{=}C{=}NR + HgS + H_2O \quad (\text{or } S + NaCl + H_2O)$$

The reactions of these reactive compounds with nucleophiles and with electrophiles are very similar to those of isocyanates and ketenes. For example, carbodi-imides react readily with water to yield the corresponding urea:

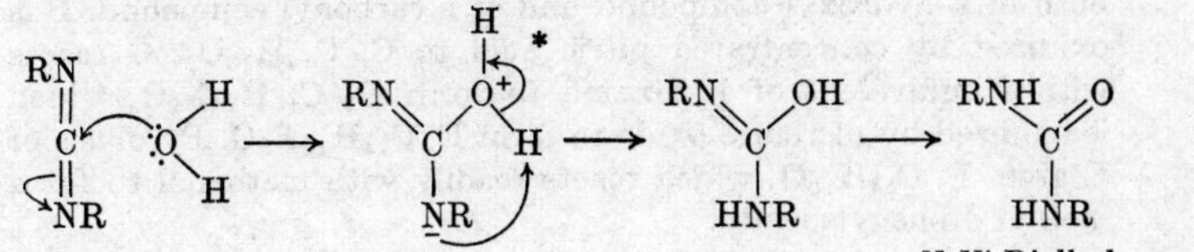

N,N′-Dialkylurea

The reaction with electrophiles is very similar. The reaction with acid is important, not because of the formation of urea, but because it is a way of preparing anhydrides under very mild conditions. It has been particularly important for the preparation of pyrophosphate esters and polyamides.

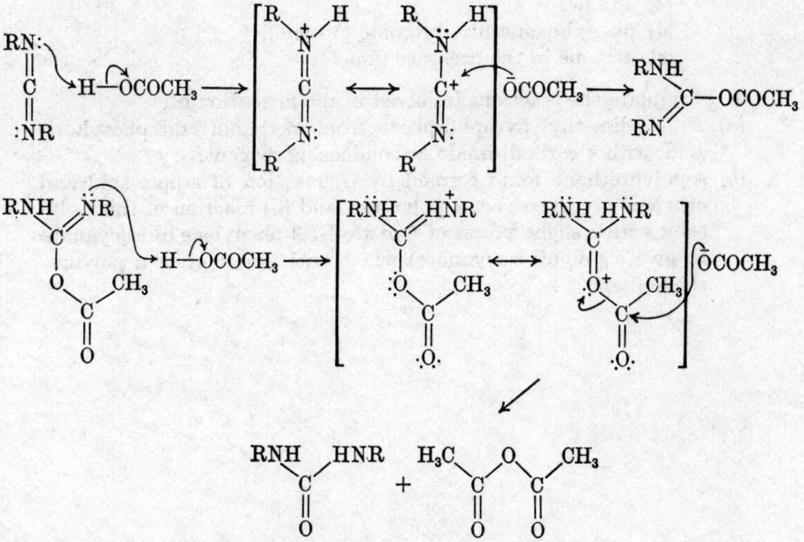

Problems

1. Elucidate the following reactions:

Compound A, C_7H_6O, reacts in the presence of potassium cyanide in ethanol to form B, $C_{14}H_{12}O_2$, which has the properties

* See footnote on page 279.

both of a hydroxyl compound and of a carbonyl compound. B is oxidized by concentrated nitric acid to C, $C_{14}H_{10}O_2$. C reacts with 1 equivalent of hydrazine to form D, $C_{14}H_{12}N_2O$, which is oxidized by mercuric oxide to form E, $C_{14}H_{10}N_2O$. Pyrolysis of E gives F, $C_{14}H_{10}O$, which reacts readily with methanol to form methyl diphenylacetate.

2. What products are formed when

(i) but-3-enal $CH_3CH{=}CHCH{=}O$

and (ii) ethylketene $CH_3CH_2CH{=}C{=}O$

react with:

(*a*) ethanol;

(*b*) hydrobromic acid;

(*c*) aniline;

(*d*) phenylmagnesium bromide (1 mole);

(*e*) acetone in the presence of acid.

3. Formulate the reactions involved in the formation of:

(*a*) P^1, P^2-dimethyl pyrophosphate from methanol and phosphoric acid, with a carbodi-imide as condensing reagent.

(*b*) A polyurethane foam formed by (i) reaction of adipic acid and ethylene glycol to give a polyester, and (ii) reaction of this polyester with a slight excess of 4-methyl-1,3-phenylene di-isocyanate to give a poly(di-isocyanate) which with water gives a polyurethane foam.

CHAPTER 22

Reactive Intermediates

Chapters 9, 10, and 11 are concerned with carbanions, carbonium ions, and carbon-free radicals. Although under special circumstances stabilized carbanions, carbonium ions, and free radicals can be prepared, these species are usually reactive intermediates occurring only transiently in a reaction process. We now have to consider three species which are neither charged, like the carbanion and the carbonium ion, nor possess an odd number of electrons, like the carbon free radical, and yet they are so reactive that they exist only transiently as reaction intermediates. These are carbenes (e.g. CH_2), nitrenes (R—N), and dehydrobenzene* (C_6H_4). We shall consider these in turn.

(a) Carbenes

Carbene as a general class name is given to neutral bivalent carbon molecules such as CH_2 or CCl_2. In these molecules there are four valency electrons from the carbon and, in the case of methylene (CH_2), two more from the two hydrogen atoms, making a total of six valency electrons. The question at once arises, how are the molecular orbitals for these molecules formed? Two arrangements seem likely, bent and linear. In the bent arrangement we can picture the carbon atom as sp^2 hybridized. Two bonding orbitals can then be formed between the hydrogen 1*s* orbitals and two of the two sp^2 orbitals of the carbon atom. This leaves us with a third sp^2 hybrid orbital and the $2p_z$ atomic orbital of the carbon atom, which have

* More correctly called didehydrobenzene because there are two hydrogen atoms less than in benzene.

different energies (i.e. they are non-degenerate). The remaining two valency electrons of the carbon will go into the third sp^2 hybrid orbital, leaving the $2p_z$ atomic orbital vacant. We have attempted to depict this below in (I). The alternative arrangement is a linear molecule in which molecular orbitals are formed between the hydrogen 1*s* orbitals and *sp* hybrid orbitals from the carbon. This leaves the carbon p_z and p_y atomic orbitals. These are of equal energy and according to Hund's rules the remaining two electrons would therefore go into these degenerate orbitals, one each, with parallel spins. We have attempted to depict this arrangement below in (II).

In the electronic arrangement (I) all the electrons are paired and the bent molecule is therefore in a 'singlet' state. In the alternative

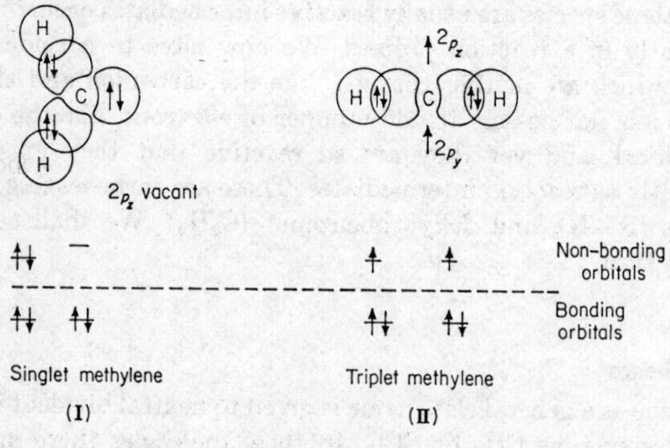

Figure 22.1.

arrangement (II) we have two electrons with parallel spins in degenerate orbitals and the linear molecule is in a 'triplet' state. Because methylene is so reactive, spectroscopic and other physical measurements on the molecule are extremely difficult, and the electronic arrangement in the methylene molecule is still a subject of controversy. The consensus of opinion at the present time is that both states occur and that both take part in chemical reactions. We shall discuss this matter further as we consider the reactions of methylene.

The two most convenient sources of methylene are the photolysis of ketene or of diazomethane:

$$CH_2{=}C{=}O \xrightarrow[2800\text{Å}]{h\nu} CH_2{:} + CO$$

$$CH_2{=}\overset{+}{N}{=}\overset{-}{N} \xrightarrow[3650\text{Å}]{h\nu} CH_2{:} + N_2$$

Both these molecules give methylene also on pyrolysis. Photolyzed or pyrolyzed by themselves, ketene and diazomethane give appreciable yields of ethylene. The ethylene is formed, not so much by the combination of two methylenes, as by the reaction of methylene with ketene or diazomethane:

$$CH_2{:} + CH_2{=}C{=}O \longrightarrow CH_2{=}CH_2 + CO$$

$$CH_2{:} + CH_2{=}\overset{+}{N}{=}\overset{-}{N} \longrightarrow CH_2{=}CH_2 + N_2$$

The reactions of methylene with other compounds depend a great deal on the method of production and on the reaction conditions. A striking reaction is that with aliphatic hydrocarbons where the methylene is inserted between a hydrogen atom and the carbon atom, e.g.:

$$CH_2{:} + CH_3CH_2CH_3 \longrightarrow CH_3CH_2CH_2CH_3 + CH_3\underset{\displaystyle CH_3}{\underset{|}{C}}HCH_3$$

Two mechanisms appear possible for this insertion reaction. Either it could be a one-step process involving three centres, or the methylene could abstract a hydrogen atom to yield a methyl radical and an alkyl radical which could then recombine:

(*a*) $CH_2{:} + H{-}C{\leqslant} \longrightarrow [H\cdots C\cdots CH_2] \longrightarrow H_3C{-}C{\leqslant}$

(*b*) $CH_2{:} + H{-}C{\leqslant} \longrightarrow CH_3\cdot \;\; \cdot C{\leqslant} \longrightarrow H_3C{-}C{\leqslant}$

Possible mechanisms of the 'insertion' of methylene.

Certainly some methyl radicals are formed in these reactions but isotopic experiments with 2-methylbut-1-ene have established that a high proportion of the methylenes are inserted in the same bond as they attack; this suggests that either a three-centre reaction is involved, as shown in (*a*), or the methyl radical shown in (*b*) is never completely free. The fine details of the mechanism of the insertion reaction are outside the scope of our present discussion: they are connected with the problem of the electronic state of methylene. When diazomethane is photolysed in the liquid phase the insertion reaction is completely random, but in the gas phase the reaction is moderately selective, especially if there is an inert gas present. Under the latter conditions the selectivity is similar to that exhibited by a free radical, i.e. methylene is inserted more readily in a tertiary hydrogen bond than in a secondary hydrogen bond and more readily in a secondary hydrogen bond than in a primary.

The second important reaction of methylene is its addition to the carbon–carbon double bond to give cyclopropane derivatives:

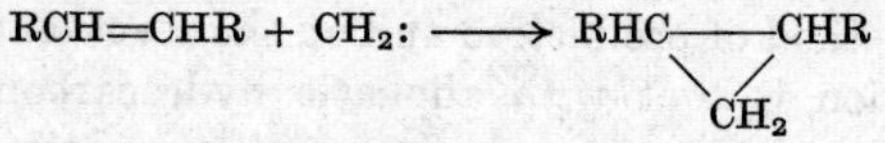

$$RCH{=}CHR + CH_2{:} \longrightarrow RHC\text{—}CHR \text{ (ring closed by } CH_2\text{)}$$

Methylene adds in an exclusively *cis*-fashion when generated by the photolysis of diazomethane either in the liquid phase or at low pressure in the gas phase (in the absence of an inert gas):

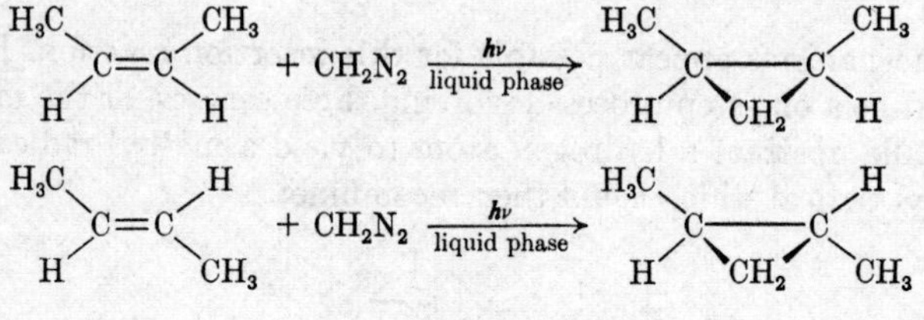

$$(H_3C)(H)C{=}C(CH_3)(H) + CH_2N_2 \xrightarrow[\text{liquid phase}]{h\nu} (H_3C)(H)C\text{—}C(CH_3)(H) \text{ (ring closed by } CH_2\text{)}$$

$$(H_3C)(H)C{=}C(H)(CH_3) + CH_2N_2 \xrightarrow[\text{liquid phase}]{h\nu} (H_3C)(H)C\text{—}C(H)(CH_3) \text{ (ring closed by } CH_2\text{)}$$

At high pressures in the gas phase or in the presence of an inert gas, the formation of a cyclopropane ring still occurs but the reaction is no longer stereospecific. These differences are currently explained in terms of the electronic state of methylene. It is assumed that methylene formed by the photolysis of diazomethane or ketene is initially in a highly energized singlet state. It is further supposed that the singlet state is the upper state and is responsible for the stereospecific addition. It is also assumed that the highly energized singlet is responsible for the indiscriminate insertion reactions. Conversion of singlet methylene into the lower-energy triplet by collision is supposed to yield a species that is more selective and adds to double bonds in a non-stereospecific fashion.

The photochemical preparation of carbenes from aliphatic diazo-compounds can proceed in high yield and in certain cases has been used for preparative purposes. Particularly interesting is the addition of ethoxycarbonylcarbene, generated by the photolysis of diazoacetic ester, to benzene:

$$C_6H_6 \xrightarrow[h\nu > 3000\text{Å}]{N_2CHCO_2Et} \left[C_6H_6{=}CH(CO_2Et) \right] \longrightarrow C_7H_7{-}CO_2Et \quad 80\text{–}90\%$$

+ isomers, e.g. $C_7H_7{-}CO_2Et$

From a preparative point of view, photochemical reactions that do not involve a chain process are very difficult to carry out on a large scale. For this reason, dihalogenocarbenes are of greater general application.

Dihalogenocarbenes are generated by various chemical processes in alkaline solution. Dichlorocarbene is particularly important and can be prepared readily from chloroform. Chloroform is attacked by concentrated alkali. The reaction is not a normal displacement but a proton-abstraction analogous to the proton abstraction by a nucleophile encountered in bimolecular elimination reactions:

$$HO^- + H{-}CCl_3 \longrightarrow H_2O + {}^-{:}CCl_3; \qquad {}^-{:}CCl_3 \longrightarrow Cl^- + {:}CCl_2$$

The trichloromethyl carbanion formed in this reaction is unstable and breaks down spontaneously to yield dichlorocarbene and the chloride anion. In the absence of other compounds the

$$\text{cyclohexene} + CHCl_3 + (CH_3)_3CO^-K^+ \longrightarrow \text{7,7-dichlorobicyclo[4.1.0]heptane } (C_6H_{10}CCl_2)$$

$$C_6H_5CH{=}CH_2 + CHCl_3 + (CH_3)_3CO^-K^+ \longrightarrow C_6H_5CH{-}CH_2{-}CCl_2 \text{ (cyclopropane)}$$

dichlorocarbene reacts with the aqueous alkali and ultimately yields carbon monoxide and formate anions. In the presence of an olefin, however, high yields of the corresponding dichlorocyclopropane can be obtained (see p. 289).

Another convenient way of preparing dichlorocarbene is by treating esters of trichloroacetic acid with alkoxides:

$$Cl_3C{-}C({=}O){-}OEt \xrightarrow{^{-}OEt} Cl_3C{-}C(OEt)_2{-}O^- \longrightarrow CCl_3^-{:} + EtO{-}C({=}O){-}OEt$$

$$CCl_3^-{:} \longrightarrow CCl_2{:} + Cl^-$$

This method often gives higher yields of dichlorocyclopropanes than do reactions using chloroform:

$$(H_3C)_2C{=}CH_2 + Cl_3CCO_2Et + (CH_3)_3CO^-K^+ \longrightarrow (H_3C)_2C{-}CH_2{-}CCl_2 \text{ (cyclopropane)}$$

The addition of dichlorocarbene to the phenoxide anion has long been known, although the mechanism of the reaction was not elucidated until fairly recently. The ultimate products of the reaction are phenolic aldehydes and the reaction is known as the Reimer–Tiemann reaction. The proportions of *ortho*- and *para*-isomers

$$C_6H_5O^- \xrightarrow{CCl_2:} \text{2-H-2-(}\bar{C}Cl_2\text{)cyclohexadienone} \;\left(\text{4-H-4-(}\bar{C}Cl_2\text{)cyclohexadienone}\right) \longrightarrow$$

$$o\text{-}{}^-OC_6H_4CHCl_2 \;\left(p\text{-}{}^-OC_6H_4CHCl_2\right) \xrightarrow[H_2O]{^-OH} o\text{-}HOC_6H_4CHO \;\left(p\text{-}HOC_6H_4CHO\right)$$

Reimer–Tiemann reaction

depend very much on the concentration and conditions. In fairly dilute solution the predominant product is the *p*-hydroxybenzaldehyde but in very concentrated solution the proportions change and *o*-hydroxybenzaldehyde (salicylaldehyde) becomes the predominant product.

The formation of isocyanides when primary amines are treated with alkaline solutions of chloroform is another reaction which has been known for a long time and is now attributed to the dichlorocarbene intermediate:

$$CHCl_3 \xrightarrow{^-OH} CCl_2: \xrightarrow{H_2\ddot{N}C_6H_5} C_6H_5\overset{+}{N}H_2\bar{C}Cl_2 \longrightarrow C_6H_5{-}\overset{+}{N}{\equiv}\overset{-}{C} + 2HCl$$

The isocyanides are unstable, extremely unpleasant-smelling substances and the treatment of a primary amine with chloroform to yield the isocyanide (carbylamine) used to be used as a diagnostic test for primary amines, the so-called 'carbylamine reaction'.

The electronic state of dichlorocarbene raises certain problems. It seems unlikely that a species produced chemically at normal temperatures would be formed in an electronically excited state, yet the dichlorocarbene has been shown to add exclusively in a *cis*-fashion. If the current hypothesis for methylene is correct, then this involves the singlet state of the carbene. In the methylene case it was argued

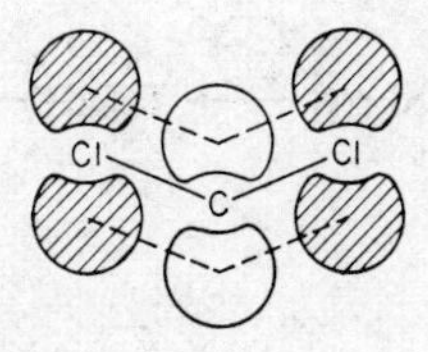

Figure 22.2. Diagram representing possible interaction between filled *p*-orbitals of the chlorine atoms (shown shaded) with the vacant $2p_z$ orbital of the carbon atom in singlet-state carbene.

that the singlet state was the upper state. Possibly the singlet and triplet states are reversed in dichlorocarbene because in the singlet state interaction between non-bonded p electrons of the chlorine atom with the vacant $2p_z$ orbital of the carbon atom becomes possible: this interaction may be sufficient to lower the energy of the singlet below that of the triplet state.

(b) Nitrenes

Neutral, univalent nitrogen compounds occur as intermediates in reactions analogous to those of carbenes. These intermediates are called nitrenes and are usually prepared by the photolysis (or pyrolysis) of organic azides:

$$R-\ddot{N}=\overset{+}{N}=\overset{-}{\ddot{N}}: \xrightarrow{h\nu} R-\ddot{N}: + \ddot{N}\equiv\ddot{N}$$

although chemical methods of preparation have also been described, e.g.:

$$R-\underset{\displaystyle H}{\underset{|}{N}}-OSO_2Ar \xrightarrow{\ddot{B}} R-\bar{N}-\bar{O}SO_2Ar \longrightarrow R-\ddot{\underset{..}{N}} + \bar{O}SO_2Ar$$

(where R = EtOOC)

There is evidence that, like carbenes, nitrenes are formed in both singlet and triplet states. The nature of R also greatly affects the reactivity of the nitrene. Four kinds of nitrene have been firmly established: aryl (R = C_6H_5), sulphonyl (R = $ArSO_2$), ethoxycarbonyl (R = EtOOC), and cyano (R = NC—)*. The reactions of nitrenes parallel those of carbenes very closely.

$$EtOOC-N_3 \xrightarrow{h\nu} EtOOC-\ddot{\underset{..}{N}} \xrightarrow{\text{cyclohexene}} EtOOC-N\text{(7-azabicyclo[4.1.0]heptane)} + \text{3-(HNCOOEt)cyclohexene}$$

(Addition) (Insertion)

$$N\equiv C-N_3 \xrightarrow{45\text{–}60^\circ} N\equiv C-\ddot{\underset{..}{N}} \xrightarrow{\text{benzene}} \text{azepine}-N-C\equiv N$$

1-Cyanoazepine

* Reactions expected to yield acyl nitrenes result in the products due to a Curtius rearrangement (see Chapter 21, p. 282).

Surprisingly an exactly analogous reaction occurs with hexafluorobenzene, giving 1-cyanohexafluoroazepine in 76% yield.

As with the carbenes the stereospecificity of addition to olefins varies with the reaction conditions, and again stereospecific addition is attributed to the singlet state of the nitrene. In insertion reactions nitrenes appear to be more selective than carbenes of related structure. In general, however, it is the similarity of the reactions of nitrenes and carbenes which should be noted, not their differences.

(c) Dehydrobenzene (benzyne)

Dehydrobenzene can be considered as being derived from benzene by the removal of two hydrogen atoms. This leaves two electrons in two sp^2 hybrid orbitals. These orbitals are not orthogonal, but, spatially, they are arranged in such a way that there is little overlap between them. We could increase the overlap by dis-

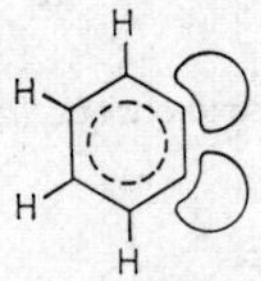

Dehydrobenzene

Figure 22.3.

torting the benzene ring but the energy we should gain by increasing the overlap would be largely offset by the energy we should lose in straining the benzene ring and distorting the π-orbital system. If we assume there is no interaction between the two sp^2 hybrid orbitals, then we have a degenerate pair of orbitals and according to Hund's rules the two electrons will go one into each orbital. Their spins will be parallel, giving us a triplet state. There is little doubt that the

Triplet

Singlet

singlet state, in which the electrons have antiparallel spins and in which there is weak interaction between the two orbitals, is the state of lower energy. The two states are conveniently illustrated in terms of resonance pictures.

At the present time it seems probable that both states occur as reactive intermediates, depending on how the dehydrobenzene is prepared. Most reactions of dehydrobenzene are in very good accord with the resonance picture of the singlet state of the molecule.*

There are basically two types of reaction in which dehydrobenzenes are formed. The former involves attack by very powerful nucleophiles on aryl halides. If chlorotoluene, for example, is treated

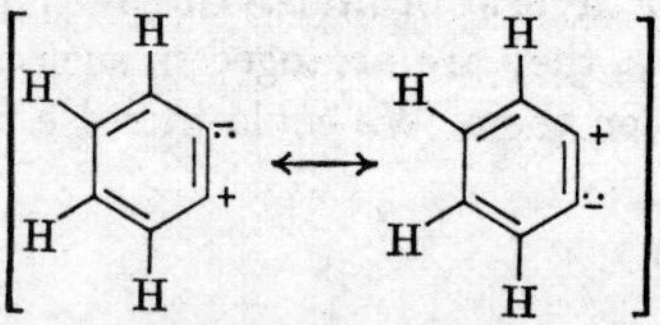

Singlet dehydrobenzene (ground state?)

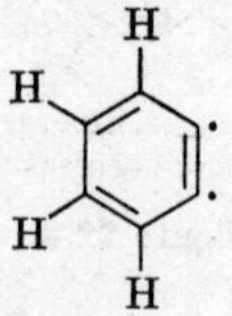

Triplet (excited state?) dehydrobenzene

Figure 22.4. Dehydrobenzene depicted by resonance theory.

*Dehydrobenzene is often written with a 'triple' bond and called 'benzyne'. The triple bond is, of course, another way of writing the singlet state, but it is misleading because it implies that the bond is analogous to those in acetylene or a nitrile. In an alkyne the bonds adjacent to the triple bond are collinear and the electron density is concentrated between the atoms. The two 'π-orbitals' are not distinguishable, and the π-electron density is completely symmetrical about the 'triple bond'. The alkyne bond is less reactive to electrophilic attack than an alkene bond and in no sense is it unstable. None of these remarks applies to dehydrobenzene. The 'third bond' in ground-state dehydrobenzene is quite different in character from the 'π-bonds' of the rest of the molecule, and the electron density is highest near the carbon atoms. The molecule is so reactive that it only exists as a transient intermediate.

with potassium amide and liquid ammonia, a mixture of *o*- and *m*-toluidine, but no *para*-isomer, is formed. Similarly, treatment of *p*-chlorotoluene yields a mixture of *m*- and *p*-toluidine, but no *ortho*-isomer, whereas treatment of *m*-chlorotoluene yields a mixture of all three toluidines (see formulae below and on p. 296).

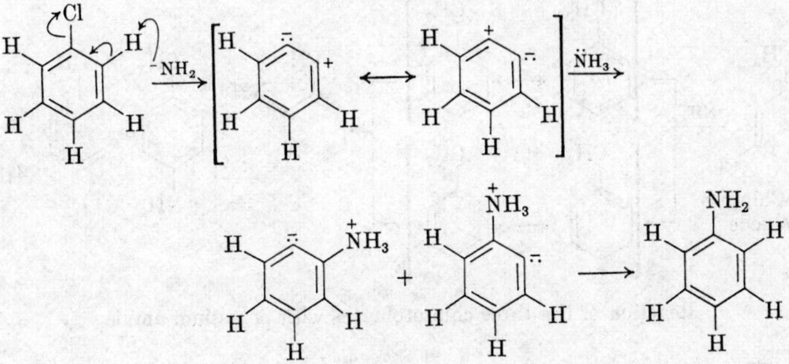

Reaction of lithium amide with chlorobenzene

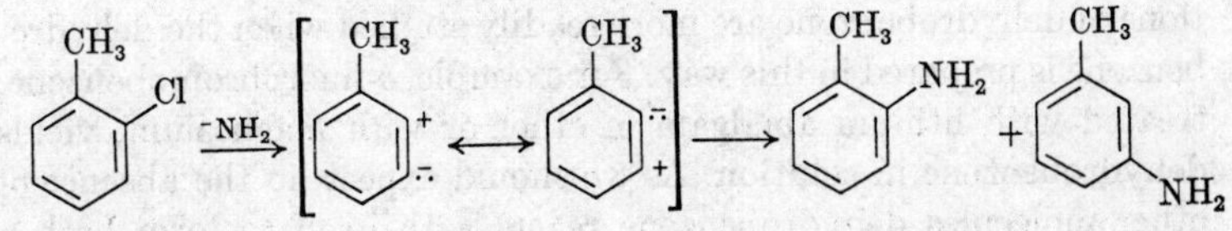

o-Chlorotoluene

Notice that this reaction is very different from the nucleophilic addition-with-elimination (nucleophilic substitution) of halogeno-benzenes activated by electron-accepting groups (see Chapter 13, p. 158). In Chapter 13 (and in Chapter 14, pp. 140–141, of Part 1) we emphasized that nucleophiles do not add to the benzene nucleus unless powerful electron-attracting or electron-accepting groups are present. With very powerful bases such as the amide anion (NH_2^-) an elimination reaction occurs to produce the transient intermediate, dehydrobenzene. Both nucleophiles and electrophiles add to dehydrobenzene, so that the overall product from the reaction of lithium amide in liquid ammonia with chlorobenzene is aniline, i.e. overall the reaction appears to be a substitution.

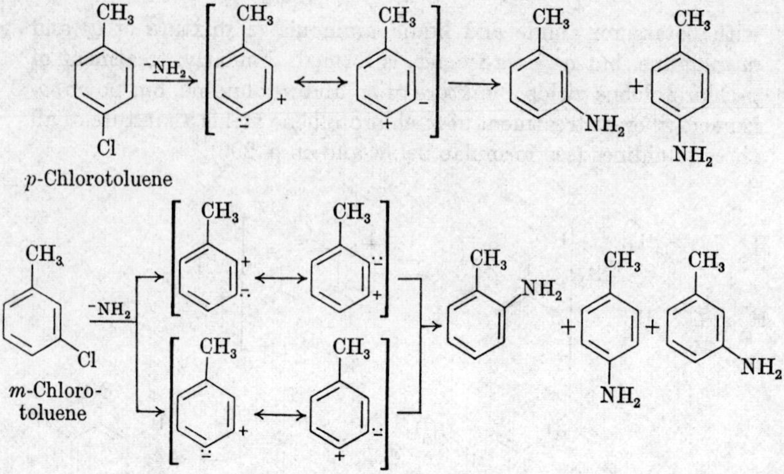

CH_3 $^{-}NH_2$ CH_3 CH_3 CH_3 CH_3

Cl NH_2 NH_2

p-Chlorotoluene

CH_3 CH_3 CH_3 CH_3 CH_3 CH_3

$^{-}NH_2$ NH_2

Cl CH_3 CH_3 NH_2

m-Chloro-
toluene NH_2

Reaction of the three chlorotoluenes with potassium amide

The second type of reaction by which dehydrobenzenes are prepared involve *ortho*-disubstituted aromatic compounds. The reactions of dehydrobenzene are more readily studied when the dehydrobenzene is prepared in this way. For example, *o*-fluorobromobenzene, treated with lithium amalgam in ether or with magnesium, yields dehydrobenzene in solution. As we should expect, in the absence of other molecules dehydrobenzene reacts with itself to form both a dimer and a trimer.

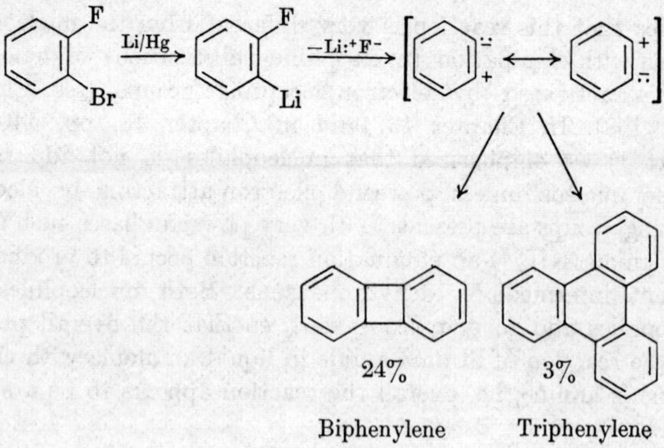

F F

Li/Hg $-Li^{+}F^{-}$

Br Li

24% 3%

Biphenylene Triphenylene

Also, as we should expect, dehydrobenzene behaves as a dienophile in a Diels–Alder or electrocyclic reaction. Thus, it will add to furan, anthracene, or tetraphenylcyclopentadienone.

A very large number of 1,2-disubstituted benzene derivatives has now been found to yield dehydrobenzene, either when heated or on photolysis. A few of the more important ones are shown in the scheme on p. 298.

Intermediates of the dehydrobenzene type are not, of course, restricted to benzene, and analogous dehydronaphthalenes have also been detected. Thus, for example, treatment of 1-chloronaphthalene with piperidine and sodamide yields a mixture of the 1- and 2-piperidyl compounds, presumably, *via* the dehydronaphthalene derivative (see p. 298).

Finally, we must briefly note that dehydrobenzenes add nucleophiles and electrophiles. Thus bromobenzene, treated with common carbanions such as enolate anions and sodamide, yields the corresponding phenyl compound. The reaction presumably involves dehydrobenzene since no reaction takes place between the carbanions and bromobenzene in the absence of a very strong base (e.g. NH_2^-).

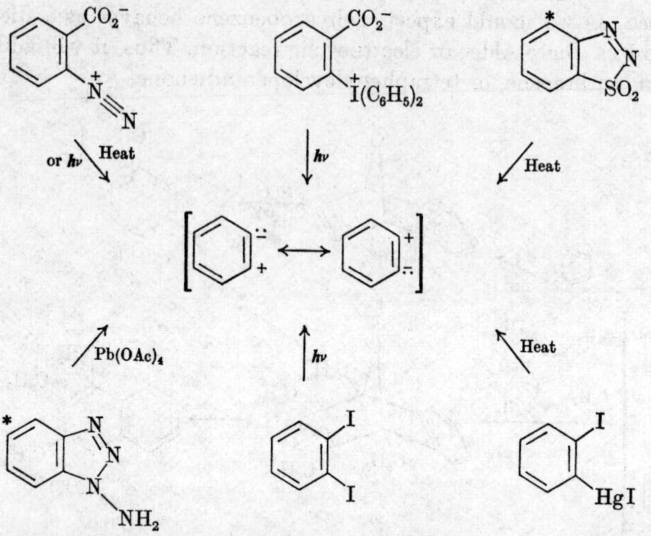

CO_2^- $\overset{+}{N}\equiv N$ — or $h\nu$ / Heat →

CO_2^- $\overset{+}{I}(C_6H_5)_2$ — $h\nu$ →

* N N SO_2 — Heat →

[dehydrobenzene resonance forms]

* N N N NH_2 — $Pb(OAc)_4$ →

I I — $h\nu$ →

I HgI — Heat →

1-Fluoronaphthalene, treated with butyl-lithium and then carbon dioxide yields a mixture of butylnaphthoic acids. A particularly interesting addition is that of triphenylphosphine and triphenylboron to dehydrobenzene to yield the internal salt or betaine (see p. 299).

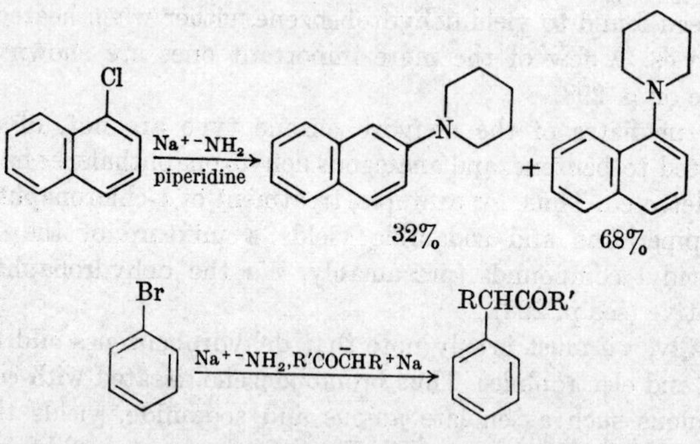

Cl $\xrightarrow[\text{piperidine}]{Na^+\,{}^-NH_2}$ N (32%) + N (68%)

Br $\xrightarrow{Na^+\,{}^-NH_2,\,R'CO\bar{C}HR\,{}^+Na}$ RCHCOR′

Dehydrobenzenes, carbenes, and nitrenes are very reactive species which exist only as transient intermediates in reaction processes.

* These compounds may give dehydrobenzene in the triplet state.

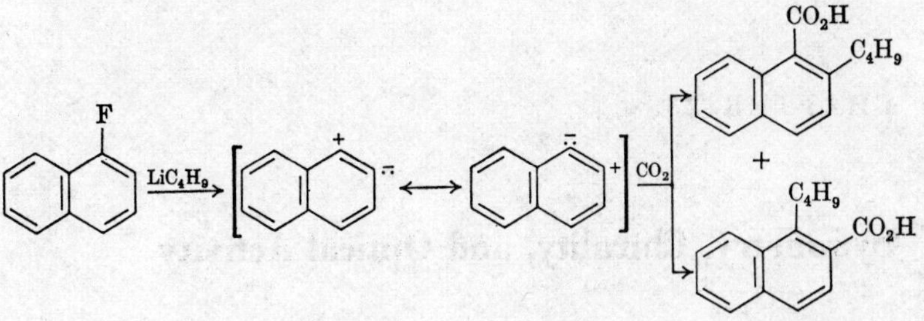

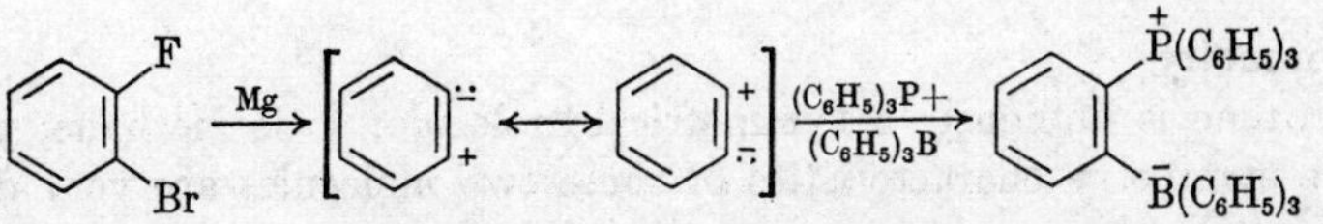

They all have a pair of electrons in a high-energy bonding orbital and have a very low-lying antibonding orbital. They are thus extremely reactive, able to combine with electrophiles, nucleophiles, or radicals. Although dehydrobenzene appears very different from methylene the two compounds have much in common.

Problems

1. What products would you expect if the following compounds are treated with strong base:

(*a*) Bromotrichloromethane in the presence of cyclohexene (phenyllithium as the base);

(*b*) *o*-chlorotoluene in piperidine solution (potassium amide as the base):

(*c*) $CH_3CClC{\equiv}CH$ (with CH_3 on the CCl carbon) in the presence of styrene (potassium *tert*-butoxide as the base).

2. Predict the outcome of the following reactions:

(*a*) $CH_2{=}C{=}O + CH_3CH_2CH_3 \xrightarrow[\text{gas phase}]{h\nu}$

(*b*) $o\text{-}C_6H_4(CO_2^-)(NH_3^+) + C_5H_{11}ONO + C_6H_5COCHN_2 \longrightarrow$

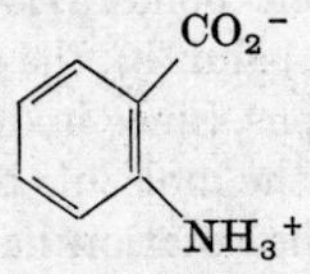

(*c*) $C_6H_5SO_2N_3 + C_6H_6 \xrightarrow{\text{heat}}$

CHAPTER 23

Symmetry, Chirality, and Optical Activity

Symmetry

Benzene is obviously a symmetrical molecule; so is methane; yet the symmetry characteristics of these two molecules are very different. In the case of more complex molecules it may not at first be obvious whether they have any symmetry or not. For this reason it

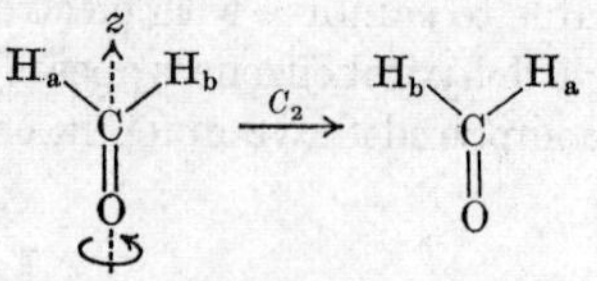

$$\underset{\text{O}}{\overset{\text{H}_a\ \ z\ \ \text{H}_b}{\text{C}}} \xrightarrow{C_2} \underset{\text{O}}{\overset{\text{H}_b\ \ \ \ \text{H}_a}{\text{C}}}$$

Figure 23.1.

is useful to define a series of *symmetry operations*. A symmetry operation is a movement of a body which transforms it into an equivalent orientation, so that every point of the body is coincident with the original site of an equivalent point (or the same point). In other words, if an observer saw the body before and after, but not during the operation, he would not be able to tell whether the symmetry operation had been performed or not.

Let us take a very simple example. Suppose we have a molecule of formaldehyde flat on the paper (see Figure 23.1) and we then rotate the molecule around the carbon–oxygen bond axis (the z axis) by 180°. If we label the hydrogen atoms (a) and (b), the effect of the rotation will be to interchange their positions. Since hydrogen atoms are indistinguishable our observer, seeing the molecule before and again afterwards, could not tell whether the rotation had occurred or not. This operation is known as a *rotation about a symmetry axis* and is denoted C_n (where n is a number, equal to $2\pi/\theta$, θ being the angle through which the molecule must be rotated to obtain an equivalent orientation). The operation we have described on formaldehyde is therefore C_2. Chloroform (see Figure 23.2) has a threefold axis coincident with the C–H bond, and the appropriate symmetry operations are denoted C_3. Benzene has a C_6 axis perpendicular to the plane of the molecule passing through its centre, and nitrogen has a C_∞ axis passing through the centres of its atoms. In a molecule such as benzene, in which a variety of rotational operations, can be performed the axis of highest n is called the *principal axis.*

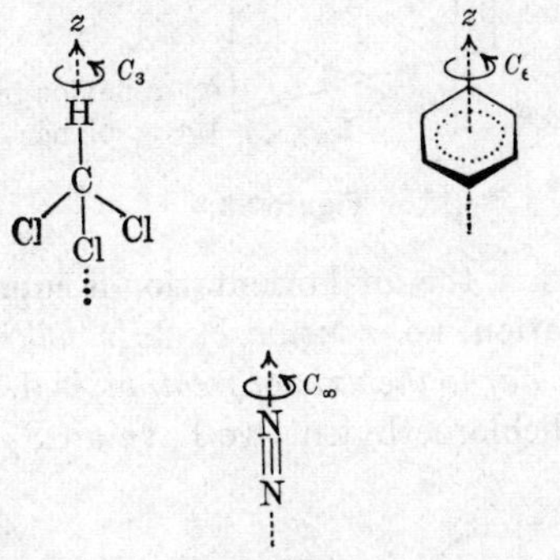

Figure 23.2.

The next symmetry operation we consider is a reflexion. If a molecule is bisected by a plane, such that each atom on one side of the plane encounters on reflexion in the plane a similar atom on the other half, we are performing another symmetry operation, a *reflexion at a plane of symmetry* designated σ. σ_h represents a reflexion in a plane perpendicular to the principal axis and σ_v represents reflexion in a plane containing the principal axis. The symmetry operations we can perform on formaldehyde are represented in Figure 23.3 (p. 302).

In very symmetrical molecules we can have another symmetry operation. In benzene, if we draw a line from any one carbon atom through the centre of the molecule, we encounter an equivalent carbon atom (i.e. in the *para*-position). Thus benzene possesses a centre of symmetry and the symmetry operation is called *inversion at the centre of symmetry* and designated i.

We now come to the most complex symmetry operation, but to organic chemists the most important one. Let us examine *trans*-1,2-dichloroethylene (Figure 23.4). If we rotate the molecule through 180° around an axis (x) coincident with the carbon–carbon bond and reflect the new orientation through a plane perpendicular

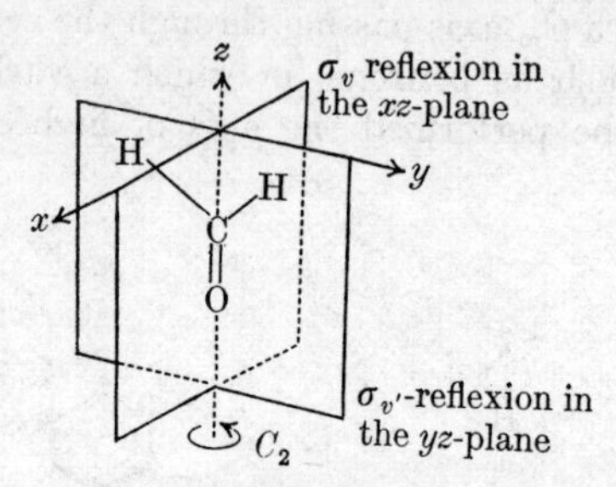

Figure 23.3

to the axis of rotation, the final orientation is superimposable on the original. The operation, i.e. *rotation by $2\pi/n$ followed by reflexion in the plane perpendicular to the axis of rotation*, is designated S_n (i.e. in the case of *trans*-dichloroethylene we have an S_2 operation).

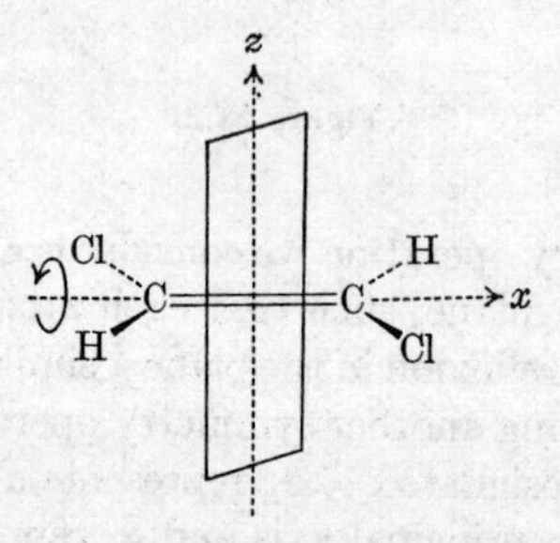

Figure 23.4.

For mathematical reasons there is a fifth symmetry operation called the *identity* operation I, which involves no movement of the molecule at all or its equivalent rotation through 360°. Obviously every object, no matter how unsymmetrical, has a symmetry element I. (Notice that $C_1 \equiv I$.)

The axes of rotation, centres of symmetry, and mirror planes are the *symmetry elements* of a molecule. The rotations, inversions, and reflexions are the symmetry operations.

Although organic chemists have long been aware of the symmetry of molecules, the formal use of symmetry has come into chemistry first through wave mechanics and then through inorganic chemistry. Symmetry is very important in wave mechanics because it is often possible to say a great deal about the wave function of a molecule solely from its symmetry without attempting to solve the appropriate wave equation. At the present time a knowledge of formal symmetry theory, though, as we shall see, very useful to an organic chemist (particularly for polycyclic molecules discussed in Part 3), is not essential. However, organic chemists will make increasing use of these ideas; hence our brief discussion of them.

Rotational Isomers and Conformation

The molecules we have considered above are all relatively rigid. Apart from small quantized vibrations the relative positions in space of the atoms making up the molecule remain virtually fixed. When we come to consider ethane, however, the situation is more complicated. As we described in Chapter 1 of Part 1 there is rotation about the carbon–carbon axis of the molecule. When the molecule is viewed along the C–C bond, the pairs of hydrogen atoms can each be in line with one other, i.e. in an 'eclipsed' form, or else the hydrogen atoms can be rearranged to be as far apart from each other as possible in a 'staggered' form. The difference between these two

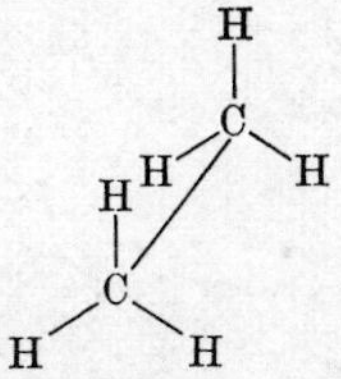

'Eclipsed' 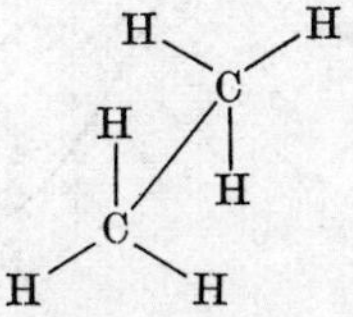 'Staggered'

extreme forms is well shown in the Newman projection formula which depicts the molecule viewed along the carbon–carbon axis (the central circle here represents a silhouette of the carbon atoms).

Clearly, there is an infinite number of rotational arrangements in between these two extreme forms. Spatial arrangements of the atoms of a molecule which result solely from the rotation about single bonds are known as *conformations*.

Eclipsed Staggered

Extreme conformations of ethane shown in Newman projection formulae, i.e. looking along the carbon–carbon axis.

Since an atom is surrounded by negative electronic charge, atoms which are not bonded to each other will mutually repel; thus the staggered form of ethane, where the hydrogen atoms are as far apart from each other as they possibly can be, will be of lower

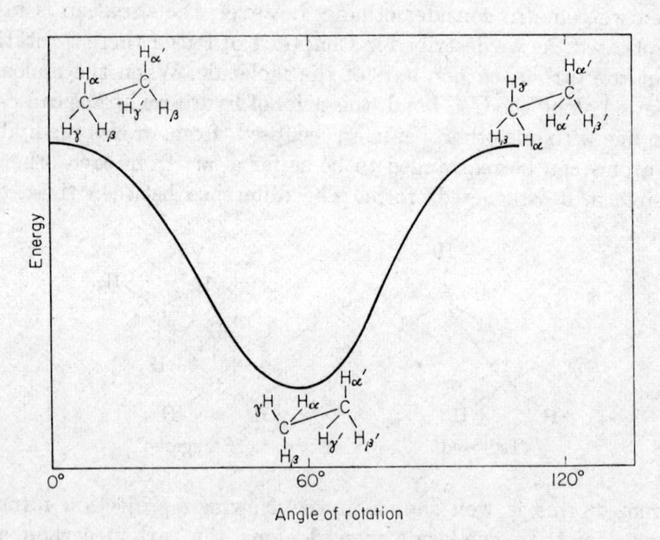

Figure 23.5. Torsional energy as a function of angular displacement.

energy than the eclipsed form where the repulsions between the hydrogen atoms will be at a maximum.*

The energy barrier to rotation in ethane is of the order of 3 kcal. mole^{-1} (cf. Figure 23.5). At normal temperatures this is insufficient to prevent rapid rotation round the carbon–carbon axis. This energy barrier can be attributed to repulsion between the electron clouds on the non-bonded hydrogen atoms (e.g. H_{α} and $H_{\alpha'}$) although the barrier is greater than can be accounted for by this explanation alone. However, non-bonded interaction between atoms and groups can be extremely important in determining the course of a reaction. Conformational effects become particularly important in cyclic

* An unambiguous but very ungainly nomenclature is used to denote rotational conformers. The relative positions of atoms a and b in a molecule a–C–C–b are denoted in terms of a 'torsional angle' θ. The term *syn* is used for

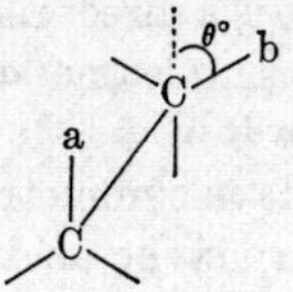

$\theta = 0° \rightarrow 90°$ and *anti* for $\theta = 90° \rightarrow 180°$. The torsion angle is subdivided further into $\theta = 0° \rightarrow 30°$, $150° \rightarrow 180°$ *periplanar* and $\theta = 30° \rightarrow 150°$ *clinal*. Thus conformations of 1,2-dichloroethane are denoted as illustrated. The nomenclature has to be elaborated further for cyclic compounds in order to distinguish $\theta = 30°$ from $\theta = 330°$ (this will be discussed in Part 3).

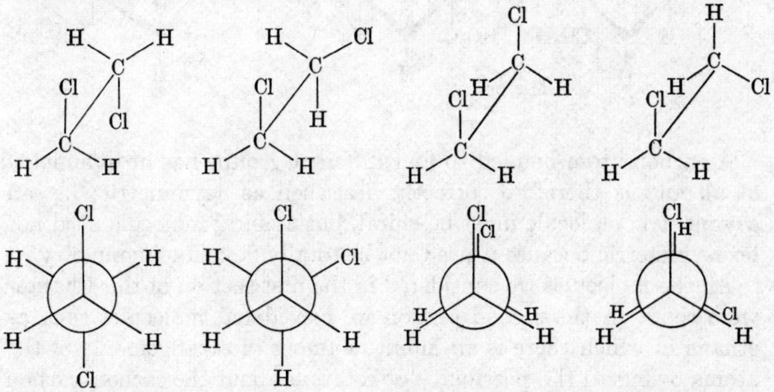

antiperiplanar synclinal synperiplanar anticlinal

compounds and this will be thoroughly discussed in Part 3. At present, we only need to be aware that aliphatic compounds can take up a variety of conformations and that some of these conformations are of lower energy than others.

Chirality and Optical Activity

In Part 1 (Chapter 15) we considered compounds which have molecules such that the mirror image of one molecule is not superimposable on that molecule, i.e. such a compound can exist in two enantiomeric forms. The separate enantiomers have the property of rotating the plane of polarized light passed through them in the gas or liquid state or through their solutions, and are said to be optically active (see Part 1, Chapter 15). The enantiomeric molecules have, like gloves, or spirals, the property of handedness. The term describing this property is chirality and a body possessing this property is called chiral. Clearly there is a direct relationship between chirality and symmetry. It is a very simple one: *a molecule is chiral if it lacks an S_n axis of symmetry, even if $n = 1$*. It is also important to appreciate that a molecule may have symmetry but lack an S_n axis. Thus *trans*-cyclopropane-1,2-dicarboxylic acid has a C_2 axis, but it lacks an S_n axis and is therefore chiral.

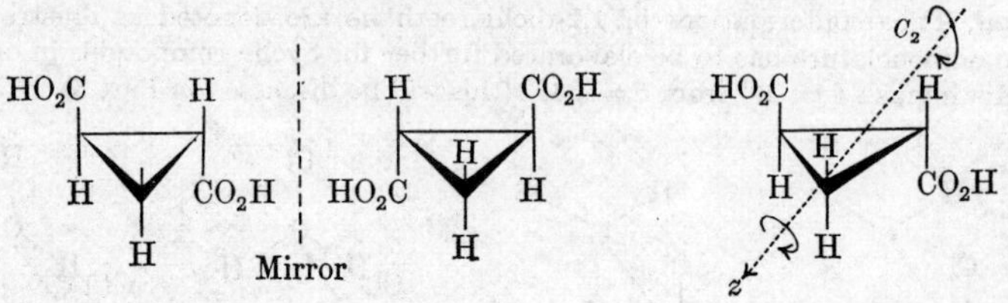

A carbon atom bonded to four different groups has no symmetry at all and is therefore correctly described as asymmetric; i.e. an asymmetric molecule must be chiral, but a chiral molecule need not be asymmetric because it need not be totally devoid of symmetry.

All the molecules we considered in the first section of this Chapter were rigid. In the second section we considered molecules such as ethane in which there is an infinite number of arrangements of the atoms owing to the possibility of rotation about the carbon–carbon bond; examination of the symmetry of such molecules is more difficult. It is often convenient to consider a molecule of this type as made up of a series of rigid units. We can consider the symmetry of the units severally and then consider what happens when we join them together. If we join together two molecular units which have symmetry by a single bond around which there is free rotation, we cannot destroy the symmetry already present, though we may add to it. Thus we cannot make a chiral molecule from two non-chiral fragments, and if we join a chiral fragment to a non-chiral fragment the resulting molecule must be chiral. Notice that these remarks *only* apply to fragments joined by single bonds around which free rotation is allowed.

The joining of two chiral fragments needs more careful consideration and instead of thinking about atoms and bonds we shall begin

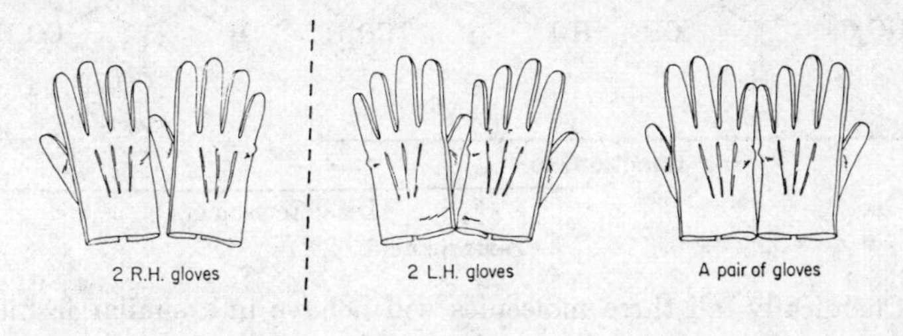

Figure 23.6.

by using the analogy we employed in Part 1, namely that of gloves. It we take two identical right-hand gloves and join them together (it doesn't matter how) we have a chiral system. The mirror image of two right-hand gloves is two left-hand gloves, and one set is not superimposable on the other. On the other hand, if we take a pair of gloves, that is a left-hand glove and a right-hand glove, and join these together, we then have a system which is no longer chiral: the mirror image of a pair of gloves is still a pair of gloves (it has an $S_1 \equiv \sigma$ element).

We must now consider a molecular system. Tartaric acid contains two asymmetric carbon atoms joined together; we can join two dextrorotatory asymmetric carbon atoms, two laevorotatory centres, or a dextro- with a laevo-rotatory centre. The first two, like two right-hand gloves or two left-hand gloves, will still be chiral and will form a pair of optical enantiomers. The last, like a pair of gloves, is no longer chiral and is not optically active. We depict this in the annexed formulae.

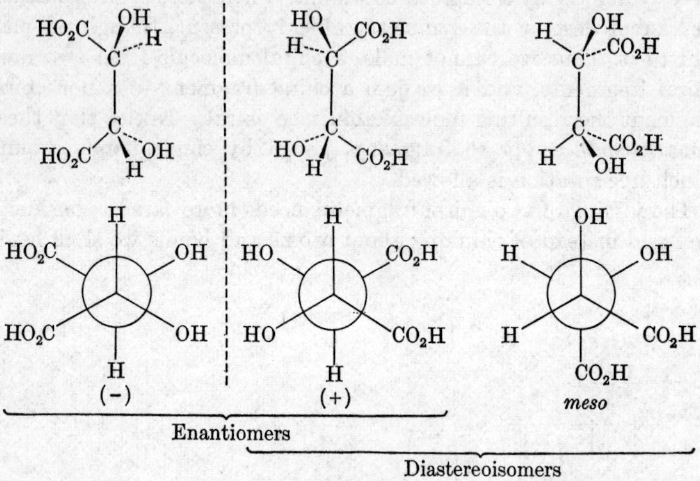

(−) (+) *meso*

Enantiomers

Diastereoisomers

Tartaric acid

Chemically, all three molecules will behave in a similar fashion. However, the two enantiomers differ only in their chirality, whereas in the 'pair-of-gloves' form, the so-called *meso*-form, we have a different species. So we should expect the physical properties of the two enantiomers to be identical saving only in their power to rotate the plane of polarized light. On the other hand, the *meso*-form, which will not rotate the plane of polarized light, should have physical properties slightly different from those of the two enantiomers. The *meso*-form is a stereoisomer of the two enantiomers. It is not an optical isomer nor is it a geometric isomer, so we must have a new word to describe the relationship between the two enantiomers and the *meso*-form. The word used is *diastereoisomer*. Enantiomers differ only in the way they rotate the plane of polarized light. Diastereoisomers, besides affecting polarized light differently (in the case of tartaric acid, the *meso*-form does not rotate the plane of polarized light at all), have different physical properties, e.g. different melting point, solubility, etc.

Now let us consider what happens when we take two non-identical centres of chirality and join them together. For gloves, this would mean taking a pair of leather gauntlets and white cotton gloves. All arrangements of one gauntlet with one cotton glove are chiral (i.e. a left-hand gauntlet and a right-hand glove, and a left-hand gauntlet and a left-hand glove, are two non-identical arrangements; similarly a right-hand gauntlet and a left-hand glove, and a right-hand gauntlet and a right-hand glove, are the enantiomeric arrangements of the first two pairs). In chemistry, a simple example of such a system would be 2,3-dibromo-3-phenylpropionic acid ($C_6H_5CHBrCHBrCO_2H$). The stereochemical forms are set out in

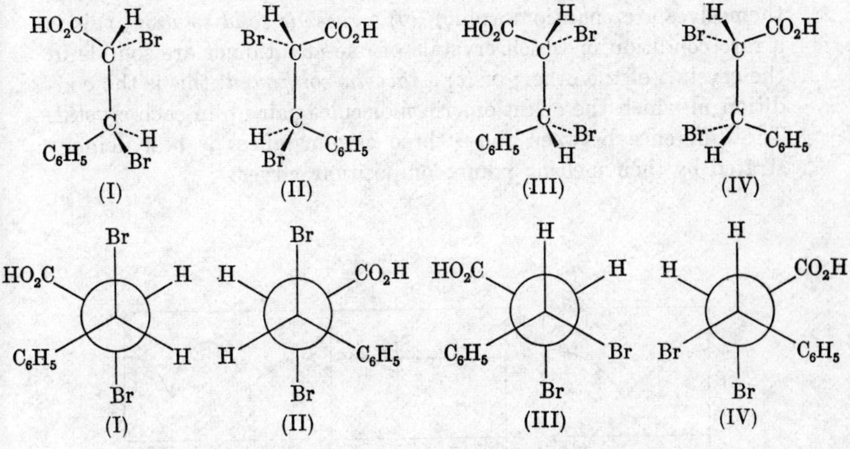

Forms of 2,3-dibromo-3-phenylpropionic acid.

the formulae I—IV. All these molecules are optically active. (II) is the enantiomer of (I); and (IV) is the enantiomer of (III). (I) is the diastereoisomer of (III) and (IV); and, similarly, (IV) is a diastereoisomer of (I) and (II). (I) and (II) have identical physical properties, saving only in their effect on polarized light. Similarly, (III) and (IV) have identical physical properties. On the other hand, the pair (I) and (II) differ somewhat in physical properties from the pair (III) and (IV).

Let us now return to consider the possible forms of tartaric acid. First, there are the two enantiomers which will have identical physical properties except for their optical activity, and then there is the *meso*-form which is optically inactive and has different physical properties from the two optical forms. There is, however, a fourth form. We saw in Part 1 that any normal chemical synthesis of an optically active molecule yields exactly equal amounts of the two enantiomers and such a mixture is called the racemic mixture. Quite often the two enantiomers will pair together in a crystal, so that a crystal of the racemic modification will differ from the crystals of either of the two pure enantiomers. In the solid state the racemic modification may exist as: (*a*) a *racemic mixture*; this consists of a mechanical mixture of exactly equal amounts of the crystals of the two enantiomeric forms (under these circumstances the crystals themselves are enantiomorphic); (*b*) a *racemic solid solution*; this is a rare condition in which crystals of one enantiomer are soluble in the crystals of the other; or (*c*) a *racemic compound*; this is the condition in which the enantiomeric molecules pair up in each crystal. The difference between these three circumstances is best demonstrated by their melting point–composition curves:

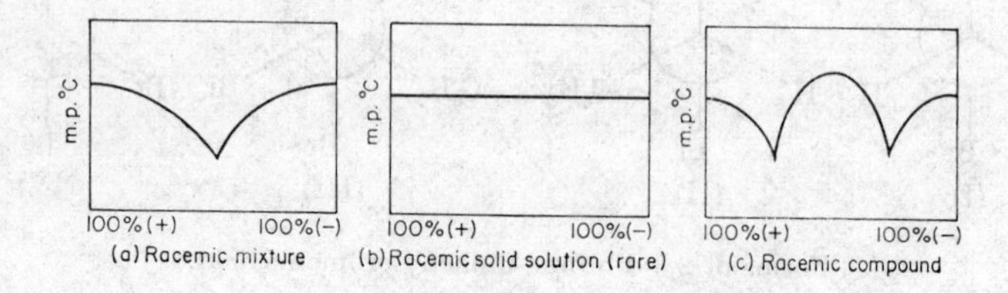

Figure 23.7. Melting points of mixtures of enantiomers.

In our diagram (*c*) the racemic compound melts higher than either of the two pure enantiomers, but racemic compounds are also known which melt lower than the optically active forms. The optically active molecules of tartaric acid do pair in the crystal, and the two active tartaric acids form a racemic compound. The physical properties of dextrorotatory, laevorotatory, racemic, and *meso*-tartaric acid are listed in Table 23.1.

Table 23.1. Physical properties of the tartaric acids

	(+) Dextro	(−) Laevo	(±) Racemic	*meso*
M.p. (°c)	170	170	206	140
$10^{-3}K_1$	1.1	1.1	1.1	0.8
$10^{-5}K_2$	5.9	5.9	5.9	1.6
Solubility (g/100 cc)	139	139	20.6	125
$[\alpha]_D^{25}$	+12°	−12°	0	0

Relative Configuration

When discussing naturally occurring organic compounds in Chapter 16 of Part 1 we described the carbohydrates, many of which have the empirical formula $(CH_2O)_n$. The most common value for n is 6 and a particularly common carbohydrate (sugar) is glucose. In solution, glucose shows the properties of an aldehyde, and when treated with sodium acetate and acetic anhydride it forms a penta-acetate, indicating that of the six oxygens five are involved in hydroxyl groups and the sixth is presumably in the carbonyl group. Kiliani treated glucose with hydrogen cyanide to form the cyano-hydrin; he then hydrolysed the cyano-group to yield the carboxylic acid and treated this acid with hydrogen iodide to yield n-heptanoic acid; in this reaction all the hydroxyl groups have been reduced to hydrogen atoms and although the yield of n-heptanoic acid is very small the fact that it is formed at all shows that glucose must consist of a linear chain of carbon atoms. The formula of glucose can therefore be written as follows:

$$CH_2OH\overset{*}{C}HOH\overset{*}{C}HOH\overset{*}{C}HOH\overset{*}{C}HOHCHO$$

Glucose (*asymmetric carbon atoms).

Notice that there are four asymmetric carbon atoms. This leads to a total of sixteen optical isomers, or eight diastereoisomers. Two of these compounds are quite common; they are known as mannose and galactose. Although these compounds differ from one another only in the relative configuration of the asymmetric carbon atoms they are quite different compounds. At first sight it would seem impossible to sort out the structure of these diastereoisomers by purely chemical techniques and yet this is exactly what Emil Fischer did in eight years between 1886 and 1894. We shall not reproduce here exactly Fischer's original argument, but our discussion is based on his treatment.

Glucose can be degraded one carbon atom at a time. There are several methods of doing this which need not concern us at present. All we need note is that it is possible to convert glucose ($C_6H_{12}O_6$) into a chemically similar five-carbon sugar called arabinose ($C_5H_{10}O_5$). (A sugar with five carbon atoms is called a pentose and that with six carbon atoms called a hexose.) There are three asymmetric carbon atoms in a pentose, giving eight optical isomers or four diastereoisomeric forms.

Our way of representing asymmetric carbon atoms is clearly too cumbersome to be employed in these more complex molecules. There is a conventional method for representing these three-dimensional structures on the two-dimensional surface of the paper, proposed by Fischer in 1891. Starting with the representation we have used throughout Parts 1 and 2 we turn the molecule so that groups a and d are inclined equally below the plane of the paper and groups b and c are projecting equally out of the plane of the paper. In the Fischer projection this is represented by the intersection of horizontal and vertical lines as shown.

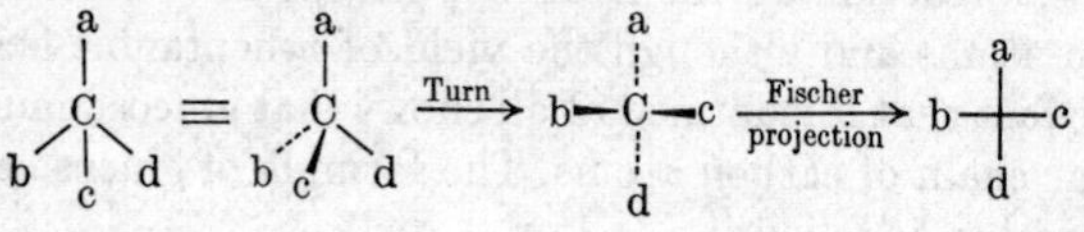

This extremely useful convention can lead us into all sorts of difficulties if we break the rules. It is vital to realize that a two-dimensional projection formula of this type can only be manipulated in the plane of the projection. Since the vertical bonds are behind the plane of the paper, the projection formula may only be rotated by 180° (not by 90°). The interchange of adjacent substituents, e.g. a and c in the projection formula, converts the formula of one enantiomer into that of the other. The value of this type of projection becomes apparent when we draw out the isomers of tartaric acid.

$$
\begin{array}{ccc}
\begin{array}{rcl} & CO_2H & \\ HO & - & H \\ H & - & OH \\ & CO_2H & \\ & (-) & \end{array}
&
\begin{array}{rcl} & CO_2H & \\ H & - & OH \\ HO & - & H \\ & CO_2H & \\ & (+) & \end{array}
&
\begin{array}{rcl} & CO_2H & \\ H & - & OH \\ H & - & OH \\ & CO_2H & \\ & \textit{meso} & \end{array}
\end{array}
$$

Fischer projections of tartaric acid.

The more centres of chirality there are in a molecule the more optically active forms we can expect. Inspection of the Fischer projection formulae will quickly show that a compound with n different chiral centres will have 2^n active forms.

We can now return to consider the configuration of glucose and, using the Fischer convention, represent the four possible pentoses ($C_5H_{10}O_5$) obtained by the degradation of glucose ($C_6H_{12}O_6$) as follows (enantiomers *not* shown):

$$
\begin{array}{cccc}
\begin{array}{rcl} & CHO & \\ H & - & OH \\ H & - & OH \\ H & - & OH \\ & CH_2OH & \\ & (V) & \end{array}
&
\begin{array}{rcl} & CHO & \\ HO & - & H \\ H & - & OH \\ H & - & OH \\ & CH_2OH & \\ & (VI) & \end{array}
&
\begin{array}{rcl} & CHO & \\ H & - & OH \\ HO & - & H \\ H & - & OH \\ & CH_2OH & \\ & (VII) & \end{array}
&
\begin{array}{rcl} & CHO & \\ HO & - & H \\ HO & - & H \\ H & - & OH \\ & CH_2OH & \\ & (VIII) & \end{array}
\end{array}
$$

Oxidized with nitric acid, arabinose gives a trihydroxyglutaric acid which is optically active. This means that arabinose must either have the structure represented by (VI) or (VIII) or be the enantiomer of (VI) or (VIII). Oxidation of (V) or (VII), or their enantiomers, would give a *meso*-trihydroxyglutaric acid:

$$
\begin{array}{c} CHO \\ | \\ (CHOH)_3 \\ | \\ CH_2OH \\ (-)\text{-Arabinose} \end{array}
\xrightarrow{HNO_3}
\begin{array}{c} CO_2H \\ | \\ (CHOH)_3 \\ | \\ CO_2H \\ \text{Trihydroxy-} \\ \text{glutaric acid} \end{array}
\qquad
\begin{array}{l} \text{Optically active.} \\ \therefore \text{Arabinose is (VI) or (VIII)} \end{array}
$$

Treated with hydrogen cyanide, arabinose gives the expected cyanohydrin, which on hydrolysis and oxidation gives two active dicarboxylic acids:

$$\underset{\text{Arabinose}}{\begin{array}{c} CHO \\ | \\ (CHOH)_3 \\ | \\ CH_2OH \end{array}} \xrightarrow{HCN} \underbrace{\begin{array}{c} CN \\ H-\!\!\!-\!\!\!\!+\!\!\!\!-\!\!\!-OH \\ | \\ (CHOH)_3 \\ | \\ CH_2OH \end{array} + \begin{array}{c} CN \\ HO-\!\!\!-\!\!\!\!+\!\!\!\!-\!\!\!-H \\ | \\ (CHOH)_3 \\ | \\ CH_2OH \end{array}} \xrightarrow[2, HNO_3]{1, H_2O}$$

$$\underbrace{\begin{array}{c} CO_2H \\ H-\!\!\!-\!\!\!\!+\!\!\!\!-\!\!\!-OH \\ | \\ (CHOH)_3 \\ | \\ CO_2H \end{array} + \begin{array}{c} CO_2H \\ HO-\!\!\!-\!\!\!\!+\!\!\!\!-\!\!\!-H \\ | \\ (CHOH)_3 \\ | \\ CO_2H \end{array}}_{\text{Both optically active}}$$

Notice that if this process were carried out with sugar (VIII), one of the dicarboxylic acids we should obtain would be a *meso*-form. Since neither of the acids was inactive, arabinose must have the configuration (VI) or be its enantiomer. Since glucose contains one more carbon atom than arabinose, glucose must either be (IX) or (X):

$$\underset{(IX)}{\begin{array}{rcl} & CHO & \\ H & + & OH \\ HO & + & H \\ H & + & OH \\ H & + & OH \\ & CH_2OH & \end{array}} \qquad \underset{(X)}{\begin{array}{rcl} & CHO & \\ HO & + & H \\ HO & + & H \\ H & + & OH \\ H & + & OH \\ & CH_2OH & \end{array}}$$

Oxidation of glucose with nitric acid gives a dibasic acid called glucosaccharic acid. The same acid can also be obtained by the oxidation of a non-identical aldohexose called gulose.

$$\underset{\text{Glucose}}{\begin{array}{rcl} & CHO & \\ H & + & OH \\ HO & + & H \\ H & + & OH \\ H & + & OH \\ & CH_2OH & \end{array}} \xrightarrow{HNO_3} \underbrace{\begin{array}{rcl} & CO_2H & \\ H & + & OH \\ HO & + & H \\ H & + & OH \\ H & + & OH \\ & CO_2H & \end{array} \rightleftarrows \begin{array}{rcl} & CO_2H & \\ HO & + & H \\ HO & + & H \\ H & + & OH \\ HO & + & H \\ & CO_2H & \end{array}}_{\text{Glucosaccharic acid}} \xleftarrow{HNO_3} \underset{\text{Gulose}}{\begin{array}{rcl} & CHO & \\ HO & + & H \\ HO & + & H \\ H & + & OH \\ HO & + & H \\ & CH_2OH & \end{array}}$$

This establishes formula (IX) or its enantiomer as the structure of glucose, since interconverting the primary hydroxy-group and the aldehyde group of formula (X) would yield the same compound. Notice that this argument is entirely based on forming *meso*-compounds, i.e. optically inactive and unresolvable compounds. These compounds must be symmetrical and by suitably working through the correct permutations the full structure of glucose was deduced. This argument shows that glucose has the configuration shown in formula (IX) or its enantiomer. No purely chemical technique will give us the absolute configuration. In glucose there are a carbonyl group and hydroxy-groups and we should expect the molecule to bite its own tail and form a cyclic acetal. This does occur (see Chapter 18) and both five- and six-membered cyclic acetals occur. Normally, glucose exists as a six-membered cyclic acetal. Notice that this introduces another asymmetric carbon atom.

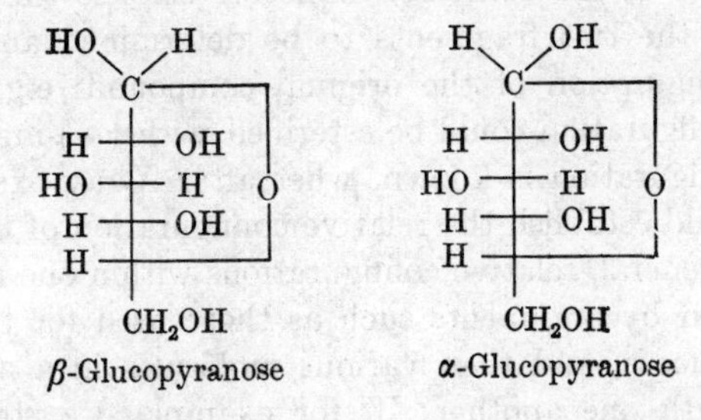

β-Glucopyranose α-Glucopyranose

Absolute Configuration

No purely chemical method can tell us the absolute arrangement of atoms in space. In Chapter 25 the various physical techniques which can be employed for determining the structures of organic molecules are discussed. One of these involves the examination of the interference patterns produced by the passage of *X*-rays through crystalline substances. Normally, the diffraction pattern produced by a molecule and its mirror image are identical. However, if one of the atoms in the molecule is electronically excited by the *X*-rays, a phase lag may be introduced and under these circumstances the diffraction pattern of the molecule and its enantiomorph will differ. Several determinations of this type have now been completed. The original one in which the sodium rubidium salt of dextrorotatory tartaric acid was examined is particularly important. The actual distribution of the groups around dextrorotatory tartaric acid is as shown on p. 316.

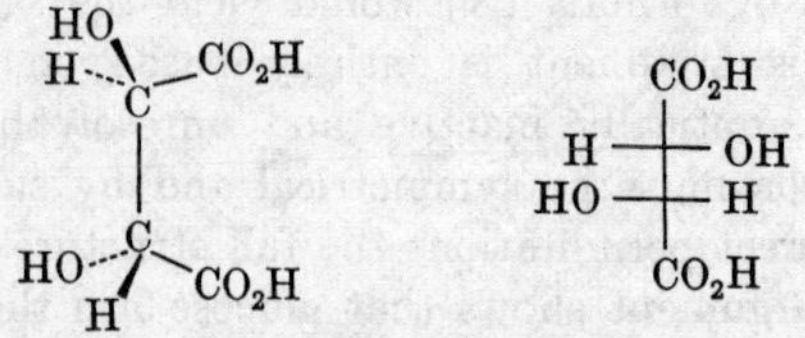

(+)-Tartaric acid: The 'absolute configuration' (real distribution of the atoms in space).

Apart from any experimental difficulties, it is not possible to make heavy-metal salts of all optically active compounds. This difficulty can sometimes be overcome by treating a compound of unknown configuration with one of known absolute configuration. *X*-ray analysis of the combined molecule enables the *relative* configuration of the two fragments to be determined and hence the absolute configuration of the original compound; e.g. an acid of unknown configuration could be esterified with (−)-menthol whose absolute configuration is known, whereafter *X*-ray examination of the ester would establish the relative configuration of the two components. In general, relative configurations within one molecule can be determined by arguments such as those used for the hydroxyl groups in glucose, and then various molecules in a series can be correlated with one another. If, for example, we start with the simplest sugar, glyceraldehyde, treatment with hydrogen cyanide will yield the corresponding cyanohydrins, in fact two of them since we introduce a new asymmetric centre. Hydrolysis, followed by oxidation with nitric acid, will yield two tartaric acids; one of these must be *meso*-tartaric acid, the other must be an optically active

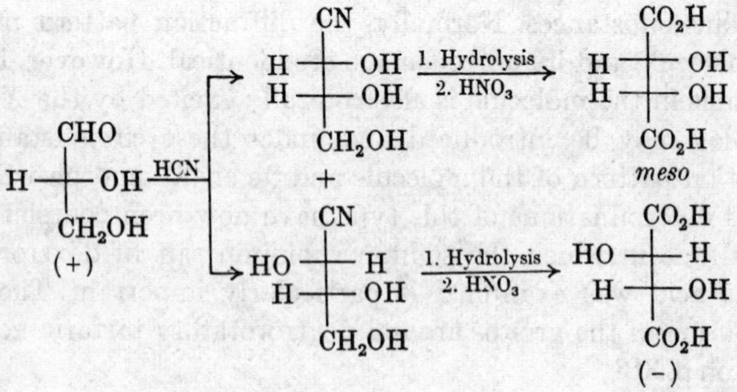

HCN

1. Hydrolysis
2. HNO_3

meso

1. Hydrolysis
2. HNO_3

(+)

(−)

form; and if we start with dextrorotatory glyceraldehyde the active tartaric acid turns out to be laevorotatory tartaric acid.

If we hydrolyse the two cyanohydrins prepared from glyceraldehyde, but then, instead of oxidizing the resulting acids with nitric acid, reduce them with sodium amalgam, two new sugars ($C_4H_8O_4$) will be formed; these are called threose and erythrose. One of these (erythrose) can also be obtained from laevorotatory arabinose by degradation.

CHO
HO — H
H — OH
CH_2OH
(−)-Threose

CHO
H — OH
CH_2OH
(+)-
Glyceraldehyde

CHO
H — OH
H — OH
CH_2OH
(−)-Erythrose

Degradation ←

CHO
HO — H
H — OH
H — OH
CH_2OH
(−)-Arabinose

These series of reactions provide together a direct stereochemical link from tartaric acid, whose absolute configuration has been determined by physical means, to glucose, whose relative configuration we deduced above. Previously, we could not tell whether the formula we drew was that of naturally occurring glucose or of its enantiomer. We can now see that the formula we used earlier in this chapter was in fact correct in an absolute sense as well as in a relative one. Many correlations of this kind have been worked out. Some, like the one we have illustrated, are limited to reactions which do not disturb the asymmetric centre; others, which we shall discuss in the next Chapter, involve the use of reactions where the configuration of the asymmetric centre is changed, but changed in a fashion which is known and understood.

Nomenclature of Stereoisomers

There are many ways in which the relative configuration of chiral centres has been named. A common and useful example is derived from the sugars threose and erythrose mentioned above. This nomenclature applies to any general system of the type CabcCabd.

(−)-Erythrose (+)-Erythrose (−)-Threose (+)-Threose

Projection formulae of the tetroses used as a basis for naming CabcCabd.

Thus, an *erythro*-form is the one in which the like groups are on the same side on the Fischer projection formula, and the *threo*-form is the one in which the like groups are on the opposite sides.

It will already be apparent that there is a large number of naturally occurring carbohydrate molecules. It turns out that the majority of naturally occurring carbohydrate molecules are related to each other in a particular stereochemical way. Besides glucose, we have already mentioned two hexoses called galactose and mannose. Other naturally occurring sugars include fructose and the pentoses ribose and xylose. Examination of the Fischer projection formulae

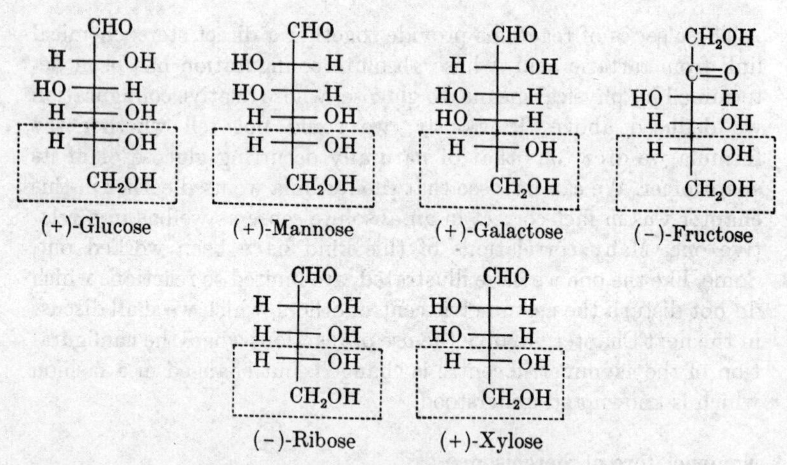

Some common naturally occurring sugars

of these seven sugars (see formulae) shows that, if they were degraded down to a triose containing the primary-alcohol end of the molecule (as glucose and mannose can be degraded down to arabinose) they would all yield the same triose, glyceraldehyde.

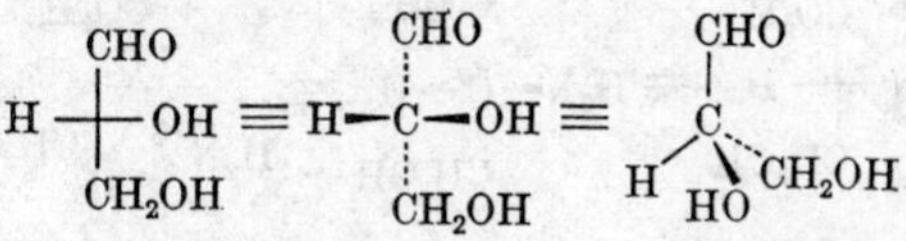

$$\begin{array}{c} CHO \\ H-\!\!\!+\!\!\!-OH \\ CH_2OH \end{array} \equiv \begin{array}{c} CHO \\ H-C-OH \\ CH_2OH \end{array} \equiv \begin{array}{c} CHO \\ C \\ H \quad HO \quad CH_2OH \end{array}$$

Glyceraldehyde, obtained by the degradation of any of these molecules, turns out to be dextrorotatory. In order to emphasize this stereochemical relationship, these sugars are all referred to as D-sugars. Thus, naturally occurring glucose is D-glucose. The D refers solely to its relationship to glyceraldehyde: it has nothing to do with its optical activity. Glucose, in fact, is dextrorotatory but naturally occurring fructose is laevorotatory and, to emphasize this, may be called D(−)-fructose. Similarly, the arabinose obtained by the degradation of glucose or mannose is laevorotatory but nonetheless it is known as D(−)-arabinose. Clearly, degradation of the enantiomer of naturally occurring glucose would yield laevorotatory glyceraldehyde, and any sugar which can theoretically be degraded to L-glyceraldehyde is known as an L-sugar (cf. gulose, p. 314).

In Part 1 (Chapter 16) we described how the structural material of animal tissues is composed of proteins, which are polyamides of high molecular weight. These polyamides are hydrolysed by acid or alkali, or also by enzymes, to α-amino-acids. Carbon atom 2 (or α) in an α-amino-acid, other than glycine, is asymmetric. It is found that the great majority of the naturally occurring α-amino-acids have the same spatial arrangements of the groups around carbon atom 2, namely, as in (XI).

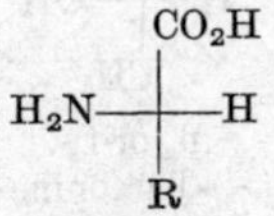

$$\begin{array}{c} CO_2H \\ H_2N-\!\!\!+\!\!\!-H \\ R \end{array}$$

(XI)

Just as the sugars are related to a three-carbon compound glyceraldehyde, so the α-amino-acids are related to the naturally occurring three-carbon amino-acid serine. The naturally occurring amino-acids are found to have the same relative configuration as laevorotatory serine, and common naturally occurring amino-acids are therefore called L-amino-acids.

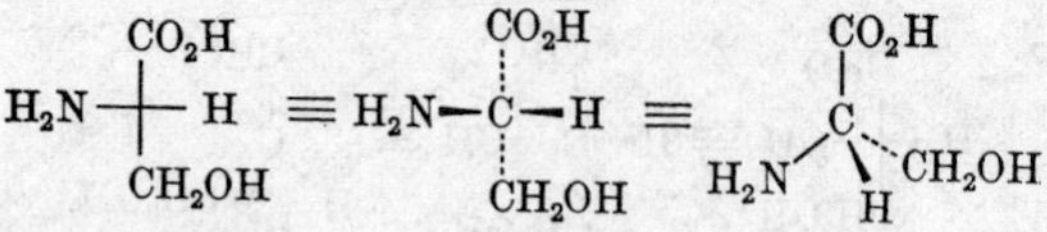

$$\begin{array}{c} CO_2H \\ H_2N - \!\!\!+\!\!\! - H \\ CH_2OH \end{array} \equiv \begin{array}{c} CO_2H \\ H_2N - C - H \\ CH_2OH \end{array} \equiv \begin{array}{c} CO_2H \\ | \\ C \\ H_2N \quad H \quad CH_2OH \end{array}$$

Thus, just as naturally occurring laevorotatory fructose is known as D(−)-fructose, so naturally occurring alanine and naturally occurring valine, both of which are dextrorotatory, are known as L-amino-acids.

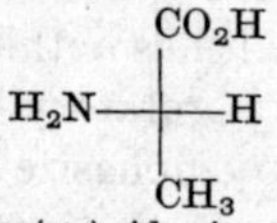

$$\begin{array}{c} CO_2H \\ H_2N - \!\!\!+\!\!\! - H \\ CH_3 \end{array} \qquad \begin{array}{c} CO_2H \\ H_2N - \!\!\!+\!\!\! - H \\ CH(CH_3)_2 \end{array}$$

L(+)-Alanine 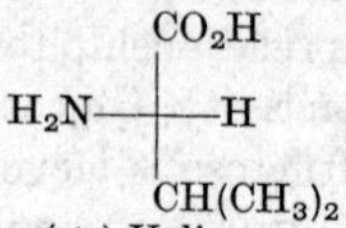 L(+)-Valine

The use of glyceraldehyde as the reference compound for the sugars, and of serine as the reference compound for the amino-acids, is extremely important and useful. The two definitions are quite independent. Their relative configurations have been established and hence the absolute configuration of serine which is correctly shown above. Compounds with both amino-groups and hydroxy-groups can, however, be confusing. For example, the hydroxy-amino-acid threonine has both a D-configuration relative to glyceraldehyde and an L-configuration referred to serine, which may be differentiated by use of the symbolism shown under the formula.

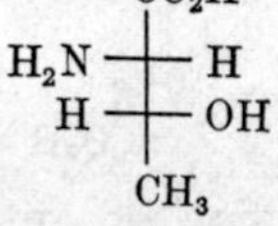

$$\begin{array}{c} CO_2H \\ H_2N - \!\!\!+\!\!\! - H \\ H - \!\!\!+\!\!\! - OH \\ CH_3 \end{array}$$

D_g or L_s
(−)-Threonine

An unambiguous method of nomenclature for specifying the absolute configuration of molecules has been devised. If a molecule contains an asymmetric carbon atom, the four atoms attached to that carbon atom are arranged in a *sequence* of decreasing atomic number. If two or more of these first atoms have the same atomic number their relative priority is chosen by comparing the atomic numbers of the second group of atoms attached to these identical first atoms. If ambiguity still persists the third set of atoms (working out from the asymmetric carbon atom) are compared and the process is continued until definite selection can be made. Having determined the sequence of the substituents attached to the asymmetric carbon atom, the molecule is now viewed from the side remote from the substituent with lowest priority. If the remaining groups in order of sequence now trace a clockwise turn, the configuration is defined as *rectus* and denoted by the symbol R. If, on the other hand, the remaining groups trace an anticlockwise turn, the configuration is called *sinister* and denoted by the symbol S. Thus, supposing we have a molecule Cabcd in which the priority of the substituents is in the order a → b → c → d. The two enantiomers are labelled as shown.

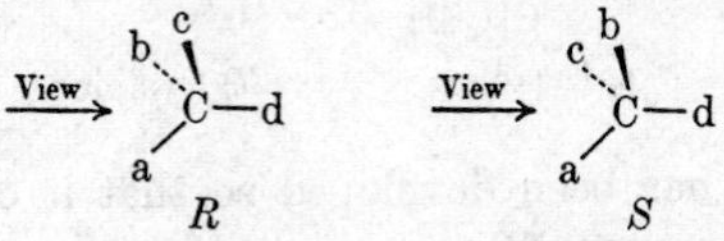

View → C(a)(b)(c)—d *R* View → C(a)(b)(c)—d *S*

Configurational nomenclature

The method of deciding the priorities a > b > c > d has been elaborated to cover all possible structures, and auxiliary rules are needed for some cases. For this Chapter we need concern ourselves only with the carbonyl group of aldehydes and acids; the effect of the auxiliary rules is that here the doubly bound oxygen atom is considered as two singly bound oxygen atoms. In this way, D-glyceraldehyde becomes (R)-glyceraldehyde, and L-serine becomes (S)-serine.

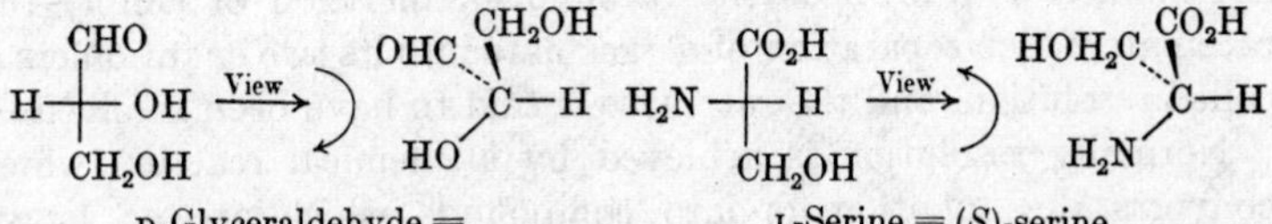

D-Glyceraldehyde ≡ (R)-glyceraldehyde L-Serine ≡ (S)-serine

This system of nomenclature has the great advantage that it is completely unambiguous with molecules containing more than one chiral centre. Each centre is examined independently and assigned a configuration and the terms, *R* or *S*, are incorporated into the systematic name. Three examples are shown below. The disadvantage of this system of nomenclature is that it does not necessarily

$$\begin{array}{c} C_2H_5 \\ H - \!\!\!+\!\!\! - Cl \\ HO - \!\!\!+\!\!\! - H \\ CH_3 \end{array} \qquad \begin{array}{c} CO_2H \\ H - \!\!\!+\!\!\! - OH \\ HO - \!\!\!+\!\!\! - H \\ CO_2H \end{array} \qquad \begin{array}{c} CO_2H \\ H_2N - \!\!\!+\!\!\! - H \\ H - \!\!\!+\!\!\! - OH \\ CH_3 \end{array}$$

3(*S*)-Chloro-2(*S*)-hydroxypentane (*R*,*R*)-Tartaric acid (2*S*,3*R*)-Threonine

correlate stereochemical series; thus the amino-acid cysteine, which is an L-amino-acid relative to serine, is (*R*)-cysteine. This system

$$\begin{array}{c} CO_2H \\ H_2N - \!\!\!+\!\!\! - H \\ CH_2SH \end{array} \equiv \begin{array}{c} HSH_2C \cdots C(CO_2H)(-H) \\ H_2N \end{array}$$

L_s-Cysteine (*R*)-Cysteine

of nomenclature has been developed so that it can be applied to complex cyclic compounds to account for all possible situations. Further sub-rules to the main sequence rule are required; they can be found in the original papers and in books relating specifically to stereochemistry.

Resolution of Racemic Forms

In Chapter 15 of Part 1 we saw that any normal chemical reaction leading to the formation of a chiral centre from achiral (i.e. non-optically active) starting materials yielded exactly equal amounts of the two enantiomers. This mixture of equal amounts of the two enantiomers is known as the racemic modification or simply the racemate.* The separation of a racemate into its two enantiomers is called resolution, and the racemate is said to have been resolved.

Normally resolution is achieved by a chemical reaction which converts the enantiomers into compounds which are no longer mirror images of each other. This will not be achieved if an achiral

* The word racemate is sometimes used (especially in German and some American literature) to mean the racemic compound as distinct from a racemic mixture which is called a conglomerate. We follow the more usual practice and use racemate to mean any racemic modification.

reagent is used. But, if the reagent is chiral, it becomes possible; a derivative, say +A, −B, is not the mirror image of −A, −B (−B being the chiral reagent or a chemical group derived from it and +A and −A being the two enantiomeric components of the racemate). Being diastereoisomers, these derivatives −A, −B and +A, −B have different physical properties, e.g. solubility, as well as different molecular rotations. In principle, the chiral reagent B may be anything that reacts readily with A. In practice, we also require that the compound A can be readily regenerated from its derivative A,B, that at least one of the pair of derivatives A,B crystallizes readily, and that the derivatives +A, −B and −A, −B differ sufficiently in solubility for fractional crystallization to be possible.

The particular class of derivative chosen for a resolution depends on the chemical nature of the substance to be resolved. Obviously, racemic acids can often be readily resolved by conversion into salts with optically active bases, and racemic bases by similar reaction with optically active acids. The salts are separated by fractional crystallization, and the acid and base are regenerated from the pure

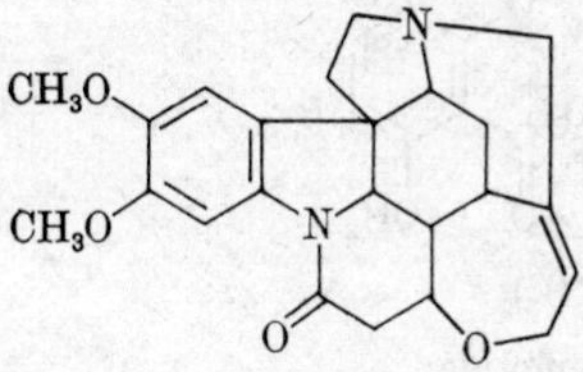

Brucine

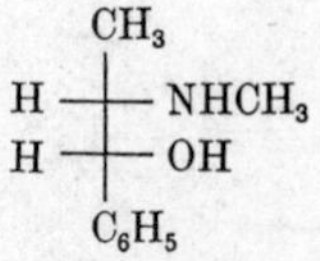

(−)-Ephedrine

Two of the naturally occurring optically active bases used for resolution.

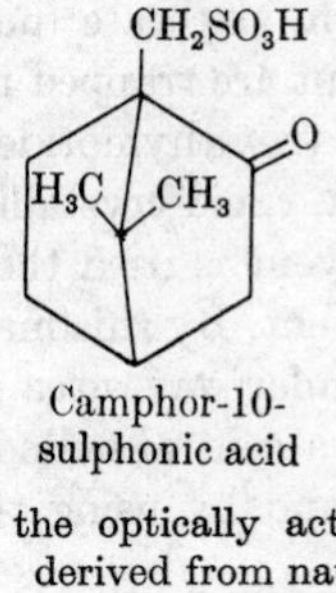

Camphor-10-sulphonic acid

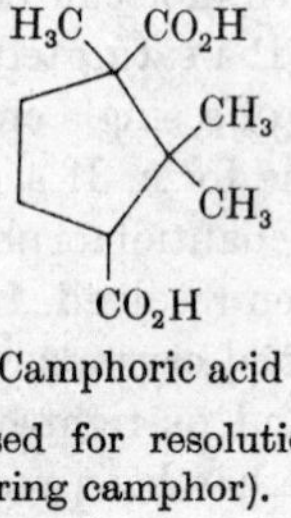

Camphoric acid

Two of the optically active acids used for resolution (both derived from naturally occurring camphor).

diastereoisomeric salts by means of alkali. Chiral reagents useful for this purpose include the naturally occurring alkaloids such as brucine and ephedrine for the acids, and camphoric or camphor-10-sulphonic acid for the base (see p. 323).

Racemic alcohols may be converted into the acid ester of a dibasic acid, which can then be resolved with an optically active base:

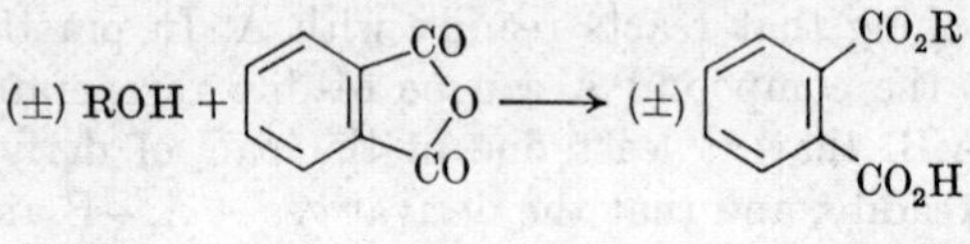

$$(\pm)\ ROH + C_6H_4(CO)_2O \longrightarrow (\pm)\ C_6H_4(CO_2R)(CO_2H)$$

Fitting a molecule into a chiral cavity can be used to effect resolution in very special cases. We saw in Chapter 18 how hydroxy-acids form lactides. A particular trimeric lactide called tri-*o*-thymotide exists as a spiral or a three-bladed propeller. In solution, the mole-

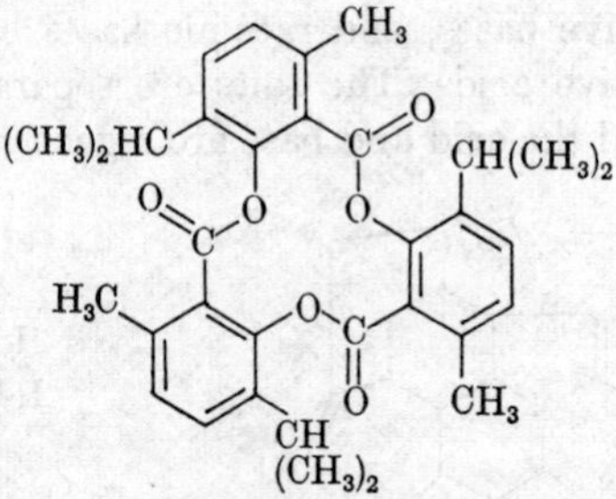

Tri-*o*-thymotide

cule can flip over from one enantiomorphic form to the other quite readily. When crystallization occurs, however, enantiomorphic crystals are formed. These crystals form clathrate (inclusion) compounds in which molecules of the solvent are trapped inside cavities in the crystal. If a saturated solution of tri-*o*-thymotide is inoculated with a crystal of a single enantiomorph, rapid crystallization occurs giving only one form. If a racemic solvent is used the crystals will trap only one enantiomorph of the solvent. By this means 2-bromobutane has been resolved. In a very similar way, urea crystallizes in hexagonal helical crystals, which have a central cylindrical channel with a left-hand or right-hand twist and by using this technique 2-chlorooctane has been resolved.

For a somewhat similar reason partial resolution may sometimes be achieved simply by passing a solution of a racemate down a column packed with cellulose. Cellulose, which is made up of glucose units joined together (Chapter 16, Part 1), possesses chirality, and the extent to which enantiomers are absorbed on cellulose may differ. It is unusual for a complete resolution to be effected in this way.

These last two examples, although of little importance in ordinary organic chemistry, are the basis of the stereospecific nature of many biological processes. Many biochemical processes take place on the surface of enzymes, which are chiral protein molecules, so that asymmetric molecules formed on their surface occur as one enantiomer only.

Eliptically Polarized Light and Fresnel's Theory of Optical Activity

In Part 1 (Chapter 15) we discussed how light travels in transverse waves. In an ordinary light wave, these transverse vectors are completely random to the direction of propagation. Any single vector may be regarded as a result of two components vibrating at right angles to each other. For example, vector OA may be regarded as the resultant of two vectors OAy and OAx. We can thus write down

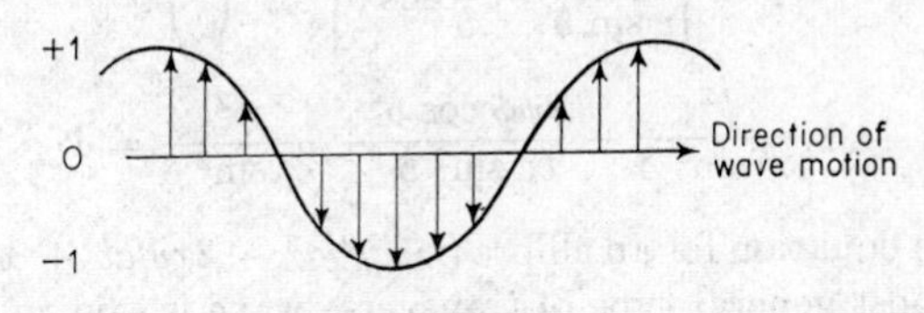

Figure 23.8

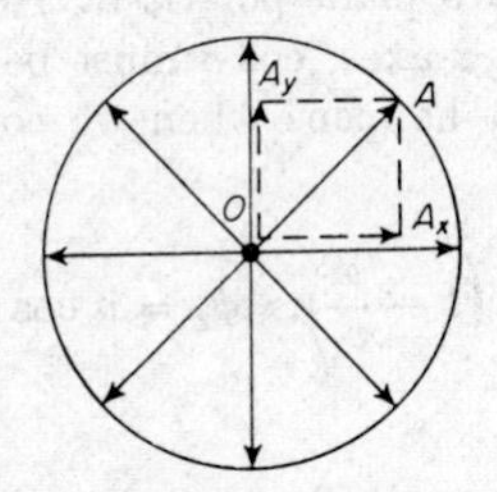

Figure 23.9

a general equation for a transverse wave where ψ represents displacements parallel to the y axis, and ϕ displacements to the z axis. v equals the velocity of the wave along the x axis. Combined, the two expressions represent a general expression for a transverse wave.

$$\psi = b \sin \frac{2\pi}{\tau}\left(t - \frac{x}{v}\right); \quad \phi = c \sin\left[\frac{2\pi}{\tau}\left(t - \frac{x}{v}\right) + \delta\right]$$

Now since $\sin(A + B) = \sin A \cos B + \cos A \sin B$, we have

$$\phi = c\left[\sin\frac{2\pi}{\tau}\left(t - \frac{x}{v}\right)\cos\delta + \cos\frac{2\pi}{\tau}\left(t - \frac{x}{v}\right)\sin\delta\right]$$

and hence

$$\frac{\phi}{c} = \frac{\psi}{b}\cos\delta + \cos\frac{2\pi}{\tau}\left(t - \frac{x}{v}\right)\sin\delta$$

$$\therefore \cos\frac{2\pi}{\tau}\left(t - \frac{x}{v}\right) = \frac{\phi}{c\sin\delta} - \frac{\psi}{b}\cos\delta$$

Remembering that $\cos^2 A = 1 - \sin^2 A$, and that $\psi/b = \sin 2\pi/\tau(t - x/v)$, we have, on squaring the above equation:

$$1 = \left[\frac{\phi}{c\sin\delta} - \frac{\psi}{b}\cos\delta\right]^2 + \left(\frac{\psi}{b}\right)^2$$

$$\therefore \frac{\psi^2}{b^2\sin^2\delta} - \frac{2\psi\phi\cos\delta}{bc\sin^2\delta} + \frac{\phi^2}{c^2\sin^2\delta} = 1$$

This is an equation for an ellipse (i.e. $y^2/a^2 - 2xy/ab + x^2/b^2 = 1$). Thus the most general type of transverse wave is said to be elliptically polarized. If $b = c$ the ellipse degenerates into a circle. If b or $c = 0$, the wave becomes plane-polarized. To put the axes of the ellipse on to the y and z axes, $\cos\delta$ must be zero, i.e. $\delta = \pm\pi/2$.

Let us now see what happens when we combine two circularly polarized waves.

$$\psi_1 = a\sin\frac{2\pi}{\tau}\left(t - \frac{x}{v_1}\right); \quad \phi_1 = a\cos\frac{2\pi}{\tau}\left(t - \frac{x}{v_1}\right)$$

and

$$\psi_2 = a\sin\frac{2\pi}{\tau}\left(t - \frac{x}{v_2}\right); \quad \phi_2 = -a\cos\frac{2\pi}{\tau}\left(t - \frac{x}{v_2}\right)$$

The first pair of equations represents a right-hand circularly polarized wave with velocity v_1, and the second pair of equations represents a left-hand circularly polarized wave with velocity v_2. If the two waves were superimposed we should obtain the expression:

$$\Psi = \psi_1 + \psi_2 = a\left[\sin\frac{2\pi}{\tau}\left(t - \frac{x}{v_1}\right) + \sin\frac{2\pi}{\tau}\left(t - \frac{x}{v_2}\right)\right]$$

Now since $\sin A + \sin B = 2\sin\frac{1}{2}(A + B)\cos\frac{1}{2}(A - B)$, we have:

$$\Psi = 2a\sin\frac{2\pi}{\tau}\left(t - \frac{x}{2}\left[\frac{1}{v_1} + \frac{1}{v_2}\right]\right)\cos\frac{\pi x}{\tau}\left(\frac{1}{v_2} - \frac{1}{v_1}\right)$$

Similarly, since $\cos A - \cos B = 2\sin\frac{1}{2}(A + B)\sin\frac{1}{2}(B - A)$

$$\Phi = \phi_1 + \phi_2 = -2a\sin\frac{2\pi}{\tau}\left(t - \frac{x}{2}\left[\frac{1}{v_1} + \frac{1}{v_2}\right]\right)\sin\frac{\pi x}{\tau}\left(\frac{1}{v_2} - \frac{1}{v_1}\right)$$

Thus the two waves combine to give a plane-polarized wave of amplitude $2a$. Dividing Ψ by Φ we get:

$$\Psi/\Phi = -\cot\frac{\pi x}{\tau}\left(\frac{1}{v_2} - \frac{1}{v_1}\right)$$

So we see that the plane of polarization rotates along the x axis (i.e. the direction of propagation), going a complete rotation in $(\pi/\tau)(1/v_2 - 1/v_1)$. If $v_1 = v_2$, the plane of polarization remains fixed.

We are very familiar with the fact that light travels at different speeds in different media. The ratio of the speed through a vacuum to the speed through a medium is, of course, the refractive index of that medium. Fresnel was the first to suggest that in an optically active medium a plane-polarized light wave was decomposed into two circularly polarized waves of slightly different velocities.

Rotatory Dispersion

The velocity of a beam of monochromatic light through a medium depends both on the nature of the medium and on the wavelength of the light. The splitting of white light into the colours of the rainbow is the most common example of this phenomenon. We cannot discuss the theory of dispersion here, and it is sufficient to say that a medium through which light passes is regarded as made up of oscillating dipoles (these are, in fact, due to the atoms and electrons making up the molecules of the medium), and dispersion is due to the difference of the interaction of light of different wavelengths with the tiny oscillators.

We have seen that Fresnel's theory attributes optical activity to the different velocities of two circularly polarized waves. It follows that the rotatory power of a substance depends on the wavelength of the polarized light being used. For many compounds the rotatory power increases progressively with diminishing wavelength, and approximately the rotatory power varies inversely as the square of the wavelength. The rotation of a large number of compounds is given more precisely by Drude's equation:

$$[\alpha] = \frac{k}{\lambda^2 - \lambda_0{}^2}$$

The two constants are known as k, the rotation constant, and λ_0, the dispersion constant. Clearly, according to our simplified picture of dispersion, the observed rotation is the sum of the effect of all the oscillators in the molecule, each of which can be regarded as providing a partial rotation. Drude's equation holds when the effect of one oscillator is predominant or when the dispersion of all the oscillators is similar. However, there are examples where the compound appears to have partial rotations of different dispersion. These led Lowry to introduce equations for (a) two partial rotations of opposite sign and unequal dispersion

$$[\alpha] = \frac{k_1}{\lambda^2 - \lambda_1{}^2} - \frac{k_2}{\lambda^2 - \lambda_2{}^2}$$

and for (b) the superposition of two different partial rotations of the same sign but unequal dispersion

$$[\alpha] = \frac{k_1}{\lambda^2 - \lambda_1{}^2} + \frac{k_2}{\lambda^2 - \lambda_2{}^2}$$

Compounds obeying the former equation he referred to as showing anomalous rotatory dispersion, and compounds obeying the latter equation he referred to as showing complex rotatory dispersion. We should, in theory, have as many terms as there are oscillators in the molecule. In practice the effect of one oscillator is likely to predominate at any particular region of the spectrum, and Drude's equation is obeyed by some very complex molecules such as sucrose.

We can depict simple, complex, and abnormal rotatory dispersion in a diagram (Figure 23.10).

Dispersion theory predicts and experiment verifies that the dispersion constant λ_0 of Drude's equation corresponds to the wavelength of the nearest absorption band and quite different phenomena are observed when polarized light with wavelengths that are partly absorbed is passed through an optically active solution. Cotton, studying solutions of potassium chromium tartrate, found that right- and left-hand circularly polarized light are absorbed to different extents. This means that a linearly polarized beam (with a wavelength within the absorption band), which can be regarded as a mixture of right and left circularly polarized light, is converted into

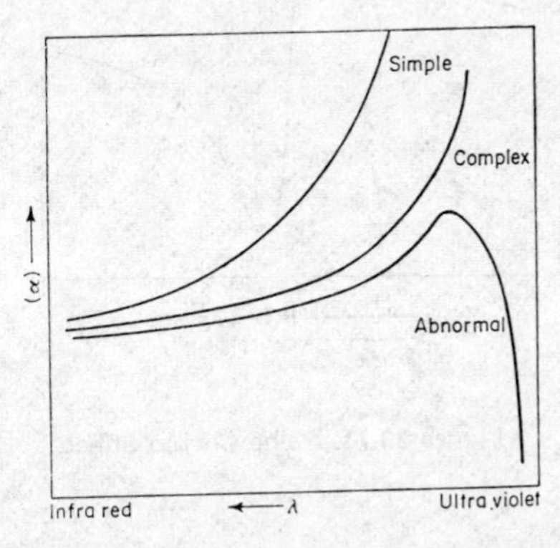

Figure 23.10. Simple, complex, and abnormal dispersion.

an elliptically polarized beam. This phenomenon is known as circular dichroism. Figure 23.11 represents the optical density and rotatory dispersion plotted against wavelength, and shows how the rotatory dispersion curve follows a very anomalous course through the absorption band. The rotatory power changes sign in the region of maximum absorption, although $[\alpha]$ does not go from $+\infty$ to $-\infty$ as predicted by an unmodified Drude equation. The ellipticity of circular dichroism is also a maximum at the same wavelength.

The actual shape of the rotatory dispersion curve through the absorbing band depends on the shape of the molecule and particularly on structural features close to the chromophore (i.e. close to the portion of the molecule associated with the absorption of the light). Rotatory dispersion curves and circular dichroism curves have been used for structural determination and particularly for the correlation of relative configuration.

Rotatory dispersion occurs only for chiral molecules. A plane of polarized light can also be rotated when a transparent non-chiral

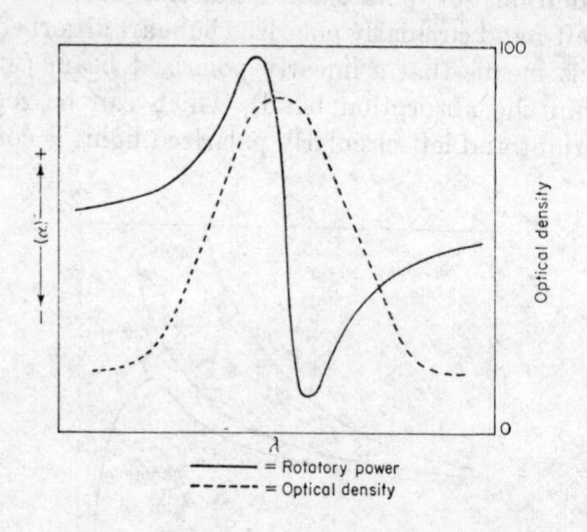

Figure 23.11. The Cotton effect.

substance is placed in a strong magnetic field. Thus if a plane-polarized beam is passed through a transparent solution such that the light travels along the lines of force, the plane of polarization is rotated. The sense of rotation is determined by the direction of the field and not by the direction of the beam of light. Thus, whereas a beam reflected back along its course through an optically active solution would show no overall rotation, a beam of plane-polarized light passed down a substance in a magnetic field and back again would show twice the rotation. The rotation depends on the strength H of the field and on the length of the light path.

$$[\alpha] = \omega l H,$$

where ω = Verdet constant, and l = length of light path.

This phenomenon of magnetic rotatory dispersion is known as the Faraday effect. The magnetic rotatory dispersion of organic compounds has been found to vary in a very regular manner with structure. Apart from early work by Perkin, however, little use has been made of the Faraday effect in structural determination.

$$[M] = \text{Molecular magnetic rotation} = \frac{M\alpha\rho_s}{M_s\alpha_s},$$

where M = molecular weight, ρ = density, α = observed rotation, and subscript 's' refers to solvent.

Problems

1. What symmetry operations can be performed on the following: (*a*) a filter ('Buchner') flask; (*b*) a conical (Erlenmeyer) flask; (*c*) CH_2Cl_2; (*d*) $CHFCl_2$;

(*e*)
C_6H_5 C_6H_5
C=C=C ;
CH_3 CH_3

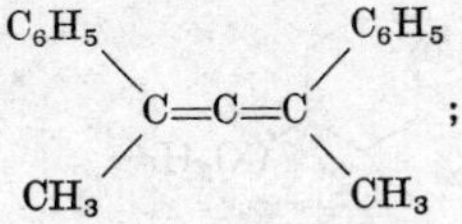

(*f*) $CHClBrSO_3H$; (*g*) HO_2CCH C $CHCO_2H$ (with ring CH_2 groups: CH_2 CH_2 above, CH_2 CH_2 below)

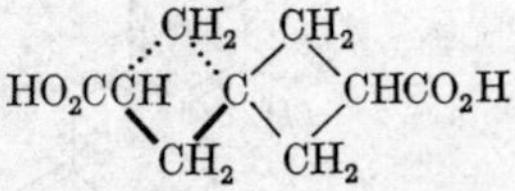

(*h*) H, H_3C, H_3C, H, C, CH_3, H, H, CH_3

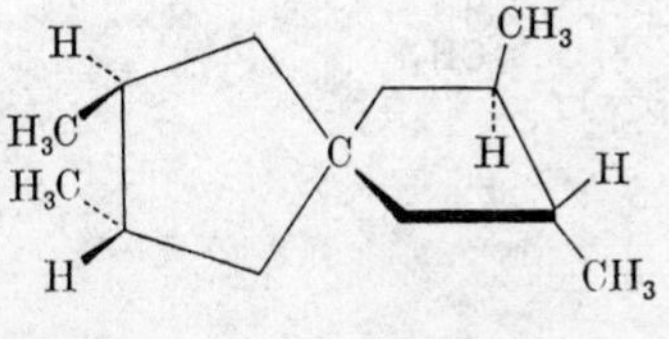

2. Deduce the configuration of the D-aldoheptose from the following observations:

The D-aldoheptose, $CH_2OH(CHOH)_5CHO$, is oxidized with nitric acid to give a *meso*-pentahydroxy-dibasic acid. If the D-aldoheptose is degraded to the D-aldohexose (i.e. loses carbon atom 1) and this hexose is then oxidized, an optically active tetrahydroxy-dibasic acid is formed. If the configuration of carbon atom 2 of this hexose is first inverted (i.e. the hexose is epimerized at C-2) and the resulting hexose is then oxidized, a *meso*-tetrahydroxy-dibasic acid is obtained.

3. Suggest structures for the following isomeric carboxylic acids $C_5H_8O_2$:

A exists in two forms, neither of which can be resolved. A can be catalytically hydrogenated to $C_5H_{10}O_2$, which can be resolved into enantiomers.

B exists in two forms, both of which are resolvable into enantiomers but cannot be hydrogenated.

C is resolvable into enantiomers and can be catalytically hydrogenated to give an acid which is also resolvable.

4. Which of the following compounds is capable of existing in optically active forms and why?

(*a*) $HOCH_2CHOHCHOHCH_2OH$; (*b*) $(CH_3)(C_6H_5)C{=}C{=}C(CH_3)(C_6H_5)$

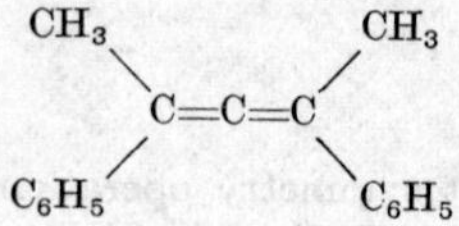

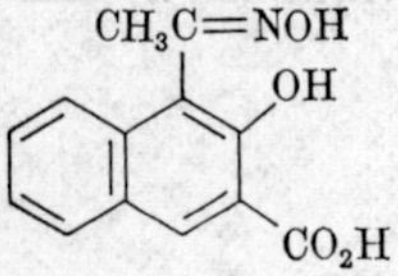

(*c*) $CH_3C{=}NOH$, OH, CO_2H (d) C_6H_5, Br^-, $\overset{+}{N}$, CO_2H

(*e*) CH_3, CH_3, CH_2CO_2H (*f*) Br, HOCH, CH_3

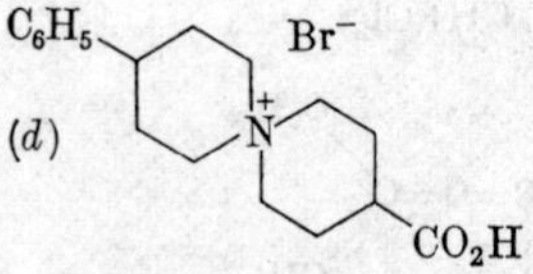

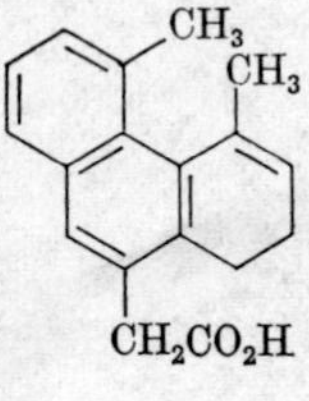

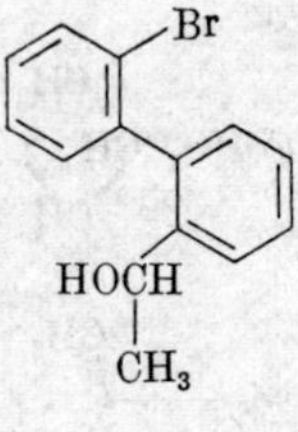

CHAPTER 24

The Stereochemical Consequences of Reactions

Throughout Parts 1 and 2 we have used formulae and diagrams that emphasize the fact that chemical reactions take place in three-dimensional space and not in two dimensions as on the plane of the paper. Much of what we shall say in the present Chapter has been implied before when various reactions have been discussed; previously, however, we did not specifically mention this stereochemical consequence. We shall now re-examine the mechanism of many reactions discussed earlier and see what steric requirements they imply.

The Stereochemical Consequences of Displacement Reactions

In Chapter 3 of Part 1 we first introduced displacement reactions and depicted a nucleophile attacking a carbon atom on the opposite side from whence the leaving group departs.* The methyl group and

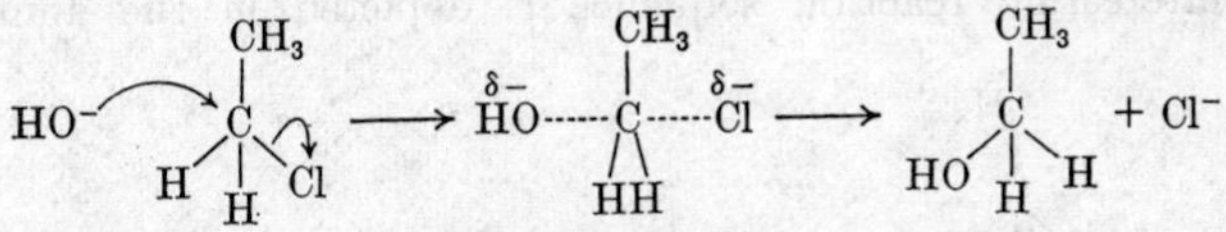

$$HO^- + CH_3CH_2Cl \longrightarrow \overset{\delta-}{HO}\cdots C(CH_3)(H)(H) \cdots \overset{\delta-}{Cl} \longrightarrow HOCH_2CH_3 + Cl^-$$

Activated complex

* In Chapter 1 of Part 1 we introduced a convention for representing tetrahedral methane as (i), where (i) is a simplified version of (ii). We have

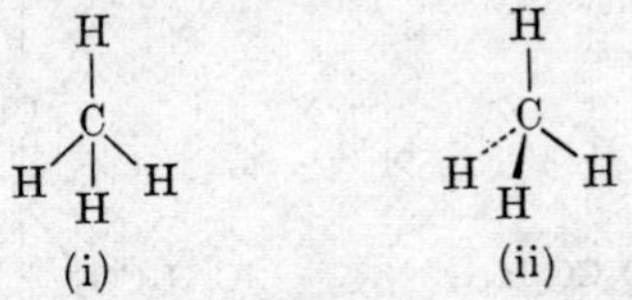

(i) (ii)

employed this convention in most formulae of the present volume, including this Chapter.

the two hydrogen atoms in the activated complex depicted lie in one plane perpendicular to the line passing through the entering hydroxyl group, the central carbon atom, and the departing chlorine atom. Such a mechanism necessarily implies an inversion of configuration at the substituted carbon atom. Suppose we take an optically active alkyl halide, (+)-2-iodo-octane, for example, and treat this with potassium iodide in acetone, the secondary iodide will be racemized (that is, will become optically inactive). This is because every such displacement involves an inversion. At the start of the reaction all the iodo-octane is dextrorotatory and each displacement involves $(+) \rightarrow (-)$; but, as the reaction proceeds and the concentration of (−)-iodo-octane builds up the reverse process, $(-) \rightarrow (+)$ also occurs. Eventually equilibrium is reached where concentration of (+)- and (−)-iodide are equal, i.e. the starting material is racemized.

$$\ddot{D} + \begin{array}{c} A \\ | \\ B-C-D \\ | \\ C \end{array} \rightleftharpoons \begin{array}{c} A \\ | \\ D-C-B \\ | \\ C \end{array} + \ddot{D}$$

$$(R) \qquad\qquad (S)$$

By using radioactive iodide and measuring the rates of both exchange and racemization, Hughes was able to show that every displacement did indeed involve an inversion, as we should expect.

An interesting reaction sequence is depicted in the annexed

$$\begin{array}{ccc}
\begin{array}{c} CH_3 \\ | \\ EtO_2C-C-OH \\ | \\ H \end{array} & \xrightarrow[\text{pyridine}]{p\text{-}CH_3C_6H_4SO_2Cl-} & \begin{array}{c} CH_3 \\ | \\ EtO_2C-C-OSO_2C_7H_7 \\ | \\ H \end{array} \\
[\alpha]_D^{20} + 11.16^\circ & & [\alpha]_D^{20} + 45.60^\circ \\
\downarrow C_6H_5COCl & & \downarrow Na^+ \ {}^-OCOC_6H_5, \ EtOH \\
\begin{array}{c} CH_3 \\ | \\ EtO_2C-C-O_2CC_6H_5 \\ | \\ H \end{array} & & \begin{array}{c} CH_3 \\ | \\ C_6H_5CO_2-C-CO_2Et \\ | \\ H \end{array} \\
[\alpha]_D^{20} - 24.60^\circ & & [\alpha]_D^{20} + 24.56^\circ
\end{array}$$

formulae. When ethyl (*S*)-lactate is esterfied with benzoyl chloride or with toluene-*p*-sulphonyl chloride, neither reaction involves the asymmetric centre; the reactions involve nucleophilic addition of the oxygen atom of the alcohol to the carbonyl group or the sulphonyl group. The reaction between a toluene-*p*-sulphonate ester and sodium benzoate is very different. Toluene-*p*-sulphonic acid is a strong acid and when the sulphonate ester is treated with sodium benzoate a nucleophilic displacement occurs. The benzoate anion displaces the toluene-*p*-sulphonate anion.

$$C_6H_5CO_2^- + EtO_2C\text{-}C(CH_3)(H)\text{-}SO_3C_7H_7 \longrightarrow C_6H_5O_2C\text{-}C(CH_3)(H)\text{-}CO_2Et + CH_3C_6H_4SO_3^-$$

Ethyl (*S*)-2-*p*-toluene-*p*-sulphyloxypropionate

Ethyl (*R*)-2-benzoyloxypropionate

Every displacement reaction in which we draw a curved arrow from a nucleophile to a carbon atom and another curved arrow from the bond joining that carbon atom to a leaving group involves inversion at the carbon atom in question. If we are using our curved arrows consistently there are no exceptions.

In Chapter 3 of Part 1 we described how certain alkyl halides, particularly those in which the halogen is attached to a tertiary group, undergo ionization to yield a transient carbonium ion and a halide anion. The resultant carbonium ion reacts extremely rapidly with any nucleophile present, and the overall consequence of the reaction appears similar to that of the true displacement reaction. The rate of the reaction depends only upon the concentration of the alkyl halide and is independent of the concentration of the anion. Such a reaction is called unimolecular and, if the nucleophile is the solvent, the reaction is termed a unimolecular solvolysis. We briefly refer to this reaction mechanism again in Chapter 10 of this volume. The carbonium ion is most stable when it is able to take up a planar trigonal configuration and whenever possible a carbonium ion takes up a planar form. Clearly, a nucleophile could approach the positive carbon atom on either side of the plane and we might predict, therefore, that unimolecular solvolysis of an optically active alkyl halide would result in racemization. In practice, the situation is a good deal more complicated than this. α-Methylbenzyl chloride undergoes unimolecular solvolysis in aqueous acetone, and if the starting halide is optically active the resultant alcohol is almost completely racemized. Most unimolecular solvolyses of optically active alkyl halides, however, yield products which are only partially racemized, the remainder being inverted. The reason for this is not difficult to understand. The alkyl halide is surrounded by the solvent which is going to behave as the nucleophile. When ionization takes place, even though the resultant carbonium ion may be strictly planar, the solvent shell surrounding it is unsymmetrical as on one side there will be the negatively charged departing anion. Thus in a reaction that involves a very short-lived carbonium ion nucleophilic attack is only possible on one side and inversion is observed. The more stable the carbonium ion the longer life it will have and the greater opportunity for nucleophiles to attack either side. α-Methylbenzyl chloride, to which we have already referred, is hydrolyzed to the corresponding alcohol in 80% aqueous acetone. The resultant alcohol is 98% racemized. In contrast, hydroloysis of 2-bromo-octane, which forms a less stable carbonium ion, under similar conditions (60% aqueous ethanol) yields an alcohol which is 34% racemized (i.e. inversion predominates). The relative extent of racemization and inversion in unimolecular solvolysis depends greatly on the reaction conditions and the nature of the solvent, but we have the general rule that a unimolecular solvolysis yields a partly racemized plus a partly inverted product.

In a few special cases neither inversion nor racemization, but complete retention of configuration, is observed. This occurs particularly with α-halogeno-carboxylic acids. In weakly alkaline media 2-bromopropionic acid undergoes unimolecular solvolysis to a lactic acid that has the same configuration about the asymmetric carbon atom as the starting halogeno-acid. Even in weakly alkaline media the carboxyl portion of the molecule will exist as the anion, and the negative charge of the carboxylate anion both facilitates the departure of the bromide anion and protects one side of the molecule from attack by the solvent (see diagram). The oxygen of the carboxylate ion probably does not become fully bound to the asymmetric carbon atom (we have represented this by a dotted 'bond' and partial charges). Such control of a reaction by a neighbouring group which becomes partially or, in some cases, wholly bound with the

$$\mathrm{CH_3C(H)(Br)CO_2^-} \longrightarrow \mathrm{[\overset{\delta-}{O}\cdots\cdots\overset{\delta+}{C}]} + \mathrm{Br^-} \xrightarrow{:OH_2} \mathrm{CH_3C(H)(OH)CO_2^-} + \mathrm{H^+}$$

(*S*)-2-Bromopropionate anion (*S*)-Lactate anion

Unimolecular solvolysis of 2-bromopropionic acid involving neighbouring-group participation and retention of configuration.

reacting centre in an intermediate is known as *neighbouring-group participation.* In concentrated alkali the mechanism of the above reaction changes and becomes bimolecular, under which conditions complete inversion of configuration occurs.

We can summarize the stereochemical consequences of replacement reactions as follows. True replacement reactions involving bimolecular displacement invariably result in inversion of configuration. Unimolecular reactions proceeding through a carbonium ion, whose overall result is one of replacement, involve concurrent racemization and inversion of configuration, except in special cases when neighbouring-group participation leads to retention of configuration.

Thus, in any particular case the stereochemical consequences of the reaction depend on the reaction mechanism (see Table 24.1).

The generality of inversion in a true displacement reaction extends to all the displacement reactions we have considered, including those

Table 24.1. Stereochemical consequences of replacement of halogen by hydroxyl

$\mathrm{R\text{–}C(H)(Hal)\text{–}CH_3}$	Bimolecular Displacement S_N2	Unimolecular solvolysis S_N1
R = n-C_6H_{13}	Inversion	Racemization + inversion
C_6H_5	Inversion	Mainly racemization + a little inversion
CO_2^-	Inversion	Retention

involving a molecular rearrangement. In Chapter 14 we showed how it was established that in the Beckmann rearrangement the displacement occurs *trans* to the carbon–nitrogen double bond (see page 167). Similarly, in Chapter 10, where we first introduced molecular rearrangements, we described the rearrangement that occurs when 2-amino-1,1-diphenylpropan-1-ol is treated with nitrous acid and we described how, starting with an optically active amino-alcohol, we obtained an optically active ketone. We commented that this confirmed our picture of a pinacol rearrangement in which the migrating group comes in behind the leaving group. In this particular experiment the configurations of the two products have been established, confirming the expected inversion.

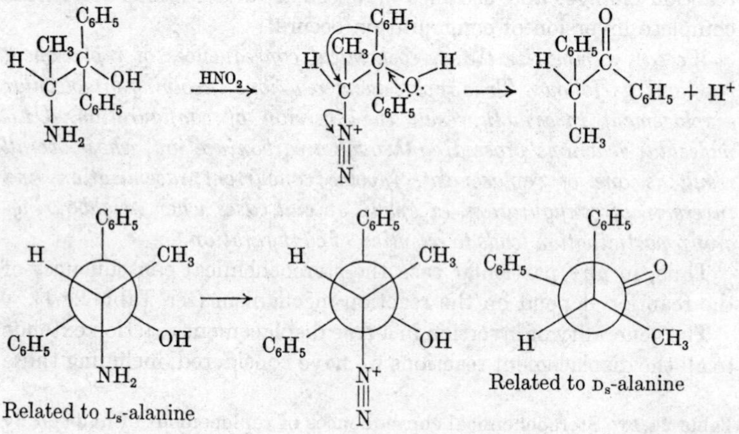

$\xrightarrow{HNO_2}$ $\longrightarrow$ $+ H^+$

Related to L_s-alanine

Related to D_s-alanine

The Stereochemical Consequences of Elimination Reactions

Elimination reactions were first introduced in Chapter 5 of Part 1. Just as there are a bimolecular mechanism and a unimolecular mechanism for displacements S_N2 and S_N1, so we saw that there are bimolecular and unimolecular mechanisms for elimination reactions ($E2$ and $E1$). We have just seen that the stereochemical consequences of bimolecular displacement are unambiguous and always the same, whereas the stereochemical consequences of unimolecular displacement are complex and depend a great deal on the nature of the compound and the reaction conditions. The situation is very similar in elimination reactions, although not quite as clear-cut: the stereochemical consequences of bimolecular elimination are usually (though not quite always) the same, whereas the stereochemical consequences of unimolecular elimination depend a great deal on the compound and the reaction conditions. Bimolecular elimination is the more common process in organic reactions occurring in solution.

Throughout Part 1 and so far in Part 2 we have depicted bimolecular elimination as shown. In Chapter 1 of Part 1 we attributed

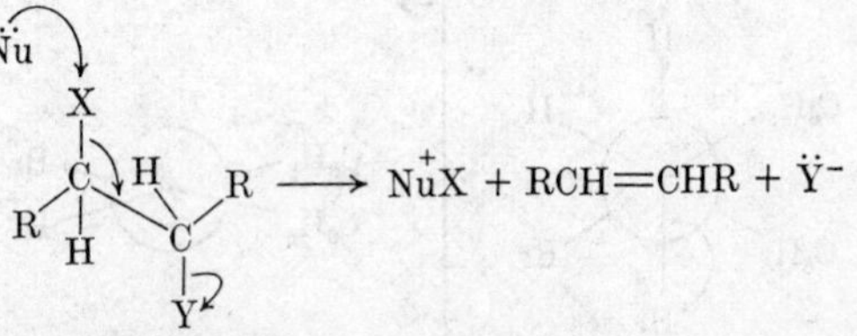

$$\longrightarrow \overset{+}{\mathrm{Nu}}\mathrm{X} + \mathrm{RCH{=}CHR} + \ddot{\mathrm{Y}}^{-}$$

the tetrahedral shape of the methane molecule to the repulsion of the electron pairs forming the carbon–hydrogen bonds and in the early Chapters of the present volume we found that this description was consistent with the ideas of quantum mechanics. In an elimination such as we have depicted above, there occurs a transfer of an electron pair from the attacking nucleophile through three bonds of the molecule on to the leaving group Y. Applying the same argument, therefore, we should expect X and Y to be as far away from each

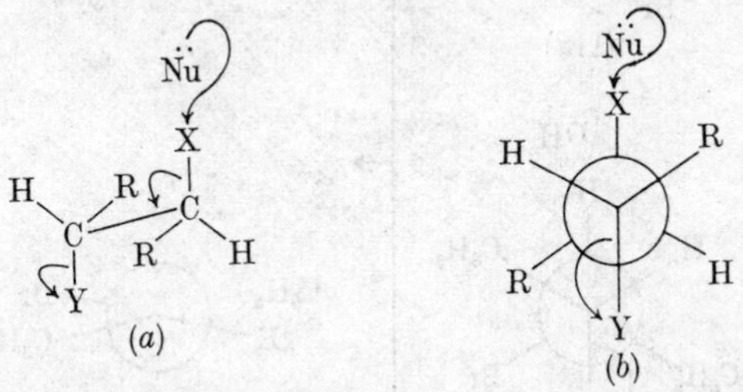

(*a*) (*b*)

Bimolecular elimination depicted in a 'three-dimensional' diagram (*a*) and in a Newman projection formula (*b*).

other as possible; this will happen when the molecule is in the skew conformation shown in which X and Y are antiperiplanar.

The steric consequences of antiperiplanar elimination are well exemplified in the classic studies of the dehydrobrominations of the

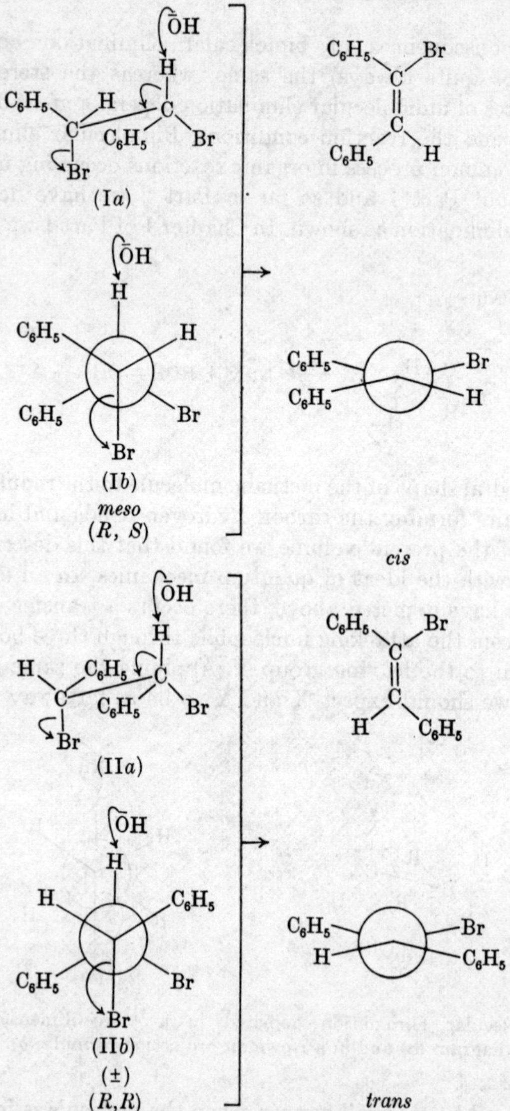

(I*a*)

(I*b*)
meso
(*R*, *S*)

cis

(II*a*)

(II*b*)
(±)
(*R*,*R*)

trans

Antiperiplanar elimination from *meso*- and (*R*,*R*)-1,2-dibromo-1,2-diphenylethane [the other enantiomorph (*S*,*S*) not depicted], (I*a*) and (II*a*) representing three-dimensional 'pictures', (I*b*) and (II*b*) representing Newman projection formulae.

two diastereomeric stilbene derivatives, *meso*- and racemic 1,2-dibromo-1,2-diphenylethane. We can depict these reactions as shown on p. 340.

The diagram shows how clearly the Newman projection formulae depict the steric situation during the course of the elimination.

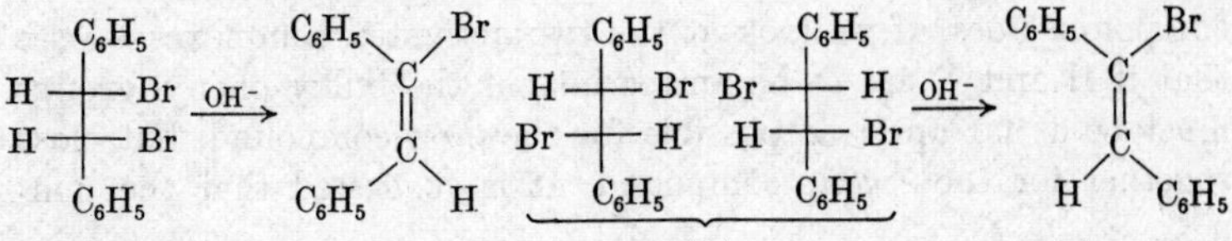

meso $\xrightarrow{OH^-}$ *cis*

(±) $\xrightarrow{OH^-}$ *trans*

The overall stereochemical consequence of the reaction may also be illustrated by using the Fischer projection formulae as above.

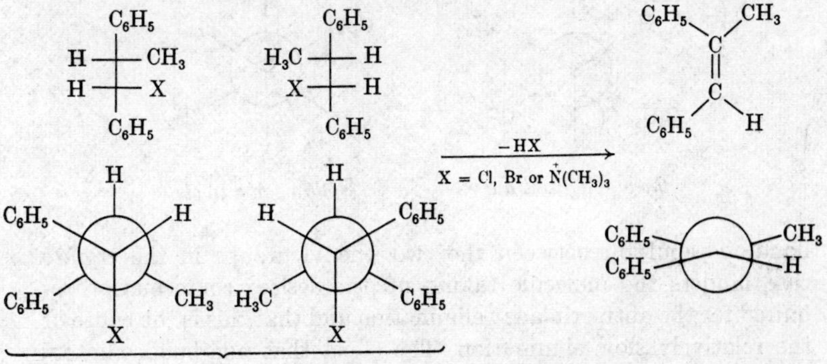

$\xrightarrow{-HX}$ X = Cl, Br or $\overset{+}{N}(CH_3)_3$

(±) *erythro* ⟶ *cis*-α-Methylstilbene

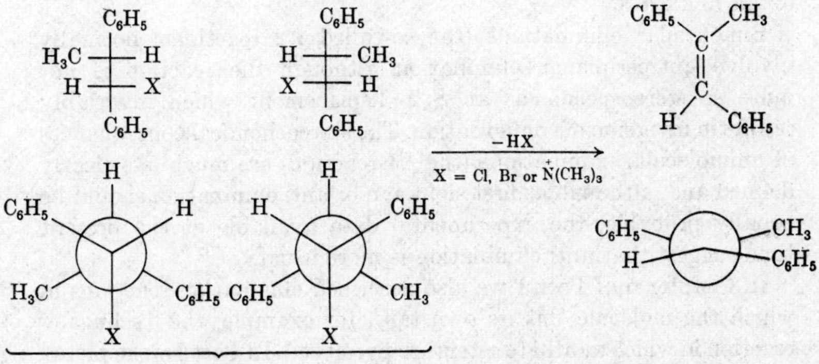

$\xrightarrow{-HX}$ X = Cl, Br or $\overset{+}{N}(CH_3)_3$

(±) *threo* ⟶ *trans*-α-Methylstilbene

Observations of this type are quite general, and base-induced elimination of 1-X-1,2-diphenylpropane gives the *trans*-α-methylstilbene from the *threo*-compound and the *cis*-α-methylstilbene from the *erythro*-compound as shown on p. 341.

In these three examples [i.e. when X = Cl, Br, or $^{+}N(CH_3)_3$], the *threo*-compound eliminates HX more rapidly than the *erythro*-compound does. If we look at the two activated complexes we see that if H and X are to be antiperiplanar the bulky phenyl groups must be as far apart as possible for the *threo*-compound, but close together for the *erythro*-compound. It is suggested that the non-

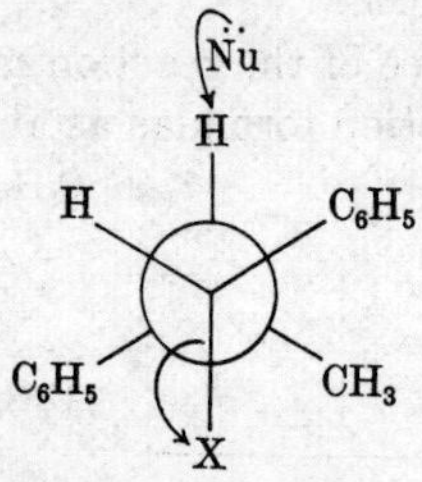

threo going to *trans* 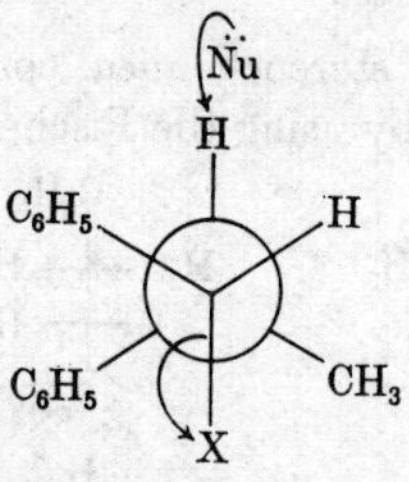 *erythro* going to *cis*

bonding repulsion between the two phenyl groups in the *erythro* case hinders the molecule taking up the desired conformation required for the antiperiplanar elimination and that this is the cause of the relatively slow elimination. The effect that non-bonded interaction can have on reaction rates will be discussed in much greater detail in Part 3.

Bimolecular eliminations (the so-called *E*2 reactions) normally involve antiperiplanar elimination although the reaction is not quite so stereospecific as an S_N2 displacement which invariably results in inversion of configuration. The stereochemical consequences of unimolecular eliminations (the *E*1 reaction) are much less clearly defined and, although at first sight syn or anti elimination should be equally probable, the experimental data available at the present time suggest that anti elimination is more usual.

In Chapter 5 of Part 1 we also described elimination reactions in which the molecule 'bit its own tail', for example, the Tschugaev reaction in which xanthate esters are pyrolyzed. In Part 1 we depicted this reaction as follows:

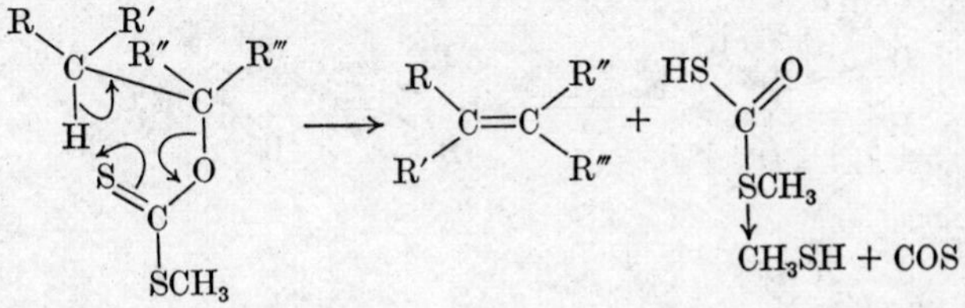

R, R′, R″, R‴ C(H)–C(O–C(=S)SCH$_3$) ⟶ RR′C=CR″R‴ + HS–C(=O)–SCH$_3$ ⟶ $CH_3SH + COS$

Note that this implies a synperiplanar elimination. Thus, diastereoisomeric xanthate esters, analogous to the compounds we have been discussing for bimolecular elimination, undergo intramolecular eliminations such that the *threo*-compound yields predominantly the *cis*-olefin and the *erythro*-compound yields predominantly the *trans*-olefin. Notice the activated complex approaches an eclipsed (synperiplanar) conformation.

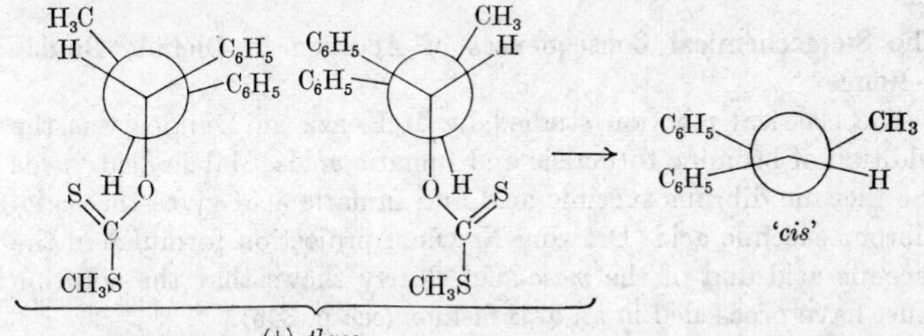

(±)-*threo* ⟶ '*cis*'

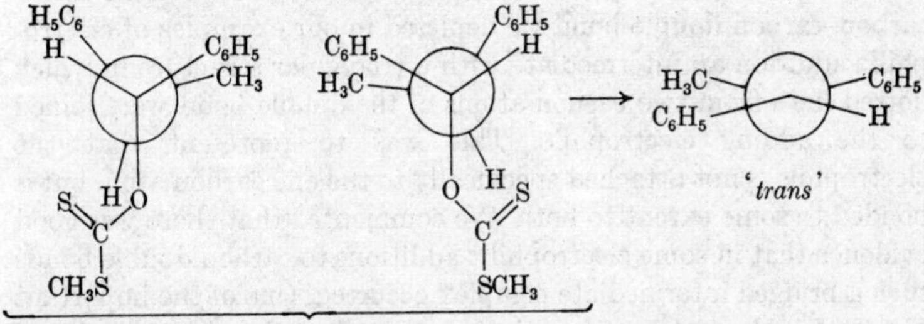

(±)-*erythro* ⟶ '*trans*'

Pyrolysis of esters and of amine oxides (the Cope elimination) likewise involves intramolecular elimination, and in both cases syn elimination predominates although pyrolysis of the esters is not completely stereospecific (see p. 344).

The stereochemical consequences of elimination reactions will be

$$\text{R(R')C(H)–C(R'')(R''')–O–C(=O)CH}_3 \xrightarrow[\text{syn}]{\text{Predominantly}} \text{RR'C=CR''R'''} + \text{CH}_3\text{C(=O)OH}$$

$$\text{R(R')C(H)–C(R'')(R''')–}\overset{+}{\text{N}}\text{(CH}_3)_2\text{–O}^- \xrightarrow{\text{syn}} \text{RR'C=CR''R'''} + (\text{CH}_3)_2\text{NOH}$$

considered in much greater detail in Part 3 when we discuss the reactions of alicyclic compounds.

The Stereochemical Consequences of Addition to Olefinic Double Bonds

The classical reaction studied by McKenzie in Dundee was the addition of bromine to maleic and fumaric acids. Maleic acid yields the racemic dibromosuccinic acid and fumaric acid gives the *meso*-dibromosuccinic acid. Drawing Newman projection formulae of the racemic acid and of the *meso*-acid clearly shows that the addition must have proceeded in an anti fashion (see p. 345).

In Chapter 6 of Part 1, when discussing addition reactions of the carbon–carbon double bond we depicted in our examples of electrophilic addition an intermediate with a triangular structure in which dotted lines from two carbon atoms of the double bond were joined to the adding electrophile. This was to represent that the electrophile is not attached specifically to the one carbon atom but is bonded to some extent to both. We commented that there was good evidence that in some electrophilic additions to carbon double bonds such a bridged intermediate complex occurred. One of the important pieces of evidence for such an intermediate is the stereospecificity of the addition of the halogens (e.g. chlorine, bromine, and iodine under ionic conditions) to yield the products resulting from anti addition. In the addition of bromine, for example, formation of the triangular intermediate with the bromine cation means that the bromine anion must then add on the other side of the molecule.

The stereochemical consequences of addition of proton acids such

H, CO_2H, C, C, H, CO_2H $\xrightarrow{Br_2}$ CO_2H, H, Br, Br, H, CO_2H + CO_2H, Br, H, H, Br, CO_2H

Maleic acid $\xrightarrow{Br_2}$ (*R,R*) + (*S,S*)

(±)-2,3-Dibromosuccinic acids

HO_2C, H, C, C, H, CO_2H $\xrightarrow{Br_2}$ CO_2H, Br, H, Br, H, CO_2H

Fumaric acid $\xrightarrow{Br_2}$ (*R,S*)

meso-2,3-Dibromosuccinic acid

as hydrogen chloride, hydrogen bromide, or sulphuric acid is far less certain. The addition of hydrogen bromide to olefins has been reported to yield the products of exclusive anti addition but to give under different circumstances the products of exclusive syn addition, and also to give a mixture of products. A mixture of products is what we might expect from the formation of a carbonium ion as

R—C(R′)=C(R″)—R‴ + Br—Br ⟶ R—C(R′)⋯Br⋯C(R″)(+)—R‴ $\bar{Br}$ ⟶ R(R′)C(Br)—C(Br)(R″)R‴

$$R(R')C{=}C(R'')(R''') + H{-}X \longrightarrow R(R')\overset{+}{C}{-}C(R'')(R''')(H)$$

Intermediate
carbonium ion

intermediate, assuming that there is free rotation about the carbon–carbon bond axis of the molecule. However, it has also been suggested that this kind of intermediate can lead to syn addition as the result of ion-pair formation. It seems reasonable to suppose that the distinction between ion-pair formation leading to syn addition and free carbonium ion leading to a mixture of products will depend very largely on the nature of the solvent, but at the present time there is insufficient experimental evidence on this subject. The occurrence of anti addition can be accommodated in the same theory invoked to explain the anti addition of the halogens, namely, the occurrence of a triangular intermediate. It is reasonable to suppose that such an intermediate would be less stable for a proton than it would be for a cation derived from an atom lower in the Periodic Table.

We can summarize the stereochemical consequences of electrophilic addition to carbon–carbon double bonds by saying that the reaction usually proceeds by an antiperiplanar process, this being generally true for the addition of the halogens, including such reactions as the addition of 'hypohalous acids' which we saw in Part 1 is initiated by the addition of a halogen cation. Addition of proton acids is far less stereospecific and depends on the compound and the reaction conditions.

The stereochemical consequences of radical addition to olefins is, at the present time, very uncertain. As we saw in Chapter 11 the addition of halogen atoms to olefins is reversible. In the presence of halogen atoms, an equilibrium between the *cis-* and *trans-*olefins is very rapidly set up. Thus the products of the reaction will not

$$Cl\cdot + R(H)C{=}C(H)(R') \rightleftharpoons R(H)\dot{C}{-}C(H)(R')Cl \rightleftharpoons Cl\cdot + R(H)C{=}C(R')(H)$$

(Free rotation)

necessarily have been those resulting from addition to the geometrical isomer with which the reaction was started. In certain circumstances there is some evidence for stereospecific radical addition in the solution phase. In the gas phase, radical addition is definitely non-stereospecific, but the situation in solution is *sub judice*.

Electrocyclic (four-centre) reactions necessarily involve a syn addition. In Chapter 6 of Part 1 the first electrocyclic reaction we discussed was the reaction of olefinic double bonds with potassium permanganate. As we should expect, maleic acid yields *meso*-tartaric acid and fumaric acid yields racemic tartaric acid. (Notice that this is the reverse of the stereochemical consequences of addition of bromine to these two acids.)

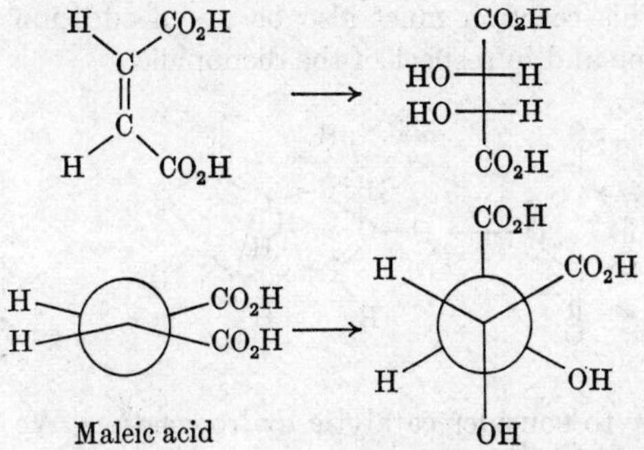

Maleic acid

meso-Tartaric acid

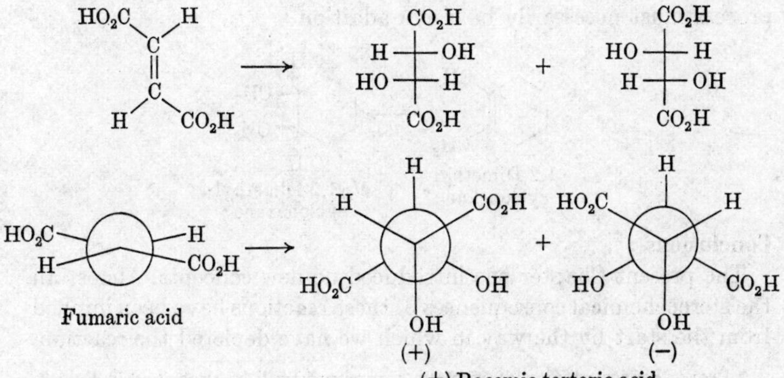

Fumaric acid

(+) (−)

(±) Racemic tartaric acid

In Part 1 we depicted such oxidation of the double bond as going through an intermediate ester, and such a reaction mechanism clearly requires syn addition, as is observed.

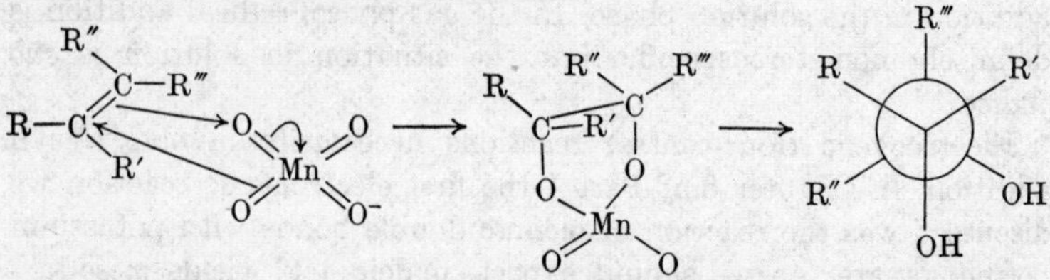

In Chapter 13 of Part 1 we discussed another type of electrocyclic reaction, namely the Diels–Alder reaction, between a conjugated diene and a suitable ethylene usually containing an acceptor group, the classical example of this reaction being that between butadiene and maleic anhydride. This reaction must also be a syn addition both in respect of the diene and in respect of the dienophile.

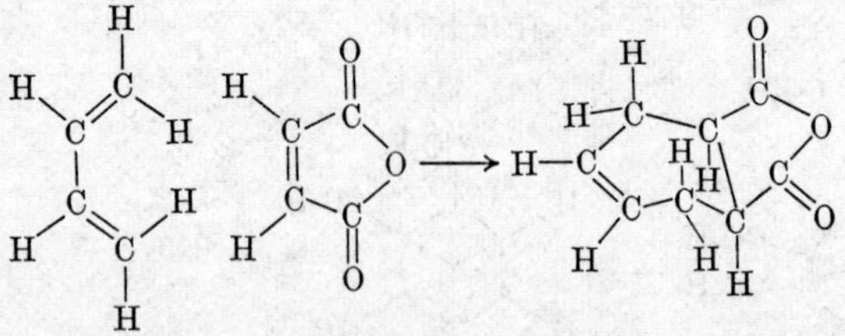

Finally, we have briefly to consider catalytic hydrogenation. We depicted this in Part 1 as involving bonding of both molecular hydrogen and olefin to the surface of the metal catalyst. Such a process must necessarily be a syn addition.

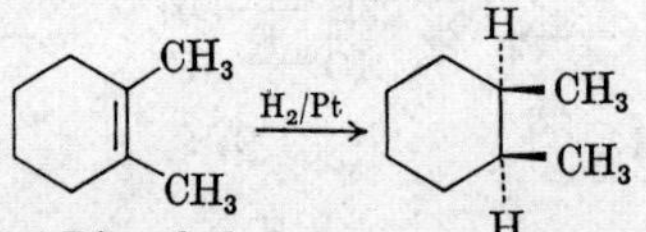

H_2/Pt

1,2-Dimethyl-cyclohexane*

cis-1,2-Dimethyl-cyclohexane*

Conclusions

The present Chapter has introduced no new concepts. Almost all the stereochemical consequences of these reactions have been implied from the start by the way in which we have depicted the reaction.

* Proper representation of alicyclic compounds will be described in Part 3.

Thus, in Chapter 3 of Part 1 we depicted a normal displacement involving an inversion of configuration at the carbon atom. We simply did not use these terms. Similarly, in Chapter 5 of Part 1, we depicted bimolecular elimination reactions proceeding by an anti-periplanar process, and tail-biting eliminations such as the Tschugaev reaction as proceeding by a synperiplanar process. All we have done in the present Chapter is to look at these reactions again and to fit them into the stereochemical definitions we developed in Chapter 23. Chemistry occurs in three dimensions, and at all times we must endeavour to think as far as we can in three dimensions, and not in two dimensions simply because we normally represent reactions on the plane of the paper.

Problems

1. In the following reaction sequences inversion of configuration is observed. Decide in each case which step involves the inversion.

(*a*) $C_6H_5CH_2CH(CH_3)(OH)$ $\xrightarrow{p\text{-}CH_3C_6H_4SO_2Cl}$ $C_6H_5CH_2CH(CH_3)(OSO_2C_7H_7)$ $\xrightarrow{Na^+\ ^-OCOCH_3}$

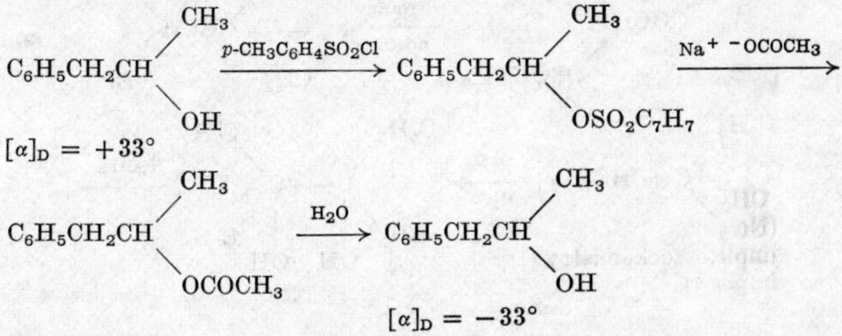

$[\alpha]_D = +33°$

$C_6H_5CH_2CH(CH_3)(OCOCH_3)$ $\xrightarrow{H_2O}$ $C_6H_5CH_2CH(CH_3)(OH)$

$[\alpha]_D = -33°$

(*b*) $HOCHCO_2H(CH_2CO_2H)$ $\xrightarrow{PCl_5}$ $ClCHCO_2H(CH_2CO_2H)$ $\xrightarrow{Ag_2O}$ $HOCHCO_2H(CH_2CO_2H)$

$[\alpha]_D = +2.9°$ $[\alpha]_D = -2.9°$

(*c*) $C(C_6H_5)(CO_2Et)(CH_3)(Cl)$ $\xleftarrow[C_5H_5N\ (excess)]{SOCl_2}$ $C(C_6H_5)(CO_2Et)(CH_3)(OH)$ $\xrightarrow{SOCl_2}$ $C(C_6H_5)(CO_2Et)(CH_3)(Cl)$

$[\alpha]_D = +120°$ (no stereochemistry implied in these formulae) $[\alpha]_D = -91°$

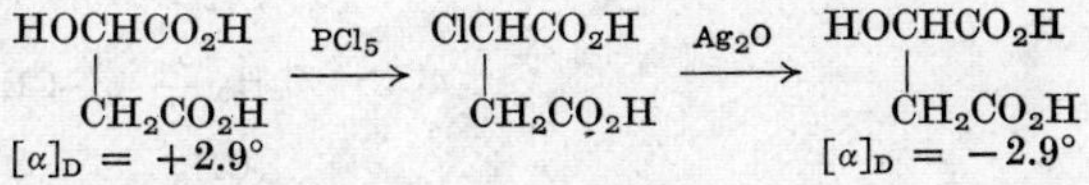

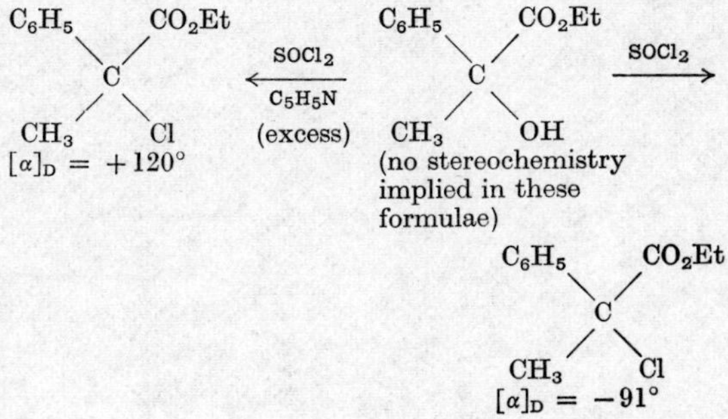

2. Predict the stereochemical structure of the products of the following reactions:

(*a*) (*cis*)-PhCH=CHPh $\xrightarrow[CCl_4]{Br_2}$ PhCHBrCHBrPh $\xrightarrow[EtOH]{KOH}$ PhCH=CBrPh

(*b*) $C_2H_5C{\equiv}CC_2H_5 \xrightarrow[\text{liq. } NH_3]{Na} C_2H_5CH{=}CHC_2H_5 \xrightarrow{KMnO_4}$ $C_2H_5CH(OH)CH(OH)C_2H_5$

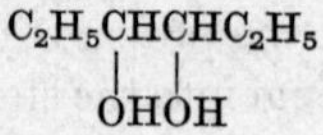

(*c*) Fischer projection: H (top); $p\text{-}CH_3C_6H_4$—, —$OCOCH_3$; C_6H_5—, —CH_3; H (bottom) $\xrightarrow{HOH}$ $C_6H_5(CH_3)CHCH(C_7H_7)(OH)$ $\xrightarrow{C_7H_7SO_2Cl}$

$C_6H_5(CH_3)CHCH(C_7H_7)(OSO_2C_7H_7) \xrightarrow[\text{acetone}]{NaBr} C_6H_5(CH_3)CHCH(C_7H_7)(Br) \xrightarrow{KOH}$

$(C_6H_5)(CH_3)C{=}CHC_7H_7$ (No stereochemistry implied) $\xrightarrow[H_2O]{OsO_4}$ $(C_6H_5)(CH_3)C(OH){-}C(C_7H_7)(H)(OH) \xrightarrow{H_2SO_4}$

$(C_7H_7)(C_6H_5)(CH_3)C{-}CHO$

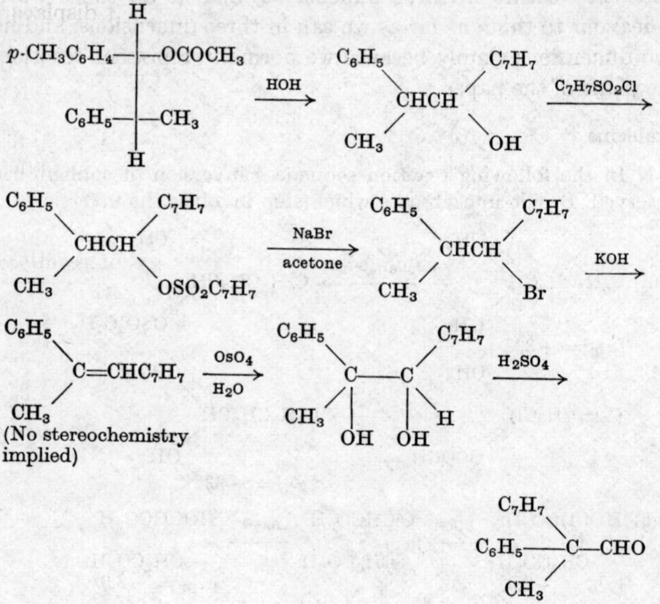

CHAPTER 25

The Determination of Structure by Physical Methods

So far, we have not considered how we know that methanol or any other organic compound has the structure we assign to it. There has been good reason for this. The complete structure of even the simplest compound such as water can only be determined by using sophisticated physical techniques. Originally, the gross structure of an organic compound, i.e. which atom is bonded to which, was largely deduced from a rationalization of its chemical reactions. The fine structure, i.e. bond lengths and bond angles, was then determined by physical means. Of recent years, physical techniques have been developed to the point where they often lead to the structure of a compound faster than a study of its chemical reactions does. The latter method is nevertheless still important and we should appreciate how it can be used.

Chemical determination of structure depends on a rationalization of observed chemical reactions, and for this we must have a large body of chemical knowledge. The absolute structure of a molecule can never be deduced from chemical reactions *ab initio*. Unfortunately, it has been common for elementary textbooks (especially British ones) to assume that they can. For example, the structure of ethanol is often 'deduced' from a series of arguments which begin as follows: 'The molecular formula is found to be C_2H_6O from analysis coupled with vapour-density measurements. Treatment of ethanol with sodium yields 0.5 mole of hydrogen per mole of ethanol. This shows that one hydrogen is bonded differently from the others. . . .' This argument is fallacious and bears no relation to the original arguments used by Kekulé. It would lead us to deduce that the two hydrogen atoms in water are differently bonded. The subsequent arguments in textbooks for the structure of ethanol usually involve treatment of ethanol with phosphorus pentachloride. These arguments are fallacious for similar reasons. Kekulé proposed the structure of methanol as we now accept it as a rationalization of chemical reactivity, particularly of the similarity of many of its reactions with those of water. The structure of ethanol followed from the similarity of the reactions of ethanol and methanol. Kekulé could not have made his proposal if it had not been for the vast accumulation of data by earlier chemists such as Gerhardt.

Any chemical argument for the structure of a molecule requires very considerable knowledge of the nature of chemical reactions. In Part 1, we showed that if we ascribed a certain structure to a molecule we could rationalize its chemical behaviour on the basis of simple ideas about the Periodic Table and electrostatics. This treatment parallels the way in which structures were originally determined. It also provides a logical and scientific argument.

No student should be asked to establish the structure of ethanol by chemical means. However, given a colourless, volatile liquid, he should, after studying its chemical reactions, be able to say that its chemical properties can be rationalized by assuming that it is an aliphatic alcohol, having based his deductions on his knowledge of the reactions of aliphatic alcohols. Then a study of the liquid's physical properties and of the physical properties of compounds prepared from it by known reactions of alcohols should enable him, by comparison with data in the literature, to establish that the unknown compound is, indeed, ethanol. Practice in the laboratory in classifying an unknown compound and then preparing derivatives in order to attempt a complete identification is a basic part of any training in organic chemistry. Originally, classification was largely by chemical tests, e.g. the presence of a carbonyl group was detected by a reaction with 2,4-dinitrophenylhydrazine, or of a primary aromatic amine by diazotization and coupling with 2-naphthol. Nowadays, these chemical tests have been supplemented, and in some cases, superseded, by the physical methods described in this Chapter.

The first step in any study of a new organic compound must be its purification; so recrystallization and distillation are the first techniques a student encounters in the laboratory, and, as his laboratory work progresses, he will meet the various forms of chromatography—column, paper, thin-layer, and gas-phase—that are used to purify organic compounds. The next step is identification of the elements present, followed by their quantitative analysis. The molecular weight is determined (e.g. from a mass spectrum) so that the molecular formula may be established. The natures of the functional groups present are next ascertained from physical observations (especially the spectroscopic methods described in the following pages) or by a study of the compound's chemical reactions. Very often, further progress requires degradation of the molecule into fragments which can be unambiguously identified and from which the structure of the starting compound can be inferred. Chapter 26 is concerned with the types of degradative methods that are employed in structural determination.

Ultimate proof of the structure of a compound can only come from physical techniques. Often, complete establishment of a structure solely by physical techniques is very laborious and usually a combination of physical and chemical methods is employed.

Virtually all regions of the electromagnetic spectrum, from X-rays to radio waves, are used in the study of the structure of organic molecules, but three regions are particularly important. These are the near-ultraviolet (200—400 mμ), the infrared (4000—600 cm^{-1}), and, in conjunction with an applied magnetic field, radio waves (60 or 100 Mc sec^{-1}). The electromagnetic spectrum spreads over a great range of wavelengths, as can be seen from Figure 25.1, so that

Table 25.1. Units of wavelength and frequency

Unit	Symbol	Value	Region used
Ångström	Å } λ	10^{-8} cm	Far ultraviolet
Millimicron	mμ } λ	10^{-7} cm	Near ultraviolet
Micron	μ } λ	10^{-4} cm }	Infrared
Reciprocal centimetres	$\bar{\nu}$ $1/\lambda$	cm^{-1} }	Infrared
Frequency*	ν	kilocycles sec^{-1} or megacycles sec^{-1}	Microwave and radio

* The internationally accepted name for the unit of frequency is the Herz (equivalent to cycle), i.e. Mc ≡ MHz.

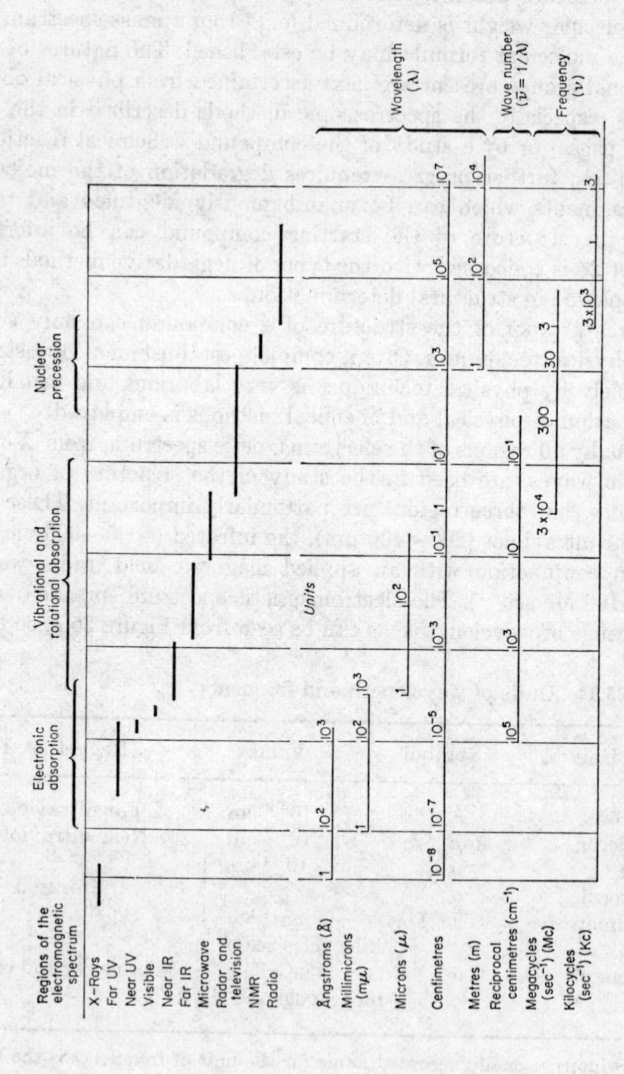

Figure 25.1. Part of the electromagnetic spectrum, and the units used in various regions.

different units are used in different regions of the spectrum. Note that units of wavelength (λ) of the electromagnetic radiation are related to units of frequency (ν) of that radiation by

$$\lambda\nu = c$$

where c is the velocity of the radiation, a constant in a particular medium (3×10^{10} cm sec^{-1} in a vacuum). For convenience, units of $1/\lambda$ (cm), i.e. cm^{-1} or 'wave number' are used in the infrared region of the spectrum. The units listed in Table 25.1 are most commonly used.

Light Absorption and Energy Changes in Molecules

The total energy of a molecule is made up of binding energy (electronic energy) and kinetic energy (vibrational and rotational energy):

$$E_{\text{total}} = \underset{\text{Binding energy}}{E_{\text{electronic}}} + \underbrace{E_{\text{vibration}} + E_{\text{rotation}}}_{\text{Kinetic energy}}$$

All molecular energy is quantized. The electronic energy levels are relatively far apart, the vibrational and rotational levels are successively closer together in energy (Figure 25.2). When a molecule absorbs a quantum of light the frequency (ν) of the light is related to the energy by the expression given in Chapter 1:

$$E = h\nu \quad (h = \text{Planck's constant})$$

Absorption of ultraviolet radiation results in a transition from one electronic level to a higher one and it may be accompanied by vibrational and rotational transitions as well.

Infrared radiation has far less energy than ultraviolet radiation. For example, a quantum of ultraviolet radiation of 2,000 Å (5×10^4 cm^{-1}) has an energy equivalent to 142 kcal mole^{-1}, while a quantum of infrared radiation of 10 μ (1,000 cm^{-1}) is equivalent to only 2.9 kcal mole^{-1}.*

Ultraviolet spectra are due to electronic transitions and give information about the electronic structure of a molecule, while

* Obtained from the relationship $E = h\nu = hc/\lambda$ (ergs molecule^{-1}) = $hc/(\lambda \times 4.8 \times 10^7)$ (kcal molecule^{-1}).

infrared spectra are due to transitions in the vibrational and rotational modes and give information about molecular architecture.

The amount of light absorbed by a medium is dependent on the intensity of the light and is proportional to the length of the light path through the medium,

$$\log_{10} (I_0/I) = kl$$

where k = extinction coefficient, I_0 = intensity of incident light, I = intensity of transmitted light, and l = thickness of absorbing medium.

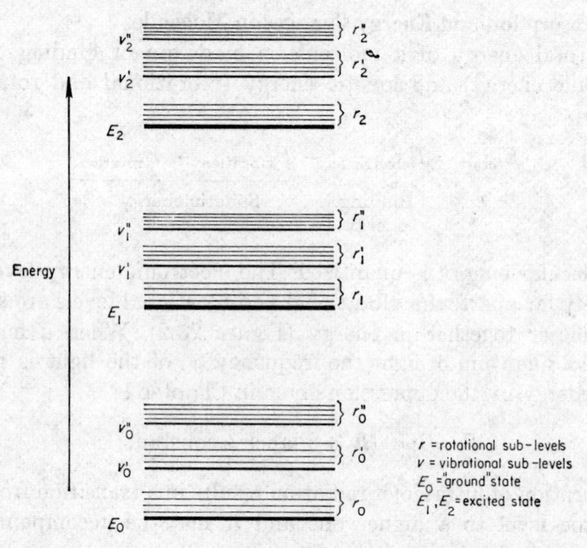

Figure 25.2. Diagram to represent energy levels in a molecule.

This relation is known as Lambert's law. If it can be assumed that there is no interaction between the molecules that absorb the light and the solvent, the amount of light absorbed is proportional to the number of absorbing molecules: this relationship is known as Beer's law. Combination of Lambert's and Beer's laws gives the following expression:

$$\log_{10} I_0/I = \epsilon cl$$

where ϵ = molecular extinction coefficient, and c = molarity.

$\text{Log}_{10}(I_0/I)$ is known as the optical density of the medium and from it can be calculated the molecular extinction coefficient

$$\epsilon = \frac{\log_{10}(I_0/I)}{c \times l}$$

Ultraviolet Spectroscopy

As the name implies, the ultraviolet region extends from the short-wave limit of the visible spectrum (i.e. violet) at about 400 mμ to the region of soft *X*-rays at about 100 mμ. Not all parts of this extensive region of the spectrum are equally accessible experimentally. Below 200 mμ, air (especially oxygen) and silica are no longer transparent, and because of the experimental techniques needed for this region it is often called the 'vacuum ultraviolet'. However, it is the longer wavelength region (200—400 mμ, the 'near ultraviolet') that is of greatest interest in organic chemistry.

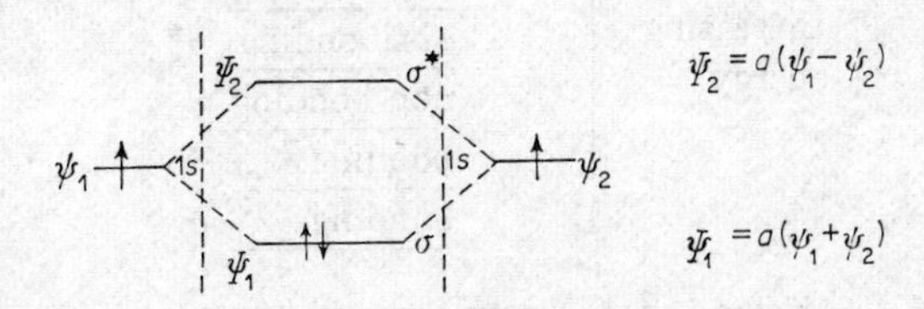

Figure 25.3. Correlation diagram showing the formation of the orbitals of a hydrogen molecule from two hydrogen atoms.

In Chapter 1 we discussed the absorption of a photon by an atom. We saw that such an absorption was accompanied by the promotion of an electron from one orbital to another of higher energy. The same applies to a molecule. In Chapter 3, for example, we saw that when two hydrogen atoms come together, their two 1*s* atomic orbitals interact to form two molecular orbitals, σ and σ^*.

In the ground state of the hydrogen molecule, both electrons are in the bonding σ molecular orbital. When a hydrogen atom absorbs light of wavelength near 110 mμ, one electron is promoted to the σ^* molecular orbital, giving rise to an excited state in which both molecular orbitals are singly occupied. Since the σ^* orbital is antibonding the electron in this orbital roughly cancels the bonding power of the electron in the σ orbital, and excited hydrogen molecules readily dissociate into atoms.

The same situation exists in more complex molecules. When methane absorbs light near 120 mμ, an electron in one of the carbon–hydrogen bonds is promoted ($\sigma \rightarrow \sigma^*$). Hydrogen and methane absorb only in the vacuum ultraviolet, and in general $\sigma \rightarrow \sigma^*$-transitions occur in this region. We saw in Chapter 3 that π molecular orbitals were of higher energy than σ orbitals, and hence the promotion of an electron from a bonding π orbital to the corresponding antibonding π^* orbital requires much less energy. Thus alkenes absorb light at much longer wavelengths (about 175–185 mμ) than alkanes, the absorption being due to $\pi \rightarrow \pi^*$-transitions. In a molecule such as hydrogen chloride, besides the two electrons associated with the σ molecular orbital, there are the non-bonding electrons of

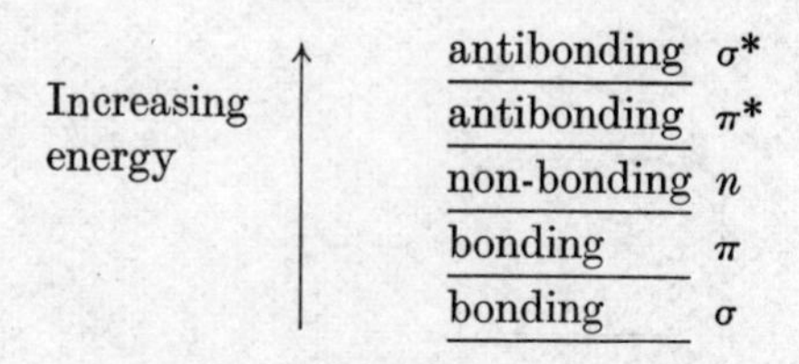

Figure 25.4*a*. Relative energies of orbitals.

the chlorine atom. Under certain circumstances, these can be promoted into the σ^* orbital. To take an example from organic chemistry, methyl iodide absorbs light near 250 mμ and this absorption is associated with the promotion of one of the non-bonding electrons of the iodine atom into the σ^* molecular orbital. This absorption is denoted $n \rightarrow \sigma^*$. Furthermore, we have groups such as >C=O, —N=O, and —N=N— in which one or both of the atoms linked in the double bond have non-bonding electrons which can be promoted into the π^* molecular orbitals. This transition, $n \rightarrow \pi^*$, represents the lowest-energy transition we observe and it can range from approximately 280 mμ (for acetone) to 650 mμ for nitroso-compounds (cf. Figure 25.4*a* and Table 25.2).

Table 25.2. Approximate ranges of electronic transitions for simple chromophores

Transition	Example	λ_{max}
$\sigma \rightarrow \sigma^*$	CH_4	120
$n \rightarrow \sigma^*$	CH_3-Cl	173
	CH_3-Br	204
	CH_3-I	259
$\pi \rightarrow \pi^*$	$C_4H_9CH{=}CH_2$	175
	Cyclohexene	183
$n \rightarrow \pi^*$	$(CH_3)_2C{=}O$	279
	$CH_3N{=}NCH_3$	340
	t-$C_4H_9N{=}O$	665

A bond causing absorption is called a *chromophore*. Compounds that absorb light in the visible region of the spectrum, e.g. nitroso-compounds, are coloured.

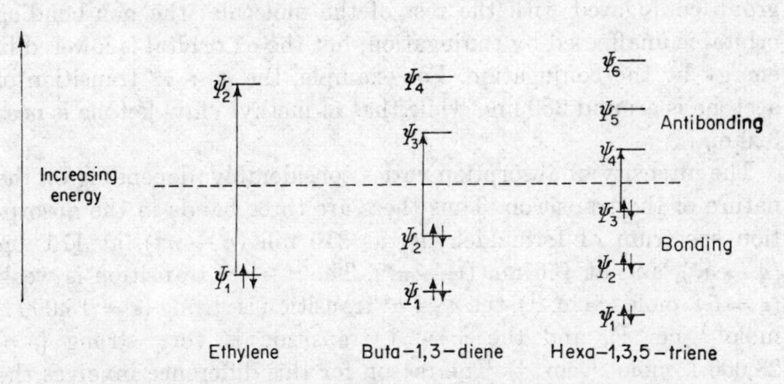

Figure 25.4*b*. Relative energies of the bonding and antibonding orbitals in molecules containing carbon–carbon double bonds.

We are familiar with the idea that there is a series of molecular orbitals of π symmetry associated with all the atoms in a chain of conjugated double bonds. Thus we saw in Chapter 3 that there are four π molecular orbitals for butadiene, two bonding and two anti-bonding. The relative energies of these orbitals, compared with the π orbitals of ethylene, are shown in Figure 25.4*b*. We see that the energy separation between the highest bonding π orbital and the lowest antibonding π^* orbital is less for conjugated molecules than for ethylene, and hence butadiene absorbs at longer wavelengths than ethylene. The longer the conjugated chain, the closer the bonding and the antibonding orbitals become and the longer the wavelength of the absorption (cf. Table 25.3).

Table 25.3. Electronic spectra of linear polyenes ($\pi \rightarrow \pi^*$)

Polyene	λ_{max}(mμ)
Ethylene	162.5
Buta-1,3-diene	217
Hexa-1,3,5-triene	251
Octa-1,3,5,7-tetraene	304

This effect will also apply to the $n \rightarrow \pi^*$ transition of a carbonyl group conjugated with the rest of the molecule: the non-bonding orbital is unaffected by conjugation, but the π^* orbital is lowered in energy by the conjugation. For example, the $n \rightarrow \pi^*$ transition of acetone is around 280 mμ, while that of methyl vinyl ketone is near 320 mμ.

The intensity of absorption varies considerably, depending on the nature of the transition. Thus there are three bands in the absorption spectrum of formaldehyde, at 310 mμ ($n \rightarrow \pi^*$), at 175 mμ ($n \rightarrow \sigma^*$), and at 156 mμ ($\pi \rightarrow \pi^*$). The $n \rightarrow \pi^*$ transition is weak ($\epsilon = 5$ l. mole^{-1} cm^{-1}), the $n \rightarrow \sigma^*$ transition is strong ($\epsilon = 18{,}000$ l. mole^{-1} cm^{-1}), and the $\pi \rightarrow \pi^*$ transition is very strong ($\epsilon = 23{,}000$ l. mole^{-1} cm^{-1}). The reason for this difference involves the orbital angular momentum, and its discussion is outside the scope of the present volume. What concerns us is that $\pi \rightarrow \pi^*$ absorption bands of conjugated molecules and the $n \rightarrow \pi^*$ absorption bands of a carbonyl group can occur in the same region of the spectrum, but that the $\pi \rightarrow \pi^*$ is 'symmetry allowed' and is strong, while the $n \rightarrow \pi^*$ is 'symmetry forbidden' and is weak, which enables the two types of transition to be distinguished in most cases.

Ultraviolet absorption bands are broad, because each electronic level has a large number of associated vibrational and rotational

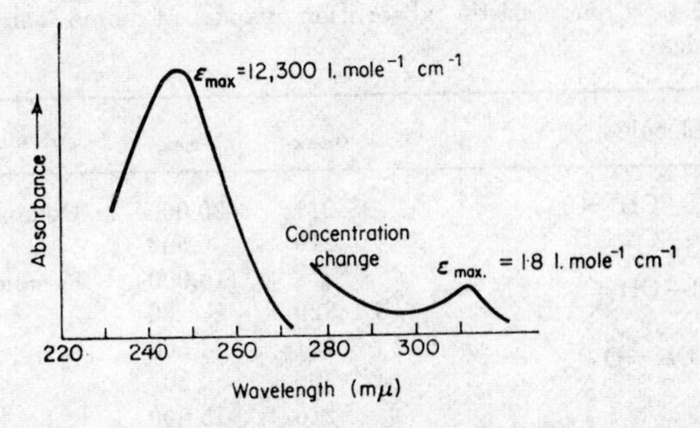

Figure 25.5*a*. The ultraviolet spectrum of mesityl oxide in methanol.

levels (see Figure 25.2). In the spectrum of mesityl oxide $(CH_3)_2C{=}CH{-}CO{-}CH_3$ shown in Figure 25.5*a*, the two absorptions $\pi \rightarrow \pi^*$ and $n \rightarrow \pi^*$ can be distinguished by their relative intensities.

Table 25.4 lists the main absorption bands of some simple conjugated molecules in the near-ultraviolet region.

Most aromatic compounds show absorption bands around 250–270 mμ with $\epsilon \approx 100$–1,000. In contrast to the ultraviolet spectra of aliphatic compounds, vibrational fine structure is sometimes observed for aromatic compounds (Figure 25.5*b*). Substituents in the benzene

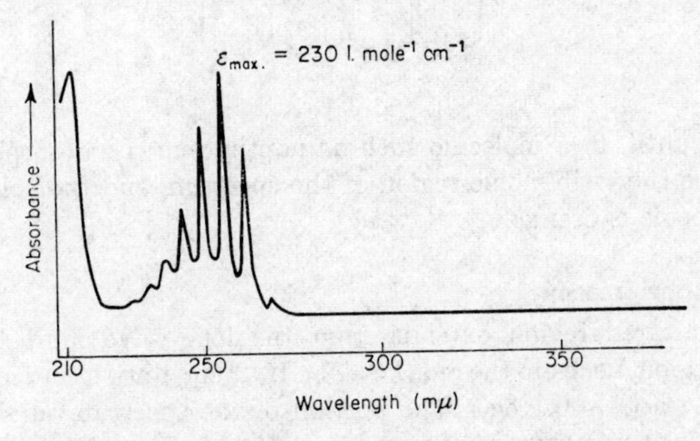

Figure 25.5*b*. The ultraviolet absorption spectrum of benzene in cyclohexane.

Table 25.4. Characteristic absorption bands of some conjugated molecules

Molecule		λ_{max}	ϵ_{max}	Solvent[a]
$CH_2{=}CH{-}CH{=}CH_2$		217	20,000	Hexane
$CH_2{=}CH{-}C{\equiv}CH$		219	6,500	,,
$CH_2{=}CH{-}CH{=}O$	{	218	18,000	Ethanol
	&	320	30	
$CH{\equiv}C{-}CH{=}O$	{	214	4,500	,,
	&	308	20	
$CH_2{=}CH{-}CO_2H$	{	210	15,000	,,
	&	245	250	

[a] With some solvents, particularly polar solvents such as ethanol, there may be some interaction between the solute and solvent molecules which can affect the position and intensity of the absorption; hence λ_{max} and ϵ_{max} may be solvent-dependent.

ring, whether acceptors or donors, increase the intensity of absorption and increase λ_{max}. In addition, if the substituent is itself a chromophore, there will be more than one absorption band (see Table 25.5).

Linear polycyclic aromatic hydrocarbons (see Chapter 21) absorb at increasingly longer wavelengths as the number of condensed rings

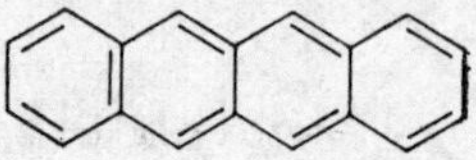

(I)

increase until, in a molecule such as napthacene (I), absorption is shifted to the visible blue region of the spectrum and the colour of the molecule is orange.

Infrared Spectroscopy

The infrared region extends from the long-wave limit to the visible region, i.e. from the red ($\lambda \sim 7 \times 10^{-5}$ cm) until it merges with microwaves ($\lambda \sim 0.1$ cm). The portion of the spectrum most used in organic chemistry ranges from 2.5×10^{-4} to 15×10^{-4} cm. This is the region of vibration-rotation spectra. Pure rotation spectra can

Table 25.5. Ultraviolet spectra of some aromatic compounds

Compound	λ_{max} (mμ)	ϵ_{max} (l. mole^{-1}cm^{-1})	Solvent
Benzene	200	8,000	Hexane
	& 255	230	
Toluene	205	7,800	,,
	& 261	300	
p-Xylene	212	8,000	,,
	& 275	490	
Styrene	224	12,000	Ethanol
	& 282	450	
Acetophenone	240	13,000	,,
	& 278	1,100	
	& 319	20	
Benzaldehyde	244	10,000	,,
	& 280	1,500	
	& 328	20	
Nitrobenzene	252	10,000	Hexane
	& 280	1,000	
	& 330	125	
Phenol	210	6,200	Water
	& 270	1,450	
Phenoxide anion	235	9,400	Aq. NaOH
	& 287	2,600	
Aniline	230	8,600	Water
	& 280	1,430	
Anilinium cation	203	7,500	Aq. acid
	& 254	160	
Acetanilide	242	14,400	Ethanol
	& 280	500	

be observed in the far-infrared region, but this is of less use for determination of organic structures and involves formidable experimental difficulties.

Let us consider a simple diatomic molecule. The bond joining the two atoms will resist extension and compression. As a model we could consider two masses connected by a spring. If the spring is stretched and then released, the masses execute simple harmonic motion about their equilibrium positions. The situation in a diatomic molecule is similar, and vibration can only occur at certain frequencies.

The vibrational energy levels are given by:

$$E_v = h\nu(v + \tfrac{1}{2})$$

where $v = 0, 1, 2, \ldots$ (the vibrational quantum number).

Notice that this means that even for the lowest vibrational level ($v = 0$) the molecule still possesses vibrational energy, known as zero-point energy (see Chapter 3).

Light waves involve oscillating electric and magnetic fields, and when light is absorbed an oscillating electric dipole is produced. To absorb near-infrared radiation, a molecule must be able to vibrate in such a way that there is a displacement of its electric charge centre, i.e. the dipole moment must change during a vibration. There is no permanent dipole in any homonuclear diatomic molecule (e.g. H_2, N_2, O_2, etc.) because of its symmetry. During vibration, charge displacement at one end is exactly balanced by a like displacement at the other end, so that the dipole moment remains zero and no infrared absorption is observed. Conversely, the bigger the change in dipole moment, the more intense the absorption. This fact becomes important when we consider complex molecules.

Polyatomic molecules have many modes of vibration. In the water molecule, for example, there are three (Figure 25.6).

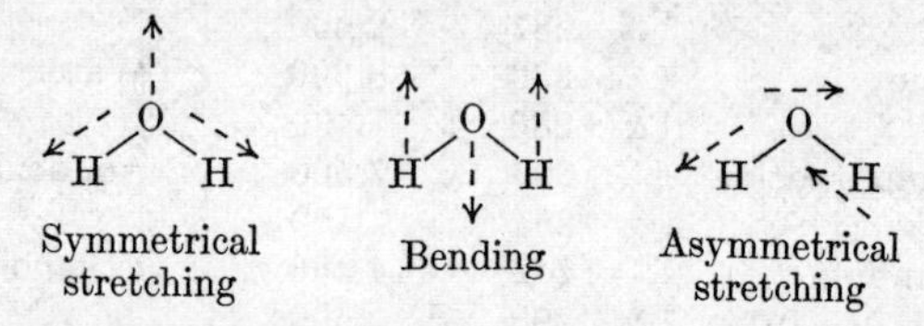

Figure 25.6. Normal vibrational modes of the water molecule.

The usefulness of infrared spectra of complex organic molecules is greatly enhanced because these complex molecules can be regarded as a heavy mass to which light masses are attached. In the infrared spectra of these molecules, absorption bands can be observed which are attributable to transitions of, for example, C—H, N—H, O—H, and C=O vibrational modes; and the wavelengths at which these absorptions occur are practically independent of the sort of molecule to which the C—H, etc., group is attached.

Bonds between different atoms absorb strongly because of the dipole associated with the bond. Thus, C—H, N—H, O—H, and C═O all absorb strongly. On the other hand, absorption due to the C—C bond is either very weak or non-existent and the absorption due to the C═C bond is only really strong in very asymmetrical molecules.

By detecting the characteristic, functional-group absorptions, observation of the infrared spectrum of a compound can, in the space of a few minutes, do the same job as a whole range of chemical reagents used for determining the presence of functional groups.

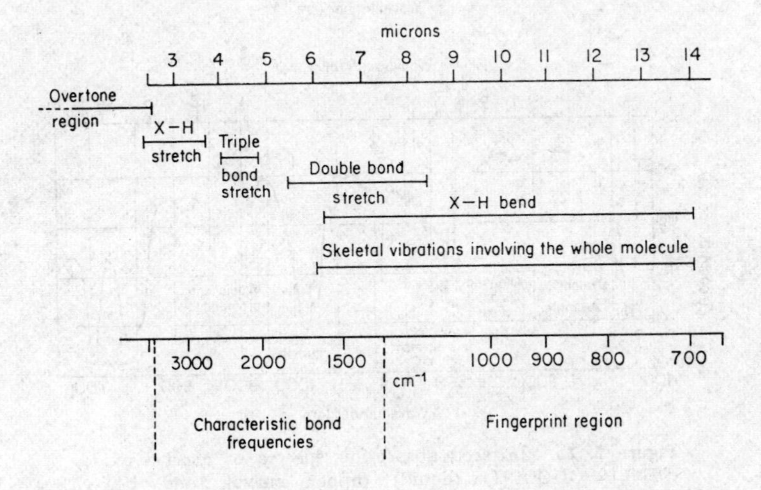

Figure 25.7a. Main regions of the infrared spectrum.

Routine examination of organic compounds is usually made with an infrared spectrometer which scans from 4,000 to 670 cm^{-1}. This range may be divided roughly into three regions as shown in Figure 25.7a, the most useful region for structural diagnosis being from 3,500 to 1,400 cm^{-1}.

The region from 1,400 to 670 cm^{-1} is where many of the absorption bands are due to vibrations of the whole molecule, i.e. not of isolated bands. This region is called the 'finger print' region as the

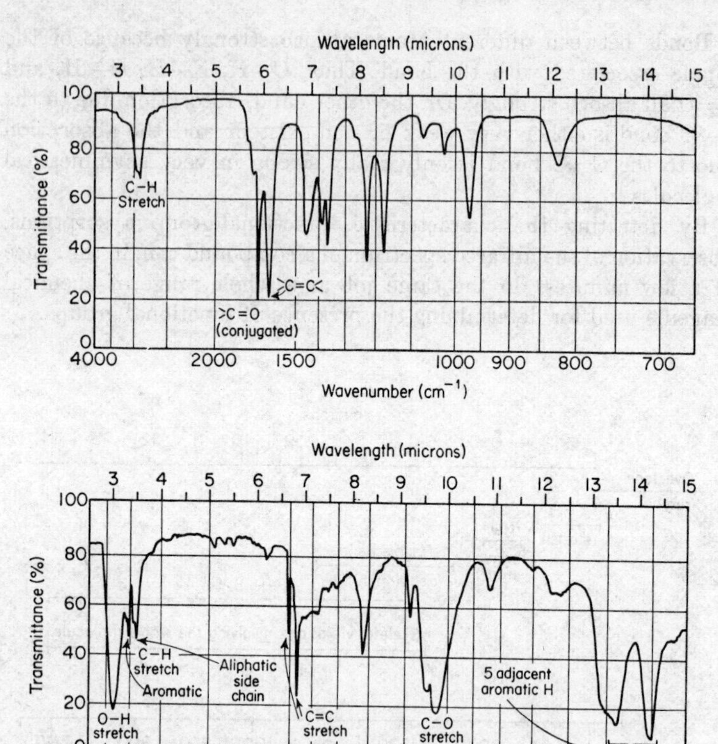

Figure 25.7b. Infrared absorption spectra of mesityl oxide, $(CH_3)_2C{=}CHCOCH_3$ (liquid) (upper curve), and benzyl alcohol, $C_6H_5CH_2OH$ (liquid) (lower curve).

observed pattern of absorptions is characteristic of individual molecules.

Correlation charts for the characteristic group frequencies are numerous, and more detailed charts may be found in many monographs on infrared spectroscopy. Table 25.6 gives some idea of the functional groups that can be detected from an infrared spectrum.

Figure 25.7b shows two infrared spectra and the assignment of characteristic group frequencies.

Table 25.6 Some characteristic infrared absorptions.

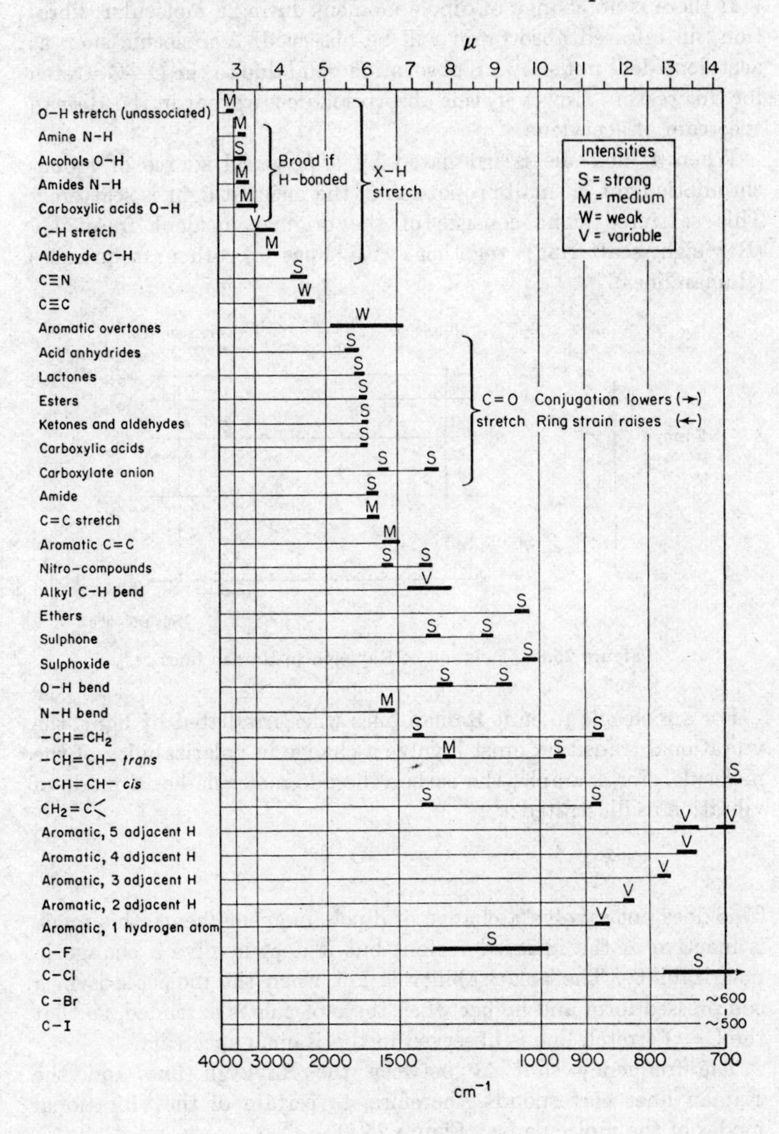

Raman Spectra

If there is no change of dipole moment during a molecular vibration, no infrared absorption will be observed. A molecule such as acetylene does not show an absorption band due to the C≡C stretch for this reason. However, this absorption does appear in the Raman spectrum of acetylene.

When a molecule is irradiated by a powerful source of monochromatic light, a small proportion of the incident light is scattered. This scattered light consists of the original incident frequency (Rayleigh scattering) together with lines of other frequencies (Raman lines).

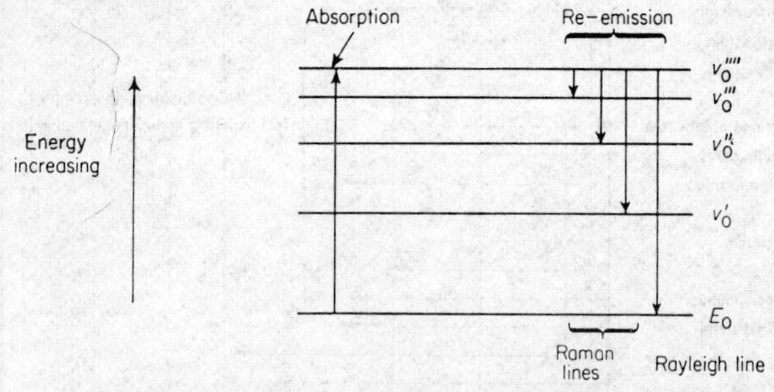

Figure 25.8. Relation of Rayleigh to Raman lines.

For a molecule to emit Raman lines when irradiated by light, the vibrational transition must involve a change in polarizability of the molecule. For example, the carbon dioxide molecule has a mode of vibration as illustrated:

$$\overset{\delta+}{\mathrm{O}}=\overset{\delta-}{\mathrm{C}}=\overset{\delta+}{\mathrm{O}}$$
$$\leftarrow \qquad \rightarrow$$

This does not involve a change of dipole moment (hence this mode is inactive in the infrared region) but it does involve a change in polarizability. The polarizability is low when the molecule is in a compressed form and higher when the molecule is extended, so that the C=O stretch line is observed in the Raman spectrum.

The frequency shift $\Delta\nu$ between the Rayleigh line and the Raman lines corresponds, therefore, to certain of the vibrational modes of the molecule (see Figure 25.8).

Energy is absorbed on irradiation of a molecule and, if the vibrational mode involves a change of polarizability, the molecule is excited to a higher vibrational energy level, from E_0 to v_0'''' in Figure 25.8. This absorbed energy can be re-emitted by transition to (*a*) the original level E_0, or (*b*) to vibrationally excited states v_0''', v_0'', and v_0'. The latter, re-emitted, radiation has slightly lower frequencies than the Rayleigh emission and the difference in frequency $\Delta\nu$ between the Rayleigh line and the Raman lines corresponds in energy to infrared frequencies.

Raman spectra, being weak emission spectra, are not easy to record since little of the irradiating light is scattered. However, comparison of infrared and Raman spectra can give information concerning the symmetry of a molecule. Evidence that the ion responsible for aromatic electrophilic nitration, the NO_2^+ ion really existed in nitrating media was obtained from Raman spectra. Spectra of nitronium salts, e.g. the perchlorate $NO_2^+ ClO_4^-$, showed a line of $\Delta\nu = 1{,}400\ cm^{-1}$ which is also observed in mixtures of concentrated nitric and sulphuric acid. The ion responsible for this line had to be linear and triatomic (like O═C═O), and the nitronium ion fitted these requirements ($O{=}\overset{+}{N}{=}O$).

Nuclear Magnetic Resonance

So far, we have not discussed the nature of the atomic nucleus and, in general, chemistry is only concerned with the extranuclear electrons. The structure of atomic nuclei is principally the concern of the nuclear physicist. However, some nuclei have a magnetic moment in consequence of their spin. Nuclear spin is governed, like electron spin, by a spin quantum number I. Nuclei which occur frequently in organic compounds are ^{12}C and ^{16}O for which $I = 0$, and ^{1}H, ^{13}C, and ^{19}F for which $I = \frac{1}{2}$.

When such a spinning nucleus is placed in a magnetic field, with the nucleus oriented at an angle θ to the magnetic field, the field will exert a force on the nucleus tending to align the spinning nucleus with the field; but, as the nucleus is spinning, the net effect will be to cause the nucleus to precess around the axis of the field with an angular velocity ω_0. This precession is analogous to the precession of a spinning gyroscope when it is allowed to topple in the earth's gravitational field.

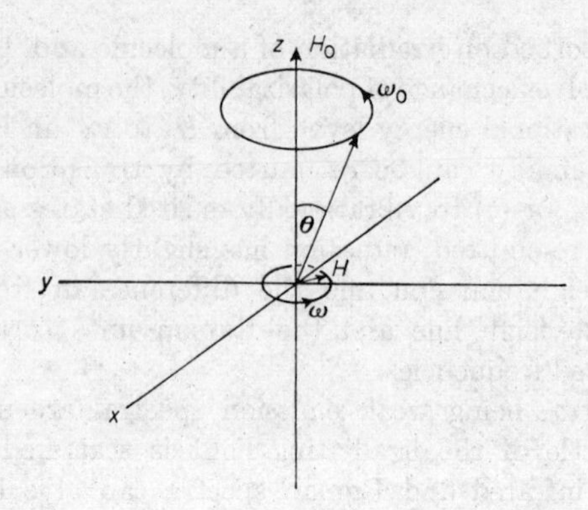

Figure 25.9. Effect of a magnetic field H_0 on a spinning nucleus.

The angular velocity ω_0 is proportional to the intensity of the applied field H_0; the constant of proportionality is known as the gyromagnetic ratio γ, whence $\omega_0 = \gamma H_0$.

If a small oscillating field H is applied perpendicular to H_0, this field H will tend to tilt the spinning nucleus towards the xy plane. It will be effective only when the oscillating field H is rotating about the axis of H_0 with the same angular velocity as the precession velocity ω_0. Under these circumstances there will be an exchange of energy between the rotating field H and the precessing nucleus. This represents a familiar resonance phenomenon, hence the name nuclear magnetic resonance (NMR).

The way in which a nucleus aligns itself with the applied magnetic field is governed by its spin. The nucleus can assume any one of $2I + 1$ orientations. For the proton, $I = \frac{1}{2}$, and there are therefore only $2(\frac{1}{2}) + 1$ = two possible orientations of the proton, correspond-

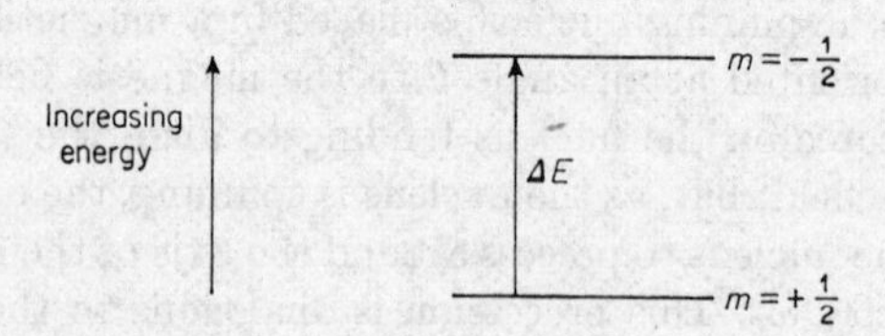

Figure 25.10. Energy levels of a proton in a magnetic field.

ing to magnetic quantum numbers of $m = +\frac{1}{2}$ and $m = -\frac{1}{2}$. One of these two orientations can be considered to be parallel to the applied field ($m = +\frac{1}{2}$) and the other antiparallel ($m = -\frac{1}{2}$). These two orientations have slightly different energies (see Figure 25.10). Excitation of the nucleus results in the transition of the nucleus to the energetically less favourable orientation. The difference in energy ΔE between the two states is proportional to the strength of the field H at the nucleus:

$$\Delta E = \gamma h H / 2\pi \qquad (1)$$

As with infrared and ultraviolet spectra, ΔE is associated with a corresponding frequency of electromagnetic radiation:

$$\Delta E = h\nu \qquad (2)$$

From equations (1) and (2) we obtain:

$$\nu = \gamma H / 2\pi \qquad (3)$$

For a hydrogen nucleus in a magnetic field of 14,000 gauss, the energy ΔE corresponds to frequencies in the short-wave radio region (60 Mc).

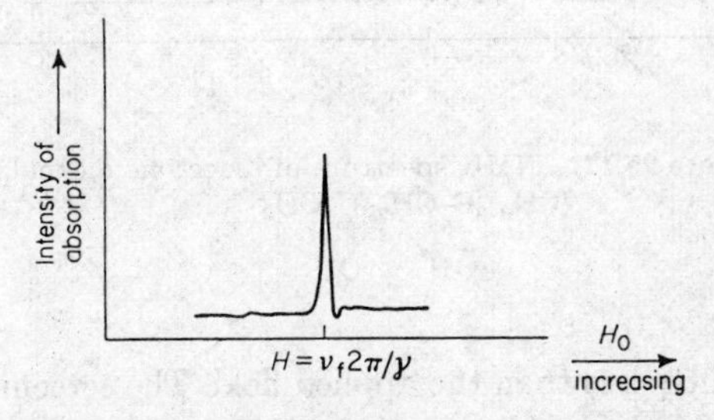

Figure 25.11. Nuclear magnetic resonance spectrum of a single nucleus.

Experimentally it is easier to vary the applied magnetic field than the additional oscillating field, so that a molecule is irradiated with a fixed frequency ν_f, and the applied magnetic field H_0 is varied until $\nu_f = \gamma H_0/2\pi$. At this point the sample absorbs energy. Any further change in H_0 will result in a decrease in absorption. An absorption spectrum can be presented as a plot of absorption against H_0 (Figure 25.11).

Absorption of energy will be observed if there is an excess of nuclei in the orientation of lower energy. The distribution of nuclei between the two levels is governed by the normal (Maxwell–Boltzmann) law: $n_1/n_2 = \exp(-\Delta E/\boldsymbol{R}T)$. The value of H_0 is made as high as possible in order to give a significant difference in population between the higher and the lower energy level.

So far we have neglected the extranuclear electrons. The electrons that surround a hydrogen atom and bind it to the rest of the molecule shield the nucleus, so that the effective field experienced by the

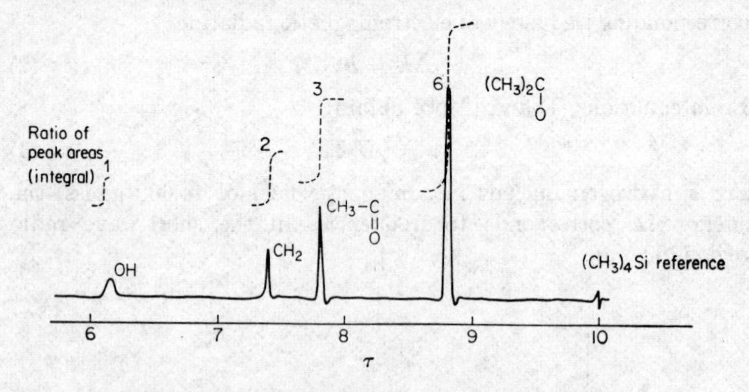

Figure 25.12. NMR spectrum of diacetone alcohol, $(CH_3)_2C(OH)–CH_2–C(=O)–CH_3$

nucleus is slightly less than the applied field. The screening effect of the electrons arises from interaction between the magnetic field of the electrons and the applied magnetic field.

The field experienced by a particular proton will therefore depend upon the electron density around the proton, and this density will depend on the atom to which the proton is attached. The resonance signals produced for differently bound hydrogen atoms occur at different field strengths. These differences in field strengths at which signals are obtained for protons in different molecular environments are called chemical shifts. As an example, let us look at the nuclear magnetic resonance spectrum of diacetone alcohol (Figure 25.12). The proton of the hydroxyl group is the least shielded of the four types of protons in the molecule (the electronegative oxygen atom withdraws electrons from the neighbourhood of the hydrogen atom of the hydroxyl group). Next appears the CH_2 group, not so strongly deshielded as the OH proton, but nevertheless, being flanked by an electron-withdrawing C═O group on one side and the C—OH group on the other, more deshielded than the methyl group attached to the C═O group. The absorption remaining is due to the protons of the pair of methyl groups at the end of the chain.

Chemical shifts are measured relative to a standard, usually the absorption of the protons in tetramethylsilane. These protons absorb at a higher frequency than the protons of the majority of organic compounds. The chemical shift δ is expressed:

$$\delta = \frac{(\nu_{\text{sample}} - \nu_{\text{reference}})}{\nu_{\text{oscillator}}} \times 10^6 \text{ cycles sec}^{-1}$$

To avoid negative values of δ, positions of NMR absorptions can also be recorded as τ values where:

$$\tau = 10 - \delta$$

In other words, τ is the chemical shift relative to a reference defined by taking $\tau = 10$ for tetramethylsilane.

The intensity of each resonance line is proportional to the number of corresponding protons in the sample, and it is usually recorded on the spectrum as a line representing the area under each peak.

As for ultraviolet and infrared spectra, a list of values (Table 25.7) can be assigned to characteristic structures—in this case, different structural environments for a proton. From the position of an absorption line in an NMR spectrum, a guess can be made as to whether the proton is in a strongly electron-deficient environment (e.g. the acidic hydrogen of a carboxylic acid) or not (e.g. the methyl group in ethanol). Aromatic protons appear at low τ values owing to the ring current which results from the action of the applied magnetic field on the closed loops of π-electrons above and below the plane of the aromatic ring (cf. Figure 25.13). This interaction produces a magnetic field which adds to the applied magnetic field experienced by the protons attached to the aromatic ring.

The values given in Table 25.7 are only approximate since individual structures affect the environment of a particular proton in different ways, but the Table can be used in conjunction with the relative peak areas to detect the presence of characteristic groups; for example, O—CH_3 protons usually appear between $\tau = 6$ and 6.4, whereas the protons of the CH_3 group in saturated chains appear around $\tau = 9$ to 9.2.

Spin–spin coupling

Nuclear magnetic resonance spectra of organic molecules are not usually as simple as illustrated in Figure 25.12. For example, the spectra of acetaldehyde and of ethyl iodide are shown in Figure 25.14.

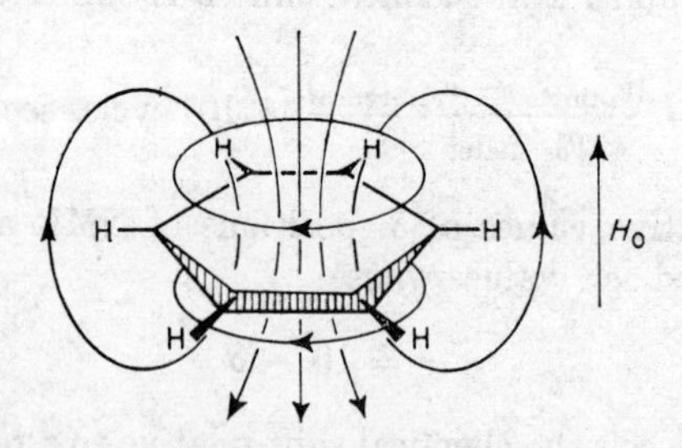

Figure 25.13. Deshielding of aromatic protons by the ring current effect.

From the integrals and the Table of chemical shifts, the absorptions can be assigned, but the question remains, why should the lines be split? To answer this question we must return to the original reason for the occurrence of NMR lines, namely, the fact that protons possess a magnetic moment due to nuclear spin (I for protons $= \frac{1}{2}$). This nuclear magnetic moment exerts an influence over its immediate surroundings, and thus the three protons of the CH_3 group in acetaldehyde are under the influence of a small magnetic field due to the spinning nucleus of the CH=O proton. When the acetaldehyde molecule is placed in an external magnetic field, the nucleus of the CH=O hydrogen may adopt one of $(2I + 1)$, i.e. two possible alignments, one with a spin quantum number

Table 25.7. Characteristic chemical shifts of protons in different environments.

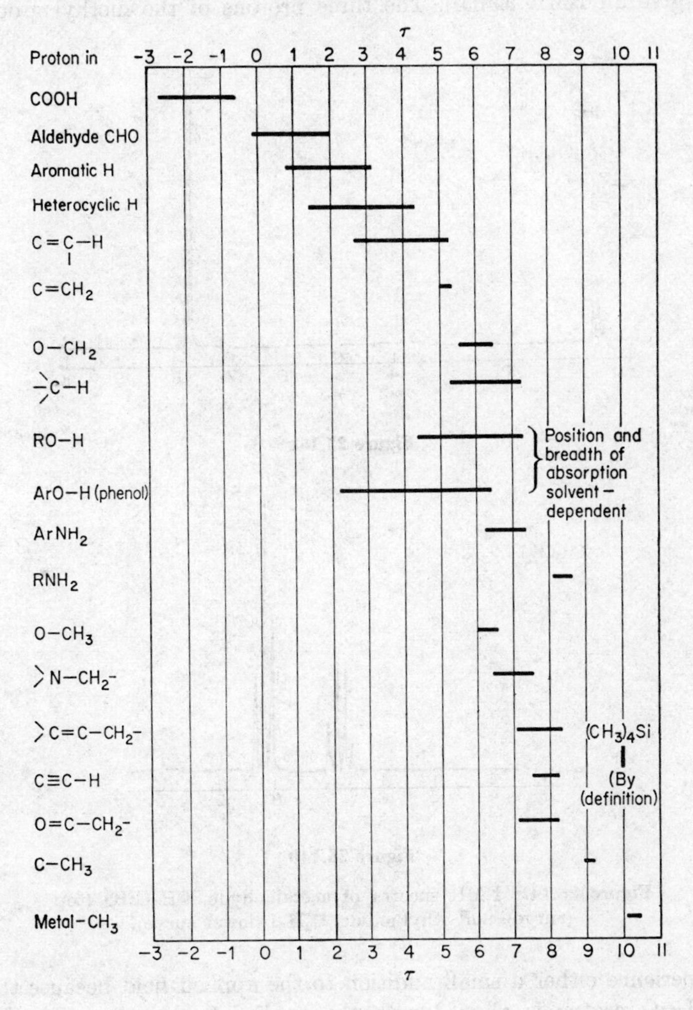

$m = +\frac{1}{2}$ and the other of slightly higher energy with $m = -\frac{1}{2}$. At ordinary temperatures, the population of these two states is nearly the same (not quite, otherwise there would be no energy absorption to give an NMR signal). The three protons of the methyl group

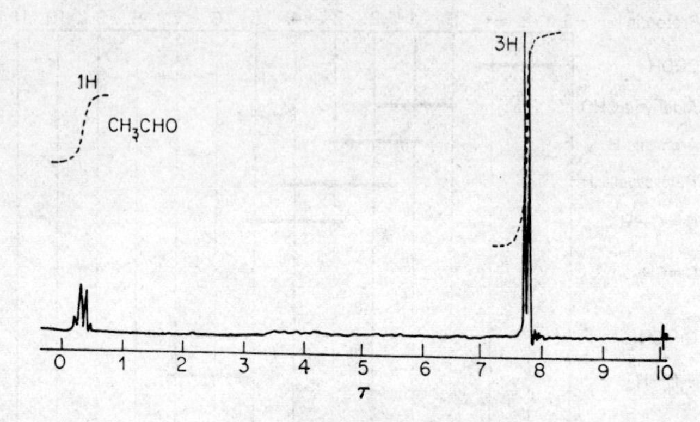

Figure 25.14*a*

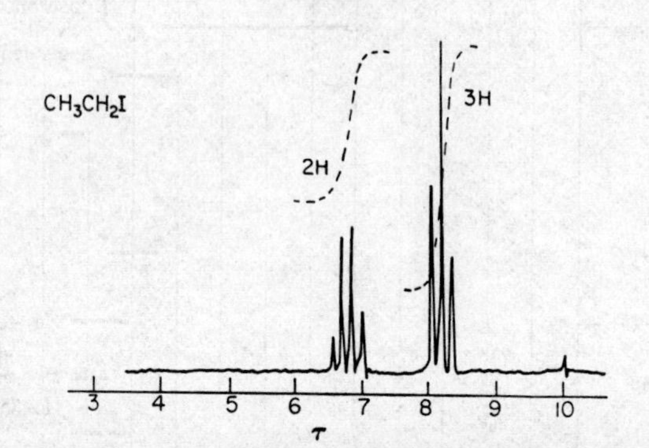

Figure 25.14*b*

Figure 25.14. NMR spectra of acetaldehyde, CH_3CHO (top (curve), and ethyl iodide, C_2H_5I (lower curve).

experience either a small addition to the applied field because the CH═O proton is aligned with the applied field—responsible for the line at $\tau = 7.83$—or a slight diminution of the applied field because the field of the CH═O proton opposes the applied field. The absorption of energy in the latter case will occur at slightly higher field ($\tau = 7.86$) (Figure 25.15*a*).

The splitting can be pictorially represented as in Figure 25.15. The splitting of the CH=O proton by the methyl protons can be accounted for by a similar argument (see Figure 25.15*b*). There is no splitting between the three protons of the methyl group, since all protons are in equivalent environments (remember that at ordinary temperature there is complete freedom of rotation about a carbon–carbon double bond).

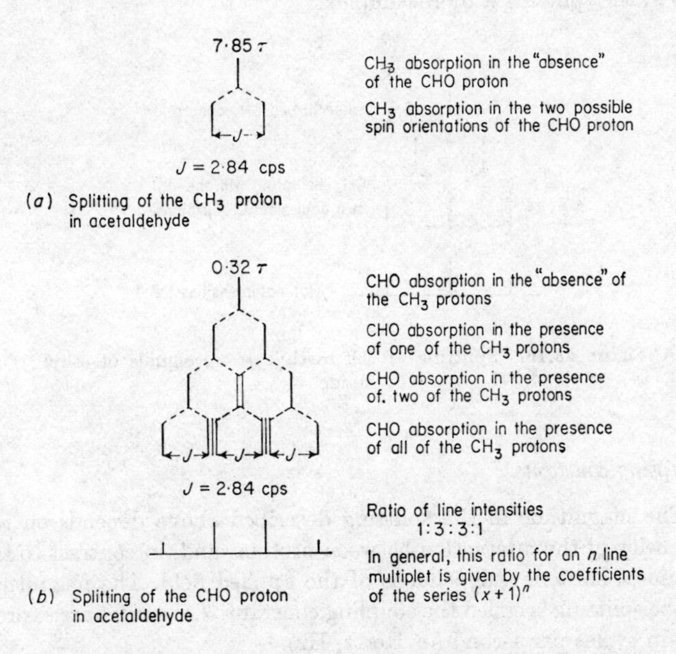

Figure 25.15. Diagram illustrating splitting of proton signals.

Splitting is most frequently encountered between protons on adjacent atoms, and there is virtually no interaction between hydrogens that are located on carbon atoms further apart unless there is a multiple bond between two of these carbon atoms.

With ethyl iodide the spectrum is one step more complex in that the CH_3 is adjacent to two protons, each of which may be aligned either with or against the applied field. The interaction may be represented as in Figure 25.16. The protons of the CH_2 group interact with the three protons of the CH_3 group and appear as a quartet (relative intensities 1 : 3 : 3: 1), like the CHO proton of acetaldehyde, with an integral equivalent to two protons.

In general, a proton in the field of n other equivalent protons will be split into $(2nI + 1)$ lines. For the proton $I = \frac{1}{2}$, so that this expression simplifies to $(n + 1)$; but for other nuclei, e.g. ^{14}N, where $I = 1$, the splitting is more complex.

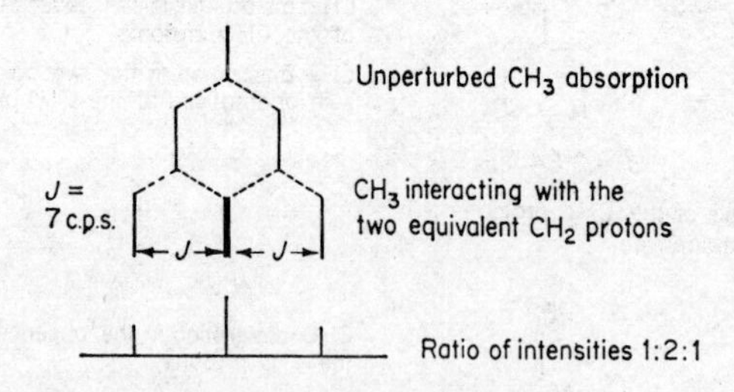

Figure 25.16. Splitting of the methyl group signals of ethyl iodide.

Coupling constants

The magnitude of the splitting described above depends on the intensity of the interaction between protons, and, in contrast to the chemical shift, is independent of the applied field. The magnitude of the splitting is called the coupling constant, J. J values are expressed in cycles per second (or Hertz, Hz).

In the spectrum of acetaldehyde, the splitting of the aldehyde proton by the methyl protons, i.e. the coupling constant $J(H, CH_3)$ ($\sim$3 cps) is the same as the coupling constant $J(CH_3, H)$ for the methyl absorption ($\sim$3 cps) since the interaction of one type of proton with another is transmitted equally either way.

The coupling of two protons in a similar environment is more complex and will not be dealt with here.

Chemical exchange

Labile protons, such as the hydroxyl proton of alcohols and phenols, which tend to exchange with neighbouring molecules in the presence of catalysts, e.g.:

$$C_2H_5O{-}H\,(H^+) \rightleftarrows C_2H_5\overset{+}{O}H_2 + H^+$$

have chemical shift values that vary according to the purity of the sample and are concentration-dependent. There will be no spin–spin splitting between such protons and neighbouring protons, provided that the rate of chemical exchange is faster than the coupling constant between these two types of protons.

Electron Spin Resonance Spectroscopy

Free radicals possess an odd electron and we know an electron has spin (Chapter 2). So, like the nucleus of a hydrogen atom, when a free radical is placed in a magnetic field, there are two possible orientations for this odd electron, with spin quantum numbers

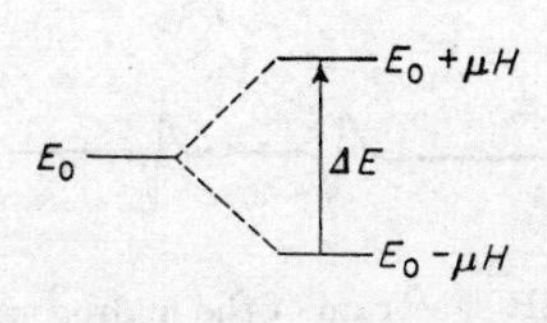

Figure 25.17. Effect of an applied magnetic field on the energy levels of an electron.

of $m = +\frac{1}{2}$ and $m = -\frac{1}{2}$, respectively. And a transition from the lower to the upper energy state may be effected by the absorption of electromagnetic radiation of suitable frequency (Figure 25.17) (where H_0 = the applied field, μ = the magnetic moment of an electron, and E_0 = the energy of the electron in the absence of the magnetic field). Absorption of energy will occur when

$$h\nu = 2\mu H_0 = \Delta E$$

The technique of measurement of this resonance is similar to that of measuring nuclear magnetic resonance. The sensitivity of this method of detecting radicals depends on the strength of the applied field, which is therefore made as large as possible. Microwave energy of the order of 1–3 cm wavelength is used, and the spectrum is scanned, as with NMR spectra, by varying the applied field. The method is extremely sensitive and concentrations of radicals as small as 10^{-10} mole $l.^{-1}$ can be detected.

In a radical, the odd electron is often located on an atom that has a nucleus with a magnetic moment, for example, hydrogen, where $I = \frac{1}{2}$. This nuclear magnetic moment will interact with the moment of the odd electron to give a splitting pattern (called hyperfine splitting), in the same way as we have described for NMR spin–spin coupling.

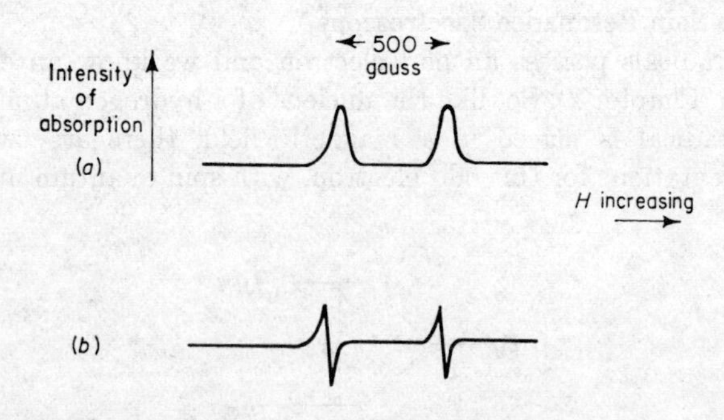

Figure 25.18. ESR spectrum of the hydrogen atom: (*a*) absorption spectrum; (*b*) derivative of the absorption spectrum.

In the hydrogen atom, the odd electron interacts with the proton to give a doublet, as shown in Figure 25.18. Electron spin resonance (ESR, for short) absorption lines are not usually displayed as shown in spectrum (*a*) of Figure 25.18, but rather as the derivative (differential) of this signal (Figure 25.18*b*).

When there is more than one proton to interact with the odd electron, the spectrum is more complex. In the case of the methyl radical there are four lines (Figure 25.19); the spacing ΔH between the lines is much less than in the spectrum of the hydrogen atom and indicates that interaction of the odd electron with the hydrogen atoms is reduced by the presence of the intervening carbon atom.

Radicals with a large number of canonical forms give complex spectra owing to the numerous protons with which the odd electron can interact, and ESR techniques can be used to determine the structures of such radicals.

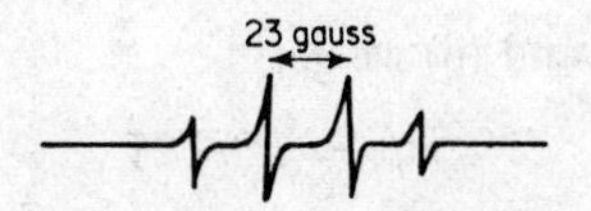

Figure 25.19. Derivative ESR spectrum of the methyl radical.

Mass Spectra

The principles underlying the production of a mass spectrum are unlike those underlying the spectra discussed so far, in that the former do not involve absorption of electromagnetic radiation. In a mass spectrometer, a compound is bombarded with electrons, which ionize and fragment the compound. The positively charged fragments are then separated and their relative abundances measured. Thus a mass spectrum of a compound is a chart showing the relative abundance of the various positive ions produced when the compound is bombarded with electrons, these ions being arranged in order of their masses.

The compound is introduced into an evacuated ionization chamber where it is bombarded with a stream of electrons:

$$e^- + XY \rightarrow XY^+ + 2e^-$$

$$XY^+ \rightarrow \underset{\text{Fragment ion}}{X^+} + Y$$

These positive ions are drawn out of the ionization chamber by an electric field V that accelerates them down the tube of the mass spectrometer. They thus acquire velocities v that depend on their masses m_1, m_2, m_3, etc.,

$$eV = \tfrac{1}{2}m_1{v_1}^2 = \tfrac{1}{2}m_2{v_2}^2 = \tfrac{1}{2}m_3{v_3}^2 = \text{etc.} \qquad \text{(i)}$$

These ions are then passed through a magnetic field where they are subject to a force eHv (H = the magnetic field strength, e the electronic charge, and v the velocity of a particular ion). The magnetic field deflects the positive ions along a circular path of radius r, and the deflecting force is balanced by the centrifugal force mv^2/r; i.e.:

$$eHv = mv^2/r \tag{ii}$$

By combining (i) and (ii) we get

$$m/e = H^2r^2/2V$$

Thus, for fixed values of the field strength H and of the accelerating potential V, ions having different mass/charge ratios (m/e) follow different paths; and, for a given value of H and V, only a selected

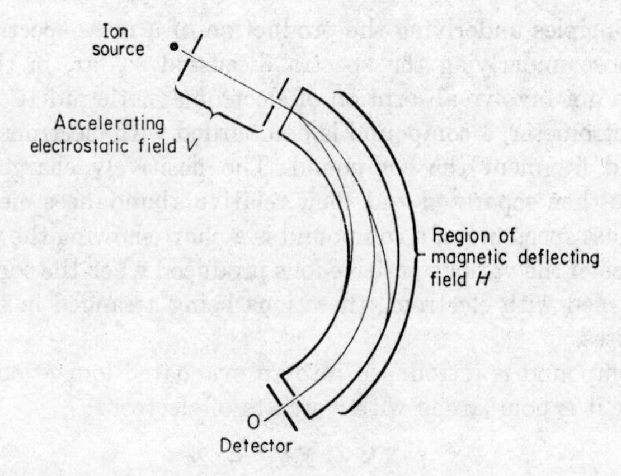

Figure 25.20. Simplified representation of a mass spectrometer.

beam of ions will pass through the detector slit (see Figure 25.20). The mass spectrum of a sample is scanned by varying either H or V.

In double-focusing instruments, the resolving power of the spectrometer is increased by passing the beam of positive ions through an electrostatic field to deflect the ions, as well as through a magnetic deflector.

The mass spectrum of an organic compound is presented in the form of a plot of relative ion abundance against the ratio, mass : charge. A typical spectrum (Figure 25.21) consists of a large number of sharp peaks, that vary in m/e ratio from the parent molecular ion formed by the initial ionization reaction:

$$XY + e^- \rightarrow XY^+ + 2e^-$$

through various ions formed by fragmentation of the parent ion:

$$XY^+ \rightarrow X^+ + Y$$

These fragment ions may break down further to give smaller fragments. In most cases the parent ion can be seen, although as with 2-methylheptane (Figure 25.21) the signal is often small. From the parent molecular ion peak, the molecular weight of the compound may be determined with an accuracy that enables a distinction to be made between formulae such as $C_{28}H_{50}$ (C, 87.0%; H, 13.0%), and $C_{29}H_{52}$ (C, 86.9%; H, 13.1%), a distinction that it would be difficult to make with certainty if combustion analysis were used to determine the percentage composition.

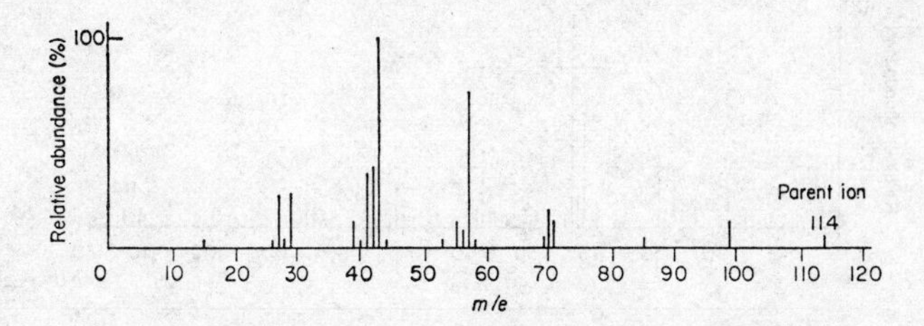

Figure 25.21. Mass spectrum of 2-methylheptane.

Ambiguities arising from different fragments of the same m/e ratio, for example, C_2H_5O and C_2H_7N (both give peaks at $m/e = 45$), can be resolved by considering the relative abundances of adjacent minor peaks due to the presence of isotopes of C, H, N, and O in the naturally occurring elements. Carbon contains about 1% of ^{13}C, nitrogen 0.4% of ^{15}N, oxygen 0.2% of ^{18}O, and hydrogen 0.02% of ^2H (D), so that most organic ion peaks show weak adjacent peaks, one or two units higher in mass, due to the presence of these isotopes. Thus a peak at $m/e = 14$ due to the $^{12}CH_2^+$ ion is accompanied by peaks at $m/e = 15$ ($^{13}CH_2^+$ and $^{12}CHD^+$) and 16 ($^{13}CHD^+$ and $^{12}CD_2^+$), as well as 17 ($^{13}CD_2^+$). The relative ratio of the first three peaks (100 : 1.11 : 0.004) can be calculated from the relative abundances of the isotopes. A peak that did not show adjacent peaks in these ratios would be due to some other molecular ion, or perhaps to a doubly charged ion of twice the mass. In our original example, the relative abundances for C_2H_5O at m/e = 45, 46, and 47 are 100 : 2.280 : 0.215, whereas for C_2H_7N they are 100 : 2.665 : 0.0228.

Molecular formulae can be determined with more reliability than described above by using a high-resolution instrument. The principal peak due to the $C_2H_5O^+$ ion actually has $m/e = 45.034$ (^1H = 1.00782; ^{12}C = 12.0000; ^{16}O = 15.9949), whereas the principal peak due to the $C_2H_7N^+$ ion actually has $m/e = 45.058$ (N = 14.0031). In a double-focusing instrument masses can be determined with an accuracy of 1 in 10,000, so that these two peaks can be distinguished.

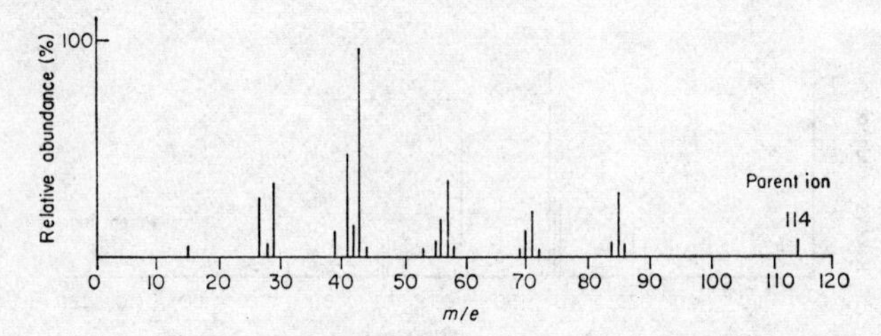

Figure 25.22. Mass spectrum of octane.

In addition to the molecular weight and molecular formula of an organic compound, mass spectra can give an indication of the structure of a molecule from the way in which it breaks down in forming the mass spectrum (the cracking pattern). Structure determination by interpretation of the cracking pattern depends largely on empirical correlations gained by studying the cracking pattern of similar groups of compounds. For example, branched-chain hydrocarbons usually break at a branch point. Thus 2-methylheptane (Figure 25.21) loses a methyl group to give an ion of $m/e = 99$. This ion is absent in the mass spectrum of octane (Figure 25.22), and the parent peak (114) is stronger in the spectrum of octane than it is in that of 2-methylheptane.

The mass spectrum of 2,2,3- and 2,2,4-trimethylpentane both show strong peaks at $m/e = 57$ due to the ion $(CH_3)_3C^+$ which arises from fragmentation at the most highly substituted carbon atom.

$$\begin{array}{c} \quad\quad\quad CH_3 \\ \quad\quad\quad | \\ (CH_3)_3C \vdots CH—CH_2CH_3 \end{array} \qquad (CH_3)_3C \vdots CH_2CH(CH_3)_2$$

2,2,3-Trimethylpentane 2,2,4-Trimethylpentane

From information of this sort, possible structures can be built up from the cracking pattern of a molecule, and there are several specialist reference books which list correlations between structure and fragmentation processes, e.g. J. H. Beynon, *Mass Spectrometry and its Applications to Organic Chemistry*, Elsevier, Amsterdam, 1963.

X-Ray Diffraction

To conclude this survey of physical methods of structure determination, we consider the most sophisticated technique, that of assigning a structure to a molecule from the diffraction pattern that is obtained when X-rays are passed through crystals of a compound.

The diffraction of X-rays by crystals depends on the fact that the atoms in crystals are arranged in a regular pattern. This regular pattern is repeated in three dimensions in the crystal, and the identical repeating unit is called the unit cell. In crystals of organic compounds, the unit cell usually contains more than one molecule arranged in a particular manner. The array of atoms acts as a three-dimensional grating for electromagnetic radiation of suitable wavelengths (0.5 to 2 Å). Each plane of atoms in a crystal partly reflects incident X-rays, so that rays reflected from different planes interact, either reinforcing or cancelling each other. The condition for reinforcement of the reflected radiation depends on the spacing (d) of the planes of the array of atoms, and on the angle (θ) and wavelength (λ) of the incident radiation (see Figure 25.23).

For the two rays 1 and 2 to reinforce one another on reflexion, the difference in path length ($xy + yz$) between rays 1 and 2 must be a whole number of wavelengths:

i.e.
$$n\lambda = xy + yz = 2xy$$
$$= 2d \sin \theta \qquad \text{(the Bragg equation)}$$

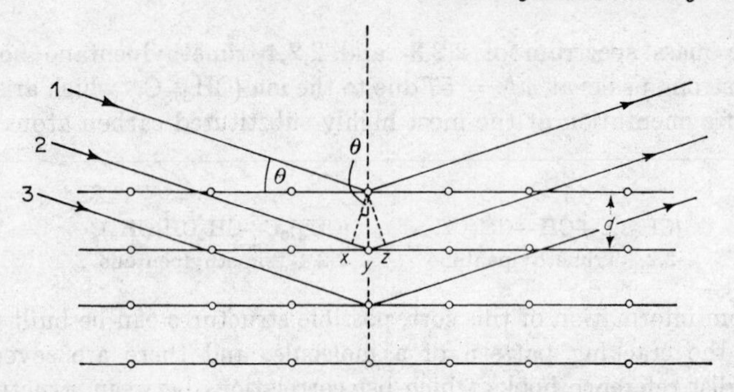

Figure 25.23. The interaction of X-rays with planes of atoms in a crystal.

In principle, it is possible to calculate d if a single crystal is placed in a beam of monochromatic X-rays and the resulting diffraction pattern and spot intensities are measured either with a photographic film or a scintillation counter. If the crystal is rotated, one should, in theory, be able to determine the direction and spacing of all the atomic planes in the crystal. During rotation (Figure 25.24) different sets of atom planes are momentarily at the critical angle θ of the Bragg equation, and reflect X-rays, thus causing a spot on the film or a response on the counter. From the pattern and intensities of the spots the symmetry of the crystal may be determined. From

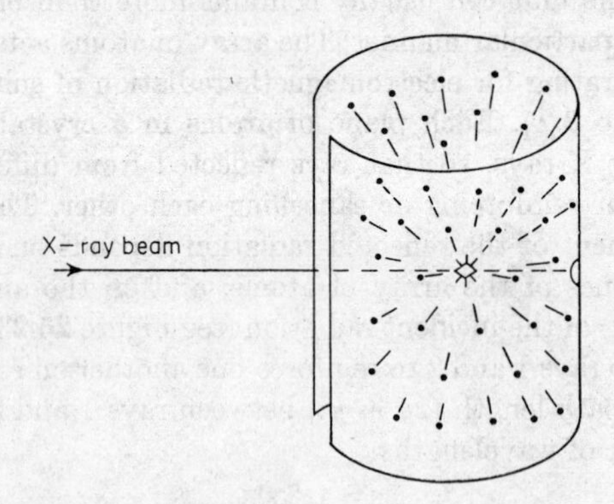

Figure 25.24. The formation of an X-ray diffraction pattern on a photographic film.

packing and symmetry considerations it is sometimes possible to determine the positions of molecules in the unit cell and sometimes also the symmetry of molecules.

The determination of the shape of the unit cell of a number of steroids by Bernal in 1932 indicated that these molecules possessed a thin lath-like structure (Figure 25.25). This knowledge was sufficient to resolve the controversy existing at that time as to the structure of these compounds.

The *X*-ray diffraction pattern can be used to give an exact location of each atom in a molecule, but examination of the pattern does not lead directly to the complete structure. The situation is similar to that encountered in wave mechanics, where the appropriate Schrödinger equation can be written down for any molecule, but it

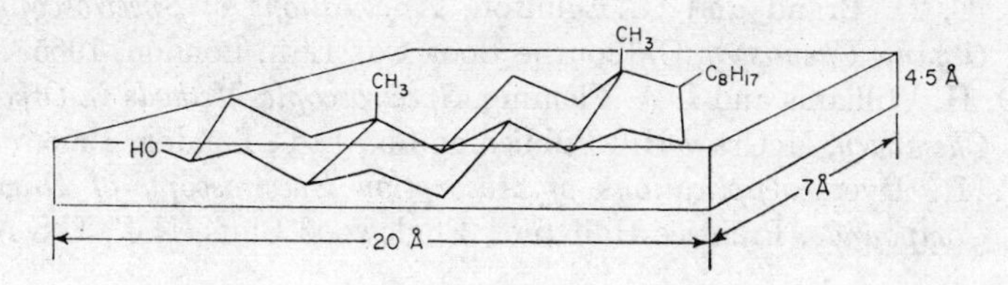

Figure 25.25. The dimensions of the cholestanol molecule from *X*-ray data.

is impossible to solve the equation completely unless approximations and intelligent guesses at the solutions are made.

In some cases the determination of molecular structure from *X*-ray data is assisted by ideas about possible structures derived from chemical degradations and from other physical methods described earlier in this Chapter. With other, more complicated, molecules, special techniques such as the 'heavy-atom method' and the isomorphous-replacement method are used. A full account of these techniques can be found in '*X-Ray Analysis of Organic Structures*', Academic Press, London 1961, by S. C. Nyburg.

Conclusions

Physical techniques have rendered inefficient the time-consuming structure determination by chemical degradation alone. At the present time a combination of various physical methods with a study of the chemical reactivity of a compound is the usual method for the rapid determination of structure.

Other physical methods, such as measurements of optical activity, dipole moment, or acid and base dissociation constant, microwave measurements of pure rotation spectra, or electron diffraction of gases, are also used for structure determination, but these are not of such general application and are therefore not discussed here. Even for the methods that have been mentioned in this Chapter, the student should refer for fuller information to some of the many books written on various physical methods of structure determination, such as:

(a) *Books on spectroscopic methods*

J. C. D. Brand and G. Eglinton, *Applications of Spectroscopy to Organic Chemistry*, Oldbourne Book Co., Ltd., London, 1965.

D. H. Williams and I. A. Fleming, *Spectroscopic Methods in Organic Chemistry*, McGraw-Hill Publishing Co., Ltd., London, 1966.

J. R. Dyer, *Applications of Absorption Spectroscopy of Organic Compounds*, Prentice-Hall Inc., Englewood Cliffs, N.J., U.S.A.

(b) *Books on physical methods in general*

M. St.C. Flett, *Physical Aids to the Organic Chemist*, Elsevier Publishing Co., Amsterdam-NewYork, 1962.

J. C. P. Schwarz (ed.), *Physical Methods in Organic Chemistry*, Oliver and Boyd, Edinburgh and London, 1964.

Students must have practice in examining spectra and deducing as much information as possible from them. Or, better, examination of chemical reactivity should accompany spectral investigation, a sample of substance as well as the relevant spectra being provided so that the student may confirm by chemical reactions the structural assignments made from the spectra.

Problem

The spectra of a compound containing carbon, hydrogen, and oxygen are shown below. From these spectra, suggest a structure for the compound.

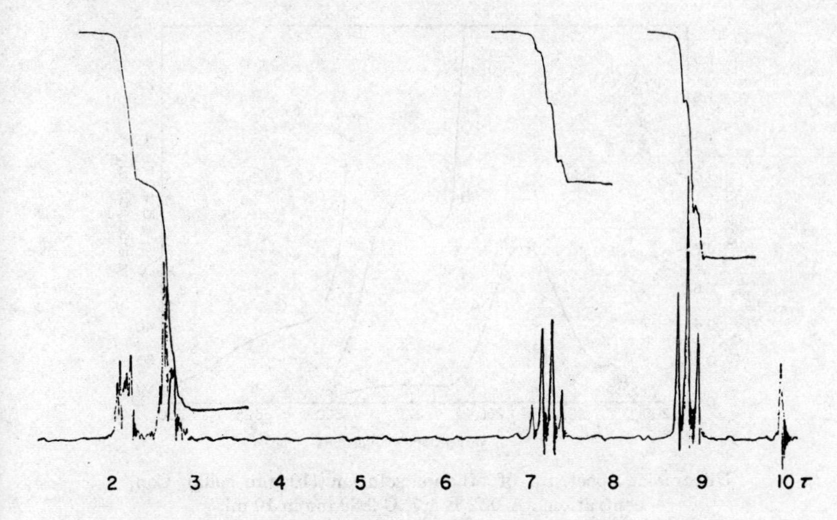

NMR spectrum of the liquid; no solvent. 60 MHz; tetramethylsilane as internal standard

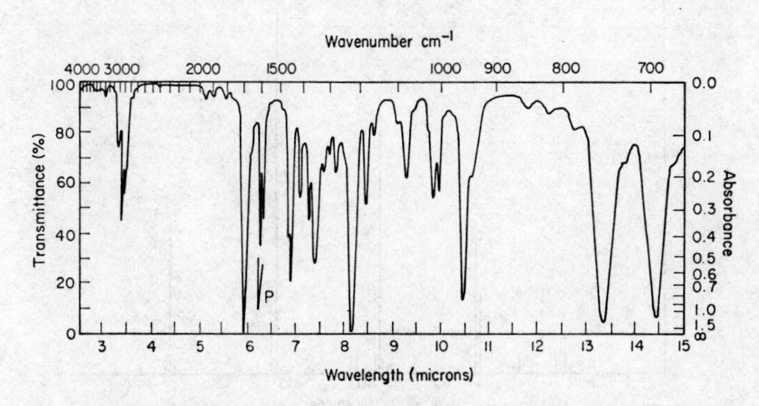

Infrared spectrum of a liquid film. P = Polystyrene reference $1603\ cm^{-1}$.

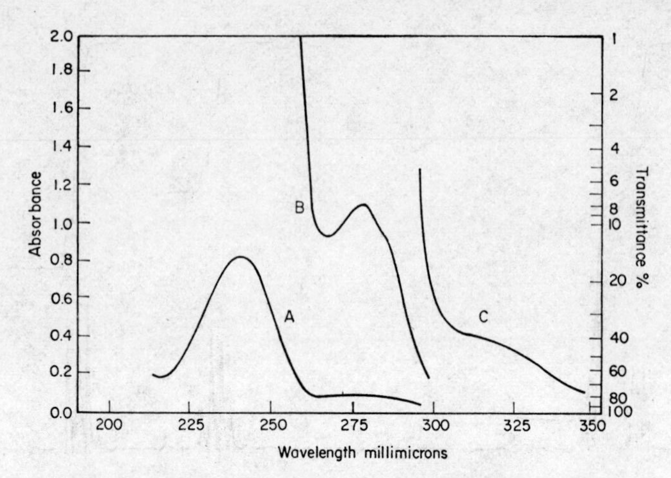

Ultraviolet spectrum of ethanol solution (10 mm cells). Concentrations: A 0.2, B 1.7, C 9.89 mg in 10 ml.

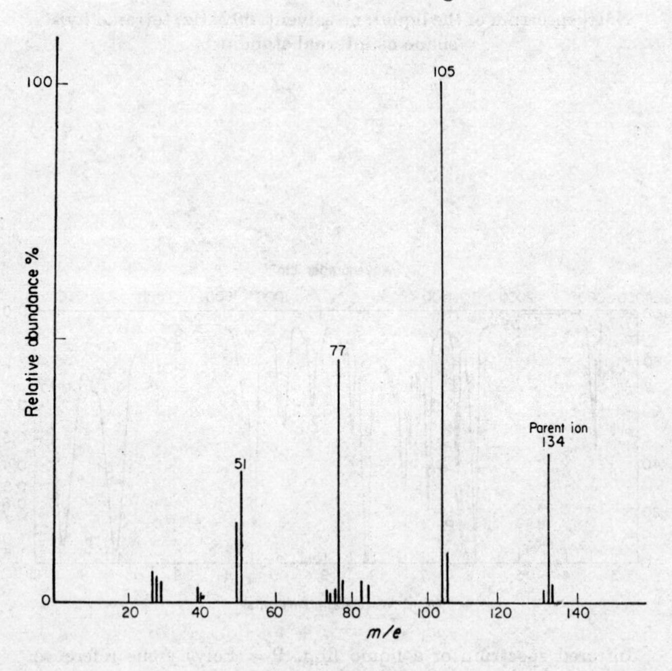

Mass spectrum.

CHAPTER 26

The Determination of Structure by Chemical Methods

Apart from *X*-ray structure analysis, it is usually impossible to determine the complete structure of a complicated organic compound solely by the physical methods described in the previous chapter. Detailed spectroscopic examination usually gives clues as to the type of structure and to the nature of the functional groups present. These clues are followed up by chemical reactions designed to break the molecule into successively smaller fragments until pieces having recognizable structure are obtained. Some of the methods most frequently used are described on the following pages. Most of them have been encountered in previous Chapters, so details of mechanism are here omitted.

Hydrolysis of Esters and Amides

Esters and amides are hydrolysed by treatment with either sodium hydroxide or mineral acid. The glycerol esters of high molecular weight present in fats and oils can be identified by such a hydrolysis, which gives glycerol and salts of the fatty acids. The fatty acids corresponding to these salts can be identified by re-esterification to the methyl esters which can be partly separated by careful fractional distillation and then identified by gas or thin-layer chromatography (cf. p. 392).

Proteins are complex polyamides of α-amino-acids. The structures of such molecules have been determined by acid hydrolysis to a mixture of α-amino-acids which can be separated by physical techniques such as paper and ion-exchange chromatography and by electrophoresis. The protein may be only partly hydrolysed by suitable choice of hydrolysis conditions, some of the amide links in

$$R'CO_2CH_2CH(O_2CR'')CH_2O_2CR''' \xrightarrow{NaOH} HOCH_2CH(OH)CH_2OH$$

Glycerol ester — Glycerol

$$+ \underbrace{R'CO_2^-Na^+ + R''CO_2^-Na^+ + R'''CO_2^-Na^+}_{\text{Fatty acid salts}}$$

$$\downarrow \text{Acidify and esterify}$$

$$R'CO_2CH_3 + R''CO_2CH_3 + R'''CO_2CH_3$$

Methyl esters of fatty acids

the protein being more susceptible to hydrolysis than others. The amino-acid at one end of the chain can be identified by treating the protein with 1-fluoro-2,4-dinitrobenzene, which serves to mark one end of the polypeptide chain (see Chapter 22).

Esters may be quantitatively hydrolysed with sodium hydroxide solution, one of the constants used as a criterion of purity of the natural fats and oils (which are mixtures of glycerol esters) being the 'saponification value'; this is defined as the number of milligrams of potassium hydroxide required to hydrolyse one gram of fat or oil.

Quantitative hydrolysis is used to estimate the number of hydroxyl groups present in a compound. The completely acetylated derivative of the hydroxy-compound is first prepared pure, and then subjected to acid hydrolysis by a non-volatile strong acid (benzenesulphonic acid). The acetic acid produced is steam-distilled from the hydrolysate and estimated. From the number of moles of acetic acid produced per mole of acetyl compound, the number of hydroxyl groups may be calculated.

Oxidation

Oxidation is the most widely used method for breaking a molecule into simpler fragments. There is a variety of oxidizing agents available for different kinds of structural demolition and the use of some of these is illustrated below.

Inorganic oxidizing acids such as nitric acid have been used extensively for bringing about general degradation of a molecule. However, its effects can be controlled by suitable choice of conditions, even with sensitive molecules; for example, it can be used to oxidize glucose to the corresponding dibasic acid, glucaric acid, without loss of carbon (see oxidation of aldehydes and alcohols, Chapter 18 of Part 1).

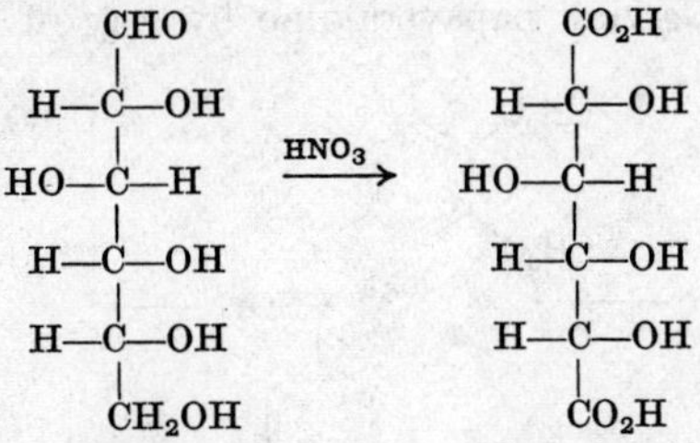

$$
\begin{array}{c}
CHO \\
H-C-OH \\
HO-C-H \\
H-C-OH \\
H-C-OH \\
CH_2OH
\end{array}
\xrightarrow{HNO_3}
\begin{array}{c}
CO_2H \\
H-C-OH \\
HO-C-H \\
H-C-OH \\
H-C-OH \\
CO_2H
\end{array}
$$

Glucose (aldehyde form) — Glucaric acid

The potential aldehyde group of an aldose is oxidized to carboxylic acid by bromine water, giving an aldonic acid, the rest of the molecule being unaffected.

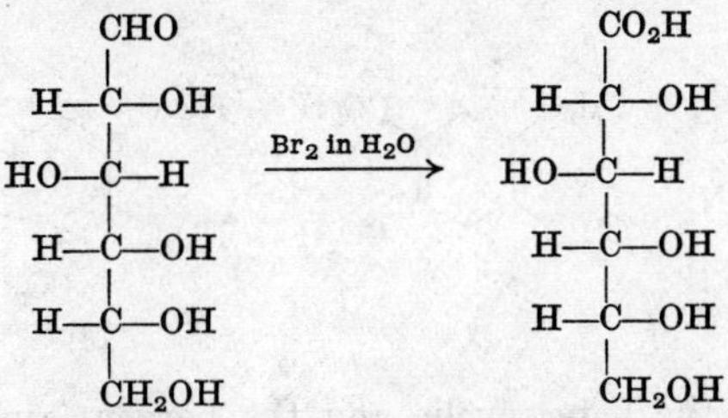

$$
\begin{array}{c}
CHO \\
H-C-OH \\
HO-C-H \\
H-C-OH \\
H-C-OH \\
CH_2OH
\end{array}
\xrightarrow{Br_2 \text{ in } H_2O}
\begin{array}{c}
CO_2H \\
H-C-OH \\
HO-C-H \\
H-C-OH \\
H-C-OH \\
CH_2OH
\end{array}
$$

Glucose (aldehyde form) — Gluconic acid

Chromium trioxide in sulphuric or acetic acid (sometimes called chromic acid), and alkaline potassium permanganate, are vigorous oxidizing agents. These two reagents oxidize aliphatic side chains but usually leave stable aromatic rings such as benzene and naphthalene intact. Nicotine, a tobacco alkaloid, gives pyridine-3-

N, N, CH_3 — Nicotine

CO_2H, N — Pyridine-3-carboxylic acid

carboxylic acid (nicotinic acid) when oxidized with chromic acid.

Papaverine, an alkaloid found in opium, contains a reactive CH_2 group, and this group is oxidized by alkaline potassium permanganate solution, giving papaveraldine by way of an intermediate

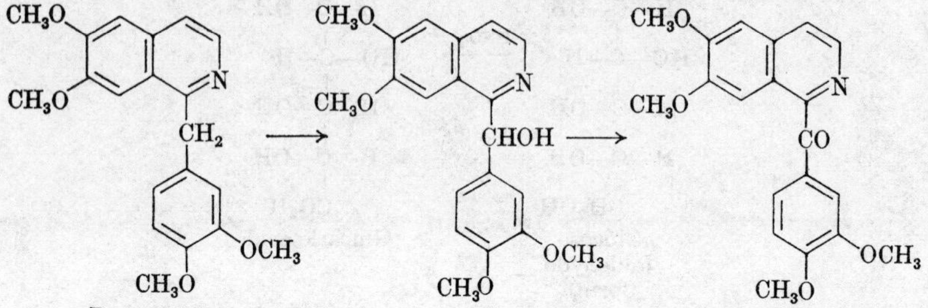

Papaverine $\longrightarrow$ (CHOH) $\longrightarrow$ Papaveraldine

secondary alcohol. Further oxidation of papaveraldine gives a mixture of carboxylic acids (**1**)–(**4**) resulting from the rupture of the

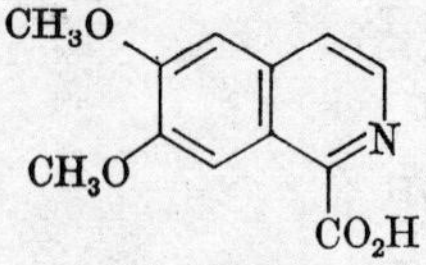

(**1**) 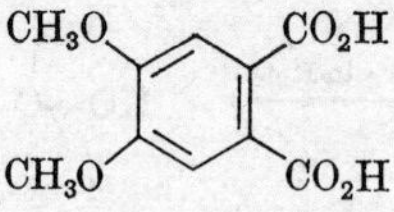 (**2**) 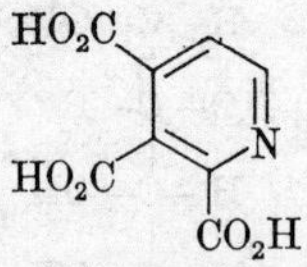 (**3**) 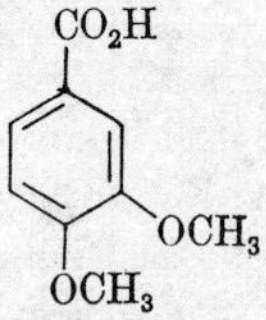 (**4**)

bonds linking the heterocyclic and the benzene ring, and partial oxidation of the heterocyclic ring system.

Aliphatic compounds, and compounds containing branched aliphatic side chains, can be completely degraded by chromium trioxide in sulphuric acid to carbon dioxide and acetic acid. Carbon dioxide is obtained from each carbon atom of the molecule except those present as methyl groups on an aliphatic chain. The latter type of group gives acetic acid, which is then steam-distilled and determined. This is known as the Kuhn–Roth method of determining the number of *C*-methyl groups in a molecule. Compounds containing CH_3—$CH(CH_3)$— or CH_3—$C(CH_3)_2$— groups give only one mole of acetic acid—usually slightly less than one—per mole of compound; thus isopropylbenzene and *tert*-butylcyclohexane each give about one mole of acetic acid. Methyl groups directly joined to aromatic rings are not oxidized to acetic acid. Toluene, for example, is oxidized under these conditions to benzoic acid. When α-tocopherol, one of the *E* group of vitamins, is first partly oxidized with chromic acid under mild conditions, an acid, $C_{16}H_{32}O_2$ (**5**) is produced, as one of the products of oxidation. This acid (**5**) when subjected to a *C*-methyl

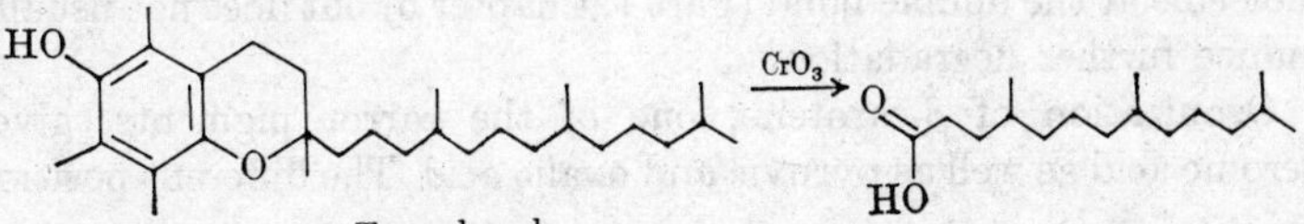

HO O $\xrightarrow{CrO_3}$ O HO

α-Tocopherol (**5**)*

determination, gave carbon dioxide and about 3 moles of acetic acid, indicating the presence of methyl groups substituted along the chain.

Saturated rings substituted with methyl groups may give much less than one mole of acetic acid per methyl group. Cholesterol, when oxidized with chromic acid in sulphuric acid, gives just less than three moles of acetic acid per mole of cholesterol (see p. 396).

For compounds containing carbon–carbon double bonds, a milder oxidizing agent such as ozone has its uses. This reagent breaks the

* Complex formulae such as (**5**) and the like are conveniently represented in a form of shorthand where a single line represents a carbon–carbon bond, plus the two carbon atoms forming the bond and all the necessary hydrogen atoms. Only hetero atoms (O, N, S, halogen, and so on) are represented by their chemical symbols, e.g.:

CH_3—CH_3 is represented as —

CH_3—CHO is represented as O

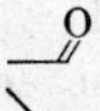

$(CH_3)_2CO$ is represented as =O

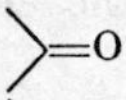

$(CH_3)_3N$ is represented as N

We are probably already familiar with this kind of symbolism in aromatic compounds; instead of writing 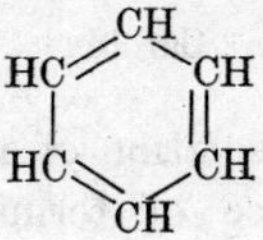 CH HC CH HC CH CH for benzene we write

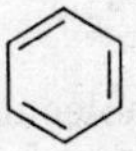

Cholesterol $\xrightarrow[\text{then distil}]{CrO_3 \text{ in } H_2SO_4,}$ $3CH_3CO_2H$ (approximately)

molecule at the double bond (Part 1, Chapter 6) but does not usually induce further degradation.

Ozonization of β-carotene, one of the carrot pigments, gives geronic acid as well as pyruvic and oxalic acid. The different position

β-Carotene $\xrightarrow{O_3}$ Geronic acid + Pyruvic acid + Oxalic acid

of the double bond in one of the two cyclohexane rings in α-carotene (another carrot pigment) is indicated by the production of iso-

α-Carotene $\xrightarrow{O_3}$ Isogeronic acid + geronic, pyruvic, and oxalic acid

geronic as well as geronic, pyruvic, and oxalic acid upon ozonolysis. The isogeronic acid is the end product of the oxidation of a 1,3-aldehydo-keto-acid.

Evidence for the acyclic nature of one end of the γ-carotene

γ-Carotene

molecule was obtained by isolation of acetone among the products of ozonolysis. The presence of terminal isopropylidene groups

(⟩=) in lycopene (responsible for the red colour of tomatoes) is

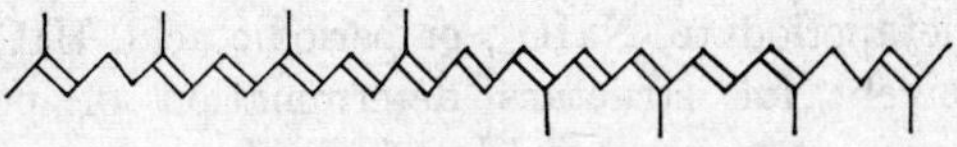

Lycopene

indicated by the formation of almost two moles of acetone per mole of lycopene oxidized.

Ozonolysis has been used extensively to determine the position of double bonds in terpenes, but misleading results are sometimes obtained. For example, the formation of formaldehyde or formic acid upon ozonolysis need not necessarily mean the presence of a terminal carbon–carbon double bond. Ozonolysis of pure geraniol, known to be entirely in the isopropylidene form (**6**), gave one mole of acetone, one of lævulaldehyde, and 0.2 mole of formaldehyde;

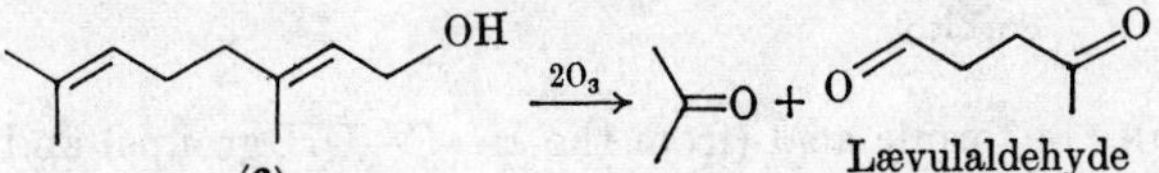

OH

(**6**) $\xrightarrow{2O_3}$ (CH₃)₂C=O + Lævulaldehyde

$+ HCO_2^- + H^+ + [CH_2O]$

0.2 mole

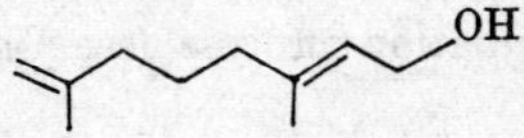

OH

(**7**)

the formaldehyde is formed during decomposition of the double ozonide and is derived from the alcoholic end of the molecule and not, as might seem at first sight, from the presence in geraniol of an isomer (**7**) with a terminal double bond.

Aqueous potassium permanganate is a powerful oxidizing agent but, at low temperature, neutral potassium permanganate solution oxidizes a carbon–carbon double bond to form a *cis* vicinal diol in rather low yield. The use of osmium tetroxide to effect this oxidation

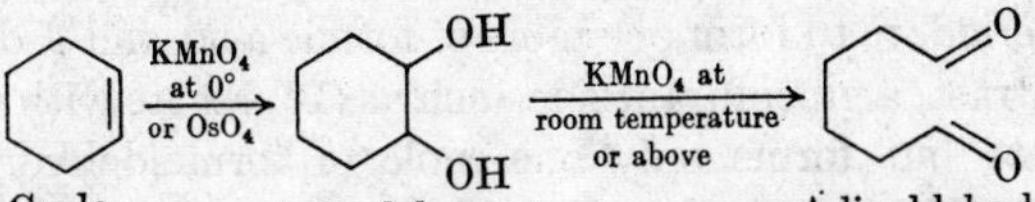

Cyclohexene $\xrightarrow[\text{or } OsO_4]{KMnO_4 \text{ at } 0°}$ *cis*-Cyclohexane-1,2-diol $\xrightarrow[\text{room temperature or above}]{KMnO_4 \text{ at}}$ Adipaldehyde (Hexanedial)

(Chapter 6 of Part 1) gives better yields of the diol. The vicinal diols thus produced are readily further oxidized.

Sodium metaperiodate, $NaIO_4$, or periodic acid, HIO_4, has been used extensively for structure determination of polyhydroxy-compounds such as the carbohydrates. Both reagent and reactant are water-soluble, and the reagent is specific for vicinal diols (see Chapter 18). A polyol such as sorbitol is oxidized quantitatively

$$\begin{array}{c} CH_2OH \\ | \\ H-C-OH \\ | \\ HO-C-H \\ | \\ H-C-OH \\ | \\ H-C-OH \\ | \\ CH_2OH \\ \text{Sorbitol} \end{array} \xrightarrow{5HIO_4} \begin{array}{c} H_2CO \\ HCO_2H \\ HCO_2H \\ HCO_2H \\ HCO_2H \\ H_2CO \end{array}$$

to four molecules of formic acid (from the H—C—OH groups) and two molecules of formaldehyde (from the $-CH_2OH$ groups). Five molecules of periodic acid are used for this oxidation, i.e. one mole per carbon–carbon bond broken, and an intermediate cyclic ion is involved in the oxidation process (see Chapter 18). In practice,

$$RCH(OH)-CH(OH)R' + IO_4^- \longrightarrow [\text{cyclic periodate ester: } RCH-O-I(OH)_2(=O)(O^-)-O-CHR'] \longrightarrow RCHO + R'CHO + IO_3^- + H_2O$$

slightly more than the theoretical quantity of periodate is usually consumed, owing to side reactions, especially if the reaction mixture is allowed to stand.

α-Methyl D-glucopyranoside (**8**) reacts quantitatively with two moles of periodate to form one mole of formic acid and a dialdehyde (**9**). In contrast, a glucofuranoside such as (**10**) reacts with two moles of periodate but forms only one mole of formaldehyde and the trialdehyde (**11**) but no formic acid (as the molecule contains no —CHOH—CHOH—CHOH— grouping).

$$\underset{(\mathbf{8})}{\begin{array}{l} CH_3O-CH- \\ H-C-OH \\ HO-C-H \quad O \\ H-C-OH \\ H-C- \\ CH_2OH \end{array}} \xrightarrow{2\,IO_4^-} \underset{(\mathbf{9})}{CH_3OCH(CHO)-O-CH(CH_2OH)-CHO} + HCO_2H$$

$$\underset{(\mathbf{10})}{\begin{array}{l} CH_3O-CH- \\ H-C-OH \\ HO-C-H \quad O \\ H-C- \\ H-C-OH \\ CH_2OH \end{array}} \xrightarrow{2\,IO_4^-} \underset{(\mathbf{11})}{CH_3-O-CH(CHO)-O-CH(CHO)-CHO} + CH_2O$$

A solution of chlorine in sodium hydroxide solution (i.e. HOCl) oxidizes the carbon atom adjacent to a carbonyl group. Methyl ketones, when oxidized in this way, lose the methyl-carbon atom and form carboxylic acids. Mesityl oxide, for example, is oxidized to 3-methylbut-2-enoic acid.

$$\underset{\text{Mesityl oxide}}{(CH_3)_2C=CH-CO-CH_3} \xrightarrow{NaOH,\ Cl_2} \underset{\text{3-Methylbut-2-enoic acid}}{(CH_3)_2C=CH-CO-OH} + CHCl_3$$

A solution of iodine in sodium hydroxide (i.e. NaOI) effects the same oxidation and the methyl group is lost as iodoform which is yellow and insoluble in water. This variation of the hypohalite oxidation was used as a qualitative test for methyl ketones and compounds which may be easily oxidized to methyl carbonyl compounds, e.g. ethanol (see Chapter 12, p. 144).

Sodium hypohalite has often been used for further oxidation of the product resulting from initial fission of a carbon–carbon double bond. α-Terpineol may be converted into the hydroxy-keto-acid (**12**) by permanganate and treatment of this product with aqueous sodium hypochlorite gives a hydroxy-acid (**13**) which is isolated as

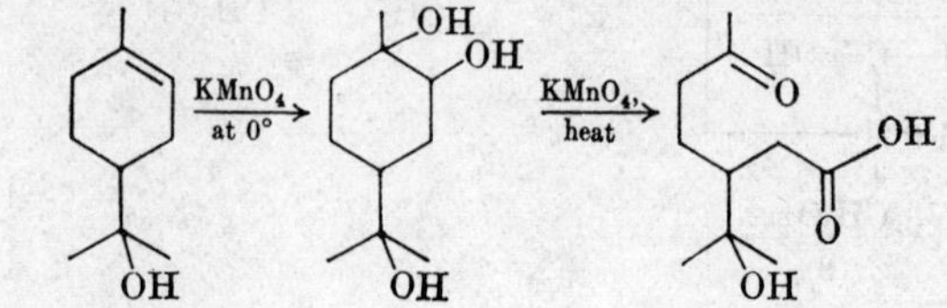

OH OH OH OH O OH O OH

α-Terpineol $\xrightarrow[\text{at } 0°]{KMnO_4}$ $\xrightarrow[\text{heat}]{KMnO_4,}$ (**12**)

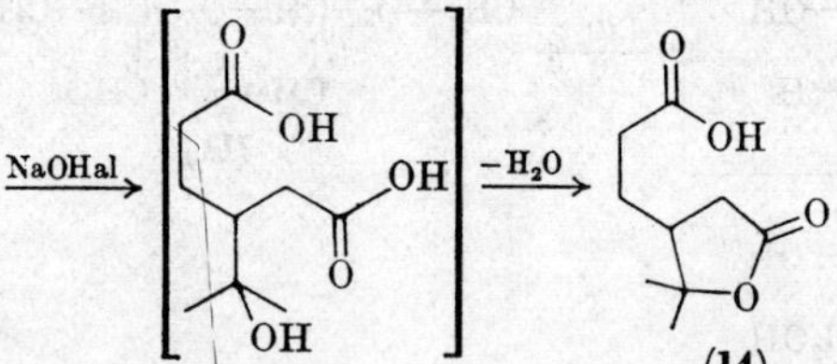

O OH OH O OH O OH O O

$\xrightarrow{NaOHal}$ [(**13**)] $\xrightarrow{-H_2O}$ (**14**)

the lactone (**14**).

If sodium hypoiodite is used as a test for methyl ketones, the formation of iodoform may not always be due to the presence of a methyl ketone in the original compound. With some α,β-unsaturated carbonyl compounds such as pulegone, iodoform is formed from acetone which results from a reverse aldol reaction of the carbonyl compound initiated by the alkaline reagent (see Chapter 9, p. 92).

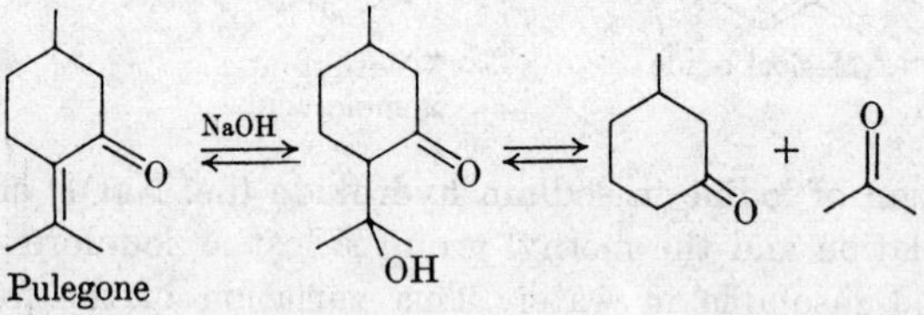

O O OH O O

Pulegone $\overset{NaOH}{\rightleftharpoons}$ $\rightleftharpoons$ +

Dehydrogenation

Information about the structure of a molecule can sometimes be obtained by dehydrogenation to an aromatic compound whose structure may be more easily determined than that of the original molecule. The reagent used to bring about this dehydrogenation is sulphur or selenium (which are reduced to H_2S and H_2Se, respectively), or sometimes palladium supported upon charcoal. Where palladium is used, the hydrogen removed is evolved as such. Temperatures between 200° and 400° are usual. Zinc dust also acts in this way and, at the same time, completely reduces hydroxy- and keto-groups; thus alizarin, a naturally occurring pigment, when submitted to distillation with zinc dust, gives anthracene. When, as often in terpenes, methyl groups are present at ring junctions, they

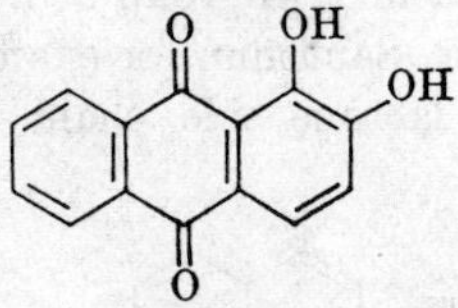

Alizarin

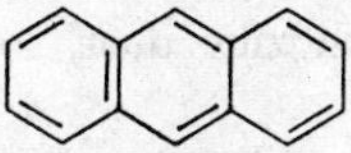

Anthracene

are lost during the reaction. Thus cyperone, a sesquiterpene ketone,

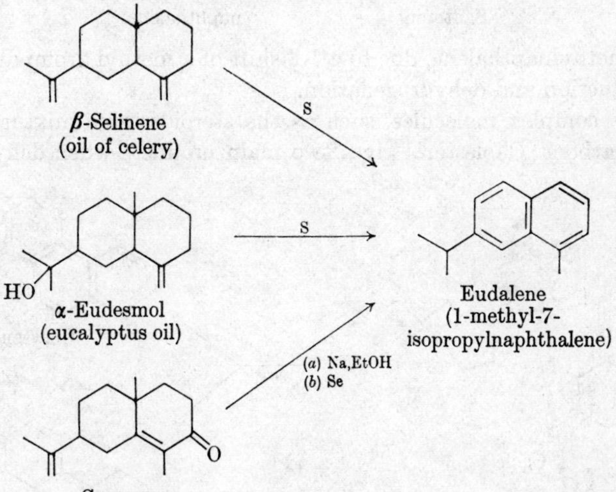

β-Selinene
(oil of celery)

α-Eudesmol
(eucalyptus oil)

Cyperone

Eudalene
(1-methyl-7-isopropylnaphthalene)

gives, after reduction of the keto-group, the same aromatic hydrocarbon, eudalene, as do the hydrocarbon β-selinene and the alcohol, eudesmol. If, however, cyperone is first treated with methylmagnesium iodide to form (**15**), dehydrogenation of this product gives

7-isopropyl-1,2-dimethylnaphthalene (**16**), the position of the new

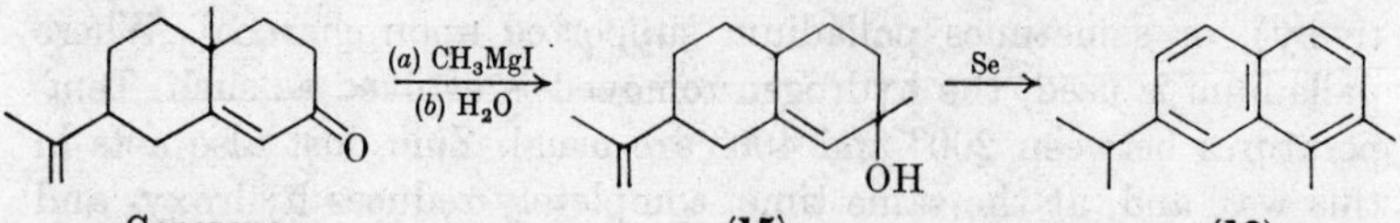

(a) CH_3MgI
(b) H_2O
Se
OH
Cyperone (**15**) (**16**)

methyl group indicating the position of the original keto-group.

Evidence obtained from such dehydrogenation experiments must be interpreted with caution. Often, a methyl group at a ring junction may migrate rather than be lost. Santonin, for example, when distilled with zinc dust, loses the lactone side chain and gives

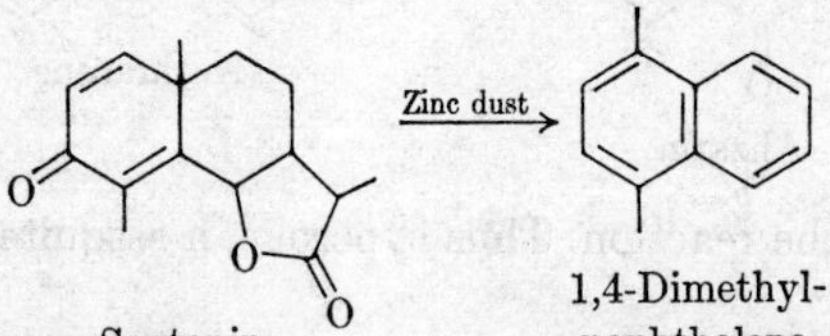

Zinc dust
Santonin
1,4-Dimethyl-naphthalene

1,4-dimethylnapthalene, due to a 1,2-shift of a methyl group during the reduction and dehydrogenation.

More complex molecules, such as the steroids, give mixtures of hydrocarbons. Cholesterol gives two main products when dehydro-

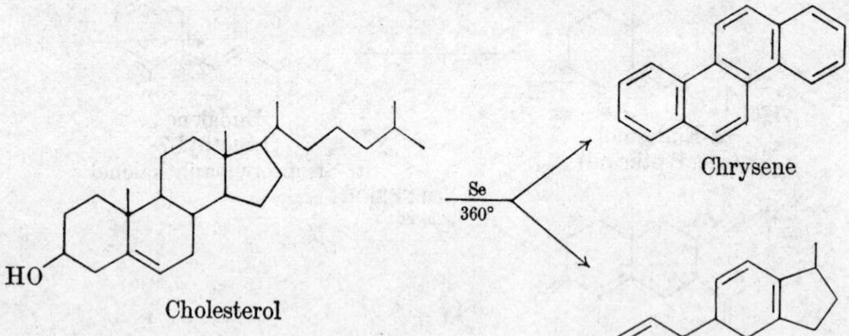

Se
360°
HO
Cholesterol
Chrysene
Diels' hydrocarbon

genated with selenium, both of which indicate the condensed nature of the saturated ring system in cholesterol.

Stepwise Degradations

Barbier–Wieland and related reactions

Another reaction that has been used in determining steroid structures is a series of reactions that can be used for shortening the carbon chain of a carboxylic acid in steps of one carbon atom at a time (see Part 1, p. 196). The method has also been used for step-

$$RCH_2CO_2H \xrightarrow[H_2SO_4]{CH_3OH} RCH_2CO_2CH_3 \xrightarrow{2\,C_6H_5MgBr} RCH_2C(OH)(C_6H_5)_2 \xrightarrow{-H_2O}$$

$$RCH{=}C(C_6H_5)_2 \xrightarrow{CrO_3} RCO_2H + O{=}C(C_6H_5)_2$$

wise degradation of long-chain branched fatty acids, the degradation being continued until a methyl ketone of known structure is obtained.

The length of the sidechain of cholic acid has been determined by first converting cholic acid into cholanic acid and submitting the ester thereof to three cycles of the degradation; the methyl ketone

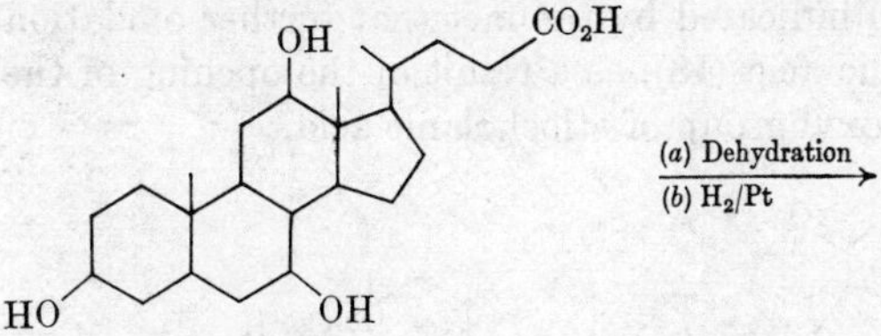

Cholic acid $\xrightarrow[(b)\ H_2/Pt]{(a)\ \text{Dehydration}}$

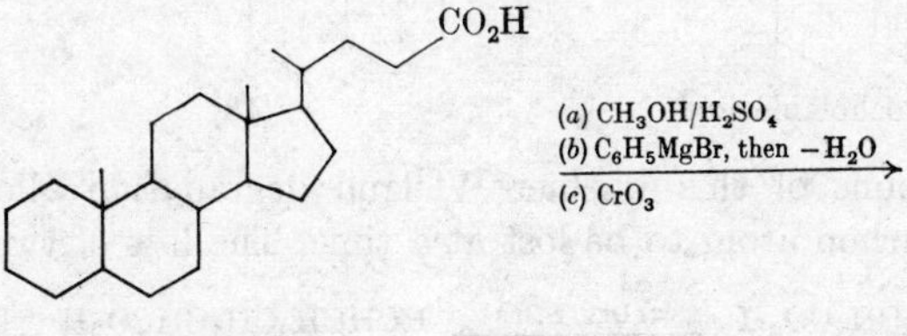

Cholanic acid $\xrightarrow[(b)\ C_6H_5MgBr,\ \text{then}\ -H_2O;\ (c)\ CrO_3]{(a)\ CH_3OH/H_2SO_4}$

Norcholanic acid $\xrightarrow[(b)\ C_6H_5MgBr,\ \text{then}\ -H_2O;\ (c)\ CrO_3]{(a)\ CH_3OH/H_2SO_4}$

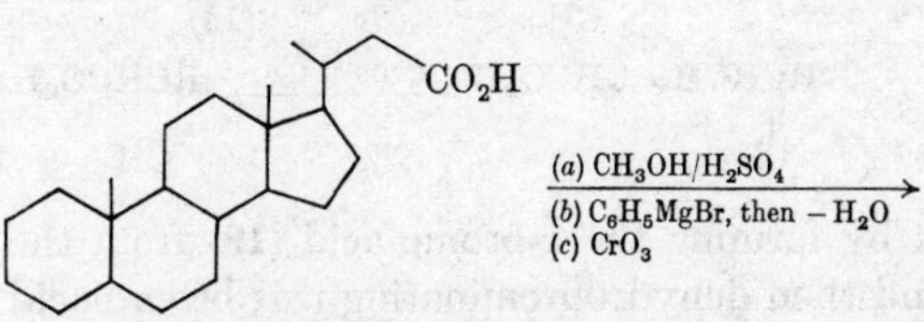

[continued on p. 404]

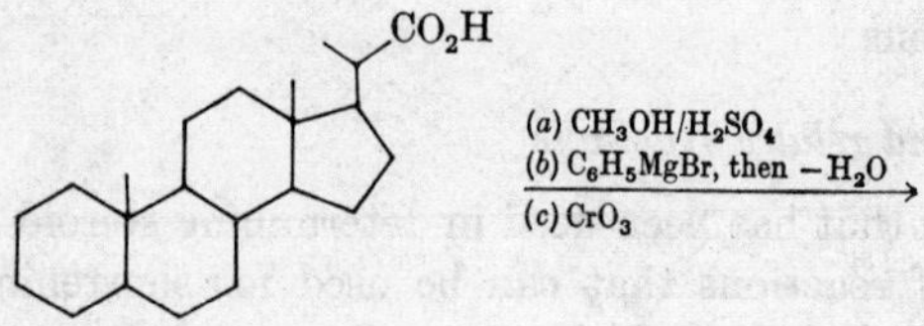

CO_2H

(a) CH_3OH/H_2SO_4
(b) C_6H_5MgBr, then $-H_2O$
(c) CrO_3

Bisnorcholanic acid

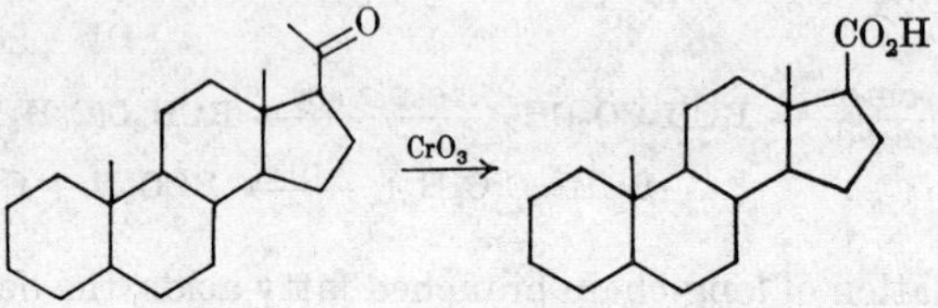

O

CrO_3

CO_2H

(17) Etiocholanic acid

(17) resulted. The formation of this methyl ketone indicated that a branch had been reached in the chain. Further chromic acid oxidation gave etiocholanic acid. That the end of the side chain had been reached was then indicated by the fact that further oxidation produced a dicarboxylic acid (**18**), as a result of the opening of the ring carrying the carboxyl group of etiocholanic acid.

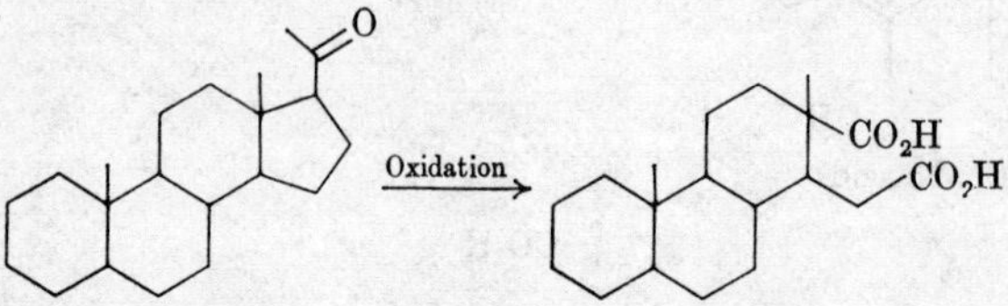

O

Oxidation

CO_2H
CO_2H

Etiocholanic acid **(18)**

Modifications of this Barbier–Wieland degradation allow more than one carbon atom to be lost at a time. The loss of two carbon

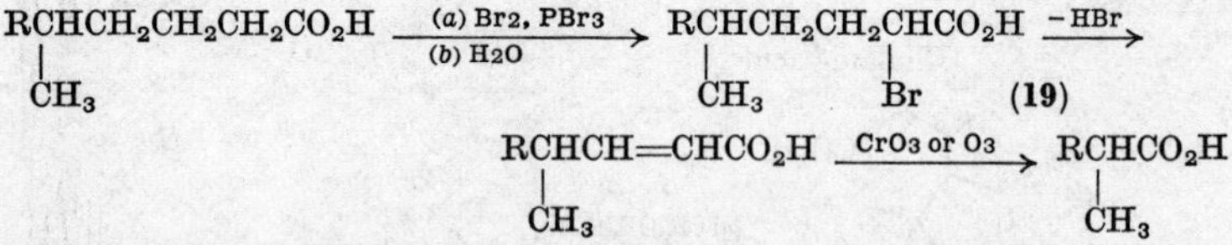

$$\underset{\displaystyle CH_3}{\underset{|}{RCH}}CH_2CH_2CH_2CO_2H \xrightarrow[(b)\ H_2O]{(a)\ Br_2,\ PBr_3} \underset{\displaystyle CH_3}{\underset{|}{RCH}}CH_2CH_2\underset{\displaystyle Br}{\underset{|}{CH}}CO_2H \ \textbf{(19)} \xrightarrow{-HBr}$$

$$\underset{\displaystyle CH_3}{\underset{|}{RCH}}CH{=}CHCO_2H \xrightarrow{CrO_3 \text{ or } O_3} \underset{\displaystyle CH_3}{\underset{|}{RCH}}CO_2H$$

atoms may be effected by forming the α-bromo-acid (**19**) from the acid to be degraded and then dehydrobrominating this bromo-acid before oxidation with chromic acid or ozone.

Three carbon atoms may be taken off by way of radical allylic

$$\underset{\displaystyle CH_3}{\underset{|}{RCH}}CH_2CH_2CO_2H \xrightarrow[(b)\ C_6H_5MgBr,\ then\ -H_2O]{(a)\ CH_3OH/H_2SO_4}$$

$$\underset{\displaystyle CH_3}{\underset{|}{RCH}}CH_2CH{=}C(C_6H_5)_2 \xrightarrow{N\text{-Bromosuccinimide}} \underset{\displaystyle CH_3}{\underset{|}{RCH}}\overset{\displaystyle Br}{\overset{|}{C}}H{=}C(C_6H_5)_2 \xrightarrow{-HBr}$$

$$\underset{\displaystyle CH_3}{\underset{|}{RC}}{=}CHCH{=}C(C_6H_5)_2 \xrightarrow{CrO_3\ or\ O_3} \underset{\displaystyle CH_3}{\underset{|}{RC}}{=}O$$

bromination with *N*-bromosuccinimide, as shown above.

Another method, known as the Varrentrap reaction, which has been used to degrade carboxylic acids, involves treatment of unsaturated acids with molten potassium hydroxide. The double bond

$$\underset{\text{Oleic acid}}{CH_3[CH_2]_7CH{=}CH[CH_2]_7CO_2H} \xrightarrow[200^\circ]{KOH} \underset{\text{Potassium palmitate}}{CH_3[CH_2]_{14}CO_2^-K^+} + CH_3CO_2^-K^+$$

migrates to the end of the carbon chain, and the resulting $\alpha\beta$-unsaturated acid undergoes oxidative fission to form the potassium salts of acetic acid and a saturated carboxylic acid having two carbon atoms less than the original acid. The reaction sequence shows how oleic acid is converted into a mixture of potassium acetate and potassium palmitate by means of this drastic reaction.

Degradations involving the Loss of Nitrogen

One method of degradation resulting in the loss of nitrogen from a molecule is known as 'exhaustive methylation'. In this, the nitrogen atom of an amine is quaternized with an excess of methyl iodide and the resulting quaternary ammonium hydroxide is pyrolysed to give a trialkylamine and an unsaturated compound. This reaction is known as a Hofmann elimination (see p. 48, Chapter 5 of Part 1, and Chapter 24 of this volume). The method is used for the study of the structure of alkaloids and cyclic bases. As an example, one of the alkaloids of hemlock, *pseudo*-conhydrine (a propylpiperidinol), forms a quaternary salt with methyl iodide which is converted into the hydroxide (**20**) by moist silver oxide or an ion-exchange column, and is then pyrolysed to form the tertiary base (**21**). This unsaturated amine can be hydrogenated to (**22**) which can be further methylated,

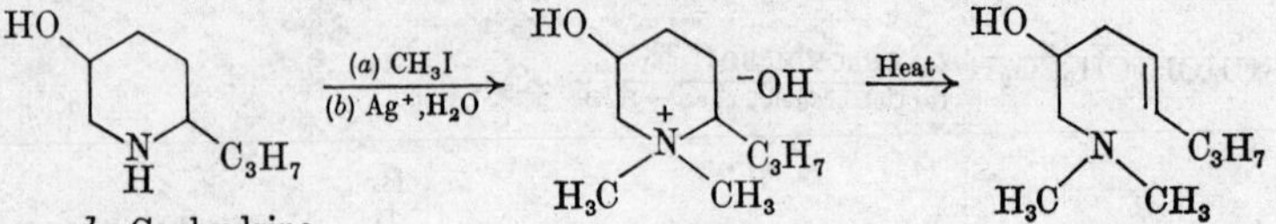
HO
C3H7
(a) CH3I
(b) Ag+,H2O
-OH
Heat
H3C
CH3

pseudo-Conhydrine (20) (21)

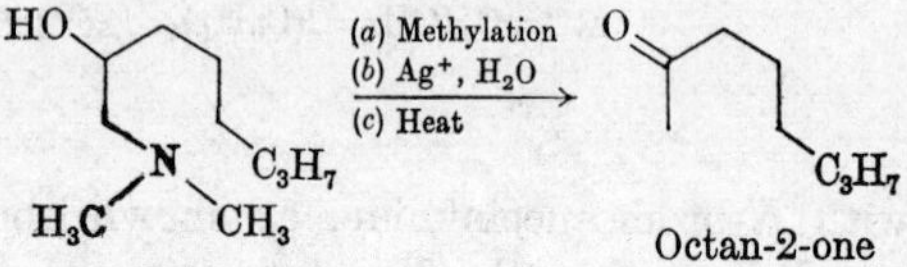
HO
(a) Methylation
(b) Ag+, H2O
(c) Heat
O
C3H7
H3C
CH3

Octan-2-one

(22)

converted into the quaternary hydroxide, and pyrolysed again, then yielding octan-2-one.

During the pyrolysis, double bonds may migrate to a position where they are more stable. This happens when piperidine (a compound originally obtained from an amide isolated from black pepper) is methylated and the quaternary hydroxide pyrolysed; penta-1,3-diene is formed after two degradation cycles rather than the unconjugated penta-1,4-diene.

Successive Hofmann degradations do not always remove nitrogen from a cyclic molecule. The tetrahydroquinoline (**23**) forms a quaternary salt with methyl iodide which does not undergo ring fission when the hydroxide is pyrolysed—it loses methanol. The quaternary hydroxide of (**24**), however, may be treated in alcohol

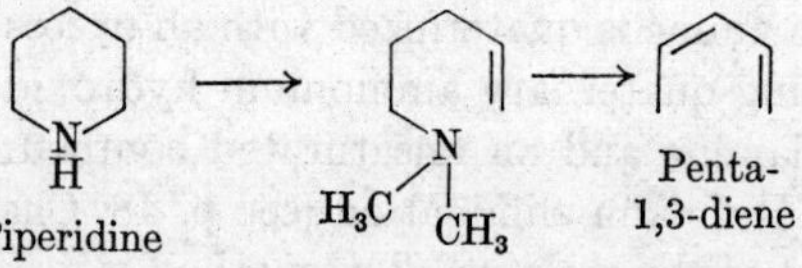
N
H
H3C
CH3

Piperidine Penta-1,3-diene

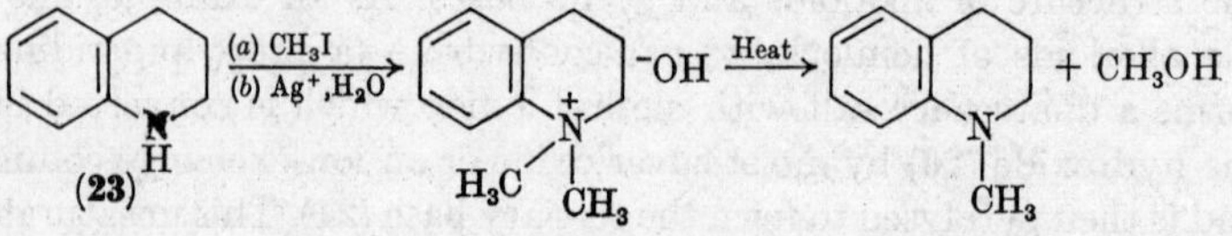
(a) CH3I
(b) Ag+,H2O
-OH
Heat
+ CH3OH
H3C
CH3

(23) (24)

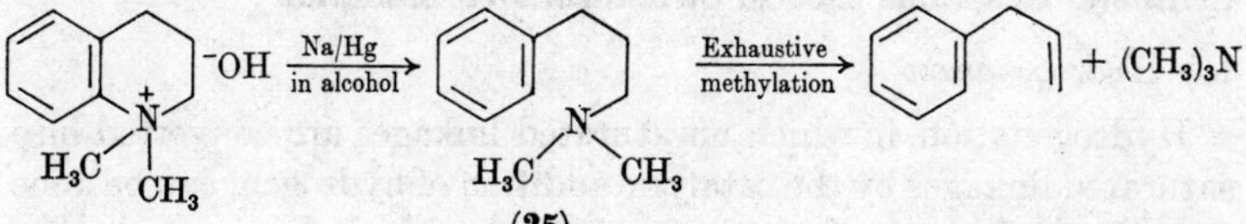

^{-}OH, H_3C, CH_3 — Na/Hg in alcohol → H_3C, CH_3 (**25**) — Exhaustive methylation → + $(CH_3)_3N$

with sodium amalgam to give a base (**25**) in which the nitrogen-containing ring is opened. This base may now be exhaustively methylated and pyrolysed, whereupon it loses trimethylamine in the usual manner.

Cyclic bases may also be made to undergo ring fission by the use of cyanogen bromide, a reagent associated with the name of von Braun. In some cases, this mode of ring fission is used in conjunction with the Hofmann elimination. Ring fission in the Hofmann elimination may occur in a different direction to that observed in the von Braun reaction. Hydrocotarnine, a minor constituent of opium,

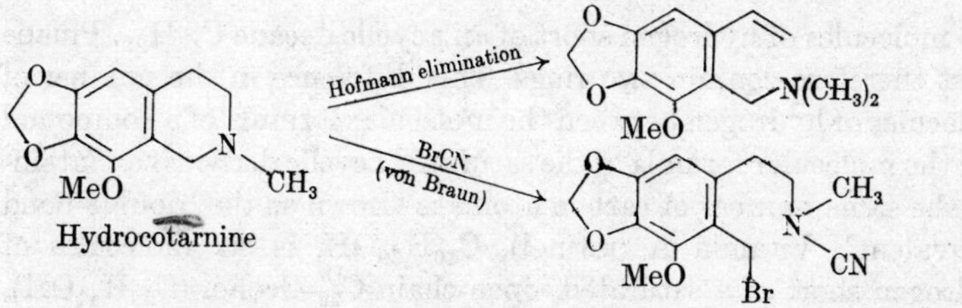

Hydrocotarnine (MeO, CH_3) — Hofmann elimination → MeO, $N(CH_3)_2$ (**26**)

Hydrocotarnine — BrCN (von Braun) → MeO, Br, CH_3, N, CN (**27**)

undergoes exhaustive methylation to form (**26**), but degradation with cyanogen bromide gives (**27**).

In the degradation of secondary amines, the ring may also be

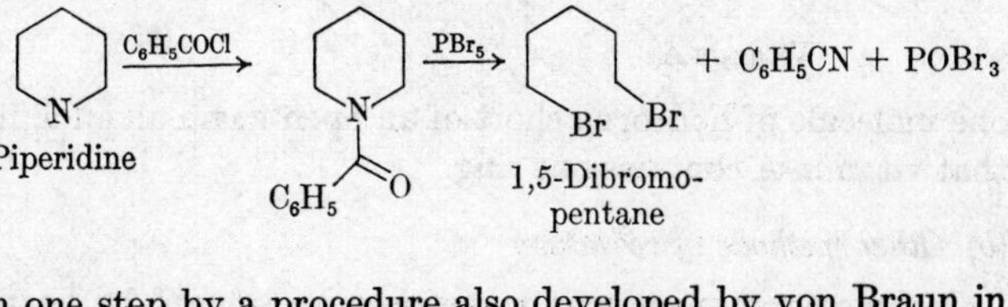

Piperidine — C_6H_5COCl → C_6H_5, O — PBr_5 → Br, Br 1,5-Dibromopentane + C_6H_5CN + $POBr_3$

opened in one step by a procedure also developed by von Braun in which the amine is treated successively with benzoyl chloride and then phosphorus pentabromide as shown above.

Synthetic Reactions used in Structural Determination

(a) *Hydrogenation*

Hydrogenation, in which unsaturated linkages are converted into saturated linkages by the catalytic addition of hydrogen, can be done quantitatively and often throws considerable light on molecular structure. Comparison of the molecular formula of the compound before and after hydrogenation indicates the number of multiple bonds and of rings in the molecule. For example, α-pinene ($C_{10}H_{16}$) reacts quantitatively with only one molecule of hydrogen to give pinane ($C_{10}H_{18}$), and thus α-pinene contains only one double bond. Pinane is saturated, although its molecular formula shows that it is

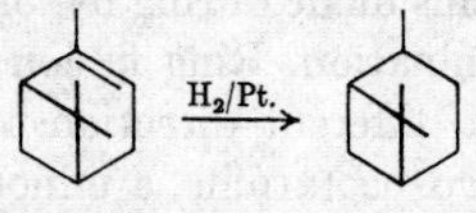

H_2/Pt.

α-Pinene Pinane

two molecules of hydrogen short of an acyclic decane $C_{10}H_{22}$. Pinane must therefore contain two rings. This difference in the number of molecules of hydrogen between the molecular formula of a compound and the molecular formula of the saturated acyclic derivative containing the same number of carbon atoms is known as the 'double-bond equivalent'. Vitamin A (retinol), $C_{20}H_{29}OH$, is six molecules of hydrogen short of a saturated, open-chain C_{20}-alcohol ($C_{20}H_{41}OH$), and thus has six double-bond equivalents. Catalytic hydrogenation produces a saturated perhydroretinol (**28**) $C_{20}H_{39}OH$, which is still

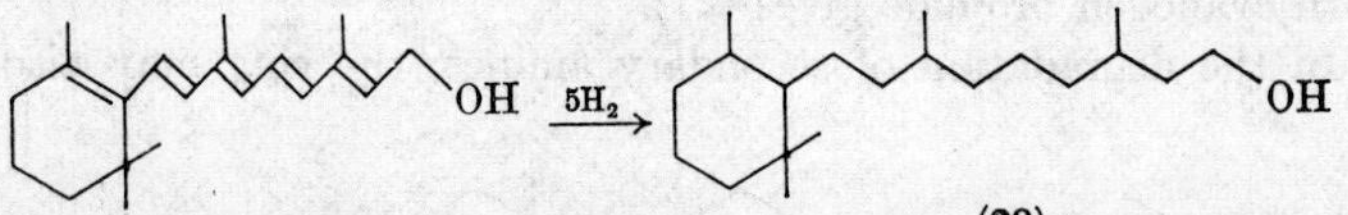

OH $5H_2$ OH

Vitamin A (**28**)

one molecule of hydrogen short of an open-chain alkanol, indicating that vitamin A contains one ring.

(b) *Other methods of reduction*

Organic compounds may sometimes be reduced by heating them with hydrogen iodide, functional groups such as hydroxyl being lost and a hydrocarbon resulting. This procedure was used at an early stage in the determination of the structure of glucose. The formation

$$\underset{\text{Glucose}}{HOCH_2[CHOH]_4CHO} \xrightarrow{HCN} \underset{\text{Glucose cyanohydrin}}{HOCH_2[CHOH]_4\underset{|}{C}HOH} \atop \quad CN$$

$$\xrightarrow[(b)\ HI]{(a)\ \text{Hydrolysis}} \underset{\text{Heptanoic acid}}{CH_3[CH_2]_7CO_2H}$$

of heptanoic acid by hydrogen iodide treatment of the acid obtained by the hydrolysis of glucose cyanohydrin was used as evidence for the straight-chain nature of this monosaccharide (see Chapter 23).

(*c*) *Ring closures*

Structural information about the geometry of a molecule and about the position of a functional group may often be obtained by observing the ease with which the compound undergoes ring formation, either with itself or with other reagents.

Thus, only one of the two possible geometric isomers of 2-aminostilbene undergoes intramolecular cyclization by means of the Pschorr reaction to form phenanthrene. This aminostilbene must

cis-2-Aminostilbene (H_2N) $\xrightarrow[(b)\ \text{Copper bronze powder}]{(a)\ HNO_2,\ H_2SO_4}$ Phenanthrene

have *cis*-geometry about the double bond.

The product (**29**) of reaction of *N*-bromosuccinimide with 3-methylbut-2-enoic acid can be hydrolysed with alkali to form a hydroxy-acid (**30**) which does not lactonize spontaneously. This

$$\underset{\text{3-Methylbut-2-enoic acid}}{(CH_3)_2C{=}CHCO_2H} \xrightarrow[\text{in } CCl_4]{N\text{-Bromosuccinimide}} \underset{(\mathbf{29})}{(CH_3)(BrCH_2)C{=}CHCO_2H}$$

$$\xrightarrow[(b)\ \text{dil. HCl}]{(a)\ NaOH} \underset{(\mathbf{30})}{(CH_3)(HOCH_2)C{=}C(H)CO_2H}$$

failure to lactonize can be taken as evidence that the methyl group *trans* to the carboxyl group reacts preferentially to the *cis*-methyl group in the radical-substitution reaction. This kind of evidence must be interpreted cautiously, however, as isomerization of the double bond may be induced by an allylic rearrangement during hydrolysis of the bromo-acid. More careful investigation of the reaction product by physical methods described in the preceding Chapter reveals that some *cis*-bromo-acid is also formed as a minor product of the bromination.

The exclusive formation of the lactone (**32**), a possible growth-nhibiting substance, shows how the stereospecific *cis*-reduction of the carbon–carbon triple bond of acid (**31**) can be demonstrated by the ease of lactonization of the reduction product. When the acetyl-

$$(CH_3)_2C(OH)-C\equiv C-CO_2H \xrightarrow[H_2]{Pd/C/BaSO_4} (CH_3)_2C<\begin{matrix} CH=HC \\ O \end{matrix}>C=O$$

(**31**) (**32**)

enic hydroxy-acid (**31**) is hydrogenated with palladium on charcoal and barium sulphate, direct formation of lactone (**32**) confirms the general observation that the carbon–carbon triple bond can be stereospecifically *cis*-hydrogenated on choice of a suitable catalyst.

In contrast, the reaction of benzaldehyde with ethyl 4-bromo-crotonate in the presence of zinc gave, after subsequent hydrolysis,

$$C_6H_5CHO + BrCH_2CH=CHCO_2C_2H_5 \xrightarrow[(b)\ Dil.\ H_2SO_4]{(a)\ Zinc} C_6H_5CH(OH)CH_2CH=CHCO_2H$$

Ethyl 4-Bromocrotonate (**33**)

a hydroxy-acid (**33**), that did not cyclize, indicating that the double bond carried the carboxyl group and the hydroxylated carbon chain in *trans*-relation to one another.

In compounds containing cyclopentane or cyclohexane rings, the size of the ring has sometimes been determined by opening the ring with oxidizing agents to form a dicarboxylic acid, and then investigating the behaviour of this acid with refluxing acetic anhydride. In Chapter 18, we described how dicarboxylic acids such as glutaric and succinic acid form anhydrides under these conditions, whereas adipic and pimelic acid give cyclic ketones (the Blanc rule).

CO_2H CO_2H $\xrightarrow[\text{Heat}]{(CH_3CO)_2O}$ O O O

Glutaric acid — Glutaric anhydride

CO_2H CO_2H $\xrightarrow[\text{Heat}]{(CH_3CO)_2O}$ =O

Adipic acid — Cyclopentanone

The ring size of ring A of the steroid cholesterol (**34**) was determined in this way by first hydrogenating it to cholestanol (**35**) and then oxidizing this saturated alcohol to a dicarboxylic acid (**36**).

C D A B HO (**34**) $\xrightarrow{Pt/H_2}$

HO (**35**)* $\xrightarrow{CrO_3}$ HO_2C HO_2C (**36**)*

This acid lost carbon dioxide and water when heated with acetic anhydride, to form a ketone (**37**), indicating that ring A was at least

* Rings C and D omitted from formulae (**35**)–(**39**) for convenience as they take no direct part in the reactions.

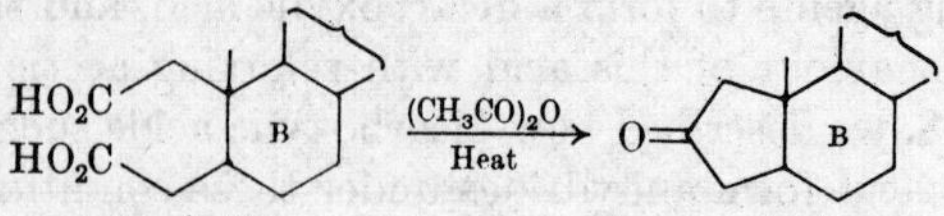

six-membered. Further oxidation of this ketone gave another dicarboxylic acid (**38**) containing one carbon atom less than the dicarboxylic acid (**36**). Refluxing the acid (**38**) with acetic anhydride gave, this time, an anhydride (**39**), indicating that ring A was indeed

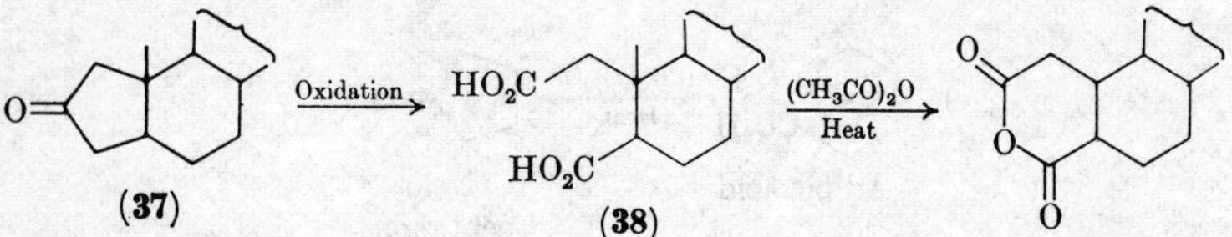

six-membered and not seven-membered, for in the latter case a ketone instead of anhydride (**39**) would have been obtained.

The same method was used to show that ring D of the steroids and related bile acids was five-membered. A ketone obtained from etiocholanic acid was oxidized and cyclized to an anhydride (the formulae show only rings C and D; cf. footnote, p. 411).

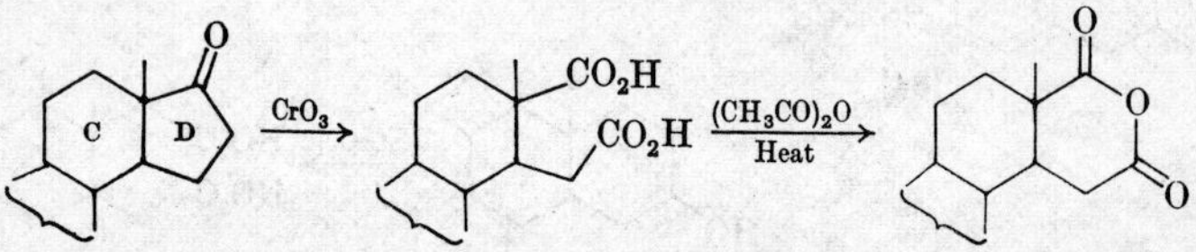

Use of the Blanc rule to determine the size of rings B and C led, however, to the erroneous conclusion that both these rings were five-membered, for it was not realized at the time that Blanc's generalizations about the reaction of dicarboxylic acids with acetic anhydride did not apply to the highly substituted dicarboxylic acids obtained when either of the two ketones (**40**) and (**41**) is oxidized.

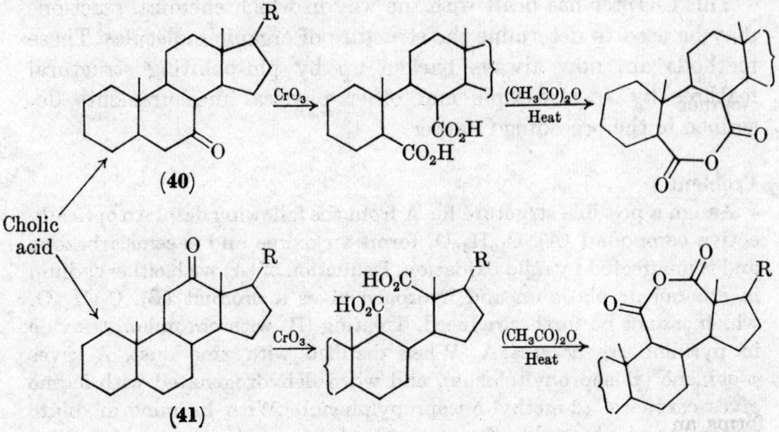

$\xrightarrow{CrO_3}$ ... CO_2H, CO_2H $\xrightarrow[Heat]{(CH_3CO)_2O}$

(**40**)

Cholic acid

$\xrightarrow{CrO_3}$ ... HO_2C, HO_2C $\xrightarrow[Heat]{(CH_3CO)_2O}$

(**41**)

(*d*) *Ring formation by Diels–Alder addition*

The presence and structure of a conjugated cyclic diene system in a molecule may be neatly demonstrated by the formation of a Diels–Alder adduct with an acetylenedicarboxylic ester, followed by pyrolysis of the adduct. By this means, the original diene system is

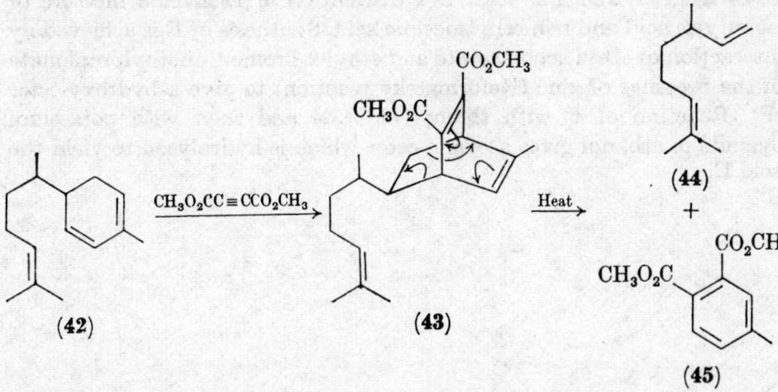

$\xrightarrow{CH_3O_2CC{\equiv}CCO_2CH_3}$ $\xrightarrow{Heat}$

(**42**) (**43**) (**44**) + (**45**)

broken up. Zingiberene (**42**), the main constituent of ginger oil, forms with methyl acetylenedicarboxylate an adduct (**43**), pyrolysis of which gives 3,7-dimethylocta-1,6-diene (**44**) and dimethyl 4-methylphthalate (**45**).

This Chapter has dealt with the way in which chemical reactions may be used to determine the structure of organic molecules. These methods are now always backed up by pin-pointing structural features by spectroscopic and other physical measurements described in the preceding Chapter.

Problem

Assign a possible structure for A from the following data. An optically active compound (A), $C_{10}H_{16}O$, forms an oxime and a semicarbazone and is unaffected by mild oxidation. Reduction of (A) with either sodium in ethanol or platinum and hydrogen gives a product (B), $C_{10}H_{18}O$, which cannot be further reduced. Treating (B) with chromium trioxide in pyridine regenerates A. When distilled with zinc dust, A gives *p*-cymene (*p*-isopropyltoluene), and when dehydrogenated with iodine gives carvacrol (2-methyl-5-isopropylphenol). With bromine in dilute mineral acid, A readily forms a monobromo-derivative; with pentyl nitrite in the presence of sodium ethoxide A forms an α-keto-oxime. Nitric acid oxidizes A to a dibasic acid, (D), $C_8H_{14}(CO_2H)_2$. This dicarboxylic acid also easily forms a monobromo-derivative. Two different half-esters are formed from D with ethanol. D is not hydrogenated with platinum and hydrogen, but forms an anhydride when heated with acetic anhydride. When subjected to further nitric acid oxidation, D forms an acid (E), $C_6H_{11}(CO_2H)_3$; further nitric acid oxidation of E gives trimethylsuccinic acid. Dry distillation of E gives a mixture of isobutyric acid and trimethylsuccinic acid. Synthesis of E is achieved by interaction of ethyl acetoacetate and ethyl 2-bromo-2-methylpropionate in the presence of zinc (Reformatsky reaction) to give a hydroxy-ester (F). Reaction of F with thionyl chloride and then with potassium cyanide in ethanol gives a cyano-ester which is hydrolysed to yield the acid E.

Answers to Problems

Chapter 1

1. $E = h\nu$; $\nu = \dfrac{c}{\lambda}$; $h = 6.6 \times 10^{-27}$ erg sec; $c = 3 \times 10^{10}$ cm sec^{-1} (see pages 1–2).

Given that $\lambda = 5.5 \times 10^{-5}$ cm.

$$\text{Hence } E = \frac{(6.6 \times 10^{-27})(3.0 \times 10^{10})}{(4 \times 10^{-5})} \text{ erg}$$
$$= 2.475 \times 10^{-12} \text{ erg}$$
$$= 1.55 \text{ ev}$$
$$= 35.7 \text{ kcal mole}^{-1}.$$

2. This is really what we have done in the case of the stretched string.

$$E_1 = \frac{h^2}{8\,\text{ml}};\quad E_2 = \frac{h^2}{2\,\text{ml}^2};\quad E_3 = \frac{9h^2}{8\,\text{ml}^2}\cdots$$

Chapter 2

1. None. Notice that as l (the length of the one-dimensional box) increases, so the energy levels tend to zero $\left(E_n = \dfrac{h^2n^2}{8\,\text{ml}^2}\right)$.

2.

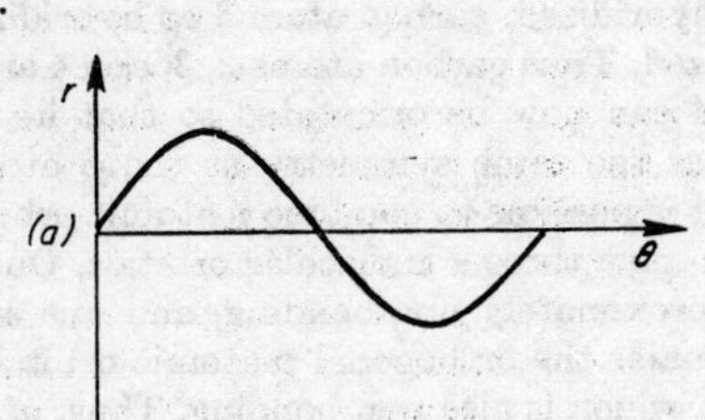

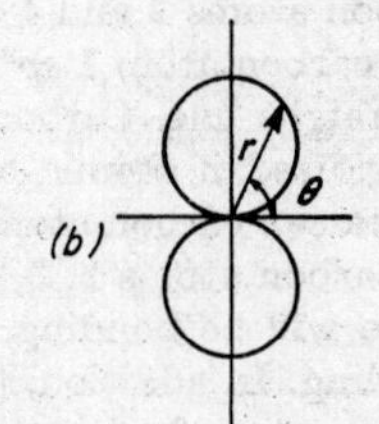

Chapter 3

1. (*a*) See Figure 3.5.

(*b*) See Figure 3.5 for $2s + 2p_z$, $2s + 2p_x =$

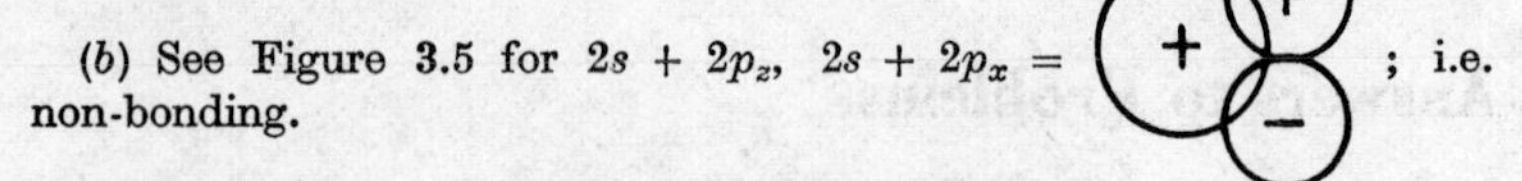

; i.e. non-bonding.

(*c*) Two orbitals of the same atom are orthogonal.

2. (*a*) $CH_2{=}CH{-}CH{=}CH_2$ discussed in detail on pages 30–32.
(*b*) $CH{\equiv}C{-}CH{=}CH_2$:

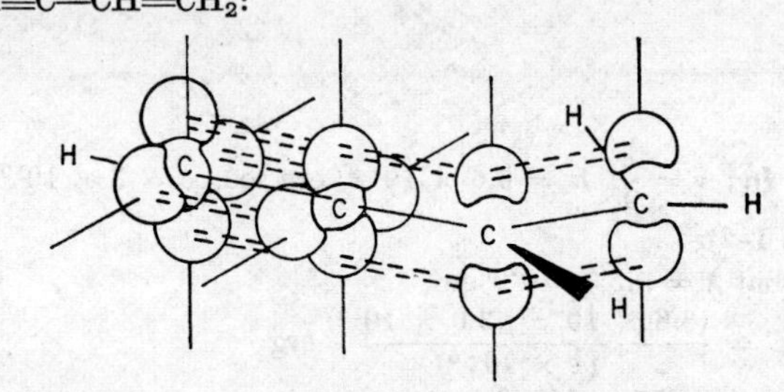

Let carbon atoms 1 and 2 be *sp* hybridized, as in acetylene, and carbon atoms 3 and 4 be sp^2 hybridized, as in ethylene. Then σ molecular orbitals can be formed between the carbon hybrid orbitals (see diagram). The remaining carbon hybrid orbitals can form σ molecular orbitals with the $1s$ atomic orbitals of the hydrogen atoms (see diagram). The p_z atomic orbitals on carbon atoms 1 and 2 can interact to form two π molecular orbitals (one bonding and one antibonding). The p_y atomic orbitals on all four carbon atoms can combine to give four π molecular orbitals, as in butadiene. For reasons that we have not discussed, the π orbitals contributing to the 'triple' bond between carbon atoms 1 and 2 have an interaction (cf. the cylindrical symmetry of the corresponding orbitals in acetylene, Figure 3.12) which will reduce the effective interaction of the p_z atomic orbitals on these carbon atoms with those of carbon atoms 3 and 4.

(*c*) $\overset{1}{C}H_3\overset{2}{C}H{=}\overset{3}{C}{=}\overset{4}{C}H_2$. Two arrangements seem possible. In both, carbon atoms 2 and 4 are sp^2 hybridized, carbon atom 3 *sp* hybridized, and carbon atom 1 sp^3 hybridized. Thus carbon atoms 2, 3, and 4 are in a straight line. Carbon atom 4 can now be orientated so that its unhybridized *p* atomic orbital has the same symmetry as the *p* atomic orbitals of carbon atom 2 (I). It is possible to combine *p* atomic orbitals on carbon atoms 2, 3, and 4 to give three π molecular orbitals. One of these will be bonding, one approximately non-bonding, and one antibonding. In addition, there remains the orthogonal *p* atomic orbital on carbon atom 3 (shown shaded), which is also non-bonding. Thus, of the

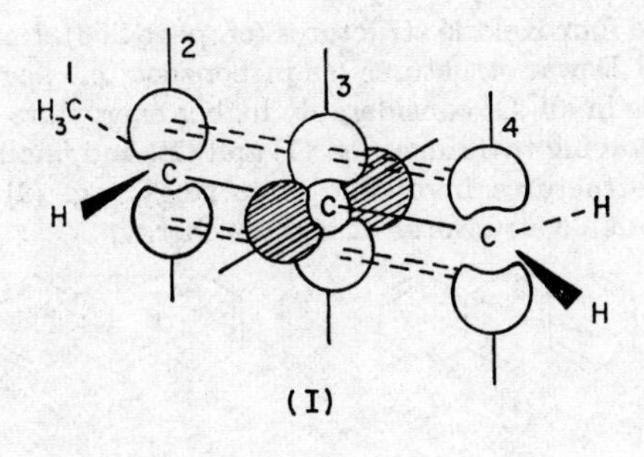

(I)

(II)

four valence electrons not associated with σ bonds, two must go into the bonding π molecular and two into one of the non-bonding orbitals (if they are both exactly non-bonding they are degenerate and according to Hund's rules each will take one electron, giving a triplet ground state for the molecule).

In the alternative arrangement (II), carbon atom 4 is orientated so that its p atomic orbital is orthogonal to the interacting orbitals of atoms 2 and 3 but has the correct symmetry to interact with the second p atomic orbital of carbon atom 3. We now have two pairs of orthogonal π molecular orbitals and the four electrons can go into the bonding orbital of each pair. These orbitals will be of higher energy than the bonding orbital of arrangement (I), but all four electrons are in bonding molecular orbitals so that the overall energy is lower. Thus we would predict that arrangement (II) would be preferred. Notice that the terminal σ bonds in this arrangement are perpendicular to each other, whereas in arrangement (I) they are all in one plane. Experiment shows these molecules assume the shape required by arrangement (II).

Chapter 4

We have to consider the ways in which we can pair the p_y atomic orbitals of the carbon atoms making up the rings. Those of lowest energy are the four Kekulé structures (cf. page 256). Next in energy are the long-bond Dewar structures (as in benzene, cf. page 38), three for each ring, nine in all. Of considerably higher energy are eight long-bond structures involving two rings, e.g. (**1**) and (**2**), and finally there are four long-bonded structures involving three rings, e.g. (**3**) and (**4**), these being the non-ionic structures of highest energy.

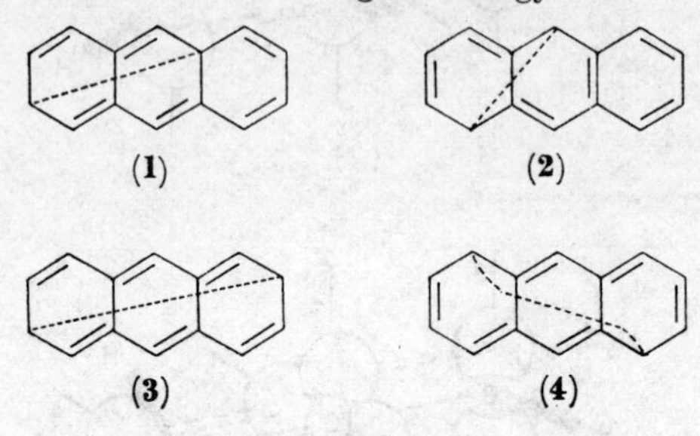

Chapter 5

Dipole moments can be represented as vectors, and the ordinary rules of vector addition apply. Thus the square of the sum of two vectors **A** and **B** inclined at an angle θ is given by $\mathbf{A}^2 + \mathbf{B}^2 + 2\mathbf{AB}\cos\theta$. For aromatic derivatives this simplifies because *ortho*-substituents are inclined at 60°, *meta* at 120°, and *para* at 180°.

o-Dichlorobenzene $\mu = 2.4$ D; *m*-dichlorobenzene $\mu = 1.6$ D; *p*-dichlorobenzene $\mu = 0$; *o*-chlorotoluene $\mu = 1.3$ D; *m*-chlorotoluene $\mu = 1.7$ D; 1,3,5-trichlorobenzene $\mu = 0$.

Chapter 7

Dissociation constants of substituted benzoic acids at 25° in water:

p-nitro (3.6×10^{-4}) > *m*-nitro (3.2×10^{-4}) > *m*-fluoro (1.4×10^{-4}) > *m*-ethoxy (8.3×10^{-5}) > *p*-fluoro (7.2×10^{-5}) > *m*-methyl (5.6×10^{-5}) > *p*-methyl (4.2×10^{-5}) > *p*-ethoxy (2.0×10^{-5}).

Chapter 8

1. Phenoxide ions couple in *ortho*- and *para*-positions (predominantly *para*). Dimethylaniline couples exclusively in the *para*-position.

2. $C_6H_5NO_2$ *meta*, $CH_3COC_6H_5$ *meta*; C_6H_5Cl *ortho* and *para*, $C_6H_5OCH_3$ *ortho* and *para*; *p*-nitrotoluene gives 2,4-dinitrotoluene, and under very mild conditions resorcinol dimethyl ether gives 2,4-dimethoxy-1-nitrobenzene.

3. $C_6H_5CHFCH_3$; CH_3CHFCH_3; $CF_3CH_2CH_2F$; $CH_3COCH_2CH_2F$; $CH_3CH_2CF_2CH_3$.

Chapter 9

1. $CH_3CH_2COCH_2CN$ $pK_a \approx 9 > CH_3CH_2NO_2$ $pK_a \approx 10 >$ $CH_3COCH_2CO_2Et$ $pK_a \approx 11 > CH_2(CO_2CH_3)_2$ $pK_a \approx 13 >$ CH_3COCH_3 $pK_a \approx 20 > C_6H_5CH_2CN > C_6H_5CH_2CH_3$ $pK_a \approx 50 >$ $(CH_3)_3CH$ [non-acidic].

2. (*a*) $CH_3CH_2CH_2COCH(C_2H_5)CO_2Et$.
(*b*) $CF_3COCH_2CO_2Et$.
(*c*) $CH_3CHOHCH_2CHO + CH_3CH{=}CHCHO$ + polymer.
(*d*) $C_6H_5CH{=}CHCOCH_3 \longrightarrow C_6H_5CH{=}CHCOCH{=}CHC_6H_5$.
(*e*)
$$\begin{array}{ccc} CH_2 & \text{———} & CHCN \\ & \diagdown\ \diagup & \\ & CH_2 & \end{array}$$

3. (*a*) $CH_3COCH(C_2H_5)CO_2Et$.
(*b*)
$$\begin{array}{ccc} CH_2 & & COCH_3 \\ | & \diagdown\ \diagup & \\ | & C & \\ | & \diagup\ \diagdown & \\ CH_2 & & CO_2Et \end{array}$$
(*c*)
$$\begin{array}{l} CH_3COCH{-}CHCOCH_3 \\ \qquad\ \ |\qquad\ \ | \\ \quad EtO_2C\quad CO_2Et \end{array}$$
(*d*) No reaction.

4. (*a*) $C_3H_7OC_2H_5$.
(*b*) $CH_3COCH_2CO_2Et$.
(*c*) $C_3H_7OC_2H_5$.
(*d*) $CH_3COCH(C_3H_7)CO_2Et$.

Chapter 10

1. (*a*) $(CH_3)_3CCH_2\overset{\overset{\large CH_3}{|}}{C}{=}CH_2 + (CH_3)_3CCH{=}C(CH_3)_2$ (see Part 1, p. 65).
(*b*) $C_6H_5C(CH_3)_3$.
(*c*) $(CH_3)_2\underset{\underset{\large OCOCF_3}{|}}{C}CH_2COCH_3 \xrightarrow{\text{MeOH}} (CH_3)_2C(OH)CH_2COCH_3$

2. (*a*) *p*-$NO_2C_6H_4COC(C_6H_5)_2CH_3$.
(*b*) $HOC(CH_3)_2CH_2CH_3$.
(*c*) $CH_3CH{=}CH_2 + (CH_3)_2CHOH$.
(*d*) $C_6H_5CH(CH_3)_2$ predominantly.

Chapter 11

1. (*a*) $C_6H_5CHBrCHBrCH_3$.
(*b*) *p*-$BrC_6H_4CHBrCHBrCH_3$ + *ortho*-compound.
(*c*) $C_6H_5CHBrCHBrCH_3 + C_6H_5CHBrCHBrCH_2Br$.
(*d*) $C_6H_5CH{=}CHCH_2Br$.

2.
$$\left.\begin{array}{l} (C_6H_5COO)_2 \longrightarrow 2C_6H_5CO_2\cdot \longrightarrow 2C_6H_5\cdot + 2CO_2 \\ C_6H_5\cdot + CF_2ClBr \longrightarrow C_6H_5Br + CF_2Cl\cdot \end{array}\right\} \text{Initiation}$$

$$
\left.\begin{array}{l}
CF_2Cl\cdot + C_2H_4 \longrightarrow CF_2ClCH_2CH_2\cdot \\
CF_2ClCH_2CH_2\cdot + CF_2ClBr \longrightarrow CF_2ClCH_2CH_2Br + CF_2Cl\cdot \\
CF_2ClCH_2CH_2\cdot + C_2H_4 \longrightarrow CF_2ClCH_2CH_2CH_2CH_2\cdot
\end{array}\right\} \text{Chain propagation}
$$

$$
\left.\begin{array}{l}
CF_2Cl\cdot + CF_2Cl\cdot \longrightarrow CF_2ClCF_2Cl \\
CF_2Cl(CH_2)_n\cdot + CF_2Cl\cdot \longrightarrow CF_2Cl(CH_2)_nCF_2Cl \\
CF_2Cl(CH_2)_n\cdot + CF_2Cl(CH_2)_n\cdot \longrightarrow CF_2Cl(CH_2)_{2n}CF_2Cl
\end{array}\right\} \text{Chain termination}
$$

Chapter 12

1. (*A*) $(C_6H_5CO)_3CH$. (*B*) $(C_6H_5CO)_2C{=}C(C_6H_5)OH$.
2. See pages 142–144.

Chapter 13

1. (*a*) $(CH_3CH(CN)CH_2CO_2Et) \longrightarrow CH_3CH(CN)CH_2CO_2Na$.

   (*b*) $C_6H_5CH(SO_3Na)CH_2COC_6H_5$.

   (*c*) $(C_6H_5)_2CHCH(C_6H_5)(NO_2)$

   (*d*) $CH_3CO(CH_2)_2CH(NO_2)CH(NO_2)(CH_2)_2COCH_3$.

   (*e*) $N(CH_2CH_2CN)_3\ [+HN(CH_2CH_2CN)_2]$

   (*f*) $CH_3CHCH(CO_2Et)_2$ with CH_3CHCO_2Et attached at the CH: $CH_3CH[CH(CO_2Et)_2]CHCO_2Et$...

$$
\left[via\ \ \begin{array}{l} CH_3CH\bar{C}HCO_2Et \\ \ \ \ \ \ \ \ | \\ CH_3C(CO_2Et)_2 \end{array} \longrightarrow \begin{array}{l} CH_3CH{-}CHCO_2Et \\ \ \ \ \ \ \ \ |\ \ \ \ \ \ \ \ | \\ CH_3C{-}CO \\ \ \ \ \ \ \ \ | \\ \ \ \ \ \ CO_2Et \end{array} \longrightarrow \right]
$$

2. No reaction; no reaction; [2,4-dinitroanisole: OCH_3, NO_2, NO_2]; same product as previous example; [2-chloro-4,6-dinitroanisole: OCH_3, O_2N, Cl, NO_2]; [2,3,5,6-tetrafluoroanisole: H, F, F, F, F, OCH_3]

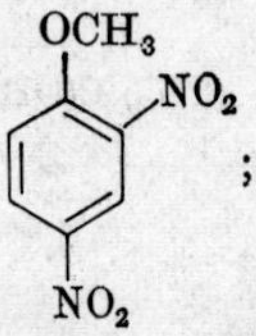

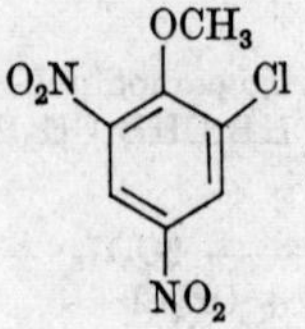

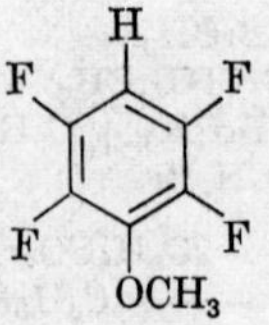

Chapter 14

$$(a)\ CH_3CH_2COCH_3 \xrightarrow{HNO_2} \underset{\displaystyle NO}{CH_3\underset{|}{C}HCOCH_3} \rightleftarrows \underset{\displaystyle HON}{CH_3\underset{\|}{C}COCH_3} \xrightarrow{NH_2OH}$$

$$(b)\ C_6H_5COC_6H_5 \xrightarrow{NH_2OH} \underset{\displaystyle NOH}{C_6H_5\underset{\|}{C}C_6H_5} \xrightarrow[2.\ H_2O]{1.\ PCl_5}$$

$$(c)\ C_6H_5NO \longrightarrow C_6H_5NH_2 \xrightarrow{C_6H_5NO}$$

Chapter 15

1. (*a*) $CH_3OC_6H_4OCH_3$; (*b*) p-$CH_3OC_6H_4CO_2CH_3$;
(*c*) o-$NO_2C_6H_4COCHN_2$ [provided diazomethane in excess],

(*d*)

```
C6H5         COCH3
    \       /
     HC — CH
    /       \
 H2C         N
    \      //
     N
```

2. All these reactions involve rearrangement:
(*a*) pinacol–pinacolone; $(CH_3)_3COCH_3$;
(*b*) Beckmann; $CH_3CONHC_6H_5$;
(*c*) Wolff; $C_6H_5CH_2CO_2CH_3$;
(*d*) Curtius; $C_6H_5CH_2NH_2$.

Chapter 16

$$\textbf{1.}\ CH_3CHO \xrightarrow{Base} \bar{C}H_2CHO \xrightarrow{HCHO}$$

$$HOCH_2CH_2CHO \xrightarrow[\text{as before}]{HCHO} (HOCH_2)_2CHCHO \xrightarrow[\text{as before}]{HCHO}$$

$$(HOCH_2)_3CHO \xrightarrow[\text{Cannizzaro reaction, slow}]{HCHO} (HOCH_2)_4C + HCO_2H$$

By-products will include methanol (simple Cannizzaro reaction), aldol and crotonaldehyde (from self-condensation of acetaldehyde), acrolein ($CH_2{=}CHCHO$) by elimination from initial product, and various compounds, especially dipentaerythritol as a result of Michael addition to acrolein. In practice the yield of pentaerythritol is about 50%.

2. (*a*)

$$\begin{matrix} CH_3 \\ \quad\diagdown \\ \quad CHBr \\ \quad\diagup \\ C_2H_5 \end{matrix} + CH_2(CO_2Et)_2 \xrightarrow{NaOEt} \begin{matrix} CH_3 \\ \quad\diagdown \\ \quad CHCH(CO_2Et)_2 \\ \quad\diagup \\ C_2H_5 \end{matrix} \xrightarrow[2.\ heat]{1.\ H_2O} \begin{matrix} CH_3 \\ \quad\diagdown \\ \quad CHCH_2CO_2H \\ \quad\diagup \\ C_2H_5 \end{matrix}$$

(*b*)

$$n\text{-}C_5H_{11}Br + CH_3COCH_2CO_2Et \xrightarrow{NaOEt} CH_3CO\overset{n\text{-}C_5H_{11}}{\overset{|}{C}}HCO_2Et \xrightarrow[2.\ heat]{1.\ Dil.\ alkali} n\text{-}C_5H_{11}COCH_3$$

(*c*)

$$Br(CH_2)_3Br + CH_2(CO_2Et)_2 \xrightarrow{2\ NaOEt} \text{cyclobutane-1,1-}(CO_2Et)_2$$

(*d*)

$$Br(CH_2)_4Br + CH_3COCH_2CO_2Et \xrightarrow{2\ NaOEt} \text{cyclopentane-1-}(COCH_3)\text{-1-}(CO_2Et)$$

(This type of reaction does not occur with acetoacetic ester and methylene tribromide; instead, a heterocyclic compound is formed.)

(*e*)

$$Br(CH_2)_3Br + 2CH_2(CO_2Et)_2 \xrightarrow{2NaOEt} \begin{matrix} CH(CO_2Et)_2 \\ \diagup \\ (CH_2)_3 \\ \diagdown \\ CH(CO_2Et)_2 \end{matrix} \xrightarrow[CH_2I_2]{2NaOEt}$$

$$\text{cyclohexane-1,1,3,3-}(CO_2Et)_4 \xrightarrow[2.\ heat]{1.\ H_2O} \text{cyclohexane-1,3-}(CO_2Et)_2$$

(*f*)

$$\begin{matrix} CH(CO_2Et)_2 \\ \diagup \\ (CH_2)_2 \\ \diagdown \\ CH(CO_2Et)_2 \end{matrix} \xrightarrow[Dieckmann]{NaOEt} \text{2-oxocyclopentane-1,1,3-}(CO_2Et)_3 \xrightarrow[2.\ heat]{1.\ H_2O} \text{2-oxocyclopentane-1,3-}(CO_2H)_2$$

3. (*a*) cyclohexanone + morpholine (HN, O) → 1-morpholinocyclohexene (N, O) $\xrightarrow[\text{2. hydrolysis}]{\text{1. } CH_3COCl}$

2-acetylcyclohexanone ($COCH_3$, O) $\xrightarrow[\text{reflux}]{\text{NaOH}}$ $CH_3CO(CH_2)_5CO_2H$ (Reverse Claisen; cf. acid hydrolysis of ethyl acetoacetate)

(*b*) $C_6H_5COCH_3 + (CH_2CO_2Et)_2 \xrightarrow{\text{Base}} (C_6H_5)(CH_3)C{=}C(CO_2Et)(CH_2CO_2Et)$ $\xrightarrow{H_2/Ni}$
(both *cis* and *trans*)

$(C_6H_5)(CH_3)CH{-}CH(CO_2Et)(CH_2CO_2Et)$

(The Knoevenagel-type condensation with diethyl succinate is known as the Stobbé reaction)

(*c*) $C_6H_5CHO + C_6H_5CH_2CO_2H \xrightarrow[\text{Et}_3\text{N reflux}]{(CH_3CO)_2O} C_6H_5CH{=}C(C_6H_5)(CO_2H)$
(70% *cis*)

(Perkin reaction, cf. page 200)

(*d*) $C_6H_5CHO + CH_3CH{=}CHCHO \xrightarrow[\text{EtOH}]{C_5H_5N} C_6H_5(CH{=}CH)_2CHO$
50%

[$C_6H_5(CH{=}CH)_4CHO$ also formed]

(Straight Claisen type condensation involving conjugated enolate anion)

(*e*) $(CH_3)_2C{=}CHCOCH_3 + CH_2(CO_2Et)_2 \xrightarrow[\text{EtOH}]{\text{NaOEt}} (CH_3)_2C(CH_2COCH_3)(CHCO_2Et\text{—}CO_2Et)$ →

cyclic diketo ester (H_3C, H_3C, C, CH_2—CO, CH_2, CH—CO, CO_2Et) $\xrightarrow[\text{2. heat.}]{\text{1. KOH}}$ 5,5-dimethylcyclohexane-1,3-dione (H_3C, CH_3, O, O)

(Michael addition, followed by Dieckmann-type condensation; finally, decarboxylation of β-keto acid.)

Chapter 17

1. (*a*) CH_3 (toluene) $\xrightarrow[H_2SO_4]{HNO_3}$ o-nitrotoluene (CH_3, NO_2) + p-nitrotoluene (CH_3, NO_2)

p-nitrotoluene $\xrightarrow{\text{1. reduce; 2. acetylate}}$ CH_3, $NHCOCH_3$

o-nitrotoluene: 1. reduce; 2. acetylate → CH_3, $NHCOCH_3$

o-nitrotoluene: Further nitration → CH_3, NO_2, NO_2 — 2,4-Dinitrotoluene

p-nitrotoluene: HNO_3, H_2SO_4, vigorous → CH_3, O_2N, NO_2, $NHCOCH_3$

p-acetotoluidide: HNO_3, H_2SO_4, mild → CH_3, NO_2, $NHCOCH_3$

o-acetotoluidide: 1. HNO_3, H_2SO_4; 2. Hydrol. → CH_3, NH_2, NO_2 + CH_3, NH_2, O_2N

1. Hydrol.; 2. diazotize; 3. Na_2HPO_2 → CH_3, O_2N, NO_2 — 3,5-Dinitrotoluene

1. Hydrol.; 2. diazotize; 3. $Na_3Co(NO_2)_6$ → CH_3, NO_2, NO_2 — 3,4-Dinitrotoluene

1. Diazotize; 2. $Na_3Co(NO_2)_6$ → CH_3, NO_2, NO_2 — 2,3-Dinitrotoluene; CH_3, NO_2, O_2N — 2,5-Dinitrotoluene

(*b*)

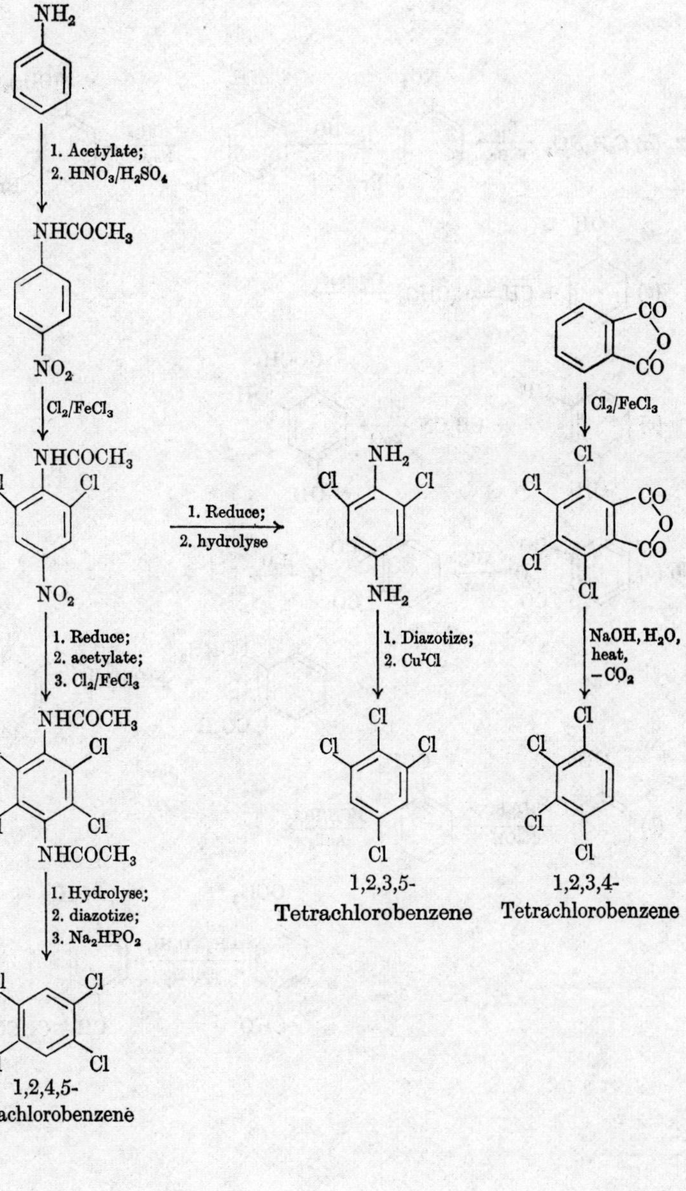

NH_2 (aniline) → 1. Acetylate; 2. HNO_3/H_2SO_4 → $NHCOCH_3$ / NO_2 → $Cl_2/FeCl_3$ → $NHCOCH_3$, Cl, Cl, NO_2

→ 1. Reduce; 2. hydrolyse → NH_2, Cl, Cl, NH_2 → 1. Diazotize; 2. Cu^ICl → Cl, Cl, Cl, Cl

1,2,3,5-
Tetrachlorobenzene

NO_2 compound → 1. Reduce; 2. acetylate; 3. $Cl_2/FeCl_3$ → $NHCOCH_3$, Cl, Cl, Cl, Cl, $NHCOCH_3$ → 1. Hydrolyse; 2. diazotize; 3. Na_2HPO_2 → Cl, Cl, Cl, Cl

1,2,4,5-
Tetrachlorobenzene

Phthalic anhydride (CO, O, CO) → $Cl_2/FeCl_3$ → Cl, Cl, Cl, Cl, CO, O, CO → $NaOH, H_2O$, heat, $-CO_2$ → Cl, Cl, Cl, Cl

1,2,3,4-
Tetrachlorobenzene

2. (*a*) $C_6H_5NO_2 \xrightarrow[FeBr_3]{Br_2}$ (NO_2, Br) $\xrightarrow{[H]}$ (NH_2, Br) $\xrightarrow[Na_2CO_3]{CH_3I}$ ($N(CH_3)_2$, Br)

(*b*) (OH) + $CH_2{=}C(CH_3)_2 \xrightarrow{\text{Catalyst}}$

(*c*) (OH, OH) + $CH_3CN \xrightarrow[ZnCl]{HCl}$ ($COCH_3$, OH, OH)

3. (*a*) (CO, O, CO) $\xrightarrow[\text{heat}]{NH_3,}$ (CO, NH, CO) $\xrightarrow{H_2O}$

($CONH_2$, CO_2H) $\xrightarrow{NaOBr}$ (NH_2, CO_2H)

(*b*) (OH) $\xrightarrow[NaOH]{(CH_3)_2SO_4}$ (OCH_3) $\xrightarrow[AlCl_3]{HCN/HCl}$

(OCH_3, CHO) $\xrightarrow[\text{pyridine}]{CH_2(CO_2H)_2}$ (OCH_3, $CH{=}CHCO_2H$)

(*c*) benzene + succinic anhydride (H_2C-CO, O, H_2C-CO) $\xrightarrow{AlCl_3}$ $C_6H_5CO-CH_2-CH_2-CO_2H$ $\xrightarrow[\text{(Clemmensen)}]{Zn/HCl}$

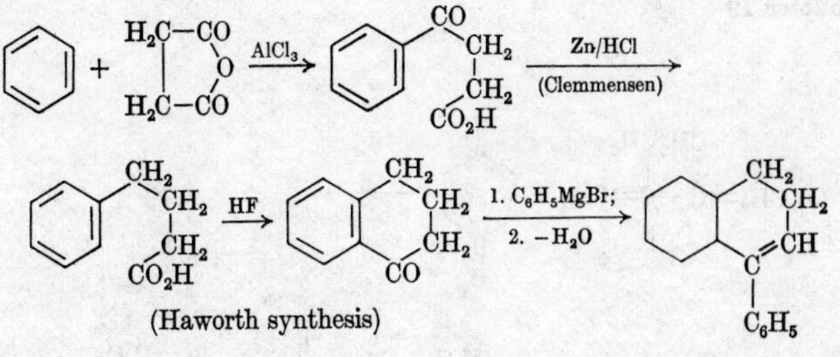

$C_6H_5CH_2CH_2CH_2CO_2H$ $\xrightarrow{HF}$ (ring ketone: CH_2, CH_2, CH_2, CO) $\xrightarrow[2.\ -H_2O]{1.\ C_6H_5MgBr;}$ (CH_2, CH_2, CH, C—C_6H_5)

(Haworth synthesis)

(*d*) toluene (CH_3) + phthalic anhydride (CO, O, CO) $\xrightarrow{AlCl_3}$

H_3C—C_6H_4—CO—C_6H_4—CO_2H (HO_2C) $\xrightarrow[H_2SO_4]{SO_3}$ H_3C-anthraquinone (O, O)

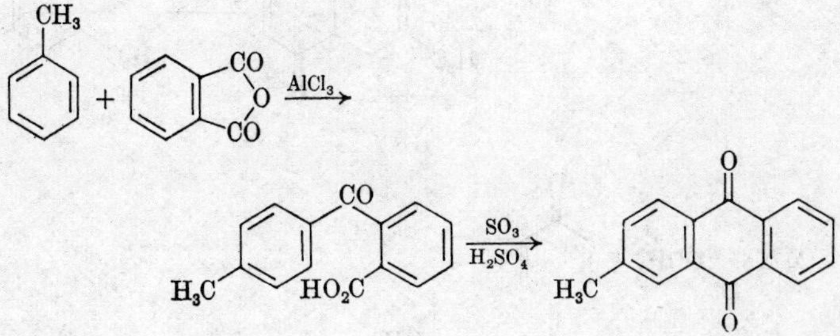

Chapter 18

1. (*a*) See pages 227–228.

(*b*) $H_3\overset{+}{N}CH_2CO_2H \underset{}{\overset{K_1}{\rightleftharpoons}} H_3\overset{+}{N}CH_2CO_2^- + H^+_{aq} \overset{K_2}{\rightleftharpoons}$

$H_2NCH_2CO_2^- + 2H^+_{aq}$

2. (*a*) 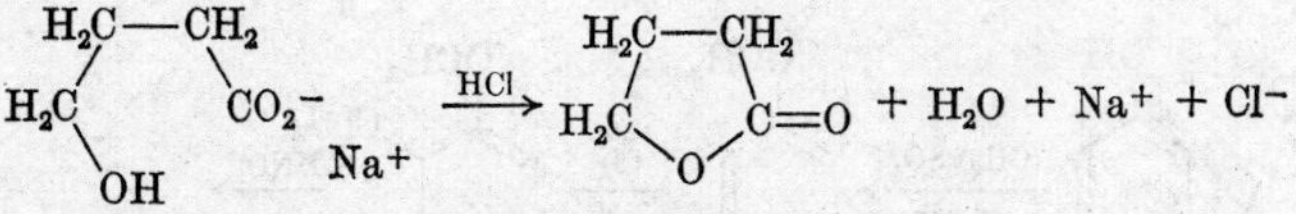 $H_2C(OH)-CH_2-CH_2-CO_2^-\ Na^+ \xrightarrow{HCl}$ (ring: H_2C-CH_2, H_2C, O, $C{=}O$) $+ H_2O + Na^+ + Cl^-$

(*b*) $(CH_2)_4(CO_2H)_2 \xrightarrow{(CH_3CO)_2O}$ (ring: H_2C, CH_2, CO, H_2C, CH_2) $+ CO_2 + H_2O$

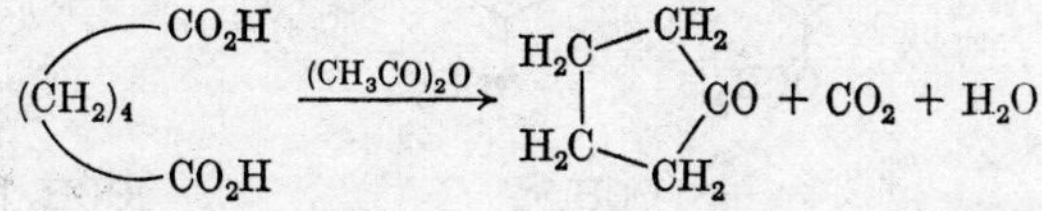

(*c*) (*A*) $HOCH_2CH_2CH(OH)CH_2CH_2OH$ (*C*) $CH_3CH(OH)-CH(OH)-CH(OH)CH_3$

(*B*) $HOCH_2CH(OH)CH_2CH_2CH_2OH$ or $HOCH_2CH(OH)CH_2CH(OH)CH_3$

Chapter 19

1. (*a*) $CH_2{=}C(CH_3){-}C(CH_3){=}CH_2$ + (p-benzoquinone) ⟶

(6,7-dimethyl-4a,5,8,8a-tetrahydro-1,4-naphthoquinone) $\xrightarrow{\text{Oxidise}}$ (6,7-dimethyl-1,4-naphthoquinone)

(*b*) $Na^+\ {}^-OCH_3$ + (p-benzoquinone) ⟶

(2,5-dimethoxyhydroquinone: OH, OCH_3, CH_3O, OH) $\xrightarrow{\text{Oxidise}}$ (2,5-dimethoxy-1,4-benzoquinone: O, OCH_3, CH_3O, O)

(*c*) (hydroquinone: OH, OH) $\xrightarrow[\text{NaOH}]{(CH_3)_2SO_4}$ (1,4-dimethoxybenzene: OCH_3, OCH_3) $\xrightarrow{Cl_2}$ (2-chloro-1,4-dimethoxybenzene: OCH_3, Cl, OCH_3) $\xrightarrow{HNO_3}$

(1-chloro-2,5-dimethoxy-4-nitrobenzene: OCH_3, Cl, O_2N, OCH_3) $\xrightarrow{\text{Morpholine}}$ (4-(2,5-dimethoxy-4-nitrophenyl)morpholine: OCH_3, N, O, O_2N, OCH_3)

2. (*a*) $2(CH_3)_2NC_6H_5 + C_6H_5CHO \xrightarrow{HCl} [(CH_3)_2NC_6H_4]_2CH$ (Leuco-base) $\xrightarrow[HCl]{PbO_2}$ Malachite Green

(*b*) $2(CH_3)_2NC_6H_5 + COCl_2 \longrightarrow [(CH_3)_2NC_6H_4]_2CO$ (Michler's ketone) $\xrightarrow[\text{dimethylaniline, } POCl_3]{\text{Excess}}$ Crystal Violet (p. 352)

(*c*) 3 mols. salicylic acid + formaldehyde $\xrightarrow[NaNO_2]{H_2SO_4}$

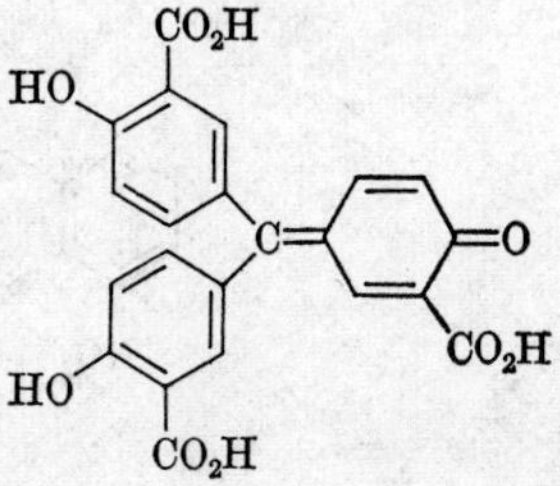

HO, CO_2H; HO, CO_2H; C=; =O; CO_2H

Chapter 20

1. (*a*) $^+NH_3\ HSO_4^-$ $\xrightarrow{\text{Heat}}$ NH_2, SO_3H

(*b*) SO_3H $\xrightarrow[H_2SO_4]{HNO_3}$ (O_2N SO_3H) + SO_3H, NO_2

S_3OH, NO_2 $\xrightarrow[H_2SO_4]{Fe}$ SO_3H, NH_2

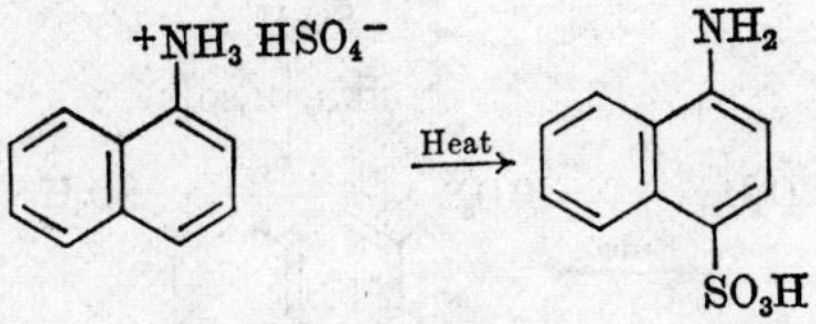

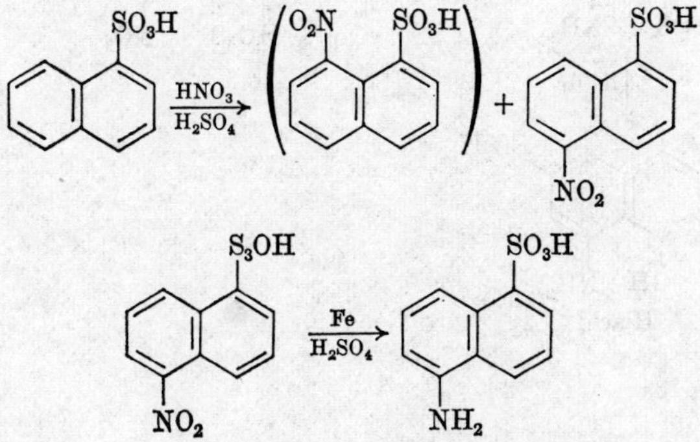

Laurent acid

(c) SO_3H $\xrightarrow[H_2SO_4]{HNO_3}$

(NO_2 SO_3H) + SO_3H NO_2

SO_3H NO_2 $\xrightarrow[H_2SO_4]{Fe}$ SO_3H NH_2

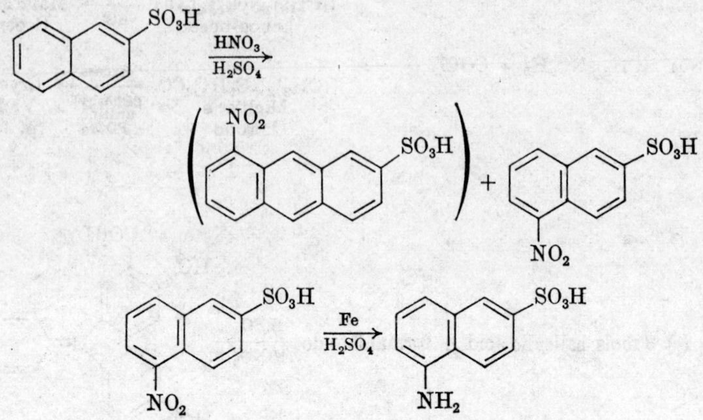

Cleve acid

(d) SO_3H $\xrightarrow[Oleum]{60\%}$ HO_3S SO_3H SO_3H

HNO_3 | H_2SO_4

HO_3S SO_3H HO_3S NO_2

$\xleftarrow{Reduce}$

HO_3S SO_3H HO_3S NH_2

NaOH

HO_3S SO_3H OH NH_2

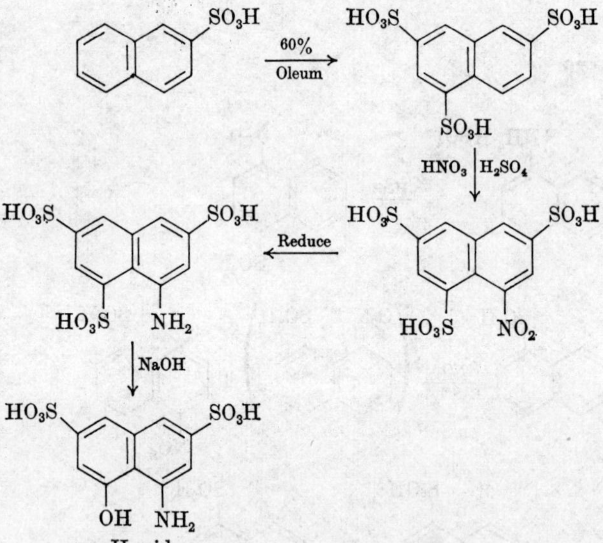

H acid

(*e*) $\overset{+}{N}H_3HSO_4^-$ $\xrightarrow[\text{Oleum}]{60\%}$

SO_3H, HO_3S, NH_2, SO_3H
(water-soluble)

+

SO_3H, NH_2, HO_3S
(water-insoluble)

Dil. H_2SO_4, boil

HO_3S, NH_2, SO_3H $\xrightarrow[\text{pressure}]{\text{Aq. NaOH}}$ HO_3S, NH_2, OH

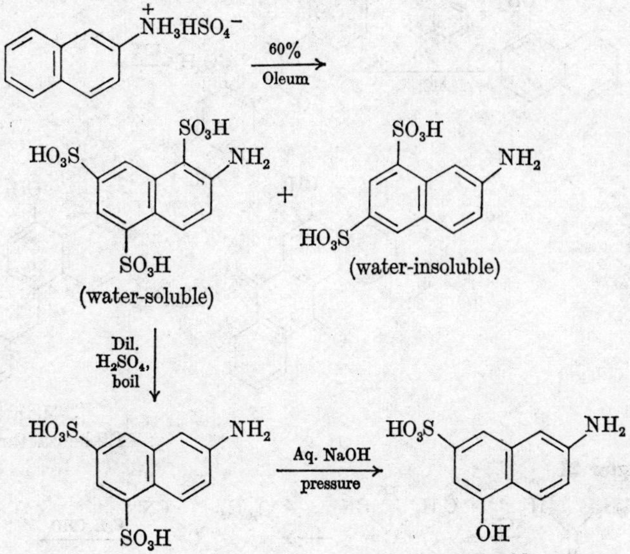

J acid

(*f*) OH $\xrightarrow{HNO_2}$ NO, OH $\xrightarrow{NaHSO_3}$ NH_2, OH, SO_3H

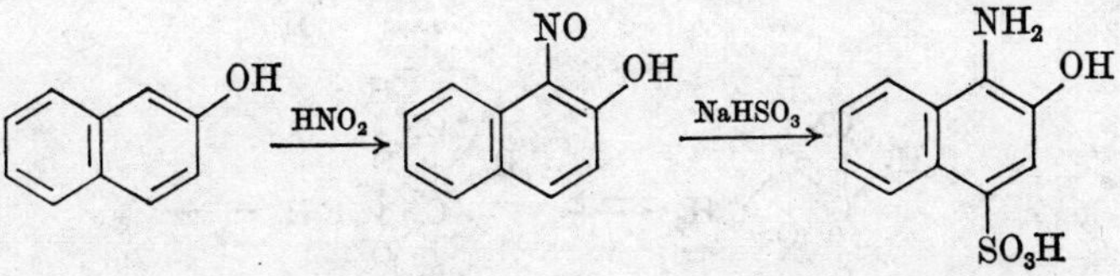

1,2,4-Acid

2. (*a*) CH_2CO_2H, H_3C, CH_3 — CHO, NO_2 (Perkin conditions) →

HO_2C, H_3C, O_2N, CH_3 → 1. Reduce; 2. diazotize; 3. Cu powder (Pschorr conditions) → HO_2C, H_3C, CH_3

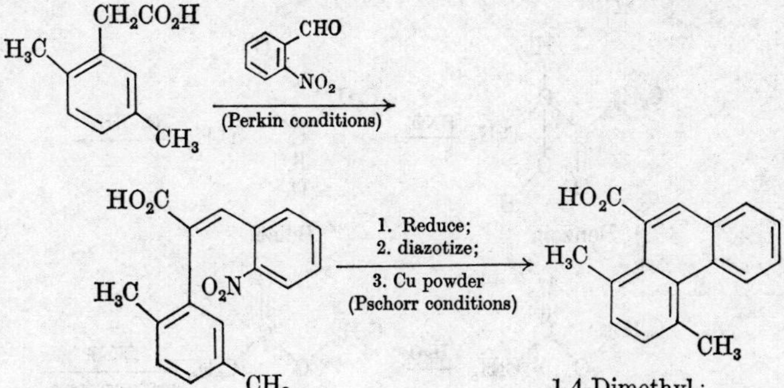

1,4-Dimethyl-
phenanthrene-10-
carboxylic acid

(b) 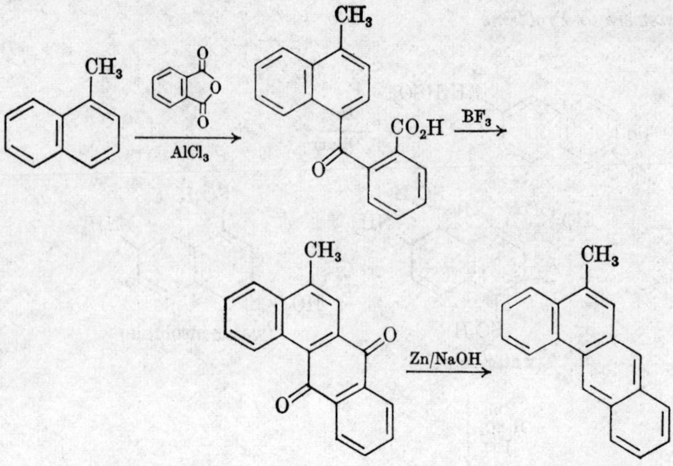 CH_3 $\xrightarrow{AlCl_3}$ CH_3 CO_2H O $\xrightarrow{BF_3}$

CH_3 O O $\xrightarrow{Zn/NaOH}$ CH_3

3-Methyl-
1,2-benzanthracene

Chapter 21

1. 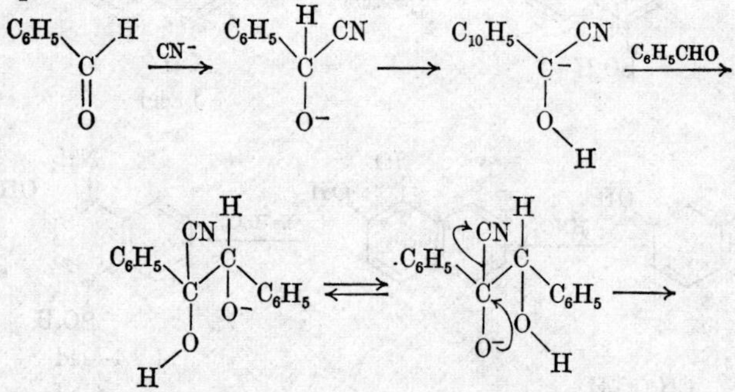 C_6H_5, H, C=O $\xrightarrow{CN^-}$ C_6H_5, H, CN, C, O^- $\longrightarrow$ $C_{10}H_5$, CN, C^-, O–H $\xrightarrow{C_6H_5CHO}$

C_6H_5, CN, H, C, C, O^-, C_6H_5, O–H $\rightleftharpoons$ C_6H_5, CN, H, C, C, C_6H_5, O^-, O–H $\longrightarrow$

C_6H_5, H, C=O, C, C_6H_5, O–H $\xrightarrow{HNO_3}$ C_6H_5, C=O, C=O, C_6H_5 $\xrightarrow{NH_2NH_2}$

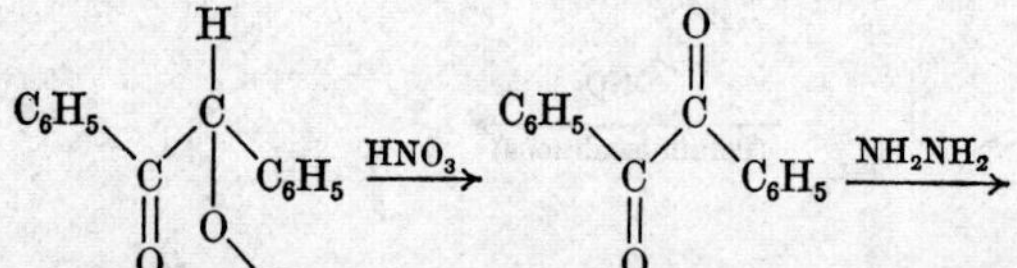

Benzoin Benzil

C_6H_5, C=O, C=N–NH_2, C_6H_5 $\xrightarrow{HgO}$ C_6H_5, C=O, C=N^+=N^-, C_6H_5 $\xrightarrow[\text{rearrangement}]{\text{Wolff}}$

$(C_6H_5)_2C{=}C{=}O \xrightarrow{MeOH} (C_6H_5)_2CHCO_2CH_3$

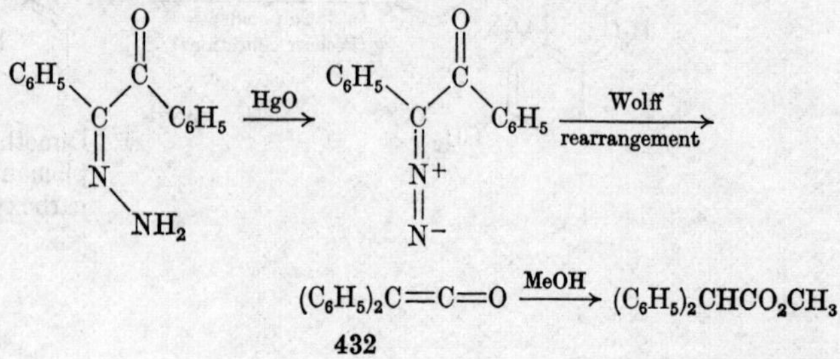

2. Reaction with

(i) $CH_3CH{=}CH{-}CH{=}O$	(ii) $CH_3CH_2CH{=}C{=}O$
(*a*) [$CH_3CH{=}CHCH(OEt)_2$ if an acid catalyst is present]	$CH_3CH_2CH_2CO_2Et$ (rapid; no catalyst required)
(*b*) $CH_3CHBrCH_2CHO$	$CH_3CH_2CH_2COBr$
(*c*) $CH_3CH{=}CH{-}CH{=}NC_6H_5$	$CH_3CH_2CH_2CONHC_6H_5$
(*d*) $CH_3CH{=}CH(C_6H_5)OH$ + $CH_3CH(C_6H_5)CH_2CHO$	$CH_3CH_2CH_2COC_6H_5$
(*e*) No reaction	$CH_3CH_2CH_2CO_2C(CH_3){=}CH_2$

Chapter 22

1. (*a*) 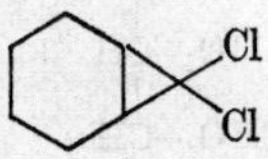

(*b*) 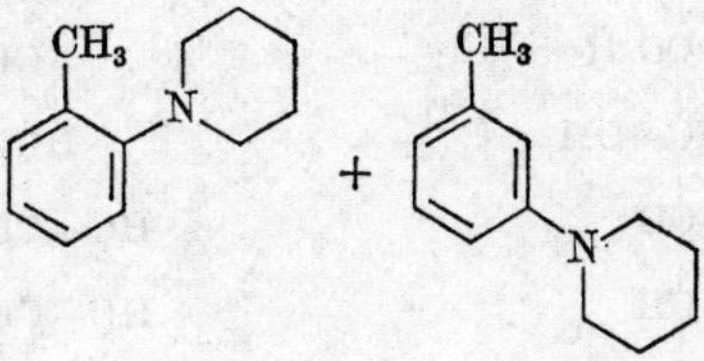

(*c*) 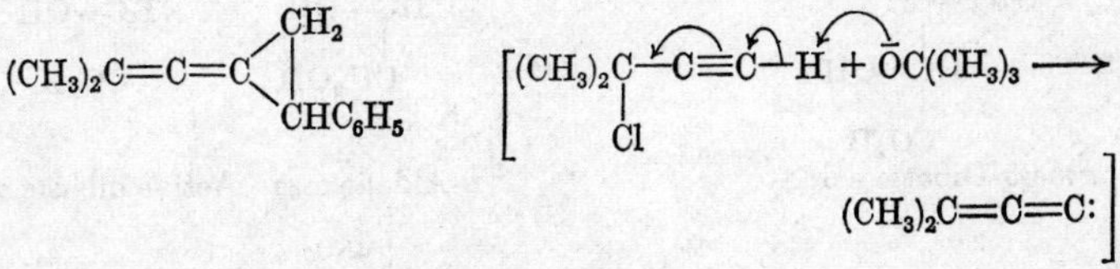

2. (*a*) $CH_3CH_2CH_2CH_3 + (CH_3)_2CHCH_3$

(*b*) 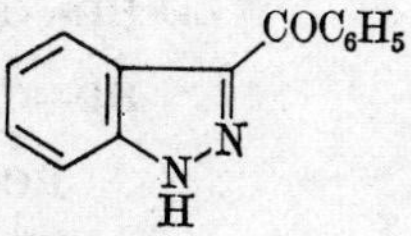

(*c*) $C_6H_5SO_2NHC_6H_5$ [$C_6H_5SO_2N_3 \longrightarrow C_6H_5\ddot{S}ON\!: + N_2$]

Chapter 23

1. (*a*) I, σ_V (*b*) I, C_∞, σ_V (*c*) $I, C_2, \sigma_V, \sigma_{V'}$ (*d*) I, σ_V
(*e*) I, C_2 (*f*) I (*g*) I, C_2 (*h*) I, C_2, S_4
(Notice that (*e*), (*f*), and (*g*) will exist in enantiomeric forms.)

2.

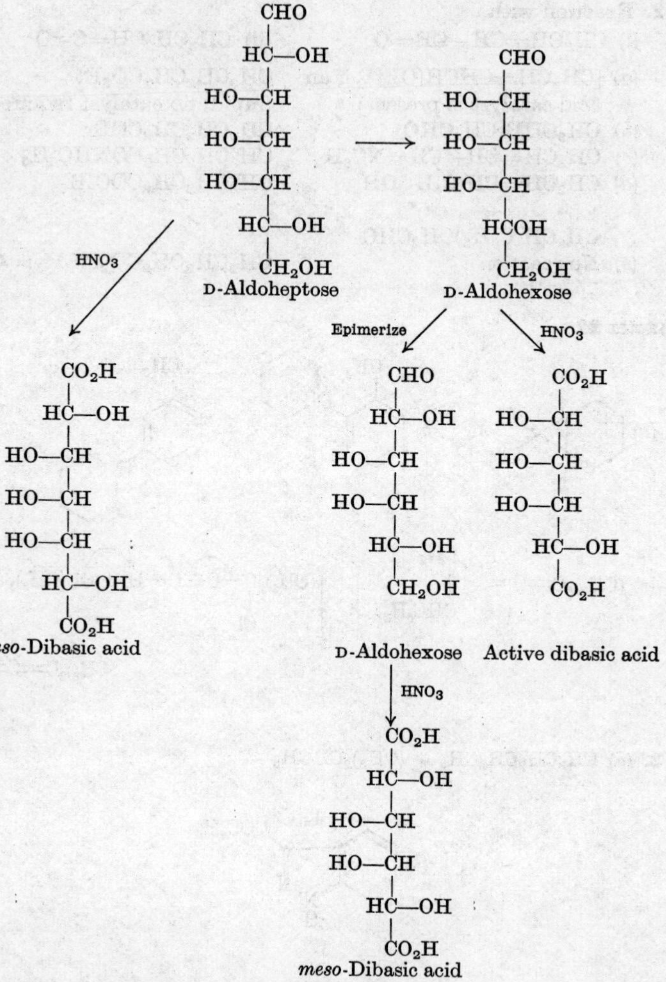

D-Aldoheptose:

$$\begin{array}{c} CHO \\ HC{-}OH \\ HO{-}CH \\ HO{-}CH \\ HO{-}CH \\ HC{-}OH \\ CH_2OH \end{array} \longrightarrow \begin{array}{c} CHO \\ HO{-}CH \\ HO{-}CH \\ HO{-}CH \\ HCOH \\ CH_2OH \end{array}$$

D-Aldoheptose → D-Aldohexose

D-Aldoheptose $\xrightarrow{HNO_3}$ *meso*-Dibasic acid:

$$\begin{array}{c} CO_2H \\ HC{-}OH \\ HO{-}CH \\ HO{-}CH \\ HO{-}CH \\ HC{-}OH \\ CO_2H \end{array}$$

meso-Dibasic acid

D-Aldohexose $\xrightarrow{Epimerize}$ D-Aldohexose; D-Aldohexose $\xrightarrow{HNO_3}$ Active dibasic acid:

$$\begin{array}{c} CHO \\ HC{-}OH \\ HO{-}CH \\ HO{-}CH \\ HC{-}OH \\ CH_2OH \end{array} \qquad \begin{array}{c} CO_2H \\ HO{-}CH \\ HO{-}CH \\ HO{-}CH \\ HC{-}OH \\ CO_2H \end{array}$$

D-Aldohexose   Active dibasic acid

D-Aldohexose $\xrightarrow{HNO_3}$

$$\begin{array}{c} CO_2H \\ HC{-}OH \\ HO{-}CH \\ HO{-}CH \\ HC{-}OH \\ CO_2H \end{array}$$

meso-Dibasic acid

3. (*A*)

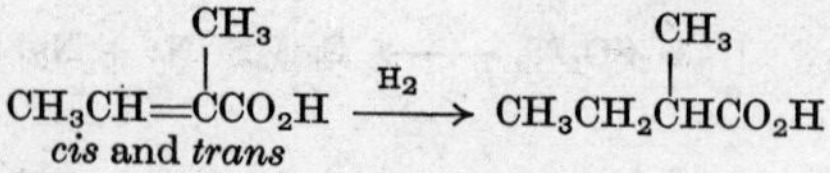

$$CH_3CH{=}\overset{\overset{CH_3}{|}}{C}CO_2H \xrightarrow{H_2} CH_3CH_2\overset{\overset{CH_3}{|}}{C}HCO_2H$$

cis and *trans*

(*B*)

$$\begin{array}{l} CH_2 \\ |\quad\diagdown \\ |\qquad CHCO_2H \\ |\quad\diagup \\ CH \\ | \\ CH_3 \end{array}$$

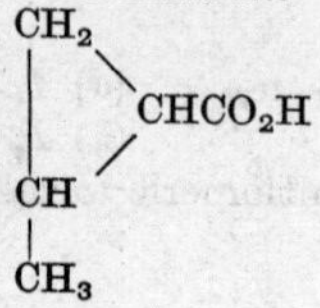

cis and *trans*

$$(C)\ \mathrm{CH_2{=}CH{-}\underset{}{\overset{\overset{\displaystyle CH_3}{|}}{C}}HCO_2H} \xrightarrow{H_2} \mathrm{CH_3CH_2\overset{\overset{\displaystyle CH_3}{|}}{C}HCO_2H}$$

4. (*a*) Two compounds with this formula are optically active; the other is a *meso*-form.

(*b*) Yes; cf. Question 1(*e*).

(*c*) Two compounds (*cis*- and *trans*-oximes), one active (see p. 167).

(*d*) Yes [cf. Question 1(*g*)].

(*e*) Yes; methyl groups are bent out of the plane of the benzene ring due to overcrowding.

(*f*) Yes; an asymmetric carbon atom.

Chapter 24

1. (*a*) $\mathrm{C_6H_5CH_2CH(CH_3)(OSO_2C_7H_7)} + \mathrm{{}^-OCOCH_3} \longrightarrow \mathrm{C_6H_5CH_2CH(CH_3)(OCOCH_3)} + \mathrm{{}^-OSO_2C_7H_7}$

(*b*) $\mathrm{HOCHCO_2H(CH_2CO_2H)} \xrightarrow{PCl_5} \mathrm{ClCHCO_2H(CH_2CO_2H)}$

(N.B. $\mathrm{\bar{O}{-}C(=O){-}C(H)(Cl)(CH_2CO_2H)} \xrightarrow{S_N1} \mathrm{\overset{\delta-}{O}\cdots\cdots C^{\delta+}(H)(CH_2CO_2H)}$ (bridged through $\mathrm{C{=}O}$) $\xrightarrow{OH_2}$ Retention of configuration)

(*c*) $\mathrm{C_6H_5(H_3C)C(CO_2Et)(OH)} \xrightarrow{\mathrm{ClOS{-}\overset{+}{N}C_5H_5\ Cl^-}} \mathrm{C_6H_5(H_3C)C(CO_2Et)(OSO{-}\overset{+}{N}C_5H_5)} + \mathrm{Cl^-} \longrightarrow \mathrm{C_6H_5(H_3C)C(CO_2Et)Cl}$

(N.B. C_6H_5, CO_2Et, H_3C — C — :Ö—H + :C̈l:—S(=O)—Cl ⟶

C_6H_5, CO_2Et, H_3C — C — O—S(=O)—Cl ⟶ C_6H_5, CO_2Et, H_3C — C — Cl

(can be isolated) Configuration retained)

2. (*a*) Ph(H)C=C(H)Ph $\xrightarrow{Br_2}$ [H; Ph—C—Br; Br—C—Ph; H] + [H; Br—C—Ph; Ph—C—Br; H] $\xrightarrow[EtOH]{KOH}$ Br(Ph)C=C(Ph)H

(*b*) EtC≡CEt $\xrightarrow[liq.\ NH_3]{Na}$ Et(H)C=C(H)Et ⟶ [H; Et—C—OH; HO—C—Et; H] + [H; HO—C—Et; Et—C—OH; H]

(*c*) [H; C_7H_7—C—$OCOCH_3$; C_6H_5—C—CH_3; H] ⟶ [H; C_7H_7—C—OH; C_6H_5—C—CH_3; H] $\xrightarrow{TsCl}$

[H; C_7H_7—C—OTs; C_6H_5—C—CH_3; H] $\xrightarrow{Br^-}$ [H; Br—C—C_7H_7; C_6H_5—C—CH_3; H] $\xrightarrow{KOH}$ $C_6H_5(CH_3)C{=}C(H)C_7H_7$ $\xrightarrow{OsO_4}$

[H; CH_3—C—OH; C_7H_7—C—OH; H] $\xrightarrow{H_2SO_4}$ CH_3—C(C_7H_7)(CHO)—C_6H_5

Chapter 25

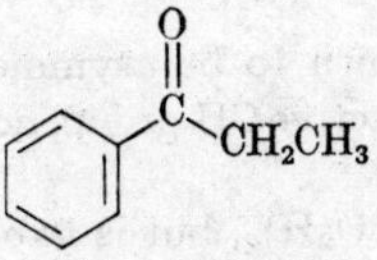

Propiophenone

The molecular weight of the compound is 134 (mass spectrum). The mass spectrum also shows that a fragment $m/e = 29$ (134 − 105), possibly $C_2H_5^+$, is readily lost from the molecule. The NMR spectrum confirms the presence of an ethyl group [a triplet centred at 8.8 τ (3H) and a quartet centred at 7.15 τ (2H). The NMR spectrum also shows aromatic protons (5H) from 2 to 3 τ, divided into two groups, 2H at ~ 2.1 τ and 3H at ~ 2.5 τ, possibly from a monosubstituted benzene ring. The mass spectrum supports this idea; the fragment appearing at $m/e = 77$ could be C_6H_5.

The infrared spectrum shows a carbonyl absorption just below 1,700 cm^{-1} (conjugated carbonyl group) and a pattern characteristic of five adjacent aromatic hydrogen atoms at 695 and 750 cm^{-1}.

The ultraviolet spectrum confirms the presence of an aromatic system [241 mμ (log ϵ 4.11) and 277 mμ (log ϵ 2.94) from curves A and B, respectively), conjugated with a carbonyl group [~ 320 mμ (log ϵ ~ 1.7) from curve *C*]. The placing of the carbonyl group in conjugation with the benzene ring also follows from a consideration of the characteristic pattern of the ethyl group shown in the NMR spectrum.

Chapter 26

A forms an oxime and semicarbazone . . . indicates C=O group.

A is not readily oxidized and is regenerated by chromium trioxide treatment of B . . . indicates a ketone.

B is formed by hydrogenation of A. So B is a saturated alcohol $C_{10}H_{17}OH$, four hydrogen atoms short of an open-chain alkanol ($C_{10}H_{21}OH$) . . . indicates a bicyclic system.

Some indications of the skeleton of A are obtained from the aromatization reactions with zinc and with iodine, which also indicates the position of the keto-group in A.

p-Cymene 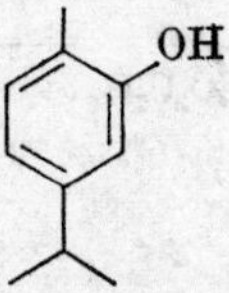 Carvacrol

Formation of a monobromo-derivative of A . . . CH group adjacent to a carbonyl group.

The dibasic acid is shown to be asymmetrical by the esterification evidence and still has a reactive CH group, now probably adjacent to one of the carboxyl groups.

D is saturated, $C_8H_{14}(CO_2H)_2$, but is two hydrogen atoms short of a dibasic aliphatic acid [$C_8H_{16}(CO_2H)_2$] and thus acid D is monocyclic.

D forms an anhydride; therefore the ring in A which was broken to form the dicarboxylic acid and carries the carboxyl group is five-membered. (In the formulae below R may be an alkyl group or hydrogen on the evidence considered so far.)

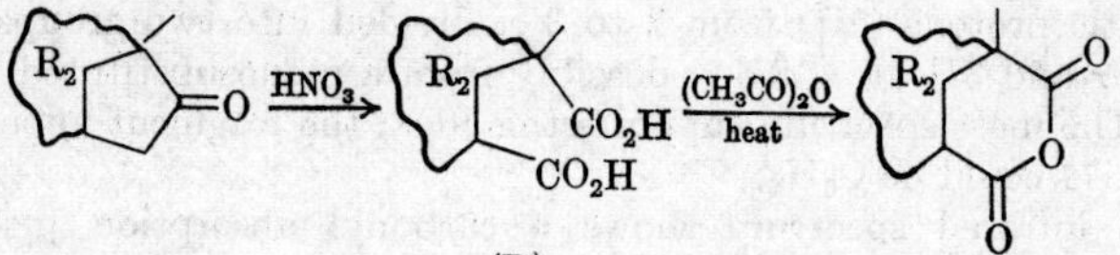

R_2 ... =O $\xrightarrow{HNO_3}$ R_2 ... CO_2H, CO_2H $\xrightarrow[\text{heat}]{(CH_3CO)_2O}$ R_2 ... O, O, O

(D)

D is oxidized to E and in the process the other ring is opened as E has the formula of an aliphatic tricarboxylic acid, $C_6H_{11}(CO_2H)_3$. During this oxidation one carbon atom has been lost.

The formation of trimethylsuccinic acid from E by oxidation and the presence of an isobutyl group in E (from the evidence of dry distillation), indicate the possible structure of E to be:

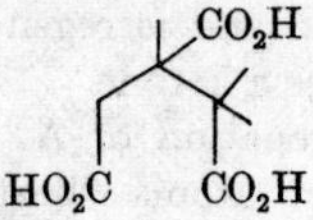

CO_2H

HO_2C CO_2H

This is confirmed by synthesis:

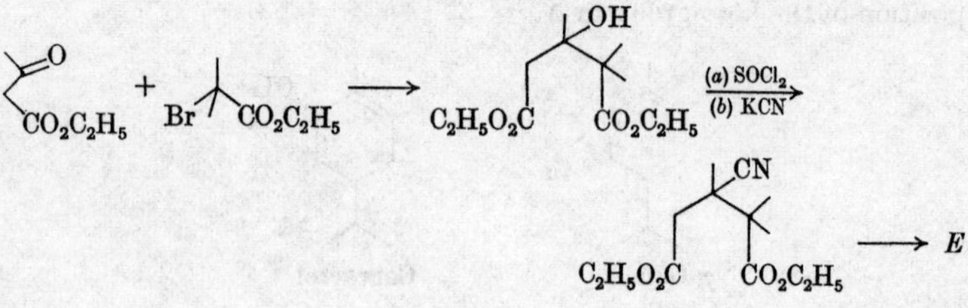

O, $CO_2C_2H_5$ + Br $CO_2C_2H_5$ $\longrightarrow$ OH, $C_2H_5O_2C$ $CO_2C_2H_5$ $\xrightarrow[(b)\ KCN]{(a)\ SOCl_2}$

CN, $C_2H_5O_2C$ $CO_2C_2H_5$ $\longrightarrow$ *E*

We can now argue that D is:

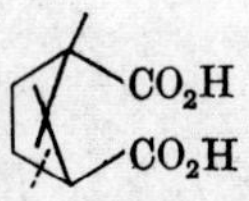

forming E by breaking the bond indicated.

Thus A has the structure:

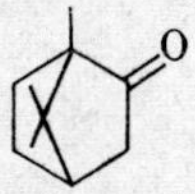

(Camphor)

and C an α-keto-oxime formed at the reactive CH_2 group, namely:

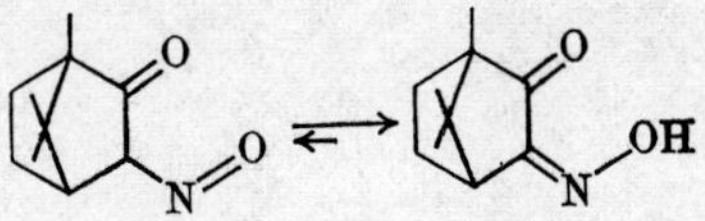

Subject Index